Energy Usage and the Environment

Energy Usage and the Environment

Editor: Ted Weyland

RC Callisto Reference

www.callistoreference.com

Callisto Reference,
118-35 Queens Blvd., Suite 400,
Forest Hills, NY 11375, USA

Visit us on the World Wide Web at:
www.callistoreference.com

ISBN: 978-1-63239-992-2 (Hardback)

Cataloging-in-Publication Data

Energy usage and the environment / edited by Ted Weyland.
        p. cm.
Includes bibliographical references and index.
ISBN 978-1-63239-992-2
1. Power resources. 2. Energy consumption. 3. Energy consumption--Environmental aspects.
4. Energy development--Environmental aspects. I. Weyland, Ted.
TJ163.2 .E54 2018
333.79--dc23

Table of Contents

Preface

In my initial years as a student, I used to run to the library at every possible instance to grab a book and learn something new. Books were my primary source of knowledge and I would not have come such a long way without all that I learnt from them. Thus, when I was approached to edit this book; I became understandably nostalgic. It was an absolute honor to be considered worthy of guiding the current generation as well as those to come. I put all my knowledge and hard work into making this book most beneficial for its readers.

Energy is an essential aspect of everyday life. Energy can be produced from renewable as well as non-renewable sources of Energy. The energy produced from non-renewable sources such as coal and petroleum produce pollution which adversely impacts our environment. Therefore, the need of the hour, is to use renewable sources of energy such as sunlight, wind, waves, etc. to meet the primary energy demands. This book attempts to understand the multiple branches that fall under the discipline of renewable energy and how such concepts have practical applications. With state-of-the-art inputs by acclaimed experts of this field, this book targets students and professionals.

I wish to thank my publisher for supporting me at every step. I would also like to thank all the authors who have contributed their researches in this book. I hope this book will be a valuable contribution to the progress of the field.

Editor

Pyrolysis of algal biomass obtained from high-rate algae ponds applied to wastewater treatment

Fernanda Vargas e Silva* and Luiz Olinto Monteggia

Institute of Hydraulic Research, Federal University of Rio Grande do Sul, Porto Alegre, Brazil

This work presents the results of the pyrolysis of algal biomass obtained from high-rate algae ponds treating sewage. The two high-rate algae ponds (HRAP) were built and operated at the São João Navegantes Wastewater Treatment Plant. The HRAP A was fed with raw sewage while the HRAP B was fed with effluent from an upflow anaerobic sludge blanket (UASB) reactor. The HRAP B provided higher productivity, presenting total solids concentration of 487.3 mg/l and chlorophyll a of 7735 mg/l. The algal productivity in the average depth was measured at 41.8 $g \cdot m^{-2}$ day^{-1} in pond A and at 47.1 $g \cdot m^{-2}$ day^{-1} in pond B. Algae obtained from the HRAP B were separated by the process of coagulation/flocculation and sedimentation. In the presence of alum, a separation efficiency in the range of 97% solid removal was obtained. After centrifugation the biomass was dried and comminuted. The biofuel production experiments were conducted via pyrolysis in a tubular quartz glass reactor which was inserted in a furnace for external heating. The tests were carried out in an inert nitrogen atmosphere at a flow rate of 60 ml/min. The system was operated at 400, 500, and 600°C in order to determine the influence of temperature on the obtained fractional yields. The studies showed that the pyrolysis product yield was influenced by temperature, with a maximum liquid phase (bio-oil and water) production rate of 44% at 500°C, 45% for char and around 11% for gas.

Keywords: high-rate algae ponds, pyrolysis, biofuels, wastewater treatment, bioremediaiton

Edited by:
Umakanta Jena,
Desert Research Institute, USA

Reviewed by:
Sandeep Kumar,
Old Dominion University, USA
Kaushlendra Singh,
West Virginia University, USA

*Correspondence:
Fernanda Vargas e Silva,
Instituto de Pesquisas Hidráulicas,
Bento Gonçalves 9500, Porto Alegre,
Rio Grande do Sul, Brazil
fervs@globo.com

Specialty section:
This article was submitted to
Bioenergy and Biofuels, a section of
the journal Frontiers in Energy
Research

Citation:
Vargas e Silva F and Monteggia LO
(2015) Pyrolysis of algal biomass
obtained from high-rate algae ponds
applied to wastewater treatment.
Front. Energy Res. 3:31.

Introduction

Biomass is considered worldwide as an important source of renewable energy, including electricity, automobile fuel, and as a source of heat for industrial equipment.

Cultures commonly used for energy production are sugarcane, corn, beans, beets, and many others. There are two main factors that define when a culture is appropriate for this process: good dry matter yield per unit of land (dry ton/ha), low area requirement for cultivation, and low costs of energy production from biomass (Dermibas et al., 2009).

However, some research has condemned the use of biofuels, associating its production with possible high food prices. Algae, among the aquatic biomass feedstocks, are considered one of the most promising sources of biofuels due to their unique characteristics. They can accumulate lipids that can be converted into biofuels, present fast proliferation, have the ability to sequester CO_2 from the atmosphere for growth and do not require agricultural land or freshwater for growth or higher water consumption, and also the whole plant matter can be used in converting biofuels processes

(Dismukes et al., 2008; Brennan and Owende, 2010; Jena and Das, 2011; Pate et al., 2011; Yanik et al., 2013; Zhou et al., 2014; Hognon et al., 2015).

Wastewater treatment associated with algae cultivation can offer an alternative way for sustainable renewable biofuels, since the large amount of freshwater needed for algae cultivation can be saved, becoming an environmentally friendly process (Zhou et al., 2014).

In a sewage treatment system, high-rate ponds are characterized by having high algal biomass generation which is an undesirable byproduct for the environment. Its presence in water bodies decreases water quality.

High-rate algae ponds are raceway-type ponds, in which water, algae, and nutrients are continually mixed. A paddle wheel generates a mean horizontal water velocity of approximately 0.15–0.3 m/s. This movement is necessary to avoid sedimentation and stratification. The maximum biomass production (mostly algae) is achieved through better use of lighting per volume. This is ensured by the low depth of the ponds and the constant movement of biomass through mechanical mixing (Nascimento, 2001; Chisti, 2007).

Usually the algal biomass productivity is determined by the measurement of solids found in the ponds.

The algal biomass production costs are mainly covered by the costs of treatment when using wastewater high-rate ponds resulting in lower environmental impacts in terms of water, energy, and fertilizer needs.

The biomass in high-rate algae ponds assimilates the nutrients needed for its growth and becomes responsible for the removal of nutrients from wastewater. This has the advantage of controlling pollution of water resources which contributes to the sustainable use of this technology on an industrial scale (Park et al., 2011; Passos et al., 2013).

The biomass separation process requires an increase in algal suspension concentration typically from 0.02 to 0.06% total suspended solids (TSS) to approximately 2 to 7% solids, which may be higher depending on the target process objective (Uduman et al., 2010).

The algal cells have reduced size, sometimes <30 μm and their density is similar to water with a low sedimentation rate, so to be successful in separation, it is necessary to aggregate the cells. Generally, the process comprises of two steps: the first involving destabilization of algal cells using coagulation followed by sedimentation or flotation. In the second step of the process, it is necessary to increase the biomass content, which is often done by filtration, centrifugation, or thermal processes (Molina Grima et al., 2003; Granados et al., 2012; Cai et al., 2013; Udom et al., 2013).

The algae cell has a negative surface charge, which prevents aggregation. This charge may be reduced or neutralized by the addition of flocculants or multivalent cations, such as cationic polymers that change the zeta potential, which is a measure of particle stability, reducing the repulsive forces. So the action of the attractive Van der Waals forces allows algae agglutination (Wessler et al., 2003; Granados et al., 2012). Salts used for this purpose should be non-toxic, low cost, and have high effectiveness at low concentrations (Molina Grima et al., 2003).

Another advantage of using coagulation/flocculation process is nutrient removal. The presence of nutrients in wastewater, particularly nitrogen and phosphorus, is a serious environmental problem and is receiving increasing attention. Nitrogen in the form of ammonia can be volatilized and cause air pollution. Phosphorus can permeate into the soil and cause damage to the underground water (Chen et al., 2012). When there are excessive levels of nutrients in the wastewater, they cause eutrophication of water sources, possibly damaging the ecosystem (Cai et al., 2013).

The algal biomass, after thickening, may reach 5–15% solid content, and, being perishable, it must be processed as soon as possible. Essential processes such as thickening and drying usually involve high operational costs. Thus, these steps are considered determining factors regarding the economical feasibility analysis of the overall process (Brennan and Owende, 2010; Uduman et al., 2010). The methods commonly used for thickening biomass are centrifugation and filtration followed by different drying techniques, such as natural, oven, spray, and fluidized bed drying.

There are three basic components in algae biomass: proteins, carbohydrates, and lipids. These oils can then be extracted and converted in to biofuels (Um and Kim, 2009).

The pyrolysis process appears to be an excellent alternative for energy conversion, it presents the advantage of using different sources of organic matter, not being limited by the lipid content, as with biodiesel production processes. The pyrolysis process is based on decomposition of organic compounds present in the total biomass under a controlled environment in the absence of oxygen and atmospheric pressure, resulting in different phases: liquid (bio-oil), gas, and solid (char). It is an endothermic reaction that occurs at a temperature of 300–700°C depending on the characteristics of the material to be pyrolyzed (Martini, 2009; Hognon et al., 2015).

Biomass pyrolysis is considered a renewable process, because biomass is turned in several gases when pyrolyzed. Carbon dioxide, one of the gases formed, is absorbed by the algae for its growth, making the process self-sustainable with no serious contribution to greenhouse effect. The relative yield of each phase generated in the process depends on operating parameters (temperature, heating rate, residence time, and flow rate of inert gas), properties of the biomass (the particle size as well as its moisture), and type of pyrolysis used (slow, fast, or flash pyrolysis) (Balat et al., 2009; Martini, 2009; Akhtar and Amin, 2012; Yanik et al., 2013; Hognon et al., 2015).

In order to obtain high yields of aqueous products, fast pyrolysis is normally used, which is characterized by higher heating rates (1000°C/min) and lower residence times of volatiles (10–20 s). In order to favor solid char formation, slow pyrolysis process with lower heating rates (5–80°C/min) and longer residence times (5–30 min) must be used (Van de Velden et al., 2010; Jena and Das, 2011; Yanik et al., 2013).

The bio-oil generated by biomass pyrolysis is generally cleaner than that from fossil fuels, due to its lower nitrogen and sulfur content. The biomass vaporizes, passes through a process of cracking and condensation, producing a dark brown liquid, consisting of a complex mixture of many different hydrocarbons. This process is most successful in fluidized bed reactors due to high heating rates, rapid devolatilization and easy control (Doshi et al., 2005; Martini, 2009).

Materials and Methods

Biomass Production

Two high-rate algae ponds were constructed in the IPH/UFRGS experimental wastewater treatment unit, at São João Navegantes Wastewater Treatment Plant, This plant is responsible for handling the sewage of the north area of Porto Alegre/RS.

The ponds were operated in closed circuit with the following dimensions: overall height: 0.9 m, length of the straight sections: 30 m, width: 5 m (at the upper edge of the slope) and surface area 320 m², as can be seen in **Figure 1**.

The high-rate algae ponds were operated under two feeding conditions: pond A was fed with raw sewage after pretreatment (screening and grit removal) and pond B was fed with effluent from an upflow anaerobic sludge blanket (UASB) reactor. In order to maximize the process of biomass production, the operating parameters of the ponds were useful depth (Hu): 0.3 m, longitudinal flow speed: 0.3 m/s, and hydraulic detention time (HDT): 3 days.

The pond samples were collected in 20 l plastic containers, directly from the body of the ponds, to provide enough biomass for the pyrolysis experiments.

In order to determine algae biomass productivity, total solids, turbidity, and chlorophyll a were measured weekly.

All experiments to determine these parameters were carried out according to Standard Methods for the Examination of Water and Wastewater [American Public Health Association (APHA) and Awwa (2005)].

Algae Separation and Thickening

Experiments of coagulation/flocculation were performed using the effluent from pond B, which showed better performance in terms of algal biomass production.

The equipment used was VELP Jar Test model F.6/S, composed of 6 jars of 2000 ml each, with agitation and controlled independently.

To evaluate the separation process and the removal of nutrients two coagulants were used, Aluminum Sulfate and Ferric Chloride and two flocculants, Sulfloc 1001 and Tanfloc SL. Their concentration ranges are shown in **Table 1**.

After the separation and removal of all the supernatant from the jars, the algae sludge was submitted to centrifugation for 20 min at 2500 rpm to obtain a sample of about 15–20% of dried solids. After

centrifugation, the biomass was dried at 105°C. Finally, the dried algae were ground in a mortar and stored separately according to the reagent used in the separation process.

Nutrient Removal

Experiments were performed to determine the concentration of nitrogen and phosphorus in effluent ponds before and after coagulation/flocculation. Thus, it was possible to determine the effect of algae upon the separation in the removal of nutrients.

Biomass Pyrolysis

The experiments obtaining biofuel via biomass pyrolysis were performed in a tubular quartz reactor, with the dimensions described in **Figure 2**.

The experiments were run in batches, to allow solid char removal. The process flow used in this work, presented in **Figure 3**, was based on Zhang et al. (2011).

In the process, the inert atmosphere was generated by nitrogen gas (1) and the heating process was provided by an external furnace (2). The condensation was performed in (3), where two condensers in series were immersed in an ice bath. The exit of non-condensable gases was in (4).

The pyrolysis reactor was fed manually with 7 g of dried and ground biomass obtained from the previous step of the process. The biomass was inserted in the reactor using an aluminum foil capsule. After it has been charged, the reactor was closed and the inert atmosphere was provided by a 0.06 l/min nitrogen gas flow.

The pyrolysis runs were started by placing the reactor in a programable tubular furnace, with a heating rate of 20°C/min. All the runs were made in two steps: heating the sample and an isothermal reaction step, maintaining the desired temperature (400, 500, and 600°C) for 60 min.

The vapors generated passed through two condensers in series, immersed in ice baths maintained at a temperature of about 0°C. At the end of each experiment, the aqueous phase generated by condensation was collected, combined, weighed, and stored. The non-condensable gases were measured by difference. Following the 60 min reaction time, the system was turned off and cooled to room temperature.

After reaching room temperature, the reactor was opened and the solid fraction (char) was collected, weighed, and stored. The

FIGURE 1 | High-rate algae ponds.

TABLE 1 | Concentration range used.

Product	FeCl$_3$ 10%	Al$_2$(SO$_4$)$_3$ 10%	Sulfloc 20%	Tanfloc 10%
Concentration range (mg/l)	200–300	100–150	250–300	50–100

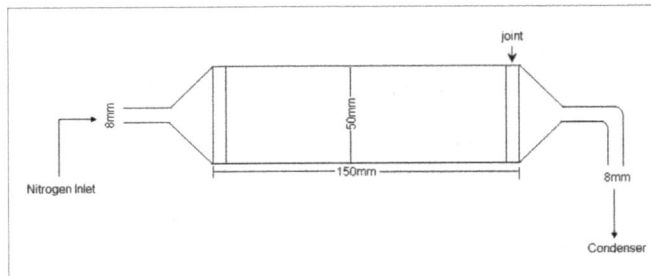

FIGURE 2 | Pyrolysis reactor.

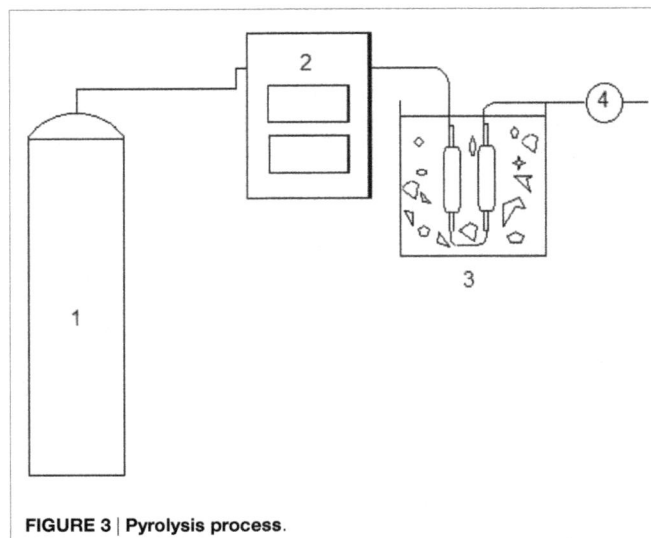

FIGURE 3 | Pyrolysis process.

reactor final mass was also determined, in order to measure the losses by wall adhesion.

The pyrolysis evaluation was performed through yields measurement. Each fraction was determined from the ratio of the weight of respective fraction to initial weight of biomass, expressed as percentage yield, according to Eq. (1).

$$\text{Yield (\%)} = \frac{\text{Fraction mass obtained after pyrolysis}}{\text{Initial algae biomass}} \times 100\% \quad (1)$$

Results and Discussions

Biomass Production

The results of solids, turbidity, chlorophyll a and productivity are shown in **Table 2**, comparing the performance of both biomass production ponds.

In the experiments, we considered the concentration of solids present in effluents and turbidity caused only by the presence of algae. From **Table 2**, it can be noted that pond B in all evaluation parameters showed higher values than those obtained from pond A. Such behavior is explained by the fact that effluent from UASB

TABLE 2 | High-rate ponds performance.

Analysis	Pond A	SD	Pond B	SD
Total solids (mg/l)	433.2	59.2	437.3	56.1
Turbidity (NTU)	41.9	8.9	63.3	13.4
Chlorophyll a (mg/l)	2338	NA	7735	NA
Productivity (g·m^{-2} day^{-1})	41.8	NA	47.1	NA

NA, not applicable.

TABLE 3 | Algae biomass productivity.

Authors	System	Biomass productivity (g·m^{-2} day^{-1})
Nascimento (2001)	HRAP	21.8
Riaño et al. (2012)	Photobioreactor	1.54
Sturm and Lamer (2011)	Open ponds	12
Terigar and Theegala (2014)	Open tanks	43.4

TABLE 4 | Crops productivity [adapted from Trzeciak et al. (2008)].

Crops	Harvest (month/year)	Biomass productivity (g·m^{-2} day^{-1})
Cotton	3	0.38
Peanut	3	0.55
Canola	3	0.60
Sunflowers	3	0.55
Dendê (*Elaeis guineensis*)	12	6.84
Mamona (*Ricinus communis* L.)	3	0.41

reactor provided low solid concentration, which facilitated higher solar irradiation in the body of the pond, an essential factor for biomass growth. Thus, the effluent selected for tests of separation, thickening, and the tests for obtaining biofuels was collected from the pond B. **Table 3** shows a comparison among biomass productivity obtained in this work and others presented in the literature.

Table 4 shows the comparison between the productivity of crops commonly used in the biofuels production.

As we can see from both tables, high-rate algae ponds can be a competitive source of biomass, with higher productivities and without need of arable land and fresh water. This system high-rate algae pond (HRAP) presents no seasonality and the biomass can be harvested all year, without competition with food crops.

Algae Separation, Thickening, and Nutrient Removal

The results of algae separation are shown in **Table 5**, based on separation efficiency related to the chemical dosage used. This table also shows the evaluation of nutrient removal for each product.

Thus, according to the results shown in **Table 5**, the biomass separated with aluminum sulfate, which was selected as the most convenient chemical due to lower dosage requirement, showed better separation and nutrient removal. The biomass was dried and crushed to be used in the pyrolysis experiments. The efficiency of N and P removal were similar when using Sulfloc 20%, but the dosage required was higher than with sulfate.

Biomass Pyrolysis

The influence of temperature (400, 500, and 600°C) on pyrolysis results are shown in **Figure 4**. According to the results, the

TABLE 5 | Removal obtained and dosage used.

Product	Ferric chloride 10%	Al sulfate 10%	Sulfloc 20%	Tanfloc 10%
Maximum separation (%)	88.4	97.9	94.5	97.5
Concentration (mg/l)	300	150	290	100
P Removal (%)	100	100	100	37.9
N Removal (%)	–	5.5	5.5	–

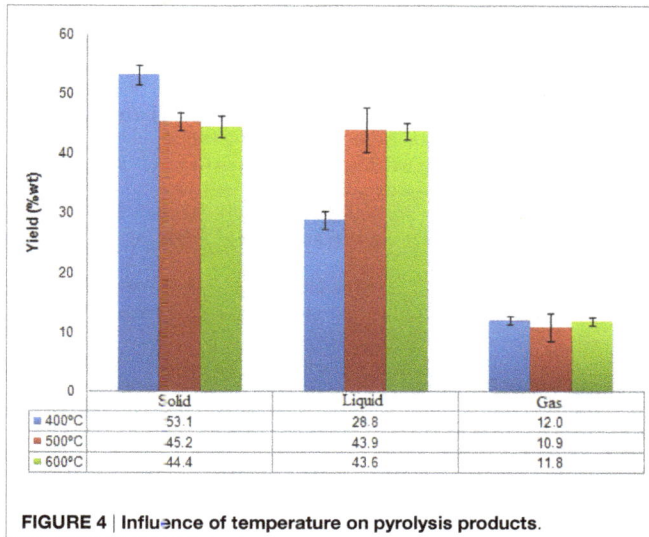

	Solid	Liquid	Gas
400°C	53.1	28.8	12.0
500°C	45.2	43.9	10.9
600°C	44.4	43.6	11.8

FIGURE 4 | Influence of temperature on pyrolysis products.

temperature of 400°C favors solid phase formation, with an average yield of 53.1%. The aqueous and gaseous phases obtained average yields of 28.8 and 12%, respectively. At 500°C, the yields for solid and aqueous phases were similar, however aqueous phase formation was slightly higher, composed of bio-oil and water. The average yield was 43.9 for solid phase 45.2 for the aqueous phase and 10.9% in the gas phase. At 600°C we can see similar yields between solid and liquid formation, 44.4 and 43.6%, respectively. The average for gas formation at this temperature was 11.8%.

The liquid phase, comprising of bio-oil and water, has a reddish brown color, with a strong and distinctive smoky smell, which confirms the information in the literature about products obtained in the pyrolysis (Jena and Das, 2011; Yanik et al., 2013).

As described in the literature, temperature plays an important role on the yield of the fractions obtained in the pyrolysis process. Studies show that temperatures between 450 and 550°C maximize the yield of bio-oil and, and at very high temperatures, secondary reactions of the volatiles may occur, thus decreasing the yield of the liquid phase, which can be seen in **Figure 4**; at 600°C, a small decrease in the aqueous phase yield was observed (Yanik et al., 2013).

For related data, we use an ANCOVA analysis (Analysis of Covariance) with a fixed factor (oven temperature) and a covariate (initial mass of algae) to identify differences in the char mass production. Five replicates were performed for each factor and the software used was SPSS version 18.

The data do not present heteroscedasticity, using the Levene test (p-value of 0.235), the tested factor was significant at a p-value of 0.01. So we went to the *post hoc* analysis, which showed a significant difference between the means of groups, the 400°C group is different from other groups and the 500°C and 600°C are not statistically different from each other.

Conclusion

In this work, the association of wastewater treatment and biofuel production through pyrolysis of algal biomass obtained in high-rate algae ponds was studied. The algal productivity, at the average depth was measured as 41.8 $g \cdot m^{-2}$ day^{-1} for pond A and as 47.1 $g \cdot m^{-2}$ day^{-1} for pond B. The algae were pyrolyzed in a tubular furnace system with external heating at different temperatures. Studies have shown that the pyrolysis process is efficient and the fractions yields are greatly influenced by temperature. Operating under mild conditions, it was possible to obtain maximum yields of 45% at 500°C for aqueous phase (bio-oil and water), 44% for char, and about 11% for gas. As we can see, through this process, it is possible to offer a promising alternative for environmental pollution control with potential economic return.

Acknowledgments

The authors acknowledge CNPq and CAPES for the financial support to this project.

References

Akhtar, J., and Amin, N. S. (2012). A review on operating parameters for optimum liquid oil yield in biomass pyrolysis. *Renew. Sustain. Energ. Rev.* 16, 5101–5109. doi:10.1016/j.rser.2012.05.033

American Public Health Association (APHA) and Awwa, W. E. F. (2005). *Standard Methods for the Examination of Water and Waste Water*, 21st Edn. American Public Health Association, 4–108; 4–147.

Balat, M., Balat, M., Kirtay, E., and Balat, H. (2009). Main routes for thermo-conversion of biomass into fuels and chemicals. Part1: pyrolysis systems. *Energy Convers. Manag.* 50, 3147–3157. doi:10.1016/j.enconman.2009.08.014

Brennan, L., and Owende, P. (2010). Biofuels from microalgae – a review of technologies for production, processing and extractions of biofuels and co-products. *Renew. Sustain. Energ. Rev.* 14, 557–577. doi:10.1016/j.rser.2009.10.009

Cai, T., Park, S. Y., and Li, Y. (2013). Nutrient recovery from wastewater streams by microalgae: status and prospects. *Renew. Sustain. Energ. Rev.* 19, 360–369. doi:10.1016/j.rser.2012.11.030

Chen, R., Li, R., Deitz, L., Liu, Y., Stevenson, R. J., and Liao, W. (2012). Freshwater cultivation with animal waste for nutrient removal and biomass production. *Biomass Bioenergy* 39, 128–138. doi:10.1016/j.biombioe.2011.12.045

Dermibas, M. F., Balat, M., and Balat, H. (2009). Potential contribution of biomass to the sustainable energy development. *Energy Convers. Manag.* 50, 1746–1760. doi:10.1016/j.enconman.2009.03.013

Dismukes, G. C., Carrieri, D., Bennette, N., Ananyev, G. M., and Posewitz, M. C. (2008). Aquatic phototrophs: efficient alternatives to land-based crops for biofuels. *Curr. Opin. Biotechnol.* 19, 235–240. doi:10.1016/j.copbio.2008.05.007

Doshi, V. A., Vuthaluru, H. B., and Bastow, T. (2005). Investigations into the control of odor and viscosity of biomass oil derived from pyrolysis of sewage sludge. *Fuel Process. Technol.* 86, 885–897. doi:10.1016/j.fuproc.2004.10.001

Granados, M. R., Acién, F. G., Gómez, C., Fernandez-Sevilla, J. M., and Molina Grima, E. (2012). Evaluation of flocculants for the recovery of freshwater microalgae. *Bioresour. Technol.* 118, 102–110. doi:10.1016/j.biortech.2012.05.018

Hognon, C., Delrue, F., Texier, J., Gateau, M., Thiery, S., Miller, S., et al. (2015). Comparison of pyrolysis and hydrothermal liquefaction of *Chlamydomonas reinhardti*. Growth studies on the recovered hydrothermal aqueous phase. *Biomass Bioenergy* 73, 23–31. doi:10.1016/j.biombioe.2014.11.025

Jena, U., and Das, K. C. (2011). Comparative evaluation of thermochemical liquefaction and pyrolysis for bio-oil production from microalgae. *Energy Fuels* 25, 5472–5482. doi:10.1021/ef201373m

Martini, P. R. R. (2009). *Conversão Pirolítica de Bagaço Residual da Indústria de Suco de Laranja e Caracterização Química dos Produtos*. Dissertação de Mestrado, PPGQ; Universidade Federal de Santa Maria, Santa Maria.

Molina Grima, E., Belarbi, E. H., Acién Fernández, F. G., Robles Medina, A., and Chisti, Y. (2003). Recovery of microalgal biomass and metabolites: process options and economics. *Biotechnol. Adv.* 20, 491–515. doi:10.1016/S0734-9750(02)00050-2

Nascimento, J. R. S. (2001). *Lagoas de Alta Taxa de Produção de Algas Para Pós-Tratamento de Efluentes de Reatores Anaeróbios*. Dissertação de Mestrado, Instituto de Pesquisas Hidráulicas, UFRGS, Porto Alegre.

Park, J. B. K., Craggs, R. J., and Shilton, A. N. (2011). Wastewater treatment high rate algal ponds for biofuel production. *Bioresour. Technol.*, 102, p. 35–42. doi:10.1016/j.biortech.2010.06.158

Passos, F., Solé, M., García, J., and Ferrer, I. (2013). Biogas production from microalgae grown in wastewater: effect on microwave pretreatment. *Appl. Energy* 108, 168–175. doi:10.1016/j.watres.2013.10.013

Pate, R., Klise, G., and Wu, B. (2011). Resource demand implications for US algae biofuels production scale-up. *Appl. Energy* 88, 3377–3388. doi:10.1016/j.apenergy.2011.04.023

Riaño, B., Hérnandez, D., and Garcia-González, M. C. (2012). Microalgal-based systems for wastewater treatment: effect of applied organic and nutrient loading rate on biomass composition. *Ecol. Eng.* 49, 112–117. doi:10.1016/j.ecoleng.2012.08.021

Sturm, B. S. M., and Lamer, S. L. (2011). An energy evaluation of coupling nutrient removal from wastewater with algal biomass production. *Appl. Energy* 88, 3499–3506. doi:10.1016/j.apenergy.2010.12.056

Terigar, B. C., and Theegala, C. S. (2014). Investigating the interdependence between cell density, biomass productivity, and lipid productivity to maximize biofuel feedstock production from outdoor microalgal cultures. *Renew. Energy* 64, 238–243. doi:10.1016/j.renene.2013.11.010

Trzeciak, M. B., das Neves, M. B., da Silva Vinholes, P., and Amaral Villela, F. (2008). Utilização de sementes de species oleaginosas para produção de biodiesel. *Inf. Abrates* 18, 30–38.

Udom, I., Zaribaf, B. H., Halfhide, T., Gillie, B., Dalrymple, O., Zhang, Q., et al. (2013). Harvesting microalgae grown on wastewater. *Bioresour. Technol.* 139, 101–106. doi:10.1016/j.biortech.2013.04.002

Uduman, N., Ying, Q., Danquah, M. K., Forde, G. M., and Hoadley, A. (2010). Dewatering of microalgal cultures: a major bottleneck to algae-based fuels. *J. Renew. Sustain. Energy* 2. doi:10.1063/1.3294480

Um, B. H., and Kim, Y. S. (2009). Review: a chance for Korea to advance algal-biodiesel technology. *J. Ind. Eng. Chem.* 15, 1–7. doi:10.1016/j.jiec.2008.08.002

Van de Velden, M., Baeyens, J., Brems, A., Janssens, B., and Dewil, R. (2010). Fundamentals, kinetics and endothermicity of the biomass pyrolysis reaction. *Renew. Energy* 35, 232–242. doi:10.1016/j.renene.2009.04.019

Wessler, R. A., Amorim, S., and Cavalli, V. (2003). *Estudo da Viabilidade Técnica e Econômica da Utilização de um Polímero Orgânico Natural Catiônico em Substituição ao Sulfato de Alumínio Convencionalmente Utilizado em Estações de Tratamento de Água (ETA'S). Artigo Técnico*. Santo André: 33 Assembleia Nacional dos Serviços Municipais de Saneamento; ASSEMAE. Available at: http://www.semasa.sp.gov.br/Documentos/ASSEMAE/Trab_29.pdf

Yanik, J., Stahl, R., Troeger, N., and Sinag, A. (2013). Pyrolysis of algal biomass. *J. Anal. Appl. Pyrolysis* 103, 134–141. doi:10.1016/j.jaap.2012.08.016

Chisti, Y. (2007). Biodiesel from microalgae. Research review paper. *Biotechnol. Adv.* 25, 294–306. doi:10.1016/j.biotechadv.2007.02.001

Zhang, B., Xiong, S., Xiao, B., Yu, D., and Jia, X. (2011). Mechanism of wet sludge pyrolysis in a tubular furnace. *Int. J. Hydrogen Energy* 36, 355–363. doi:10.1016/j.ijhydene.2010.05.100

Zhou, W., Chen, P., Min, M., Ma, X., Wang, J., Griffith, R., et al. (2014). Environment-enhancing algal biofuel production using wastewaters. *Renew. Sustain. Energ. Rev.* 36, 256–269. doi:10.1016/j.rser.2014.04.073

Conflict of Interest Statement: The authors declare that the research was conducted in the absence of any commercial or financial relationships that could be construed as a potential conflict of interest.

2

Optimizing the Critical Factors for Lipid Productivity during Stress Phased Heterotrophic Microalgae Cultivation

P. Chiranjeevi[1,2] and S. Venkata Mohan[1,2]*

[1] *Bioengineering and Environmental Sciences (BEES), CSIR-Indian Institute of Chemical Technology (CSIR-IICT), Hyderabad, India,* [2] *Academy of Scientific and Innovative Research (AcSIR), India*

Edited by:
Bo Hu,
University of Minnesota, USA

Reviewed by:
Jianguo Zhang,
University of Shanghai for
Science and Technology, China
Cong Li,
Ovivo Water, USA

*Correspondence:
S. Venkata Mohan
vmohan_s@yahoo.com

Specialty section:
This article was submitted to
Bioenergy and Biofuels,
a section of the journal
Frontiers in Energy Research

Citation:
Chiranjeevi P and Venkata Mohan S
(2016) Optimizing the Critical
Factors for Lipid Productivity during
Stress Phased Heterotrophic
Microalgae Cultivation.
Front. Energy Res. 4:26.

Microalgae-derived biodiesel production is one of the promising and sustainable platform. The effect of selected stress factors (pH, temperature, salinity, and carbon supplementation) on microalgal lipids and carbohydrate production during heterotrophic mode of operation was studied using design of experimental (DOE) methodology (Taguchi approach) with variation at four levels ($2^1 \times 4^4$). Experiments were performed with allegorical batch experimental matrix (16 experimental trails). All the selected factors showed marked influence on the lipid production, whereas temperature and carbon concentration showed major influence on the carbohydrate synthesis. Interesting, relatively higher total lipid production (55% of DCW) was obtained from Experimental no. 6 (pH: 6; salinity: 1 g/l; temperature: 20°C; carbon concentration: 30 g/l). Relatively good neutral lipid fraction (13.6%) was observed with Experimental no. 8: pH: 6; salinity: 5 g/l; temperature: 30°C; carbon concentration: 1 g/l. Good carbohydrate synthesis (262 mg/g biomass) was observed with Experiment no. 3 (pH: 4; salinity: 2 g/l; temperature: 30°C; carbon concentration: 15 g/l). Fatty acid methyl esters (FAME) analysis the presence of higher number of saturated fatty acids (C12:0 to C24:0) in experimental setups 6 and 8, favoring the biodiesel properties.

Keywords: biomass, chlorophyll, mixotrophic, nutritional mode, biodiesel, FAME, data envelopment analysis (DEA)

INTRODUCTION

Microalgae cultivation is attracting renewed interest due to its ability to produce diverse photosynthetic products, including lipids. Microalgae can be cultivated either by autotrophic mode (Dayananda et al., 2007; Liu et al., 2011; Venkata Mohan et al., 2011) fixing CO_2 in the presence of sunlight or by heterotrophic mode (Brennan and Owende, 2010; Garcia et al., 2011; Devi and Venkata Mohan, 2012) using organic compounds as energy and carbon sources or by mixotrophic mode using both organic compounds and inorganic CO_2 (Qiao and Wang, 2009; Chen et al., 2011; Devi et al., 2012; Cong and Lu-Kwang, 2014; Cong et al., 2016). At present, inherent limitations encountered with microalgae cultivation is its low lipid/biomass production, which invariably impedes its scale-up operation.

To enhance lipid synthesis, strategies by altering the nutrient regime and cultivation conditions are generally employed. Factors, such as nutrient stress, light, temperature, CO_2, and salinity, have been explored to enhance lipid accumulation in microalgae (Merchant et al., 2012; Sharma et al., 2012; Chandra et al., 2014; Venkata Mohan et al., 2015; Chiranjeevi and Venkata Mohan, 2016). Stress has a critical role on the lipid synthesis during microalgae cultivation. The nutrient stress affects the growth of algae as well as lipid productivity and its profile. Under adverse/stress conditions, microalgae tend to accumulate neutral lipids to protect cells from photooxidation (Adams et al., 2013; Zhang et al., 2013). Among the different modes of cultivation, heterotrophic operation offers several advantages over the others, including the elimination of light requirement, and lipid yields (Devi et al., 2012; Venkata Mohan et al., 2015). Moreover, in heterotrophic mode of operation, microalgae are influenced by the stress factors facilitating synthesis of higher lipid along with significant substrate degradation (Devi et al., 2012; Cong et al., 2016).

In order to study the effect of heterotrophically induced nutrient stress on the process, design of experimental (DOE) methodology was employed. The focal objective of this investigation is to study the methodological application of Taguchi orthogonal array (OA) experimental design (DOE) to optimize selected stress factors, viz., pH, salinity, temperature, and carbon supplementation. Factorial-based DOE methodology by Taguchi OA approach merges statistical and engineering techniques (Taguchi, 1986; Venkata Mohan et al., 2005, 2007). Mixotrophic microalgae cultivation was studied to achieve higher biomass productivity using Taguchi DOE methodology by optimizing eight factor (Chiranjeevi and Venkata Mohan, 2016).

Analysis of the experimental data using analysis of variance (ANOVA) provides information about statistically significant factors and their optimum levels.

EXPERIMENTAL METHODOLOGY

Design of Experimental Methodology

Taguchi's DOE methodology was used by selecting important factors, viz., pH, salinity, temperature, and carbon concentration, whose variation will have a critical effect on the heterotrophic lipid synthesis and carbohydrates production (Table 1). Four levels of factor variations were selected, which represent the experimentation size of 16 with an array matrix of M-16 (Table 2).

Microalgae

Mixed microalgae culture collected from Nacharam Lake (Pedda Cheruvu), Hyderabad, during the pre-monsoon season was used as parent inoculum. The culture was washed twice with water and pelletized (2.1 g with a concentration of 0.2 g/l) by centrifugation (3000 rpm; 10 min at 30°C) to remove associated debris and restored in rectangular plastic tubs (36 cm × 24 cm × 12 cm) exposed to diffused sunlight. Domestic sewage [(DS) pH, 7.8; COD, 220 mg/l; VFA, 165 mg/l; BOD, 120 mg/l; total alkalinity, 140 mg/l; chlorides, 175 mg/l; nitrates, 115 mg/l] was used as feedstock for microalgal cultivation. After the consistent amount

of biomass (cell density) was achieved, this culture was used as inoculum for the experimental study.

Experimental Details

The combinatorial of 4 factors at 4 levels (Table 1) with 16 experimental variations (M-16) in batch mode (Table 2), during stress phase (SP) was performed. In accordance with the designed experiments, the 250 ml of modified growth medium (as per design, Table 1) was sterilized (autoclaved for 20 min at 121°C and 1.05 kg/cm steam pressure) prior to the inoculation (10% v/v; OD, 0.1) to avoid the contamination. Growth phase (GP) was operated for a period of 8 days in mixotrophic mode [BG11 media, pH: 8.2, temperature: 32°C, glucose 1.5 g/l and light 4000 lux]. After 8 days of growth, the cultures were harvested, and the resulting biomass was used as inoculum for the SP of the experiments operated in heterotrophic mode (Table 2). All the experimental setups were placed on a temperature shaking incubator (120 rpm). During operation (GP and SP), biomass, pigment analysis (chlorophyll a and b) along with total carbohydrate concentration were estimated every alternate day. Lipid analysis was performed at initial (before the GP) and end of GP and SP. In both the phases of operations, to avoid bacterial contamination, antibiotic (ampicillin: 0.2 g/l) was added to each experimental setup once in every alternate day. All the experiments were carried out in triplicates, and the average of three independent operations was presented and discussed.

Analysis

Microalgae biomass was monitored by measuring OD at 600 nm and chlorophyll a and b were measured at 647 and 664 nm, respectively. Algal cells were disrupted by sonication (40 kHz for 2 minutes) followed by centrifugation (5000 rpm for 5 minutes at 28°C), and the dissolved carbohydrates in supernatant of the solution were estimated by Anthrone method. COD, nitrates, phosphates, and pH were analyzed as per standard procedure (APHA, 1998).

TABLE 1 | Selected factors and assigned levels.

S. no.	Factor	Level 1	Level 2	Level 3	Level 4
1	pH	4	6	8	10
2	Salinity (g NaCl/l)	0	1	2	5
3	Temperature (°C)	20	25	30	35
4	Carbon supplementation (g glucose/l)	0	1	15	30

TABLE 2 | Orthogonal array (OA) of designed experiments with output parameters.

Experiment number	Factors				Lipid productivity (% of dry biomass)	Neutral lipid (%)	Total carbohydrates (mg/g biomass)
	1	2	3	4			
1	1	1	1	1	14.8	4.7	144
2	1	2	2	2	25.6	12.3	28
3	1	3	3	3	22.1	11.3	262
4	1	4	4	4	17.7	4.2	40
5	2	1	2	3	49.1	10.9	28
6	2	2	1	4	55.0	9.4	146
7	2	3	4	1	10.6	4.3	28
8	2	4	3	2	34.5	13.6	123
9	3	1	3	4	45.1	5.7	235
10	3	2	4	3	32.9	10.5	28
11	3	3	1	2	20.6	5.4	157
12	3	4	2	1	14.6	7	28
13	4	1	4	2	23.2	6.2	28
14	4	2	3	1	23.5	12.9	192
15	4	3	2	4	20.5	10	36
16	4	4	1	3	16	9.1	188

Lipid Extraction and Derivatization of FAME

At the end of GP (2.28 g/l) and SP (1.25–10.38 g/l), the biomass was separated (5000 rpm; 5 min; 28°C), and the resulting biomass after solar drying was powdered by blending. The blended powder was further sonicated (4 kHz; 2 min), and lipids was extracted using chloroform and methanol (2:1) as solvents (modified Bligh and Dyer method). Hexane was used for neutral lipid extraction (Devi et al., 2012). Furthermore, followed by centrifugation, lipid layer was transferred to pre-weighed round bottom flask, and the total and neutral lipids were determined gravimetrically in terms of percentage dry cell weight (% DCW). Lipid productivity (%) was calculated based on the ratio of total lipid extracted to dry weight of algal biomass. The resulted lipid was transesterified (using a methanol–sulfuric acid mixture) to fatty acid methyl esters (FAME). FAME composition was detected by gas chromatography (Nucon-5765) using FID with capillary column [30 mm (0.25 mm × 0.25 μm); Valcobond] using nitrogen carrier gas (1 ml/min). The oven temperature was maintained initially at 140°C (for 5 min), later increased to 240°C (ramp rate of 4°C/min; 10 min). The temperature of injector and detector were maintained at 280 and 300°C, respectively, with a split ratio of 1:10. The FAME composition was compared with the standard (C4- C24:1) (SUPELCO).

Data Analysis

The data derived from the experiments was analyzed employs "bigger is better" performance characteristics using Qualitek-4 (Nutek Inc.) software.

RESULTS AND DISCUSSION

Growth Phase

Mixotrophic mode of cultivation was used for microalgae cultivation during GP. Good increment in biomass growth [1.2 g/l (0 days) to 2.28 g/l with biomass productivity of 28.5 mg/l/day (8th day)] and marginal improvement in lipid productivity [total/neutral lipids 7.8/1.5 to 12.6/4.4% (8th day)] was observed. In agreement to the biomass growth pattern, the total chlorophyll (TC) also increased linearly from 1.3 to 23.6 μg/ml by the end of GP. The organic fraction (carbon) and nutrients (nitrogen and phosphate) contributed to the accumulation of carbohydrates (total, 40 mg/g biomass) and proteins (5.3 mg/ml) at the end of the GP operation.

Stress Phase

Selected Factors Influence on Lipid Synthesis

The influence of the selected factors on stress-induced heterotrophic microalgae cultivation showed marked effect on the lipid synthesis and profile. The magnitude of the difference between average effects (L2 − L1) represents the relative influence of the selected factors on the process. The average effect of the factors and resulting interaction on the total and neutral lipids synthesis is depicted in Tables 3–5. The bigger the difference, the stronger is the influence (ignoring the negative value). The experiments output showed a marked variation in the lipid production as well as on its profile. Among the four factors, pH showed a stronger

TABLE 3 | Main effects of selected factors on total lipids production.

S. no.	Factor	Level 1	Level 2	Level 3	Level 4	L2 − L1
1	pH	20.049	37.299	28.299	20.799	17.25
2	Salinity (NaCl g/l)	33.049	34.25	18.45	20.7	1.201
3	Temperature (°C)	26.6	27.449	31.299	21.1	0.849
4	Carbon supplement (glucose g/l)	15.875	25.974	30.024	34.574	10.099

TABLE 4 | Main effects of selected factors on neutral lipids synthesis.

S. no.	Factor	Level 1	Level 2	Level 3	Level 4	L2 − L1
1	pH	8.125	9.549	7.15	9.549	1.423
2	Salinity (NaCl g/l)	6.875	11.274	7.75	8.475	4.398
3	Temperature (°C)	7.15	10.05	10.875	6.3	2.9
4	Carbon supplement (glucose g/l)	7.224	9.375	10.45	7.324	2.15

TABLE 5 | Main effects of selected factors on carbohydrates production.

S. no.	Factor	Level 1	Level 2	Level 3	Level 4	L2 − L1
1	pH	1.185	0.812	1.12	1.11	−0.373
2	Salinity (NaCl g/l)	1.087	0.982	1.207	0.947	−0.102
3	Temperature (°C)	1.587	0.3	2.029	0.31	−1.297
4	Carbon supplement (glucose g/l)	0.98	0.839	1.264	1.142	−0.142

effect on the total lipid synthesis, while temperature showed the least influence (pH > carbon supplementation > salinity > temperature). On the contrary, salinity showed a stronger effect on the neutral lipids synthesis, while pH showed least influence (salinity > temperature > carbon supplementation > pH).

The effect of pH on the total lipid synthesis was significant. Acidic redox microenvironment (pH 6) illustrated higher lipid synthesis. The favorable pH range for microalgae cultivation varies between 6 and 9, depending on the metabolic mode of cultivation and strain used. Heterotrophic mode cultivation between pH 6 and 7 was reported to be optimum (Kumar et al., 2010). The CO_2/organic carbon uptake by algae associates with an increase in biomass yield but leads to a decrease in pH that affects the microalgae physiology toward lipid synthesis (Belkin and Boussiba, 1991). Experiment no. 6 (pH 6, 30 g COD/l, 2 g NaCl/l, and 20°C) documented relatively higher lipid (total) productivity (55% of DCW; 21.62 mg/l/day) (**Figure 1A**). During cultivation of microalgae, the decrement in the pH from 6 to 4.8 was noticed. Acidic pH facilitates the diffusion of protons required for activation of hexose/proton symporter system for transport of glucose molecules from ambient cell surface to internal cytoplasm to take part in glycolysis pathway and subsequent lipid synthesis (Morales-Sánchez et al., 2013). With this experimental condition, protein and carbohydrates concentrations of 9.4 mg/ml and 146 mg/g DCW were observed, respectively. Subsequent to pH, carbon concentration (glucose) showed significant control on the

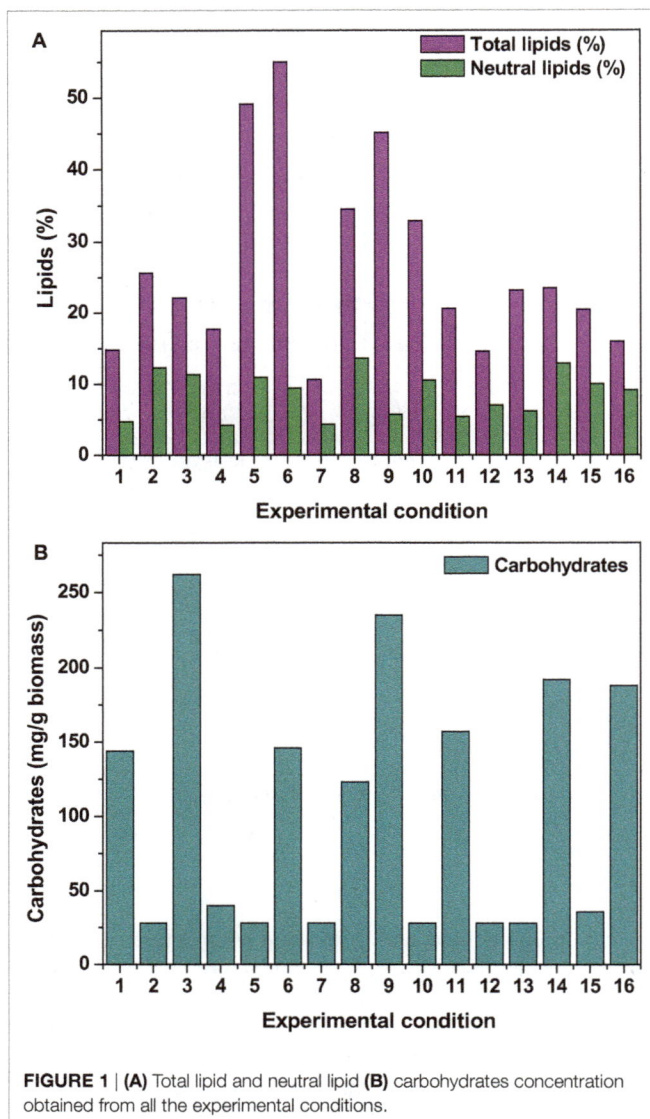

FIGURE 1 | (A) Total lipid and neutral lipid (B) carbohydrates concentration obtained from all the experimental conditions.

total lipid synthesis. Carbon in the form of glucose influenced the lipid accumulation (up to certain concentration). Glucose loading of 30 g/l at level 4 showed higher lipid productivity (55% of DCW; Exp no. 6). Carbon at optimal concentration during the SP increased the metabolic rate, manifesting glycolysis which also might influence the lipid synthesis (Chandra et al., 2014). Maximum biomass productivity (39.37 mg/l/day) with COD removal efficiency of 66% was observed with this experimental variation. Glucose absorption consumes less energy compared to CO_2 uptake by photosynthesis (Garcia et al., 2011). TC concentration [9.19 µg/ml (chl a, 6.61 µg/ml; chl b, 2.58 µg/ml)] also correlated with the biomass production.

Contrary to the total lipids, salinity (5 g/l NaCl) showed marked effect on the neutral lipids production [13.6% of DCW (total, 36% of DCW)] as documented with Exp no. 8 (1 g COD/l; pH 6; 30°C). Salinity-induced stress influences both physiological and biochemical mechanisms associated with the lipid synthesis (Kalita et al., 2011). Exposure of microalgae to hyper saline conditions alter their metabolism to uptake and export ions through the cell membrane, and the resulting stress proteins accumulated facilitates increment in the lipids content to adapt to the extreme environment (Talebi et al., 2013). Lipid induction phase operated under salt stress showed production of higher number of saturated fatty acid (SFA) methyl esters with the improved fuel properties along with increased lipid production (Venkata Mohan and Devi, 2014). The temperature at level 1 (20°C) and level 3 (30°C) showed favorable lipid synthesis. Cultivation at 30°C registered noticeable improvement in the lipid production, especially neutral lipids. The change in the cultivation temperature from optimal range invariably effects the composition of membrane lipids to maintain the normal function of microalgae (Subhash et al., 2014). With Exp no. 8, the pH increased from 6 to 7.1 during cultivation with COD removal efficiency of 86.2% and biomass productivity of 11.43 mg/l/day, were neutral lipid productivity of 3.88 mg/l/day [TC, 8.13 µg/ml (Chl a/Chl b, 5.61/2.51 µg/ml)]. The protein and carbohydrates concentration was found to be 7.68 mg/ml and 106 mg/g DCW, respectively.

Lipids Profile

The increase in cultivation temperature resulted in increased distribution of lipids in the form of SFAs (Boussiba et al., 1987). Temperature-induced stress showed influence on increment in the neutral lipid content illustrating feasibility toward good biodiesel properties with higher SFAs to unsaturated fatty acids (USFA) ratio (Subhash et al., 2014). Similar trend was observed in the present experiment with the presence of stearic (C18:0) and palmitic (C16:0) acids (**Figure 2**). The presence of long-chain fatty acids from C12:0 to C20:0 were observed in the Exp no. 8 due to the combined effect of salinity and temperature. The concentrations of SFAs were high with palmitic acid (C16:0) (50.8%), margaric acid (C17:0) (2.1%), and stearic acid (C18:0) (10.1%) compared to USFA [linoleic acid (C18:2) (6.8%); linolenic acid (C18:3) (2.8%)]. Fatty acids, including palmitic (40.7%) and myristic acids (4.2%), are known to have potential antibacterial and antifungal activities. Modification of fatty acids composition is one of the mechanisms used by algae to regulate osmotic balance and maintain membrane fluidity under the altered conditions (Adams et al., 2013). Low temperature induces the synthesis of more USFA to speed up the lipids desaturation as a recompensation measure to maintain the cell membrane fluidity (Perez-Garcia et al., 2011). Operating at lower temperature gives rise to more intracellular molecular oxygen and consequently improves the activities of desaturases and elongases that are involved in the biosynthesis of USFA (Chen and Chen, 2006). The presence of long-chain fatty acids from C12:0 to C24:0 were also observed due to the combined effect of pH and high carbon content.

Factor Influence on Carbohydrates Accumulation

Among the studied stress factors, temperature showed marked effect on the carbohydrate accretion, while salinity showed the least influence (temperature > pH > carbon supplementation > salinity). Nutrient limitation facilitates the accumulation of carbohydrates due to distinct reduction in cell growth (Ball et al., 1990). Temperature at level 3 (30°C) showed high carbohydrate accretion (262 mg/g biomass; Exp no. 3; 4 pH, 2 g

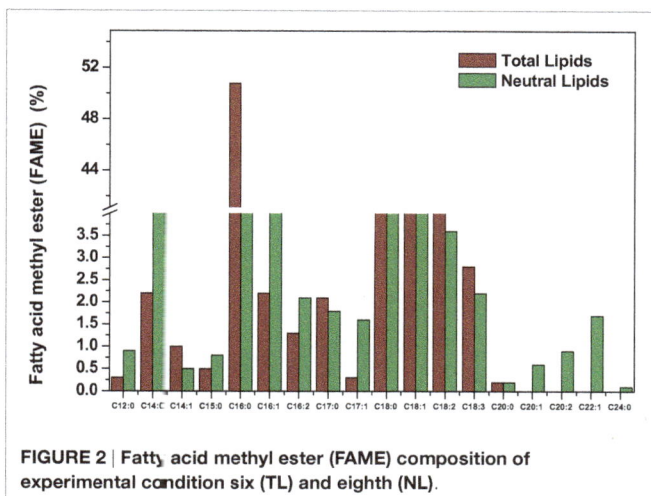

FIGURE 2 | Fatty acid methyl ester (FAME) composition of experimental condition six (TL) and eighth (NL).

TABLE 6 | Automated test for presence of interaction factors.

S. no.	Interacting factor pairs (based on SI)	Columns	SI (%)	Reserved column	Optimum levels
1	Salinity (NaCl) × temperature (°C)	2 × 3	71.73	1	[2,1]
2	pH × temperature (°C)	1 × 3	18.8	2	[2,1]
3	pH × carbon supplement	1 × 4	14.75	5	[2,4]
4	Salinity (NaCl) × carbon supplement	2 × 4	7.09	6	[2,4]
5	Temperature (°C) × carbon supplement	3 × 4	5.85	7	[1,4]
6	pH × salinity (NaCl)	1 × 2	5.51	3	[2,2]

NaCl/l; 15 g COD/l) than level 1 (20°C), level 2 (25°C), and level 4 (35°C) (**Table 2**; **Figure 1B**). Next to the temperature, carbon load (level 3; 15 g COD/l) showed feasibility to higher accretion of carbohydrate with substrate degradation of 66%. Carbohydrate accumulation improves by increasing the carbon load during the SP of microalgae cultivation (Giordano, 2001; Xia and Gao, 2005; Chandra et al., 2014). Under nutrient limiting conditions and with good supply of carbon, the protein content of microalgae function as nitrogen source, and the carbohydrate content may increase considerably (Ball et al., 1990). Salinity at level 3 (2 g/l NaCl) showed good carbohydrates accumulation. Accumulation of carbohydrates takes place as a response to an immediate NaCl shock (Zheng et al., 1997). Salt stress facilitates production of reactive oxygen species, which inhibit RuBisCO activity and lead to photoinhibition that concurrently decreases the biomass growth (Neale and Melis, 1989; Murata et al., 2007). High NaCl concentration also inactivates the ATP-synthase resulting protein synthesis inhibition in microalgae (Allakhverdiev et al., 2005). With Exp no. 3, the pH varied marginally from 4 to 3.9 during cultivation with COD removal efficiency of 66% and biomass productivity of 19.0 mg/l/day [TC, 7.3 µg/ml (*Chl a/Chl b*, 5.61/1.69 µg/ml)]. The protein concentration was found to be 10.71 mg/ml.

Maximum substrate degradation (based on COD removal efficiency) of 66, 86, and 66% was observed at the end of SP in experiments 6, 8, and 3, respectively. Apart from the lipid synthesis, the substrate utilized by the microalgae also associates with the biomass/carbohydrate production (6.37/146, 1.83/123, and 3.9/262 g/l/mg/g) during the heterotrophic microalgae cultivation under starvation phase. Maximum chlorophyll concentration was observed at the end of the starvation phase (Exp no. 6: 9.19 µg/ml; Exp no. 8: 10.9 µg/ml; Exp no. 3: 7.3 µg/ml) along with good protein concentration (Exp no. 6: 9.46 mg/ml; Exp no. 8: 7.68 mg/ml; Exp no. 3: 10.71 mg/ml).

Factor Interactions
Evaluating factors' interaction facilitates understanding the complexity of the process. The estimated interactions based on the severity index (SI) help to know the influence of two individual factors at various levels (**Table 6**). Salinity with temperature showed highest SI (71.73%), followed by pH and temperature (18.8%), pH and carbon supplementation (14.75%) (**Figure 3**). Temperature showed high level of interaction with pH (71%). The interaction between carbon and temperature (SI, 5.8%) showed a marginal influence on carbohydrate production. Interaction between salinity and pH (SI, 5.5%) has no significant influence on lipid and carbohydrate production.

Analysis of Variance
It is evident from ANOVA that all the factors and interactions considered in the experimental design had statistically significant at 95% confidence limit (*F*-ratios). The experimental degree of freedom (DOF) is 15 (factors-DOF, 3) (**Tables 7–9**). Salinity had the maximum contribution (29.12%) on total lipid production, followed by pH (27.97%) and carbon supplementation (27.49%) at the individual level (**Table 7**) (**Figure 3A**). The temperature had minimal impact on the lipid synthesis (6.33%). On the whole, salinity, pH, and carbon supplementation at their individual levels contributed over 95% of the influence on the lipid production, indicating that these factors play crucial role on lipid synthesis. The temperature had the maximum contribution (32.82%) on neutral lipid synthesis at the individual level, followed by salinity (23.08%) and carbon supplementation (14.56%) (**Table 8**) (**Figure 3B**). pH showed minimal impact on the production process (5.84%). Overall, temperature, salinity, carbon supplementation, and pH at their individual levels contributed over 76.3% of the influenced factors in an escalation of neutral lipid synthesis. The temperature showed maximum contribution (85.25%), followed by carbon supplementation on carbohydrate accumulation at the individual level (**Table 9**) (**Figure 3C**). On the whole, temperature and carbon supplementation at their individual levels contributed over 86.013% of the influence on carbohydrates production.

Optimum Conditions for Maximum Productivity
Optimum conditions during SP of operation to achieve higher lipid (total) productivity from microalgae by the selected factors contribution is shown in **Table 10**. The optimum operating conditions

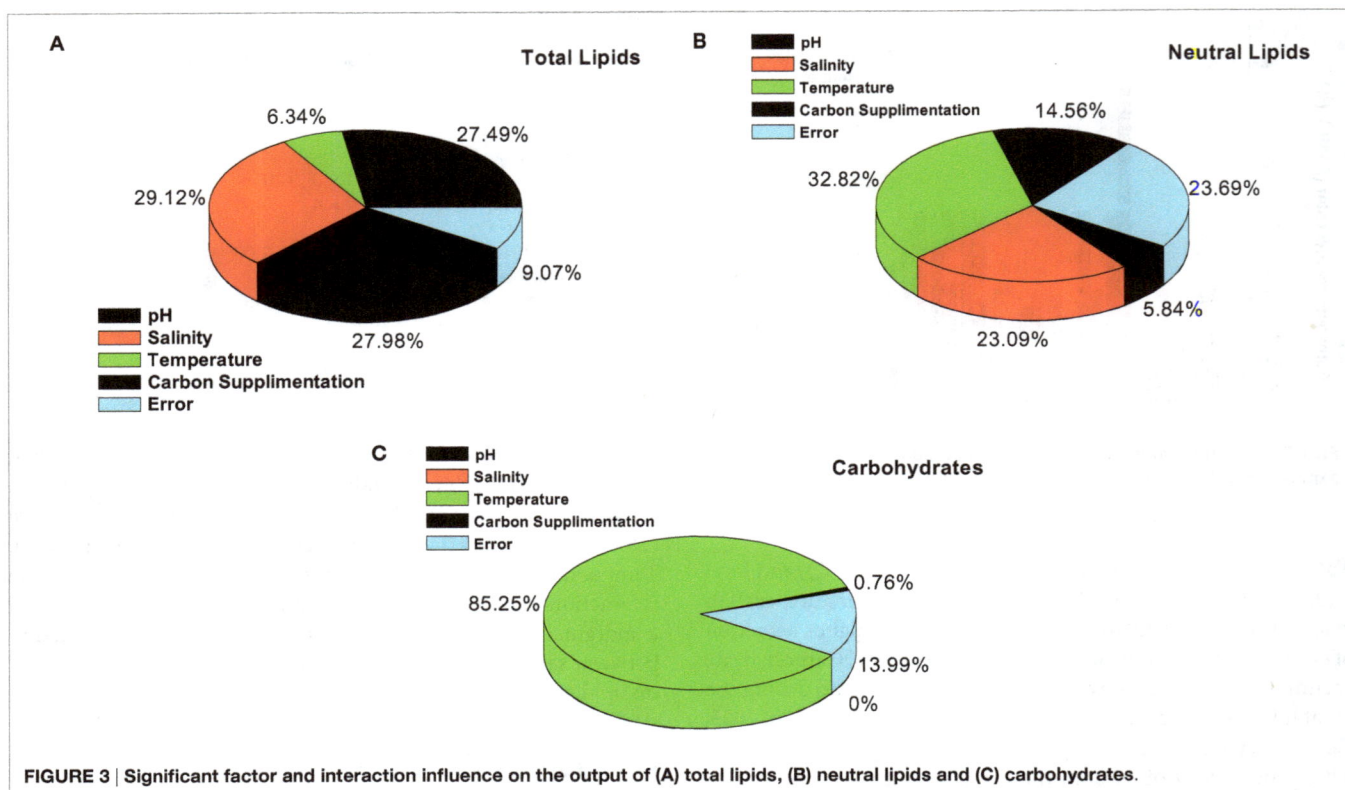

FIGURE 3 | Significant factor and interaction influence on the output of (A) total lipids, (B) neutral lipids and (C) carbohydrates.

TABLE 7 | Analysis of variance (ANOVA) for total lipids production.

S. no.	Factor	DOF	Sum of squares	Variance	F-ratio	Pure sum	Percent
1	pH	3	775.686	258.562	16.416	728.437	27.97
2	Salinity (NaCl g/l)	3	805.427	268.475	17.046	758.177	29.12
3	Temperature (°C)	3	212.247	70.749	4.492	164.997	6.337
4	Carbon supplement (glucose g/l)	3	762.986	254.328	16.147	715.736	27.49
	Other/error	3	47.249	15.749			9.075
	Total	15	2603.597				100

TABLE 8 | Analysis of variance analysis (ANOVA) for neutral lipids synthesis.

S. no.	Factor	DOF	Sum of squares	Variance	F-ratio	Pure sum	Percent
1	pH	3	16.513	5.51	2.233	9.129	5.843
2	Salinity (NaCl g/l)	3	43.476	14.492	5.873	36.074	23.088
3	Temperature (°C)	3	58.681	19.56	7.927	51.28	32.819
4	Carbon supplement (glucose g/l)	3	30.156	10.052	4.074	22.755	14.563
	Other/error	3	7.40	2.47			23.69
	Total	15	156.25				100

TABLE 9 | Analysis of variance (ANOVA) for carbohydrates production.

S. no.	Factor	DOF	Sum of squares	Variance	F-ratio	Pure sum	Percent
1	pH	3	0.331	0.11	0.996	0	0
2	Salinity (NaCl g/l)	3	0.163	0.054	0.489	0	0
3	Temperature (°C)	3	9.436	3.145	28.346	9.103	85.251
4	Carbon supplement (glucose g/l)	3	0.414	0.138	1.244	0.081	0.762
	Other/error	3	0.331	0.11			13.98
	Total	15	10.678				100

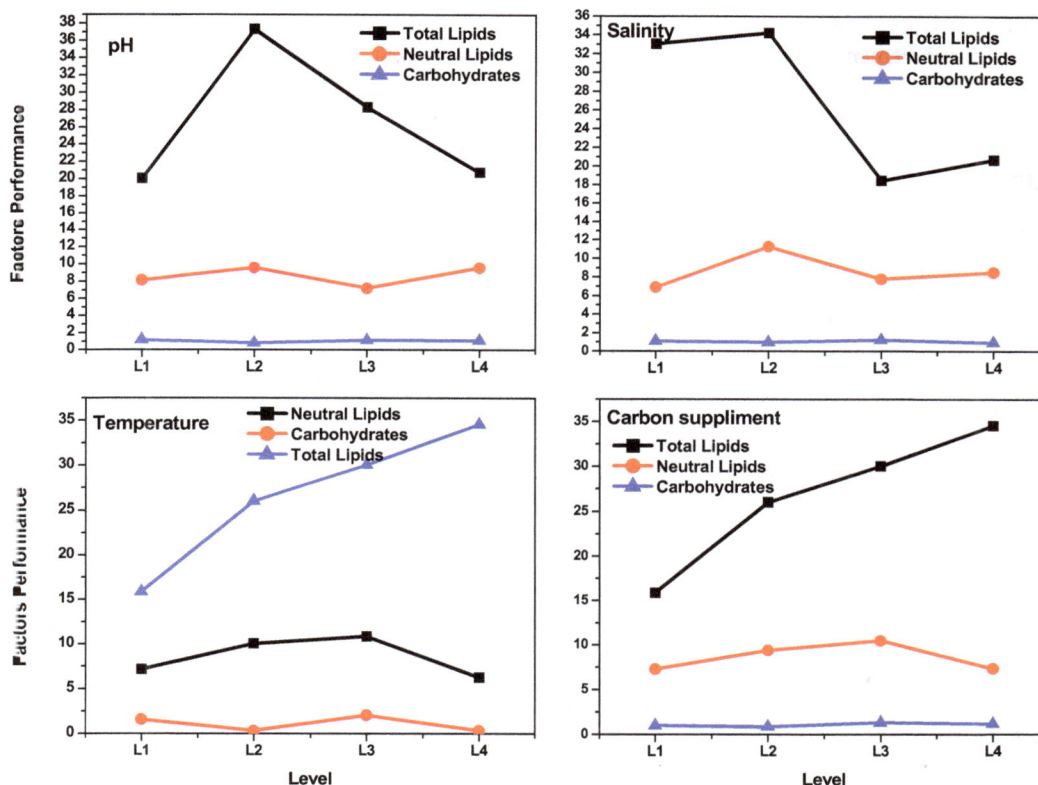

FIGURE 4 | Individual factors performance at different levels.

TABLE 10 | Optimum conditions and their contribution on total lipids.

S. no.	Factor	Level description	Level	Contribution
1	pH	6	2	10.687
2	Salinity (NaCl g/l)	1	2	7.637
3	Temperature (°C)	30	3	4.687
4	Carbon supplement (glucose g/l)	30	4	7.962
Total contribution from all factors				30.97
Current grand average of performance				26.61
Expected result at optimum condition				57.58

TABLE 11 | Optimum conditions and their contribution on neutral lipids.

S. no.	Factor	Level description	Level	Contribution
1	pH	6	2	0.956
2	Salinity (NaCl g/l)	1	2	2.681
3	Temperature (°C)	30	3	2.281
4	Carbon supplement (glucose g/l)	15	3	1.856
Total contribution from all factors				7.77
Current grand average of performance				8.59
Expected result at optimum condition				16.36

showed the requirement of acidic microenvironment (pH 6) with 1 g NaCl/l, 20°C, and 30 g COD/l, which resulted in the increment of lipid (total) production from 14.6 to 57.6% (from 26.6%) (**Figure 4**). A total contribution of factors at optimized conditions was 30.97% for lipid production with a maximum contribution of 10.68% from pH. Salinity was specifically found to be the most significant factor influencing neutral lipid production (**Table 11**). The optimum operating conditions showed the requirement of acidic microenvironment (pH 6) with 1 g NaCl/l, 30°C, and 15 g COD/l, which resulted in the increment of neutral lipid from 8.6% to 16.4% (**Figure 4**). The temperature was found to be the most important factor influencing carbohydrate production (**Table 12**).

TABLE 12 | Optimum conditions and their contribution on carbohydrates.

S. no.	Factor	Level description	Level	Contribution
1	pH	4	1	0.128
2	Salinity (NaCl g/l)	2	3	0.15
3	Temperature (°C)	30	3	0.973
4	Carbon supplement (glucose g/l)	15	3	0.208
Total contribution from all factors				145.9
Current grand average of performance				105.6
Expected result at optimum condition				251.5

The favorable operating conditions showed the requirement of acidic microenvironment (pH 4) with 2 g NaCl/l, 30°C, and 15 g COD/l results in the increment of carbohydrate production from 140 to 251.5 mg/g biomass (**Figure 4**). Carbohydrate profile showed the presence of 35% of xylose and 20% of arabinose.

DISCUSSION

Data envelopment analysis (DEA) technique was used to analyze the relative performance of the system based on the output parameters viz., lipid productivity (total and neutral) and carbohydrates production. DEA offers an easy approach to measure, compare the performance, and analyze the results obtained by interpreting the output of a given system based on the relative efficiency (not on absolute efficiency) using graphical approach (Venkata Mohan et al., 2008). The relative efficiency of any given system indicates the extent to which other systems can improve its performance. Total and neutral productivity and carbohydrates production were considered as three output parameters corresponding to input parameter (pH) (**Table 13**; **Figure 5**). The relative efficiency of the system can be calculated by comparing the current performance of the system to the best possible performance that the system could be reasonably expected to achieve as per the Eq. 1, where, X represents the length of the line from the origin to the point obtained by plotting two ratios of the system and Y denotes length of the line from the origin through the point obtained by the system to the efficient frontier.

$$\text{Relative efficiency} = [(X/Y) \times 100] \quad (1)$$

Experiment 6 (carbon concentration 30 g/l; temperature 20°C; salinity 1 g/l; pH 6), experiment 3 (carbon concentration 15 g/l; temperature 30°C; salinity 2 g/l; pH 4) and experiment 2 (carbon concentration 1 g/l; temperature 25°C; salinity 1 g/l; pH 4) yielded maximum efficiency (100%) among the experimental variations studied with respect to total lipids, neutral lipids and carbohydrates productivities. In all the three cases of data analyses by DEA considering pH as input, experimental points on the efficient frontier lines (relative efficiency 100%) illustrated the efficacy of acidophilic pH (6–4) with salinity (1 or 2 g/l), carbon supplemented with high glucose concentration and temperature in the range of 20–30°C. About six of the experimental variations studied are above 60% of the relative efficiency. Experiments above 96% relative efficiency showed positive influence of acidophilic cultivation.

As shown in **Table 14**, microalgae cultivation at pH 6 increased lipid productivity with more number of SFAs in both experiments 6 and 8. Influence of salinity was observed markedly on increment in palmitic acid (C16:0) fraction. Microalgae can survive in extreme microenvironments; a minor shift in pH in external environment can change their metabolic activity (Varshney et al., 2015). A change in acidophilic environment from pH 6 to 4, microalgae showed transition of biochemical pathway from lipid to carbohydrates synthesis when experimental conditions 6/8 and 3 were compared. Studies also reveal that at acidophilic conditions showed increased basal level of heat shock protein (HSP), higher levels of the antioxidative enzymes, ascorbate peroxidase (APX) and which is thought to be an adaptive mechanism to extreme acidic environment (Gerloff-Elias et al., 2006; Garbayo et al., 2007). The acidophiles also showed increased production of the antioxidants lutein and β-carotene, respectively (Garbayo et al.,

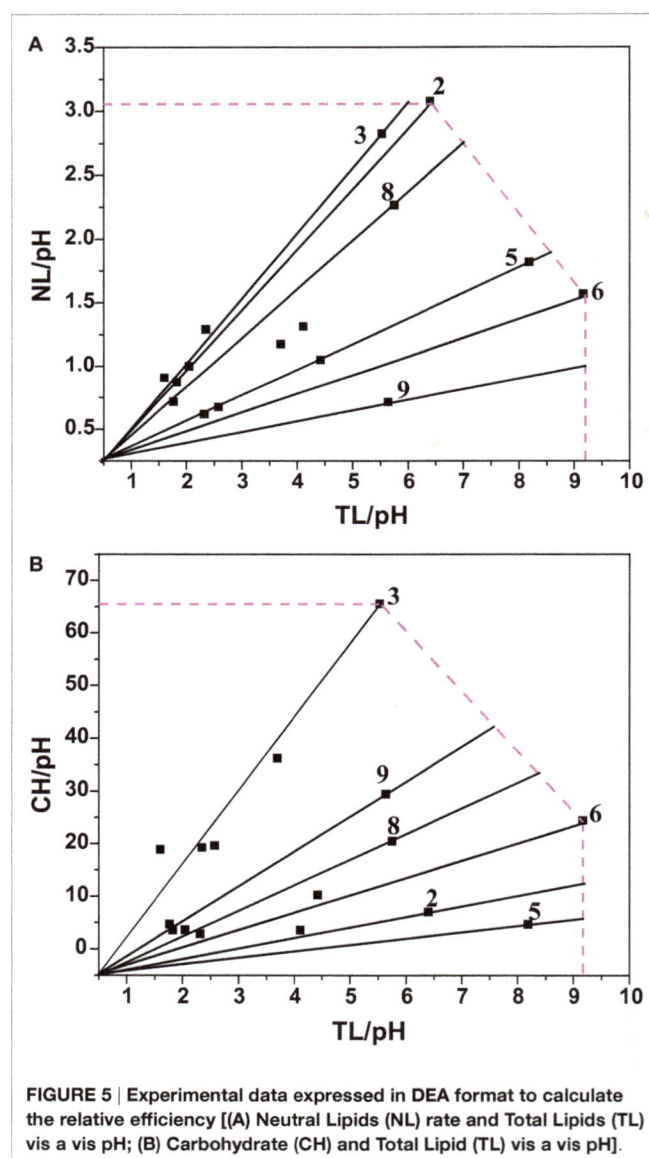

FIGURE 5 | Experimental data expressed in DEA format to calculate the relative efficiency [(A) Neutral Lipids (NL) rate and Total Lipids (TL) vis a vis pH; (B) Carbohydrate (CH) and Total Lipid (TL) vis a vis pH].

TABLE 13 | Consolidated data of DEA analysis.

TL vs NL		TL vs Carbohydrates	
Experiment number	Relative efficiency (%)	Experiment number	Relative efficiency (%)
6	100	6	100
2	100	3	100
5	96.8	2	90.1
3	91.0	5	71.9
8	82.1	9	70.6
9	62.5	8	66.1

TABLE 14 | Consolidate optimum conditions toward biomass and lipid production obtained from the study.

Experiment number	Experimental condition				Total/neutral lipids (% DCW)	Biomass production (g/l)	Lipid productivity (mg/l/day)	SFAs	UFAs		Carbohydrates (mg/g biomass)
	pH	Salinity (g/l)	Temperature (°C)	Carbon supp (g/l)					MUFA	PUFA	
Experiment No. 6	6	1	20	30	55.0/9.4	6.3	21.62	7 No. (C16:0; 40.7%)	5 No. (C18:1; 21.4%)	4 No. (C18:2; 3.6%)	146
Experiment No. 8	6	5	30	1	34.5/13.6	1.8	3.88	6 No. (C16:0; 50.8%)	3 No. (C18:1; 19.4%)	3 No. (C18:2; 6.8%)	123
Experiment No. 3	4	2	30	15	22.1/11.3	3.9	4.21				262
Total lipids (Table 10)[a]	6	1	30	30	57.5						251
Neutral lipids (Table 11)[a]	6	1	30	15	16.3						
Carbohydrates (Table 12)[a]	4	2	30	15							

[a]Optimum condition.

2008; Vaquero et al., 2012). Increase in carbon concentration and salinity favored high number of mono unsaturated fatty acids (MUFA). In experiment No. 6, more number of polyunsaturated fatty acids (PUFAs; structural lipids) with major fractions of oleic acid (C18:1) and linoleic acid (C18:2) was observed, which basically contributes to the high biomass productivity. Higher lipid fraction (55% of DCW) and biomass productivity (39.3 mg/l/day) were obtained with high carbon, minimal salinity, and low temperature conditions (Experiment No. 6).

CONCLUSION

Design of experimental methodology based on Taguchi approach illustrated the specific function of selected factors on microalgae lipid synthesis and carbohydrate production during stress-induced heterotrophic cultivation. Among the four factors, pH, salinity, and carbon concentration showed significant influence on the lipid synthesis. This study reported a high total lipids concentration (55% of DCW; 21.6 mg/l/day), observed in high carbon, minimal salinity, and low temperature conditions. A maximum neutral lipids concentration (13.6% of DCW; 3.8 mg/l/day) was observed in high salinity, low carbon conditions at

acidophilic microenvironment. Acidophilic microenvironment, moderate salinity, and high carbon concentration illustrated significant influence on the carbohydrates synthesis. The FAME analysis documented good biodiesel properties with a higher number of SFAs, especially palmitic acid (C16:0) followed by MUFA and PUFA with major fractions of oleic acid (C18:1) and linoleic acid (C18:2). The study documented feasibility of heterotrophic mode of microalgae cultivation toward regulating higher lipid productivity, along with waste treatment.

AUTHOR CONTRIBUTIONS

All the authors have equally contributed to the manuscript and mutually agreed for submission of the work.

ACKNOWLEDGMENTS

The authors wish to thank Director, Council for Scientific and Industrial Research (CSIR)-IICT, Hyderabad, for encouragement. Authors acknowledge funding from CSIR in the form of Network project – BioEn (CSC-0116). PC acknowledges CSIR for providing research fellowship.

REFERENCES

Adams, C., Godfrey, V., Wahlen, B., Seefeldt, L., and Bugbee, B. (2013). Understanding precision nitrogen stress to optimize the growth and lipid content tradeoff in oleaginous green microalgae. *Bioresour. Technol.* 131, 188–194. doi:10.1016/j.biortech.2012.12.143

Allakhverdiev, S. I., Nishiyama, Y., Takahashi, S., Miyairi, S., Suzuki, I., and Murata, N. (2005). Systematic analysis of the relation of electron transport and ATP synthesis to the photodamage and repair of photosystem II in *Synechocystis*. *Plant Physiol.* 137, 263–273. doi:10.1104/pp.104.054478

APHA. (1998). *Standard Methods for the Examination of Water and Wastewater*, 20th Edn. Washington, DC: American Public Health Association/American Water Works Association/Water Environment Federation.

Ball, S. G., Dirick, L., Decq, A., Martiat, J. C., and Matagne, R. F. (1990). Physiology of starch storage in the monocellular alga *Chlamydomonas reinhardtii*. *Plant Sci.* 66, 1–9. doi:10.1016/0168-9452(90)90162-H

Belkin, S., and Boussiba, S. (1991). Resistance of spirulina platensis to ammonia at high pH values. *Plant Cell Physiol.* 32, 953–958.

Boussiba, S., Vonshak, A., Cohen, Z., Avissar, Y., and Richmond, A. (1987). Lipid and biomass production by the halotolerant microalga *Nannochloropsis salina*. *Biomass* 12, 37–47. doi:10.1016/0144-4565(87)90006-0

Brennan, L., and Owende, P. (2010). Biofuels from microalgae – a review of technologies for production, processing, and extractions of biofuels and co-products. *Renew. Sustain. Energ. Rev.* 14, 557–577. doi:10.1016/j.rser.2009.10.009

Chandra, R., Rohit, M. V., Swamy, Y. V., and Venkata Mohan, S. (2014). Regulatory function of organic carbon supplementation on biodiesel production during growth and nutrient stress phases of mixotrophic microalgae cultivation. *Bioresour. Technol.* 165, 279–287. doi:10.1016/j.biortech.2014.02.102

Chen, G. Q., and Chen, F. (2006). Growing phototrophic cells without light. *Biotechnol. Lett.* 28, 607–616. doi:10.1007/s10529-006-0025-4

Chen, M., Tang, H., Ma, H., Holland, T. C., Ng, K. Y., and Salley, S. O. (2011). Effect of nutrients on growth and lipid accumulation in the green algae

Dunaliella tertiolecta. Bioresour. Technol. 102, 1649–1655. doi:10.1016/j. biortech.2010.09.062

Chiranjeevi, P., and Venkata Mohan, S. (2016). Critical parametric influence on microalgae cultivation towards maximizing biomass growth with simultaneous lipid productivity. *Renew. Energy.* doi:10.1016/j.renene.2016.03.063

Cong, L., and Lu-Kwang, J. (2014). Conversion of wastewater organics into biodiesel feedstock through the predator-prey interactions between phagotrophic microalgae and bacteria. *RSC Adv.* 4, 44026–44029. doi:10.1039/C4RA06374K

Cong, L., Suo, X., and Lu-Kwang, J. (2016). Cultivation of phagotrophic algae with waste activated sludge as a fast approach to reclaim waste organics. *Water. Res* 91, 195–202. doi:10.1016/j.watres.2016.01.021

Dayananda, C., Sarada, R., Usha Rani, M., Shamala, T. R., and Ravishankar, G. A. (2007). Autotrophic cultivation of *Botryococcus braunii* for the production of hydrocarbons and exopolysaccharides in various media. *Biomass Bioenergy* 31, 87–93. doi:10.1016/j.biombioe.2006.05.001

Devi, M. P., and Venkata Mohan, S. (2012). CO_2 supplementation to domestic wastewater enhances microalgae lipid accumulation under mixotrophic microenvironment: effect of sparging period and interval. *Bioresour. Technol.* 112, 116–123. doi:10.1016/j.biortech.2012.02.095

Devi, M. P., Venkata Subhash, G., and Venkata Mohan, S. (2012). Heterotrophic cultivation of mixed microalgae for lipid accumulation and wastewater treatment during sequential growth and starvation phases. Effect of nutrient supplementation. *J. Renew. Energy.* 43, 276–283. doi:10.1016/j.renene.2011.11.021

Garbayo, I., Domínguez, M. J., Vigara, J., and Vega, J. M. (2007). Effect of abiotic stress on Chlamydomonas acidophila viability. *Appl. Microbiol.* 1, 184–189.

Garbayo, I., Cuaresma, M., Vílchez, C., and Vega, J. M. (2008). Effect of abiotic stress on the production of lutein and β-carotene by Chlamydomonas acidophila. *Process Biochem.* 43, 1158–1161. doi:10.1016/j.procbio.2008.06.012

Garcia, O. P., Bashan, Y., and Puente, M. E. (2011). Organic carbon supplementation of sterilized municipal wastewater is essential for heterotrophic growth and removing ammonium by the microalgae *Chlorella vulgaris*1. *J. Phycol.* 47, 190–199. doi:10.1111/j.1529-8817.2010.00934.x

Gerloff-Elias, A., Barua, D., Mölich, A., and Spijkerman, E. (2006). Temperature- and pH-dependent accumulation of heat-shock proteins in the acidophilic green alga Chlamydomonas acidophila. *FEMS Microbiol.* 56, 345–354. doi:10.1111/j.1574-6941.2006.00078.x

Giordano, M. (2001). Interactions between C and N metabolism in *Dunaliella salina* cells cultured at elevated CO_2 and high N concentrations. *J. Integr. Plant Biol.* 158, 577–581. doi:10.1078/0176-1617-00234

Kalita, N., Baruah, G., Goswami, R. C. D., Talukdar, J., and Kalita, M. C. (2011). *Ankistrodesmus falcatus:* a promising candidate for lipid production, its biochemical analysis and strategies to enhance lipid productivity. *J. Microbiol. Biotechnol. Res.* 1, 148–157.

Kumar, A., Ergas, S., Yuan, X., Sahu, A., Zhang, Q., and Dewulf, J. (2010). Enhanced CO_2 fixation and biofuel production via microalgae: recent developments and future directions. *Trends Biotecnol.* 28, 371–380. doi:10.1016/j. tibtech.2010.04.004

Liu, L., Huang, J., Sun, Z., Zhong, Y., Jiang, Y., and Chen, F. (2011). Differential lipid and fatty acid profiles of photoautotrophic and heterotrophic *Chlorella zofingiensis:* assessment of algal oils for biodiesel production. *Bioresour. Technol.* 102, 106–110. doi:10.1016/j.biortech.2010.06.017

Merchant, S. S., Kropat, J., Liu, B., Shaw, J., and Warakanont, J. (2012). TAG, you're it *Chlamydomonas* as a reference organism for understanding algal triacylglycerol accumulation. *Curr. Opin. Biotechnol.* 23, 352–363. doi:10.1016/j. copbio.2011.12.001

Morales-Sánchez, D., Tinoco-Valencia, R., Kyndt, J., and Martinez, A. (2013). Heterotrophic growth of *Neochloris oleoabundans* using glucose as a carbon source. *Biotechnol. Biofuels* 6, 100. doi:10.1186/1754-6834-6-100

Murata, N., Takahashi, S., Nishiyama, Y., and Allakhverdiev, S. I. (2007). Photoinhibition of photosystem II under environmental stress. *Biochim. Biophys. Acta* 1767, 414–421. doi:10.1016/j.bbabio.2006.11.019

Neale, P. J., and Melis, A. (1989). Salinity-stress enhances photoinhibition of photosynthesis in *Chlamydomonas reinbardtii. J. Plant Physiol* 134, 619–622. doi:10.1016/S0176-1617(89)80158-0

Perez-Garcia, O., Escalante, F. M. E., de-Bashan, L. E., and Bashan, Y. (2011). Heterotrophic cultures of microalgae: metabolism and potential products. *Water Res.* 45, 11–36. doi:10.1016/j.watres.2010.08.037

Qiao, H., and Wang, G. (2009). Effect of carbon sources on growth and lipid accumulation in *Chlorella sorokiniana* GXNN01. *Chin. J. Oceanol. Limnol.* 27, 762–768. doi:10.1007/s00343-009-9216-x

Sharma, K. K., Schuhmann, H., and Schenk, P. M. (2012). High lipid induction in microalgae for biodiesel production. *Energies* 5, 1532–1553. doi:10.3390/ en5051532

Subhash, G. V., Rohit, M. V., Devi, M. P., Swamy, Y. V., and Venkata Mohan, S. (2014). Temperature induced stress influence on biodiesel productivity during mixotrophic microalgae cultivation with wastewater. *Bioresour. Technol.* 169, 789–793. doi:10.1016/j.biortech.2014.07.019

Taguchi, G. (1986). *Introduction to Quality Engineering: Designing Quality in to Products and Processes.* Tokyo: Asian Productivity Organization.

Talebi, A. F., Tabatabaei, M., Mohtashami, S. K., Tohidfar, M., and Moradi, F. (2013). Comparative salt stress study on intracellular ion concentration in marine and salt-adapted freshwater strains of microalgae. *Nat. Sci. Biol.* 5, 309–315.

Varshney, P., Mikulic, P., Vonshak, A., Beardall, J., and Wangikar, P. P. (2015). Extremophilic micro-algae and their potential contribution in biotechnology. *Bioresour. Technol.* 184, 363–372. doi:10.1016/j.biortech.2014.11.040

Vaquero, I., Ruiz-Domínguez, M. C., Márquez, M., and Vílchez, C. (2012). Cu-mediated biomass productivity enhancement and lutein enrichment of the novel microalga Coccomyxa onubensis. *Process Biochem.* 47, 694–700. doi:10.1016/j.procbio.2012.01.016

Venkata Mohan, S., Chandrasekhara Rao, N., Krishna Prasad, K., Muralikrishna, P., Sreenivasa Rao, R., and Sarma, P. N. (2005). Anaerobic treatment of complex chemical wastewater in a sequencing batch biofilm reactor: process optimization and evaluation of factors interaction using the Taguchi dynamic DOE methodology. *Biotechnol. Bioeng.* 90, 732–745. doi:10.1002/bit.20477

Venkata Mohan, S., and Devi, M. P. (2014). Salinity stress induced lipid synthesis to harness biodiesel during dual mode cultivation of mixotrophic microalgae. *Bioresour. Technol.* 165, 288–294. doi:10.1016/j.biortech.2014.02.103

Venkata Mohan, S., Devi, M. P., Mohanakrishna, G., Amarnath, N., Babu, M. L., and Sarma, P. N. (2011). Potential of mixed microalgae to harness biodiesel from ecological water-bodies with simultaneous treatment. *Bioresour. Technol.* 102, 1109–1117. doi:10.1016/j.biortech.2010.08.103

Venkata Mohan, S., Lalit Babu, V., and Sarma, P. N. (2008). Effect of various pretreatment methods on anaerobic mixed microflora to enhance biohydrogen production utilizing dairy wastewater as substrate. *Bioresour. Technol.* 99, 59–67. doi:10.1016/j.biortech.2006.12.004

Venkata Mohan, S., Rohit, M. V., Chiranjeevi, P., Chandra, R., and Navaneeth, B. (2015). Heterotrophic microalgae cultivation to synergize biodiesel production with waste remediation: progress and perspectives. *Bioresour. Technol.* 184, 169–178. doi:10.1016/j.biortech.2014.10.056

Venkata Mohan, S., Sirisha, K., Sreenivasa Rao, R., and Sarma, P. N. (2007). Bioslurry phase remediation of chlorpyrifos contaminated soil: process evaluation and optimization by Taguchi design of experimental (DOE) methodology. *Ecotoxicol. Environ. Saf.* 68, 252–262. doi:10.1016/j.ecoenv.2007.06.002

Xia, J. R., and Gao, K. S. (2005). Impacts of elevated CO_2 concentration on biochemical composition, carbonic anhydrase, and nitrate reductase activity of freshwater green algae. *J. Integr. Plant Biol.* 47, 668–675. doi:10.1111/j.1744-7909.2005.00114.x

Zhang, Y. M., Chen, H., He, C. L., and Wang, Q. (2013). Nitrogen starvation induced oxidative stress in an oil-producing green alga *Chlorella sorokiniana* C3. *PLoS ONE* 8:69225. doi:10.1371/journal.pone.0069225

Zheng, S. Z., Wang, D. Y., Zhang, Q., and Shen, X. W. (1997). Chemical studies on the polysaccharide from *Dunaliella salina* (I). *J. Northwest Normal Univ.* 33, 93–95.

Conflict of Interest Statement: The authors declare that the research was conducted in the absence of any commercial or financial relationships that could be construed as a potential conflict of interest.

3

Net-Energy Analysis of Integrated Food and Bioenergy Systems Exemplified by a Model of a Self-Sufficient System of Dairy Farms

Edited by:
Luís Alexandre Duque Moreira De Sousa,
Public Research Centre Henri Tudor, Luxembourg

Reviewed by:
Ramesh Lekshmana,
Dr. MGR Educational and Research Institute, India
Michae Carbajales-Dale,
Global Climate and Energy Project, USA

*Correspondence:
Hanne Østergård
haqs@kt.dtu.dk

†Present address:
Piotr Oleskowicz-Popiel,
Institute of Environmental Engineering, Poznan University of Technology, Poznan, Poland;
Jens Ejbye Schmidt,
Institute Center for Energy (iEnergy),
Masdar Institute of Science and Technology, Abu Dhabi,
United Arab Emirates

Specialty section:
This article was submitted to Energy Systems and Policy,
a section of the journal
Frontiers in Energy Research

Mads Ville Markussen[1], Siri Pugesgaard[2], Piotr Oleskowicz-Popiel[1†], Jens Ejbye Schmidt[1†] and Hanne Østergård[1]*

[1]Department of Chemical and Biochemical Engineering, Technical University of Denmark, Kgs Lyngby, Denmark,
[2]Department of Agroecology, Aarhus University, Tjele, Denmark

Agriculture is expected to contribute in substituting of fossil fuels in the future. This constitutes a paradox as agriculture depends heavily on fossil energy for providing fuel, fodder, nutrients, and machinery. The aim of this paper is to investigate whether organic agriculture is capable of providing both food and surplus energy to the society as evaluated from a model study. We evaluated bioenergy technologies in a Danish dairy-farming context in four different scenarios: (1) vegetable oil based on oilseed rape, (2) biogas based on cattle manure and grass-clover lays, (3) bioethanol from rye grain and whey, and (4) a combination of (1) and (2). When assessing the energetic net-contribution to society from bioenergy systems, two types of problems arise: how to aggregate non-equivalent types of energy services and how to account for non-equivalent types of inputs and coproducts from the farming? To avoid the first type, the net output of liquid fuels, electricity, useful heat, and food were calculated separately. Furthermore, to avoid the second type, all scenarios were designed to provide self-sufficiency with fodder and fertilizer and to utilize coproducts within the system. This approach resulted in a transparent assessment of the net-contribution to society, which is easy to interpret. We conclude that if 20% of land is used for energy crops, farm-gate energy self-sufficiency can be achieved at the cost of 17% reduction in amount of food produced. These results demonstrate the strong limitations for (organic) agriculture in providing both food and surplus energy.

Keywords: dairy farms, self-sufficiency, net-energy, vegetable oil, biogas, bioethanol, organic farming

Citation:
Markussen MV, Pugesgaard S, Oleskowicz-Popiel P, Schmidt JE and Østergård H (2015) Net-Energy Analysis of Integrated Food and Bioenergy Systems Exemplified by a Model of a Self-Sufficient System of Dairy Farms.
Front. Energy Res. 3:49.

INTRODUCTION

Agricultural production depends on fossil energy for providing fuels, fertilizer, pesticides, fodder, and machinery (Østergård et al., 2010; Pelletier et al., 2011). Consequently, food production is vulnerable to fluctuating and rising oil prices (Neff et al., 2011). At the same time, agriculture is expected to contribute in substitution of depleting fossil energy sources for the society (Farrell et al., 2006; EU, 2009; Cherubini, 2010). This constitutes a paradox. In addition, the production of biofuels from energy crops will require replacing land for food with land for energy. For instance, it has been shown that even drastic land use changes may only provide 7–20% of total Danish energy supply in economically feasible scenarios (Callesen et al., 2010).

The limitations of biofuels as an energy source for the contemporary industrial economy may be considered in the light of history. Before the era of fossil fuels, agriculture was together with forestry the main source of net-energy in the society by providing food for human labor, fodder for draft animals, and biomass for heating (Odum, 2007; Hall et al., 2009). During the nineteenth and twentieth centuries, coal, oil, and natural gas took over as society's main source of net-energy and agriculture was industrialized in the way that farms were supplied with oil- and industrial-based inputs. Farmers no longer needed to produce fodder for draft animals because they could import oil to power their machinery. On-farm production of organic fertilizers was substituted by import of commercial fertilizers produced by the use of fossil energy, and manual or mechanical weeding was substituted by applying fossil fuel-based pesticides (Hall et al., 1986; Conforti and Giampietro, 1997). Altogether, the productivity per hectare was boosted with the consequence that food supply systems now uses 4–10 times more fossil energy than the food energy they produce (Heller and Keoleian, 2003; Markussen and Østergård, 2013), i.e., agriculture became a net-energy sink. If agriculture should play a significant role in the future energy system, then the first milestone to be achieved would be to become net-energy neutral, e.g., self-sufficient with fuels.

Organic agriculture has taken the first step in reducing the dependency of fossil energy by using neither fossil-based fertilizer nor pesticides (IFOAM, 2014). However, although the omission of pesticides and mineral fertilizers reduces the external energy use for organic agriculture, organic production still depends on fossil fuels for both fuel and electricity (Dalgaard et al., 2001; Halberg et al., 2008). The next step in making agriculture into a net-energy provider is to implement strategies that increase farm energy output. In this respect, it is important to pay attention to the various non-equivalent energy carriers, which are needed in agriculture (i.e., liquid fuels, electricity, and heat) and in which way they can be provided by biomass. Scenarios for energy production have been investigated for different organic farming systems (Halberg et al., 2008; Karpenstein-Machan, 2001; Fredriksson et al., 2006; Østergård and Markussen, 2011; Oleskowicz-Popiel et al., 2012). When comparing such scenarios, two types of problems typically arise: (1) How to aggregate non-equivalent energy data, i.e., joules of energy of different quality providing different services like electricity, liquid fuel, and heat (Giampietro, 2006)? This problem is often either neglected by simply balancing inputs and outputs in joules disregarding their lack of substitutability (Fredriksson et al., 2006; Pugesgaard et al., 2013) or solved by converting to monetary units (Karpenstein-Machan, 2001) or by applying assumptions regarding how much fossil fuel is replaced in the surrounding society (Halberg et al., 2008). (2) How to compare scenarios with different types and amounts of inputs and/or outputs? In a system perspective, changing one component of the system implies that many other components are affected. This problem has often been addressed by system expansion (Halberg et al., 2008) or allocation based on heating value or economical values of output (Fredriksson et al., 2006). Common

to most approaches to the two types of problems is that they summarize the net-energy balance of the specific system in one number and thus facilitate direct comparison between different systems. However, they also result in a severe loss of information regarding the qualitative nature of the output, which may make results problematic to explain and interpret (Giampietro, 2004; Giampietro et al., 2006).

The aim of this work is to assess the possibilities of contemporary dairy farms to become net-energy providers at the same time as producing food. Based on a Danish organic dairy farm model system for a specific land area, four scenarios for bioenergy production are analyzed. The four scenarios are designed to address the two types of problems outlined above. The first problem is approached by balancing energy consumption and production of liquid fuels, electricity, and useful heat separately. The second problem is approached by designing all scenarios such that the farms are self-sufficient with fertilizers and fodder, and all coproducts from the farms are used within the farming system as fodder or fertilizer. In this way, the consumption and production of the three energy products and food can be compared among the scenarios.

Technologies for Producing Bioenergy in Organic Agriculture

The bioenergy technologies of interest for our scenarios are the production of vegetable oil from oilseed rape, biogas from manure, and grass-clover and bioethanol from grain and whey. Their specific characteristics in relation to our design of scenarios are described below.

Producing vegetable oil from oilseed rape to be used in modified diesel engines is a simple and inexpensive process (Karpenstein-Machan, 2001), which is practically independent of scale in terms of effectiveness. A trade-off of this technology is that oilseed rape, especially in organic agriculture under Danish growing conditions, is considered a problematic crop due to high fertilizer needs and to high risk of pests. The latter risk increases with increasing density, and therefore no more than 10% land should be used for oilseed rape (Halberg et al., 2008). The residual oilcakes can be used as fodder on the farm making it useful for dairy farms.

Biogas is useful in organic agriculture because at the same time it produces energy and an effluent that can be applied as fertilizer (Rehl and Müller, 2011; Johansen et al., 2013). Grass-clover has been suggested as feedstock for biogas in organic agriculture (Stinner et al., 2008; Pugesgaard et al., 2013) as clover is a nitrogen-fixing plant that contributes with extra nitrogen to the other crops in the rotation. During the anaerobic digestion process, part of the organic N in the feedstock is mineralized implying that the ammonia content of the digestate is higher than in the feedstock (Halberg et al., 2008). Therefore, using digested plant biomass and manure as a fertilizer may increase the yields of crops as compared to using undigested biomass (Stinner et al., 2008; Pugesgaard et al., 2013). The methane in the biogas can be converted to electricity or used as motor fuel. In practice, however, biogas may not be a preferable motor fuel on the farm because the biogas is produced throughout the year, and the motor fuel is needed in seasonal peaks for crop management. If the biogas is stored as a gas, it takes up enormous space, and if it is stored as compressed gas, it takes large investment in pressure-proofed storage (Fredriksson et al., 2006). The biogas could be utilized continuously in a combined heat and power facility (CHP), producing hot water and electricity to the grid (Karpenstein-Machan, 2001). However, depending on the specific location of the biogas plant it may be difficult to utilize the potential useful heat. Another possibility is to upgrade the biogas and distribute it via the natural gas grid or use it as fuel in the transportation sector (Ahlgren et al., 2010).

Bioethanol has the advantage of being a liquid fuel that to some extend can be handled in the existing infrastructure and supplement gasoline as a fuel for some combustion engines. However, it has limited use in agriculture since most existing machinery and trucks run on diesel, which cannot easily be substituted by ethanol. Another disadvantage is that distilling of bioethanol is energy intensive and should preferably be done in large-scale facilities capable of reusing thermal energy and utilizing waste heat (Gan and Smith, 2011). Starch-based feedstock for bioethanol production is often wheat or maize, but at a dairy farm, returned whey combined with a cereal, e.g., rye, may be used for energy production (Kádár et al., 2011). The residuals from the bioethanol process dried distillers grains with solubles (DDGS) can be used at the farm as fodder.

MATERIALS AND METHODS

In this study, we model farm-level production and consumption of energy and analyze a reference scenario and four bioenergy scenarios (vegetable oil, biogas, bioethanol, and combined vegetable oil and biogas) which aim at increasing energy output. The farm model accounts for flows of energy and material based on published data for input and output parameters. The model consists of a farm community of 10 identical dairy farms of 100 ha each (specified below) representing full time organic dairy farms on loamy soil in accordance to Pugesgaard et al. (2013). However, to isolate the net-energetic contributions of these farms, the otherwise representative farms are modified to have no import of fodder and fertilizer (**Figure 1**), and the number of animals is adjusted accordingly to match on-farm fodder production.

In the reference scenario, the only output is cheese and animals for slaughtering. Whey resulting from the cheese production is returned to the farms for fodder or as feedstock in the bioethanol scenario. In the bioenergy scenarios, either 10 or 20% of land is allocated to bioenergy feedstock production (**Figure 2**) to consider a range, which is within reach (Pugesgaard et al., 2013). It is assumed that the average distance from each farm to the shared dairy and biogas facility is 5 km. The assumption is arbitrary but is considered as a best case based on the structure of Danish organic agriculture. The main components of the model are production and consumption of liquid fuels, electricity, heat, fodder, and food from a self-sufficient farming system on a specific land area. Each of these energy services is balanced separately to emphasize the lack of substitutability between them. They are all accounted in joules to be able to compare the numbers:

(1) Liquid fuel (i.e., diesel, vegetable oil, and bioethanol). Diesel is used for crop cultivation and for transportation of different substances specified in each scenario.

(2) Electricity. Electricity is used for powering the livestock houses and the bioenergy facilities.

(3) Useful heat. The only consumption of heat takes place at the biogas plant itself as the livestock houses are unheated and potential energy requirements for drying cereals not are considered. Housing for the people at the farm is outside the system boundary.

(4) Fodder. Fodder is modeled in Scandinavian Fodder Unit (SFU) which is defined as "12 MJ of metabolizable energy or the equivalent to the fodder value in 1 kg barley" (Dalgaard et al., 2001). In all scenarios, production and consumption is balanced such that all fodders are used on the farm. This implies that the number of cows and amount of produced milk are adjusted in each scenario according to the amount of produced fodder. The comparison between the scenarios is still possible as it is self-sufficient agricultural systems covering a specific land area and with different energy production which are compared

(5) Food. Food is accounted in food energy from a human nutritional perspective. Production and consumption is balanced based on UN's recommendation for a healthy life, which is a daily intake of minimum 8.8 MJ (2100 kcal) (United Nations, 2014).

Notably, indirect energy requirements to provide inputs and capital investments are omitted in the analysis. In addition, energy costs in the dairy process are not accounted for as the system boundaries are at the farm gate including the shared biogas facility.

The scenarios have different N-dynamics. This dynamic is not considered in our model since nutrients are not a limiting factor due to the availability of animal manure and a large grass-clover area for fodder (see the nutrient balance in Pugesgaard et al., 2013). Using digested plant biomass and manure as a fertilizer may increase the yields of crops as compared to undigested biomass (Stinner et al., 2008; Pugesgaard et al., 2013). In this study, we assume that crop yields are the same independently of whether the manure is treated or not.

Finally, for consistency best-case assumptions, i.e., assumptions that are favorable for increasing production and reducing consumptions have been chosen whenever possible.

The Reference Scenario
In this section, the farm model for the reference scenario (**Figure 1**) is described in terms of data sources and assumptions and consumption and production of energy and materials.

Crops
Data for full-time organic dairy farms on loamy soil have been chosen for the analysis in accordance to Pugesgaard et al. (2013). The mix of crops grown (**Table 1**) corresponds to the average mix for Danish organic dairy farms in 2006 on non-irrigated loamy soil based on data from the Danish annual farm account statistics (StatBankDanmark, 2007). This gives 10–30% better

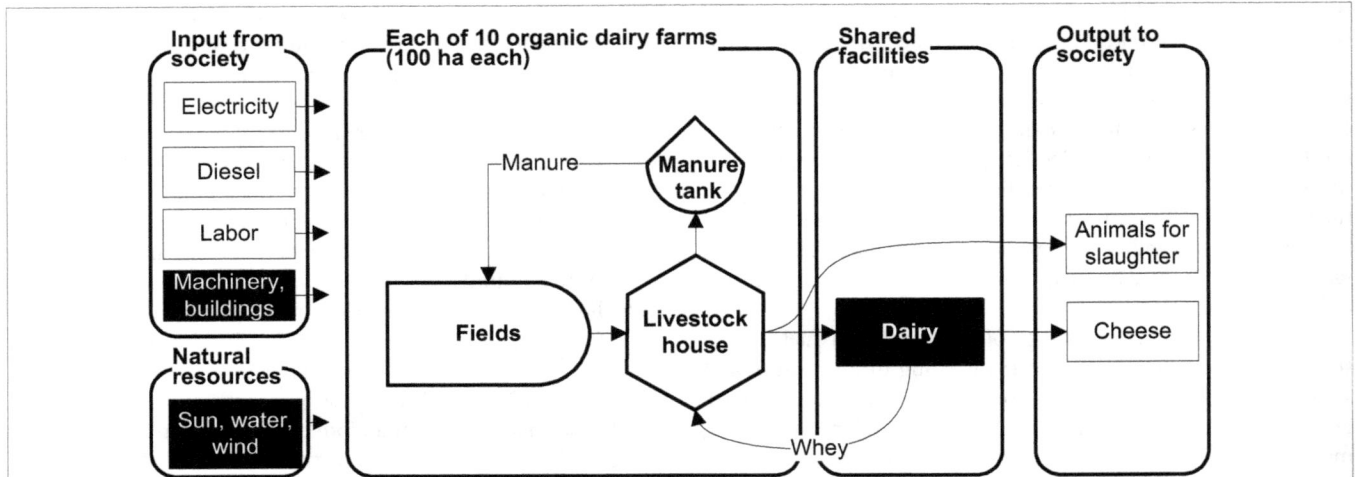

FIGURE 1 | Material and energy flows in the reference scenario. The system consists of 10 identical farms and a shared dairy. Based on the natural resources and input from the society, each farm produces fodder for the livestock. The cows produce manure, animals for slaughter and milk for the dairy. The dairy then produces cheese and whey that are fed back to the cows as fodder. The manure is used as fertilizers on the fields. The dairy is included as a black box because energy and labor use in this process is not accounted for, neither is indirect energy use for the production of machinery and buildings. For further explanation, see text (see The Reference Scenario).

yield than the more representative sandy soil type in Denmark, but it has been chosen as a best-case assumption. The total fodder production from grazed areas makes up 20% of the total fodder production measured in SFU, which corresponds to the amount of time that the livestock is spending grazing per year according to Danish regulation (specified below). Remaining grass-clover ley is cut four times per year. Yields in SFU and dry matter (DM) are given in **Table 2**. All crops are used as fodder, and all straws are used for livestock management.

Livestock

The livestock diet consists of 78% whole crop (grass-clover, maize, barley, and peas), 16% cereals and peas, and 6% whey from the cheese production with a fodder value of 15 kg whey per SFU (Møller et al., 2005). No specific calculations have been done on the protein intake of the livestock. The milk production per cow per year is 7969 kg milk corresponding to an average organic dairy farm cow (StatBankDanmark, 2007). The consumption of fodder is calculated according to Olesen et al. (2006). It is assumed that each milk-producing cow has one offspring per year. The production of meat is estimated to 195 kg of meat per year per milk-producing cow (StatBankDanmark, 2007).

Milk is transported in average 5 km to the shared dairy where it is converted to cheese and the whey is transported back to the farms and utilized as fodder for the livestock. The number of milk-producing cows is calculated as the maximum herd size possible based on the fodder produced on the farms including the whey from the cheese production. The results of combination of data and the assumptions above are found in **Tables 3** and **4**.

Manure

The production of manure is estimated according to the Danish standards that take into account the yield of milk and the

consumption of fodder (Poulsen, 2008). Based on this, each lactating cow produces 19.7 ton/year of manure, and each calf of 0–6 months produces 1.9 ton/year and young animals and of 6–28 months produces 6.5 ton/year. A fraction of the manure corresponding to the time the cows spend outside is assumed to be excreted at the grazing fields and is not available for collection. It is assumed that lactating cows are grazing 8 h/day half of the year, meaning that 17% of year they are outside. Calves of 0–6 months are outside 25%, and calves of 6–28 months are outside 45% of the year (according to Danish regulation of organic dairy cows). All manures collected in the livestock house are spread on the fields as a fertilizer. The results of combination of data and the assumptions above are found in **Table 3**.

Energy Consumption and Production

Energy input from society is calculated in terms of electricity, diesel, and labor. Electricity use is 6.6 GJ/year per milk-producing unit (one cow + one heifer) (Refsgaard et al., 1998). The diesel use for crop production (**Table 2**) is calculated based on the Danish Agricultural Advisory Service (2008), which prescribes the fields operation needed for each crop, and Dalgaard et al. (2001), which estimate the diesel use for the field operations per hectare or per ton of harvested biomass. Diesel use for mowing of whole crop peas, once per year, and grass-clover, four times per year, is based on 5 l/ha. Chopping and handling is calculated based on the weight of the whole crop, where the weight is calculated according to the DM content of ensilaged crops (Møller et al., 2005). Manure collected in the livestock house is assumed to be spread equally on all fields with 14.3 ton/ha. For this, 0.3 l diesel is used per ton of manure. Diesel use for transporting milk to the dairy and whey back to the farm is included with the value of 1 MJ/t km as a transport of liquids in truck exclusive empty return (Berglund and Börjesson, 2006).

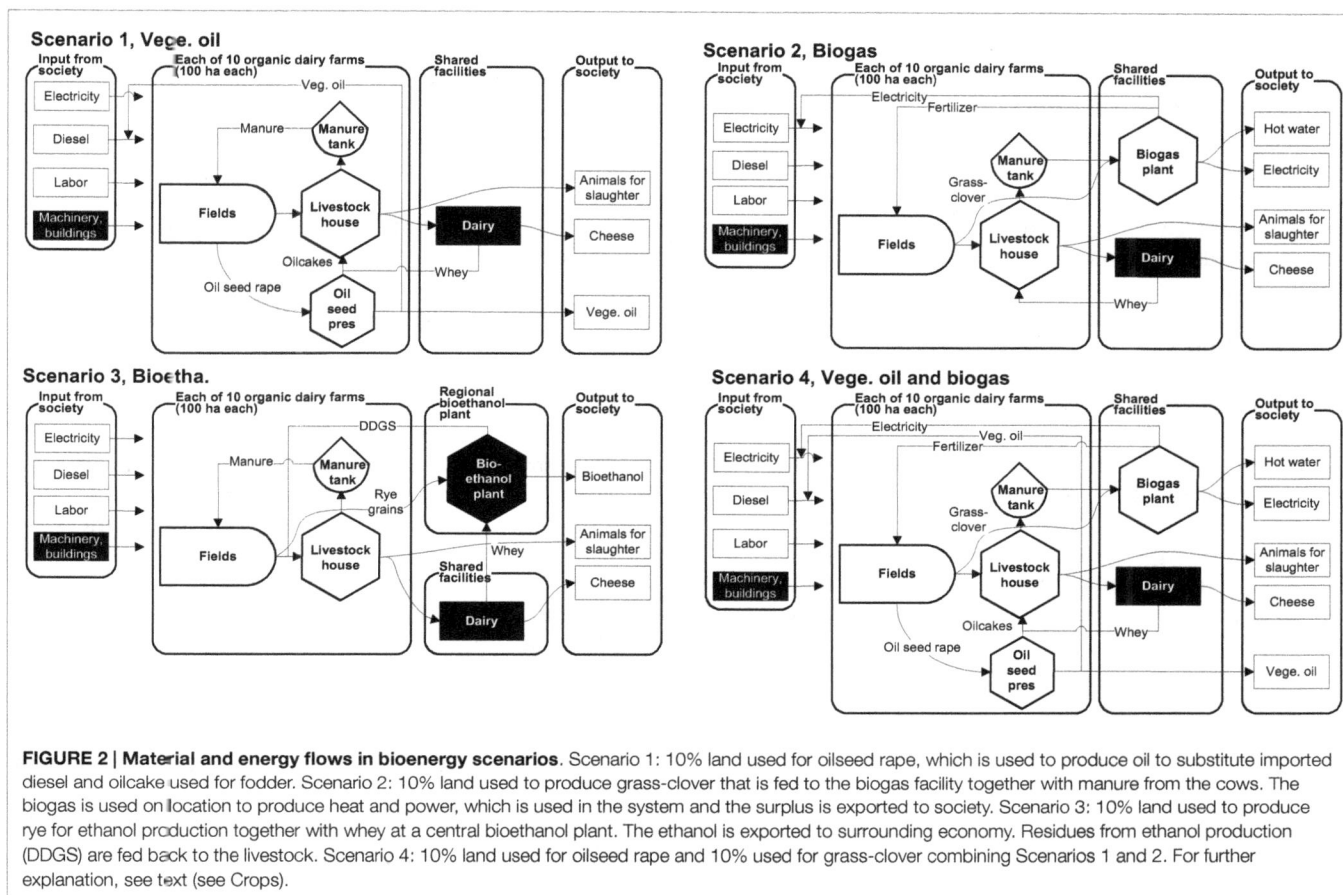

FIGURE 2 | Material and energy flows in bioenergy scenarios. Scenario 1: 10% land used for oilseed rape, which is used to produce oil to substitute imported diesel and oilcake used for fodder. Scenario 2: 10% land used to produce grass-clover that is fed to the biogas facility together with manure from the cows. The biogas is used on location to produce heat and power, which is used in the system and the surplus is exported to society. Scenario 3: 10% land used to produce rye for ethanol production together with whey at a central bioethanol plant. The ethanol is exported to surrounding economy. Residues from ethanol production (DDGS) are fed back to the livestock. Scenario 4: 10% land used for oilseed rape and 10% used for grass-clover combining Scenarios 1 and 2. For further explanation, see text (see Crops).

TABLE 1 | Mix of crops in percentage of total area in each scenario.

Crops	Reference scenario[a]	Scenario 1 (Vege. oil)[b]	Scenario 2 (Biogas)[b]	Scenario 3 (Bioetha.)[b]	Scenario 4 (Vege. oil and biogas)[b]
Spring barley	5.3	4.8	4.8	4.8	4.2
Winter wheat	4.6	4.1	4.1	4.1	3.7
Spring wheat	2.2	2.0	2.0	2.0	1.8
Winter rye	0.9	0.8	0.8	0.8	0.7
Winter triticale	2.7	2.4	2.4	2.4	2.2
Oats	5.4	4.9	4.9	4.9	4.3
Peas	0.4	0.4	0.4	0.4	0.3
Peas – whole crop	2.9	2.6	2.6	2.6	2.3
Maize – whole crop	5.4	4.9	4.9	4.9	4.3
Barley – whole crop	10.5	9.5	9.5	9.5	8.4
Grass-clover	27.1	24.4	24.4	24.4	21.7
Grass-clover, 70% grazed	20.0	18.0	18.0	18.0	16.0
Permanent grass	12.6	11.3	11.3	11.3	10.1
Oilseed rape (energy crop)	0	10.0	0	0	10.0
Grass-clover (energy crop)	0	0	10.0	0	10.0
Winter rye (energy crop)	0	0	0	10.0	0

[a]The average mix of crop for Danish organic dairy farms on loamy soil according to StatBankDanmark (2007).
[b]The bioenergy scenarios have 10 (1–3) or 20% (4) energy crops but maintain the same relative mix of fodder crops.

The input of labor is accounted for as the amount of food energy needed to feed the labor force. It is assumed that each farm employs two workers full time.

Output to society is calculated in terms of food energy. Cheese is evaluated as the food energy in milk minus the food energy in whey (Møller et al., 2005). Food energy value of meat is estimated based on the 1050 kcal/kg of live weight cows (Syrstad, 1993).

Bioenergy Scenarios

In four alternative scenarios, different strategies aiming at increasing energy production by means of vegetable oil, biogas, and bioethanol are explored. Flows of materials and energy for the four scenarios are illustrated in **Figure 2**. In Scenarios 1–3, 10% of the arable land is used for producing energy crops, and in Scenario 4, 20% is used for energy crops (**Table 1**). On the remaining land, the relative mix of fodder crops is the same as in the reference scenario. The milk and manure productions per cow are also kept constant, but the number of cows is reduced according to the reduced fodder production (**Table 3**). Energy use in cultivation and livestock management and food output is calculated in the same way as in the reference scenario except that it is adjusted according to the new mix of crops (**Table 1**) and number of animals (**Table 3**).

Scenario 1 – Vegetable Oil

Each farm uses 10% land to produce oilseed rape. The seeds are used in a farm scale cold press system that produces vegetable oil to substitute diesel as well as oilcakes, which are used as fodder (**Figure 2**). Average yield of the oilseed rape is 2200 kg/ha for oilseed rape (85% DM) on fertile loamy soil (Danish Agricultural Advisory Service, 2008). Oil yield is 33% of oilseed yield, and lower heating value is 37 MJ/kg corresponding to 34 MJ/l. The vegetable oil is assumed to substitute diesel on a 1:1 J basis (Halberg et al., 2008). Fodder value of oilcakes is 1.1 SFU/kg (Møller et al., 2005). Electricity for the oil press is included as 1.1 kWh/30 kg seeds (Jørgensen and Dalgaard, 2004). The straw is used for other purposes at the farms, and it is assumed that no extra labor is needed at the farms to produce the vegetable oil.

Scenario 2 – Biogas

The 10 farms share a biogas facility to take reduced costs. The average distance between biogas facility and each farm is 5 km. The biogas reactor is fed with mixture of manure and grass-clover from 10% of land. The biogas is used to generate electricity and useful heat on location. Part of it is used in the system, and the surplus is exported to the society. The biogas effluent is spread on the fields as fertilizer (**Figure 2**). Transport of manure from the 10 farms and transport of the biogas effluent back to the farms require 1 MJ/t km (Berglund and Börjesson, 2006), and the weight of the digested feedstock is assumed to be equal to the weight of manure and grass-clover, respectively. Transport of grass-clover is accounted for as 0.7 MJ/ton km (Berglund and Börjesson, 2006). The biogas yield from manure and grass-clover is 6.2 and 10.6 GJ per dry ton, respectively (Berglund and Börjesson, 2006). The use of electricity and heat is 26 and 190 MJ, respectively, per ton manure feedstock and 92 and 540 MJ, respectively, per ton grass-clover feedstock (Börjesson and Berglund, 2006). The CHP unit is assumed to convert the raw biogas to electricity and heat with an efficiency of 32 and 55%, respectively (Jungbluth et al., 2007). It is assumed that one person full time is employed at the shared biogas facility.

Scenario 3 – Bioethanol

The 10 farms use rye grains from 10% land and whey from the dairy to produce bioethanol at a regional bioethanol production plant outside the system. The residue from the distilling process, DDGS, is fed back to the livestock as a fodder (**Figure 2**). Energy use at the bioethanol production plant is not accounted for as it is outside the system boundary. The bioethanol yield is calculated based on the starch content in rye grains, 64% of

TABLE 2 | Yield and diesel use per hectare per year for crops included in study.

Crops	Yields, DM (ton/ha)[a]	Yield (1000 SFU/ha)[a]	Diesel use (l/ha)[b]	SFU/l diesel
Spring barley	3.26	3.61	83	44
Winter wheat	4.25	5.14	86	60
Spring wheat	3.29	3.99	81	49
Winter rye	4.76	5.57	90	62
Winter triticale	5.13	6.10	92	67
Oats	4.36	3.97	86	46
Peas	2.36	3.03	79	39
Peas – whole crop[c]	4.60	4.00	69	58
Maize – whole crop[c]	9.60	8.00	115	70
Barley – whole crop[c]	6.50	5.00	79	63
Grass-clover	8.76	7.30	83	88
Grass-clover, 70% grazed	7.15	6.50	26	253
Permanent grass	2.40	2.00	5	409
Oilseed rape	1.87	3.73	63	59

[a]Adapted from Kádár et al. (2011).
[b]See text (Energy Consumption and Production) for elaboration.
[c]Yields based on crop with undersown grass-clover.

TABLE 3 | Inputs to livestock production for the entire system of 1000 ha.

Scenario	Fodder crops (1000 SFU)	Whey for fodder (1000 SFU)	Fodder residues (1000 SFU)	Milk-producing cows	Electricity use (TJ)
Reference	5560	359	0	730	4.82
1 (Vege. oil)	5004	334	163[a]	680	4.49
2 (Biogas)	5004	324	0	660	4.36
3 (Bioetha.)	5004	0[b]	237[c]	650	4.29
4 (Vege. oil and biogas)	4448	300	163[a]	610	4.03

[a]Oil cakes.
[b]The whey is used for ethanol fermentation together with rye.
[c]DDGS.

Net-Energy Analysis of Intergrated Food and Bioenergy Systems Exemplified by a Model of a Self-Sufficient...

23

TABLE 4 | Products (other than fodder crops) that stay within each farm [in brackets] and products that are transported 5 km to or from shared facilities in each scenario.

Scenario	Bioenergy crop[a]	Manure collected	Bioenergy residue[b]	Milk	Whey
Reference	0	[14.30]	0	5.82	5.2
1 (Vege. oil)	[0.22]	[13.32]	[0.15]	5.42	4.9
2 (Biogas)	2.50	12.93	15.43	5.26	4.7
3 (Bioetha.)	0.56	[12.73]	0.19	5.18	0[c]
4 (Vege. oil and biogas)	[0.22] and 2.50	11.95	[0.15] and 14.45	4.86	4.4

Numbers in 1000 ton/year for the entire system of 1000 ha.
[a]*No feedstock, oilseed rape, grass-clover, rye, and combined oilseed rape and grass-clover, respectively.*
[b]*No residue, oil cakes, biogas effluent, DDGS, and biogas effluent and oil cakes, respectively.*
[c]*The whey is used for ethanol fermentation together with rye.*

TABLE 5 | Annual diesel use for farming and transportation for the entire system of 1000 ha for each scenario.

Scenario	Diesel use in farming (TJ)	Diesel use for transport (TJ)	Transport as pct. of total (%)
Reference	2.31	0.06	2.3
1 (Vege. oil)	2.30	0.05	2.2
2 (Biogas)	2.40	0.20	7.7
3 (Bioetha.)	2.39	0.05	2.1
4 (Vege. oil and biogas)	2.39	0.19	7.3

DM, and the lactose content in whey, 77.5% of DM (Møller et al., 2005). Based on the stoichiometric mass balance, ethanol yield is assumed to be 0.51 g/g starch and 0.54 g/g lactose. All starch and lactose are assumed to be converted to ethanol. The fodder value of the DDGS is 34% of rye grain fodder value (Bentsen et al., 2007) plus the fodder value of whey protein (Møller et al., 2005). Transport of grains from the farms to a collection point 5 km away and transport of DDGS from that point and back to the farms require 0.7 MJ/t km (Berglund and Börjesson, 2006). The weight of DDGS is assumed to be 34% of rye grains. No extra labor is assumed.

Scenario 4 – Vegetable Oil and Biogas

The 10 farms share a biogas facility that is fed with grass-clover from 10% of the fields (similar to Scenario 2), and each farm uses another 10% of land to produce oilseed rape that is used in a low-tech farm scale oilseed press (similar to Scenario 1).

RESULTS AND DISCUSSION

The aim of this study is to investigate the limitations and potentials for agriculture in providing both food and energy and the trade-off between growing fodder or energy crops. In four alternative scenarios, different strategies aiming at increasing energy production by means of vegetable oil, biogas, and bioethanol are modeled. The scenarios are described in Section "Materials and Methods" and illustrated in **Figure 2**. The modeling is based on a

TABLE 6 | Inputs and outputs from the vegetable oil production process for the entire system of 1000 ha.

	Scenarios 1 and 4 (Unit/year)	Unit
Inputs		
Rape seed (85% DM)	220	ton
Electricity[a]	0.03	TJ
Outputs		
Oil[a]	2.71	TJ
Oil cakes[a]	163	1000 SFU

[a]*See text (Scenario 1 – Vegetable Oil).*

number of critical assumptions as outlined in Section "Materials and Methods." However, because conservative estimates and best-case assumptions have been used regarding, e.g., yields, transport distances, and conversion efficiencies, the resulting energy balances are expected to demonstrate what may be possible but in many cases, the requirements could be much more demanding.

Energy Consumption

Diesel is used for field operations and for transport between farms and shared facilities. There is a notable difference between the scenarios in the amount of products that are transported (**Table 4**) and in the corresponding use of diesel for transportation (**Table 5**). Biogas production has the highest impact as it involves transportation of large quantities of high water content feedstock and residues. In Scenario 2, 15,430 ton of biogas feedstock is transported to the biogas plant, and the same amount of biogas residue is transported back (**Table 4**). This means that each day throughout the year, more than 40 ton should be transported each way by tractors and trailers. In Scenario 4, 14,450 ton is transported each way. For this reason, the diesel use for transportation is significantly higher in these scenarios, making up 7.7 and 7.3%, respectively, of total diesel use at the farm. In scenarios without biogas, the diesel use for transportation of milk and whey (reference scenario and Scenario 1) and also grain for ethanol production (Scenario 3) is 2.3, 2.2, and 2.1%, respectively, of total diesel use in the farming system (**Table 5**).

Electricity is primarily used for livestock production, and the consumption is proportional to the number of cows (**Table 3**). In the bioenergy scenarios, additional 0.03 TJ of electricity is required for vegetable oil production (**Table 6**), 0.57 and 0.54 TJ, respectively, for biogas production in Scenarios 2 and 3 (**Table 7**) and nothing for bioethanol production (**Table 8**). Heat is only used for the biogas production.

Energy Balance

The inputs and outputs to each of the bioenergy production processes were summarized in **Tables 6–8**. These tables quantify feedstock and residue production in the specific scenarios in addition to the energy consumption and production. The main results of this study are the gross energy consumptions and productions (**Figure 3**) and the resulting net-energy output of each of the three energy services and food for each of the five scenarios (**Table 9**).

TABLE 7 | Inputs and outputs from the biogas production process for the entire system of 1000 ha.

	Scenario 2 (Unit/year)	Scenario 4 (Unit/year)	Unit
Inputs			
Cattle manure (11% DM)	12.93	11.95	1000 ton
Grass-clover (34.4% DM)	2.50	2.50	1000 ton
Electricity use in process[a]	0.57	0.54	TJ
Heat used in process[a]	3.81	3.62	TJ
Intermediate			
Raw biogas[a]	17.87	17.20	TJ
Outputs			
Residue (15% DM)	15.43	14.45	1000 ton
Electricity generation[a]	5.72	5.50	TJ
Useful heat production[a]	9.83	9.46	TJ

[a]See text (Scenario 2 – Biogas).

TABLE 8 | Inputs and outputs from the bioethanol production process for the entire system of 1000 ha.

	Scenario 3 (Unit/year)	Unit
Inputs		
Whey (5.8% DM)	4.66	1000 ton
Rye grain (85% DM)	0.56	1000 ton
Outputs		
Ethanol (90% water)[a]	4.42	TJ
DDGS (90% DM)[a]	237	1000 SFU

[a]See text (Scenario 3 – Bioethanol).

Liquid Fuel

The production of liquid fuels consists of vegetable oil in Scenarios 1 and 4 and bioethanol in Scenario 3. Using 10% of land for oilseed rape as energy, crop produces 2.71 TJ of vegetable oil. If the produced oil replaces purchased diesel, this results in a net output in Scenarios 1 and 4 of 0.36 and 0.14 TJ, respectively (**Table 9**), corresponding to 15 and 5% of the diesel used within the system for field operations and transport. This result is in line with Karpenstein-Machan (2001) who found that using 10% of land for oilseed rape as an energy crop would be sufficient for achieving self-sufficiency with liquid fuels for German dairy farms. In a Danish context, 10% of land for oilseed rape could produce 50–60% of the diesel required at a cash-crop farm (Halberg et al., 2008). In our dairy farm model, we have lower diesel consumption due to the high share of permanent grass and grazed areas, which demand less consumption of diesel in their production (**Table 2**).

A Swedish study of motor fuel self-sufficiency in a cash-crop farm concluded that allocating 9.3, 5.9, and 3.8% of farm land to oilseed rape for biodiesel, winter wheat for ethanol, and ley for biogas, respectively, would be sufficient to meet the on-farm fuel demand (Fredriksson et al., 2006). The energetic values of our results are essentially in agreement with these results. However, in our study, we do not consider the possibility of using biogas as a motor fuel or using ethanol to substitute diesel or to be produced on small scale suitable for a farm community (see Technologies for Producing Bioenergy in Organic Agriculture).

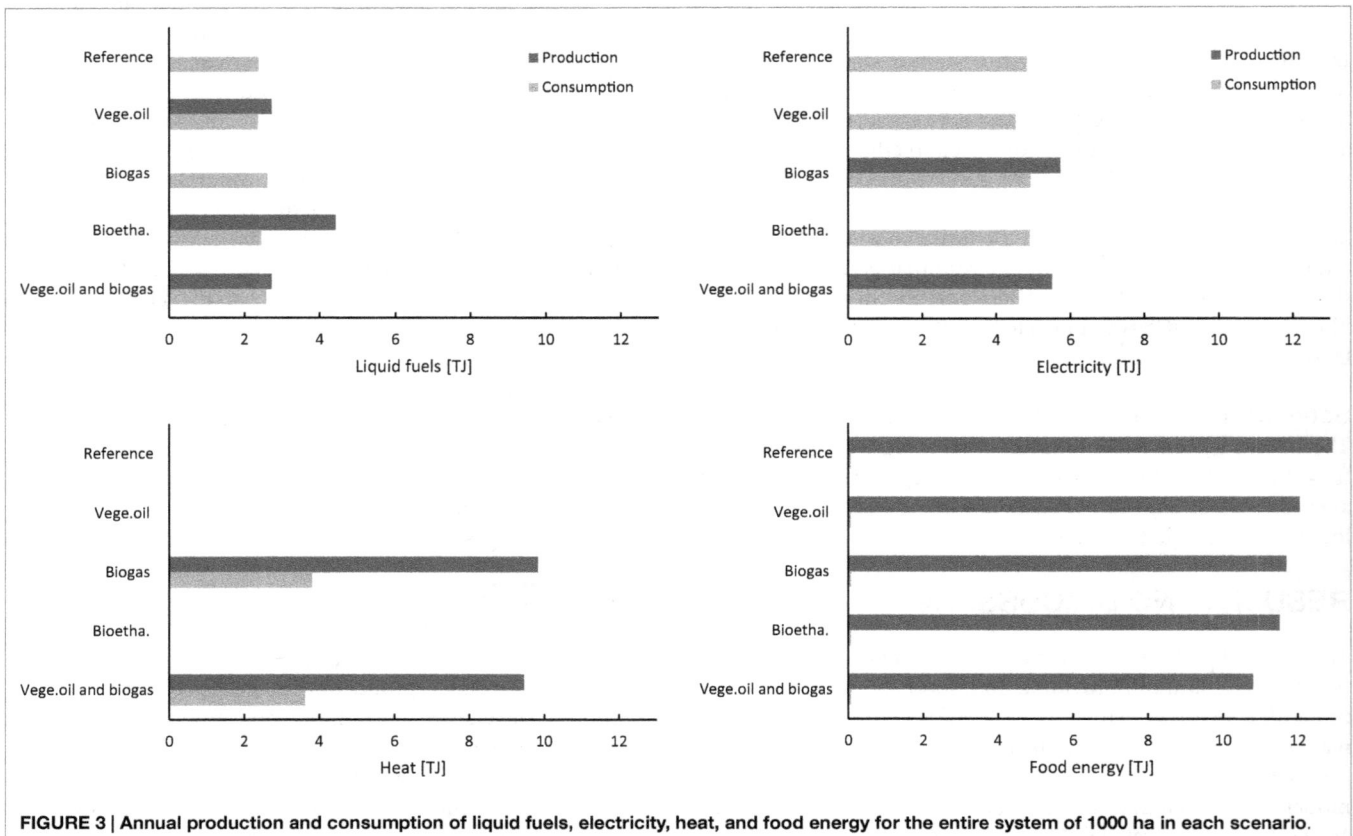

FIGURE 3 | Annual production and consumption of liquid fuels, electricity, heat, and food energy for the entire system of 1000 ha in each scenario.

TABLE 9 | Annual net output of liquid fuels, electricity, heat, and food energy for the entire system of 1000 ha in each scenario.

Scenarios	Energy crop (crop area)	Liquid fuels (TJ)	Electricity (TJ)	Heat (TJ)	Food energy (TJ)
Reference	–	−2.37	−4.82	0	12.86
1 (Vege. oil)	Oilseed rape (10%)	0.36	−4.52	0	11.98
2 (Biogas)	Grass-clover (10%)	−2.60	0.79	6.02	11.62
3 (Bioetha.)	Winter rye (10%)	(1.98)[a]	−4.29	0	11.44
4 (Vege. oil and biogas)	1 + 2 (20%)	0.14	0.91	5.84	10.73

[a]The ethanol is produced outside the system boundaries and not used locally on the farms.

In Scenario 3, bioethanol is produced in a regional ethanol plant based on whey and rye grains from 10% of the land. The ethanol produced is equivalent to 4.42 TJ, resulting in a net output of liquid fuels of 1.98 TJ (**Table 9**). This is the highest gross production of liquid fuel (**Figure 3**), but it also implies additional reduction in food production (discussed below). When comparing Scenario 3 to the other scenarios, it is important to take into account that energy use for ethanol production is not included. The production is at a regional ethanol plant outside the system boundaries so it depends on the existence of such a plant. Data from literature for ethanol production from grains indicate that the production of 1 MJ ethanol requires 0.27 MJ steam or 0.36 MJ of natural gas and 0.027 MJ of electricity (Bentsen et al., 2007). For Scenario 3, this would imply an additional consumption of 0.12 TJ of electricity and 1.60 TJ of natural gas. Besides, more energy would be needed to transport whey and rye to the regional plant and DDGS back to the collection point.

Electricity and Useful Heat

Electricity and useful heat are produced only in the two scenarios with biogas production (Scenarios 2 and 4; **Figure 3**). Producing biogas from manure and grass-clover from 10% of the area gives a net electricity output of 0.79 or 0.91 TJ in Scenarios 2 and 4, respectively (**Table 9**). This corresponds to 16 and 20%, respectively, of the electricity used in the systems to power the livestock house and to operate the biogas facility.

Energy Self-Sufficiency

In this study, the imperative of energy self-sufficiency has been applied as a way to model the potential energetic contribution from agriculture. The aim was to produce quantitative results, which are straightforward to interpret by avoiding allocation and systems expansion and quantitative normalization of qualitative differences.

The results show that using 10% of land for oilseed rape production (Scenario1) can make the system more than self-sufficient with liquid fuel and using 10% land for grass-clover biogas feedstock (Scenario 2) can make the system more than self-sufficient with electricity and heat. Only Scenario 4 achieves self-sufficiency with both electricity and fuel, and it produces a small energy surplus, i.e., 0.14 TJ of vegetable oil

and 0.91 TJ of electricity (**Table 9**). These results are in line with a study of a German livestock system, which showed that energy self-sufficiency could be achieved by using 18% land for energy crops with the combination of vegetable oil and anaerobic codigestion of energy crops and manure (Karpenstein-Machan, 2001). However, the results are in contrast to the previous Danish livestock model farming system with biogas production analyzing similar scenarios as in our study (Pugesgaard et al., 2013). They concluded that manure and grass-clover from 10% of the area could alone make the farm into a net-energy producer (also considering consumption of diesel). This discrepancy is mainly due to that liquid fuel, useful heat, and electricity were added in their energy balance, and thus that they implicitly assumed that these energy qualities can substitute each other, which is not the case. In Scenario 2, we would reach a similar result if the net-energy balance was calculated as the sum of net production of liquid fuel, electricity, and heat. In this case, the net-energy output ends up in a positive balance (4.21 TJ).

In a study of a Danish organic cash-crop system with 20% of land grown with grass-clover for biogas and 20% of land grown with oilseed rape for vegetable oil, twice the amount of liquid fuel needed on the farm was produced as well as a large surplus of useful heat and electricity (Østergård and Markussen, 2011). However, our conclusion that it requires about 20% land to produce a net-energy output differs significantly from a study of a Danish cash-crop system which was shown to be a "net-energy producer" by using alone 10% grass-clover as energy crop for biogas (Halberg et al., 2008). In that study, the energy balance was calculated based on how much fossil fuel the produced electricity and useful heat could replace in the economy. Such an approach implicitly answers the question: how much fossil fuel can be replaced in the economy if one farm changes its cropping system? It thus assumes that everything outside the system of interest stays the same, e.g., that the demand for energy in the economy is constant and that it will be met. Consequently, that approach is suitable for short-term outlooks of incremental changes. If, as in our case, the aim is to assess the possibilities of agriculture to provide both food and energy in a future where supply of fossil fuel is limited, then the assumption that everything else stays the same is less useful; there may not be any fossil fuel consumption to replace.

The energy self-sufficiency that is obtained in Scenario 4 should be seen in perspective of the energy requirements that was not included in the model, namely energy required upstream for producing and maintaining machinery and buildings and energy required downstream for processing the food and distributing it to consumers. Energy use for the construction of machinery and buildings is particularly relevant for the biogas plant, which requires a significant investment of energy. Furthermore, other studies show that energy consumption at the farm only constitutes from 20 to 30% of the total energy requirement in the food system (Heller and Keoleian, 2003).

Food Production

The primary product of a farming system is food (**Figure 3**), which is directly correlated to the fodder production. The production of fodder from crops, whey, and bioenergy residues and the resulting number of milk-producing cows in each scenario are summarized in **Table 3**. In Scenario 1, the fodder production is only 7% smaller than in the reference scenario even though the area for fodder crops is reduced by 10%. This is due to the fodder value of the oil cake residue, which corresponds to 1630 SFU/ha oilseed rape used for bioenergy. In Scenario 2, there are no fodder residues, and, therefore, exactly 10% less fodder is produced when compared to the reference scenario. In Scenario 3, the fodder production is 11.5% lower than in the reference scenario. This is because rye and whey are used for ethanol production instead of fodder, and the lactose in the whey is consumed during the fermentation. Finally in Scenario 4, the fodder production is reduced by 17% (i.e., the combination of Scenarios 1 and 2) when compared to the reference scenario.

The net output of food for each scenario (gross food production minus food consumed by workers) is shown in **Table 9**. In Scenarios 2 and 4 with biogas, there are 21 workers employed compared to 20 workers in the other scenarios. Each of these workers is assumed to need 8.8 MJ of food energy per day (United Nations, 2014). The food output thus corresponds to that each person within the system produces food to support 201, 187, 173, 178, and 159 persons for the five scenarios, respectively.

Another study of a 100% self-sufficient and manually managed mini farm of 372 m^2 (4000 ft^2) that grows grain crops, root crops, and vegetables and produces eggs showed that each farmer could support up to five additional people with a complete diet (Schramski et al., 2011). Compared to this, Scenario 4 is impressively 30 times more productive per labor input, but also needs 40 times more land per person supported (2994 versus 74 m^2). In addition, our system needs input of machinery and other industrial inputs, which are not needed in the manually managed farm.

CONCLUSION

The methodological approach of assessing bioenergy technologies in the context of a farming system that is self-sufficient with fodder and fertilizers and that utilizes all coproducts within the system is useful for providing a clear picture of limitations and possibilities of agriculture as energy provider. The approach circumvents the problem of comparing scenarios with different types of inputs and outputs. In this way, the agricultural system is seen in isolation without taking into account input of manure or fodder, i.e., agricultural production outside the system boundaries. Furthermore, balancing different types of energy separately is a useful way to avoid the problem of adding non-equivalent energy data. For these two reasons, the comparison of different scenarios for production of fodder, livestock, and bioenergy from a specific area is straightforward to interpret.

According to our farm model, a community of dairy farms of 1000 ha in total can be self-sufficient with fodder and fertilizer as well as liquid fuels, electricity, and heat. This is possible if 10% land is used for oilseed rape for vegetable oil production and 10% for grass-clover used codigestion with manure in a biogas plant. In addition, the system produces a surplus of heat, which may be used in nearby buildings. This strategy reduces the food production with approximately 17% when compared to the reference scenario. However, even if 20% land is used for energy crops, the net output of energy is marginal and in any case insufficient to provide energy for, e.g., downstream processing and distributions of the produced food or any other activities in the surrounding economy. Overall, it seems unlikely that (organic) agriculture can contribute significantly in powering an industrialized economy, as we know it today, without devoting an unacceptable large share of the land for energy crops. However, a net-energy neutral (or even slightly positive) agriculture will save energy resources for other uses.

AUTHOR CONTRIBUTIONS

MM did all the calculations in collaboration with HØ, and MM was the prime writer of the manuscript. SP and PO-P contributed with data and drafting the text for the agricultural part and the technical part, respectively. JS and HØ developed the project and contributed to drafting the text.

ACKNOWLEDGMENTS

This study has partly been within the BioConcens project, which is linked to the International Centre for Research in Organic Food Systems (ICROFS) and funded under the research program: Research in Organic Food and Farming, International Research Co-operation and Organic Integrity.

REFERENCES

Ahlgren, S., Bernesson, S., Nordberg, Å, and Hansson, P. (2010). Nitrogen fertiliser production based on biogas – energy input, environmental impact and land use. *Bioresour. Technol.* 101, 7181–7184. doi:10.1016/j.biortech.2010.04.006

Bentsen, N. S., Felby, C., and Ipsen, K. H. (2007). Energy balance of 2nd generation bioethanol production in Denmark. Available at: http://www.tekno.dk/pdf/projekter/p09_2gbio/ClausFelby/p09_2gbio%20Bentsen%20et%20al%20(2006).pdf

Berglund, M., and Börjesson, P. (2006). Assessment of energy performance in the life-cycle of biogas production. *Biomass Bioenergy* 30, 254–266. doi:10.1016/j.biombioe.2005.11.011

Börjesson, P., and Berglund, M. (2006). Environmental systems analysis of biogas systems – Part 1: fuel-cycle emissions. *Biomass Bioenergy* 30, 469–485. doi:10.1016/j.biombioe.2005.11.014

Callesen, I., Grohnheit, P. E., and Østergård, H. (2010). Optimization of bioenergy yield from cultivated land in Denmark. *Biomass Bioenergy* 34, 1348–1362. doi:10.1016/j.biombioe.2010.04.020

Cherubini, F. (2010). The biorefinery concept: using biomass instead of oil for producing energy and chemicals. *Energy Conver. Manage.* 51, 1412–1421. doi:10.1016/j.enconman.2010.01.015

Conforti, P., and Giampietro, M. (1997). Fossil energy use in agriculture: an international comparison. *Agric. Ecosyst. Environ.* 65, 231–243. doi:10.1016/S0167-8809(97)00048-0

Dalgaard, T., Halberg, N., and Porter, J. R. (2001). A model for fossil energy use in Danish agriculture used to compare organic and conventional farming. *Agric. Ecosyst. Environ.* 87, 51–65. doi:10.1016/S0167-8809(00)00297-8

Danish Agricultural Advisory Service. (2008). *Budget Estimates for Organic Production 2008 [in Danish; Original Title "Budgetkalkuler 2008"]*. Aarhus: Landbrugsforlaget.

EU. (2009). *Directive 2009/28/EC of the European Parliament and of the Council of 23 April 2009 on the Promotion of the Use of Energy from Renewable Sources and Amending and Subsequently Repealing Directives 2001/77/EC and 2003/30/EC*. Official Journal of the European Union. L 140, 16–62

Farrell, A. E., Plevin, R. J., Turner, B. T., Jones, A. D., O'Hare, M., and Kammen, D. M. (2006). Ethanol can contribute to energy and environmental goals. *Science* 311, 506–508. doi:10.1126/science.1121416

Fredriksson, H., Baky, A., Bernesson, S., Nordberg, O., Norén, O., and Hansson, P. (2006). Use of on-farm produced biofuels on organic farms – evaluation of energy balances and environmental loads for three possible fuels. *Agric. Syst* 89, 184–203. doi:10.1016/j.agsy.2005.08.009

Gan, J., and Smith, C. T. (2011). Optimal plant size and feedstock supply radius: a modeling approach to minimize bioenergy production costs. *Biomass Bioenerg.* 35, 3350–3359. doi:10.1016/j.biombioe.2010.08.062

Giampietro, M. (2004). *Multi-Scale Integrated Analysis of Agroesosystems.* Boca Raton: CRC-press.

Giampietro, M. (2005). Comments on "the energetic metabolism of the European union and the United States" by Haberl and colleagues – theoretical and practical considerations on the meaning and usefulness of traditional energy analysis. *J. Ind. Ecol.* 10, 173–185. doi:10.1162/jiec.2006.10.4.173

Giampietro, M., Allen, T. F. H., and Mayumi, K. (2006). The epistemological predicament associated with purposive quantitative analysis. *Ecol. Complex.* 3, 307–327. doi:10.1016/j.ecocom.2007.02.005

Halberg, N., Dalgaard, R., Olesen, J. E., and Dalgaard, T. (2008). Energy self-reliance, net-energy production and GHG emissions in Danish organic cash crop farms. *Renew. Agr. Food Syst.* 23, 30–37. doi:10.1017/S1742170507002037

Hall, C., Balogh, S., and Murphy, D. (2009). What is the minimum EROI that a sustainable Society must have? *Energies* 2, 25–47. doi:10.3390/en20100025

Hall, C. A. S., Cleveland, C. J., and Kaufmann, R. (1986). *Energy and Resource Quality: The Ecology of the Economic Process.* New York, NY: Wiley.

Heller, M. C., and Keoleian, G. A. (2003). Assessing the sustainability of the US food system: a life cycle perspective. *Agric. Syst.* 76, 1007–1041. doi:10.1016/S0308-521X(02)00027-6

IFOAM. (2014). *The principles of Organic Agriculture, International Federation of Organic Agriculture Movements.* Available at: http://www.ifoam.bio/en/organic-landmarks/principles-organic-agriculture 2014

Johansen, A., Carter, M. S., Jensen, E. S., Hauggard-Nielsen, H., and Ambus, P. (2013). Effects of digestate from anaerobically digested cattle slurry and plant materials on soil microbial community and emission of CO2 and N2O. *Appl. Soil Ecol.* 63, 36–44. doi:10.1016/j.apsoil.2012.09.003

Jørgensen, U., and Dalgaard, T. (2004). *Energy in Organic Agriculture [In Danish, original title "Energi i økologisk jordbrug"].* Foulum: Danish Research Center in Organic Farming.

Jungbluth, N., Chudacoff, M., Dauriat, A., Dinke, F., Doka, G., Faist Emmenegger, M., et al. (2007). *Life Cycle Inventories of Bioenergy.* Ecoinvent Report No. 17. Dübendorf: Swiss Centre for Life Cycle Inventories.

Kádár, Z., Christensen, A. D., Thomsen, M. H., and Bjerre, A. (2011). Bioethanol production by inherent enzymes from rye and wheat with addition of organic farming cheese whey. *Fuel* 90, 3323–3329. doi:10.1016/j.fuel.2011.05.023

Karpenstein-Machan, M. (2001). Sustainable cultivation concepts for domestic energy production from biomass. *Crit. Rev. Plant Sci.* 20, 1–14. doi:10.1080/20C13591099164

Markussen, M. V. and Østergård, H. (2013). Energy analysis of the Danish food production system: food-EROI and fossil fuel dependency. *Energies* 6, 4170–4186. doi:10.3390/en6084170

Møller, J., Thøgersen, R., Helleshøj, M. E., Weisbjerg, M. R., Søegaard, K., and Hvelplund, T. (2005). *Feed Material Table 2005 – Composition and Nutritive Value of Feeds for Livestock [In Danish, original title "Fodermiddeltabel 2005 – Sammensætning og Foderværdi af Fodermidler til kvæg"].* Available at: https://www.landbrugsinfo.dk/Kvaeg/Foder/Sider/Fodermiddeltabel_2005.aspx

Neff, R. A., Parker, C. L., Kirschenmann, F. L., Tinch, J., and Lawrence, R. S. (2011). Peak oil, food systems, and public health. *Am. J. Public Health* 101, 1587–1597. doi:10.2105/AJPH.2011.300123

Odum, H. T. (2007). *Environment, Power, and Society for the Twenty-First Century: The Hierarchy of Energy.* New York, NY: Columbia University Press.

Olesen, J. E., Schelde, K., Weiske, A., Weisbjerg, M. R., Asman, W. A. H., and Djurhuus, J. (2006). Modelling greenhouse gas emissions from European conventional and organic dairy farms. *Agric. Ecosyst. Environ.* 112, 207–220. doi:10.1016/j.agee.2005.08.022

Oleskowicz-Popiel, P., Kádár, Z., Heiske, S., Klein-Marcuschamer, D., Simmons, B. A., Blanch, H. W., et al. (2012). Co-production of ethanol, biogas, protein fodder and natural fertilizer in organic farming – evaluation of a concept for a farm-scale biorefinery. *Bioresour. Technol.* 104, 440–446. doi:10.1016/j.biortech.2011.11.060

Østergård, H., and Markussen, M. V. (2011). "Energy Self-sufficiency from an Emergy Perspective Exemplified by a Model System of a Danish Farm Cooperative," in *Emergy Synthesis 6, Theory and Application of the Emergy Methodology*, eds M. T. Brown and S. Sweeney (Gainesville, FL: The Center for Environmental Policy, Department of Environmental Engineering Sciences, University of Florida), 311–322.

Østergård, H., Markussen, M. V., and Jensen, E. S. (2010). "Challenges for sustainable development in a biobased economy," in *The Biobased Economy*, eds H. Langeveld, M. Meeusen, and J. Sanders (Washington, DC: Earthscan Publications Ltd), 33–48.

Pelletier, N., Audsley, E., Brodt, S., Garnett, T., Henriksson, P., Kendall, A., et al. (2011). Energy intensity of agriculture and food systems. *Annu. Rev. Environ. Resour.* 36, 223–246. doi:10.1146/annurev-environ-081710-161014

Poulsen, H. D. (2008). *Standard Figures for Manure – 2008 [In Danish, Original Title "Normtal for Husdyrgødning – 2008"].* Faculty of Agricultural Sciences Aarhus University. Available at: http://anis.au.dk/fileadmin/DJF/Anis/normtal2008.pdf

Pugesgaard, S., Olesen, J., Jørgensen, U., and Dalgaard, T. (2013). Biogas in organic agriculture – effects on productivity, energy self sufficiency and greenhouse gas emissions. *Renew. Agr. Food Syst.* 29, 28–41. doi:10.1017/S1742170512000440

Refsgaard, K., Halberg, N., and Kristensen, E. S. (1998). Energy utilization in crop and dairy production in organic and conventional livestock production systems. *Agric. Syst.* 57, 599–630. doi:10.1016/S0308-521X(98)00004-3

Rehl, T., and Müller, J. (2011). Life cycle assessment of biogas digestate processing technologies. *Resour. Conserv. Recy.* 56, 92–104. doi:10.1016/j.resconrec.2011.08.007

Schramski, J. R., Rutz, Z. J., Gattie, D. K., and Li, K. (2011). Trophically balanced sustainable agriculture. *Ecol. Econ.* 72, 88. doi:10.1016/j.ecolecon.2011.08.017

StatBankDanmark. (2007). *Accounts Statistics for Agriculture Organic Holdings, Financial Results and Balance by Type and Account Items (1996–2009).* Available at: http://www.statistikbanken.dk

Stinner, W., Möller, K., and Leithold, G. (2008). Effects of biogas digestion of clover/grass-leys, cover crops and crop residues on nitrogen cycle and crop yield in organic stockless farming systems. *Eur. J. Agron.* 29, 125–134. doi:10.1016/j.eja.2008.04.006

Syrstad, O. (1993). Evaluation of dual-purpose (milk and meat) animals. *World Animal Rev.* 77, 56–59.

United Nations. (2014). *What is Hunger? United Nations World Food Programme – Fighting Hunger Worldwide.* Available at: http://www.wfp.org/hunger/what-is

Conflict of Interest Statement: The authors declare that the research was conducted in the absence of any commercial or financial relationships that could be construed as a potential conflict of interest.

4

Determination of Terpenoid Content in Pine by Organic Solvent Extraction and Fast-GC Analysis

Anne E. Harman-Ware[1]*, Robert Sykes[1], Gary F. Peter[2] and Mark Davis[1]

[1] National Bioenergy Center, National Renewable Energy Laboratory, Golden, CO, USA, [2] School of Forest Resources and Conservation, University of Florida, Gainesville, FL, USA

Terpenoids, naturally occurring compounds derived from isoprene units present in pine oleoresin, are a valuable source of chemicals used in solvents, fragrances, flavors, and have shown potential use as a biofuel. This paper describes a method to extract and analyze the terpenoids present in loblolly pine saplings and pine lighter wood. Various extraction solvents were tested over different times and temperatures. Samples were analyzed by pyrolysis-molecular beam mass spectrometry before and after extractions to monitor the extraction efficiency. The pyrolysis studies indicated that the optimal extraction method used a 1:1 hexane/acetone solvent system at 22°C for 1 h. Extracts from the hexane/acetone experiments were analyzed using a low thermal mass modular accelerated column heater for fast-GC/FID analysis. The most abundant terpenoids from the pine samples were quantified, using standard curves, and included the monoterpenes, α- and β-pinene, camphene, and δ-carene. Sesquiterpenes analyzed included caryophyllene, humulene, and α-bisabolene. Diterpenoid resin acids were quantified in derivatized extractions, including pimaric, isopimaric, levopimaric, palustric, dehydroabietic, abietic, and neoabietic acids.

Keywords: fast-GC, pyrolysis-molecular beam mass spectrometry, cell wall chemistry, renewable materials, biofuels, bioproducts, biomaterials

Citation:
*Harman-Ware AE, Sykes R, Peter GF
and Davis M (2016) Determination of
Terpenoid Content in Pine by Organic
Solvent Extraction and Fast-GC
Analysis.
Front. Energy Res. 4:2.*

INTRODUCTION

Renewable chemicals, including fuels, solvents, fragrances, flavors, and pharmaceutical compounds, can be generated or extracted from renewable biomass sources. Many types of compounds can be generated from different components of the biomass, e.g., ethanol from the carbohydrate fraction or biodiesel from the lipid components in biomass. Other biomass components, including lignin and oleoresin excretions from conifers, have also been used as a source of renewable chemicals. For example, pine oleoresin is used to generate turpentine, a solvent and source of synthetic platform chemicals such as α-pinene (Palmer, 1943; Beglinger, 1958). Terpenoids, naturally occurring organic compounds derived from isoprene units, are the primary constituents of pine oleoresin (Palmer, 1943; Bohlmann and Keeling, 2008; Rodrigues-Corrêa et al., 2012). Monoterpenes (C_{10}), sesquiterpenes (C_{15}), and diterpenoid resin acids (C_{20}), the main terpenoids found in pine oleoresin, are a

Abbreviations: BSTFA, N,O-Bis(trimethylsilyl)trifluoroacetamide; GC/FID, gas chromatography/flame ionization detector; LTM MACH, low thermal mass modular accelerated column heater; PLW, pine lighter wood; PS, pine sapling; py-MBMS, pyrolysis-molecular beam mass spectrometry.

valuable source of chemicals with many industrial applications (Beglinger, 1958; Martin et al., 2002; Monteiro and Veloso, 2004; Bohlmann and Keeling, 2008; Harvey et al., 2009; Rodrigues-Corrêa et al., 2012). Turpentine is composed of monoterpenes and is used in the flavor and fragrance industry (Bohlmann and Keeling, 2008; Rodrigues-Corrêa et al., 2012). Other uses for terpenoid compounds have been found in pharmaceutical, cosmetics, and polymer industries (Beglinger, 1958; Bohlmann and Keeling, 2008; Rodrigues-Corrêa et al., 2012). Monoterpenes and sesquiterpenes have also been investigated as potential sources of renewable fuel (Monteiro and Veloso, 2004; Harvey et al., 2009; Peralta-Yahya et al., 2011; Renninger et al., 2011; Meylemans et al., 2012; Rodrigues-Corrêa et al., 2012; Hellier et al., 2013; Vallinayagam et al., 2014).

Efforts have been made to understand the genetic, ecological, and physicochemical processes behind the production and accumulation of terpenoids in pine and other feedstocks (Nerg et al., 1994; Manninen et al., 2002; Martin et al., 2002; Bojovic et al., 2005; Schmidt et al., 2011; Achotegui-Castells et al., 2013; Susaeta et al., 2014). For this type of research, it is important to accurately measure the terpenoid content in biomass to be able to compare the variables potentially affecting terpenoid production and accumulation. With the wide variety of uses and applications of terpenoid components as well as the biological variability associated with their formation, development of rapid analytical methods used to characterize terpenoids and measure their abundance in the biomass is becoming increasingly important.

Currently, analytical methods primarily use hexane or other non-polar solvents to extract terpenoid components from pine and other biomass sources and GC/MS to identify and quantify the components of the extract (Lewinsohn et al., 1993; Manninen et al., 2002; Bojovic et al., 2005; Thompson et al., 2006; Varming et al., 2006; Ormeño et al., 2007, 2010; Zhao et al., 2010; Achotegui-Castells et al., 2013). There have been other methods and solvents used to extract and quantify terpenoids in biomass, including accelerated solvent extraction, dynamic headspace analysis as well as the use of methyl tert-butyl ether and derivatization agents during extraction (Martin et al., 2002; Varming et al., 2006; Fojtová et al., 2008; Zhao et al., 2010). Several terpenoids are available to be used as standards for quantitation, whereas the abundance of others are estimated based on assumption of response factors relative to internal standards (Martin et al., 2002; Varming et al., 2006; Ormeño et al., 2007, 2010; Fojtová et al., 2008; Zhao et al., 2010; Achotegui-Castells et al., 2013). However, the more polar resin acids, such as abietic, dehydroabietic, neoabietic, palustric, pimaric and isopimaric acid, are not accounted for in their entirety from hexane (or other non-polar) extractions, especially without derivatization prior to GC analysis, and constitute large fractions of pine oleoresin (Martin et al., 2002; Keeling and Bohlmann, 2006). Accelerated solvent extractions require the use of high temperatures and pressures and reported methyl tert-butyl ether extractions are tedious and time consuming (Martin et al., 2002; Fojtová et al., 2008). Additionally, most methods require multiple types or steps of extractions to remove different types of terpenoids or utilize equipment, such as shaker tables and Soxhlet extractors. Currently, there is not a high-throughput technique that successfully extracts and accurately measures both the polar and non-polar most abundant terpenoid contents in pine or other biomass samples.

The goal of this investigation was to develop a rapid, accurate, screening method using an optimized solvent system to quantify the abundant terpenoids in pine biomass. This technique can be used to compare biomass genetic transformations, biological variation, and the effects of physical and chemical treatments on plants in order to screen for desired genetic constructs and treatments. The method minimizes the use of specialized equipment such as a Soxhlet apparatus or shaker table and does not require multiple extraction steps or labor-intensive treatment of the biomass, such as grinding. Implementation of this method allows rapid screening of pine samples based on terpenoid content; which directly affects the potential of the pine to be used as a source of renewable fuels and chemicals.

We evaluated the effects of the solvent type, time, and temperature used in the extractions. In an effort to develop a more rapid method, a low thermal mass modular accelerated column heater (LTM MACH) GC/FID, also known as ballistic or fast-GC, was used to analyze the extracted components using a 3.5 min GC method. Fast-GC is a method that has been used to successfully and rapidly analyze many types of samples, including petroleum products, natural oils, and plant extracts (Luan and Szelewski, 2008; Firor, 2011). Pyrolysis-molecular beam mass spectrometry (py-MBMS), a technique used to study biomass components based on their pyrolyzate profiles,(Evans and Milne, 1987; Sykes et al., 2009) was also used to supplement the analysis of the terpenoid components of the pine samples before and after extractions in order to quantify extraction efficiency.

MATERIALS AND METHODS
Preparation of Calibration Standards
Terpenoid calibration standards were obtained from Sigma Aldrich, Alfa Aesar, CanSyn Chem. Corp., and purified bisabolene was provided by the Joint BioEnergy Institute, Emeryville, CA, USA. Calibration standards (5–400 µg/mL) were prepared in hexane/acetone (v/v, 1:1) and hexadecane was used as an internal standard (1 mg/mL in hexane/acetone (1:1) stock solution). **Figure 1** shows the terpenoid compounds that were calibrated in this method. Trimethylsilyl derivatives of resin acids (pimaric, levopimaric, isopimaric acid, neoabietic acid, abietic acid, palustric acid, dehydroabietic acid) were prepared by making serial dilutions of the resin acids in hexane/acetone (1:1) and derivatizing each standard individually using N,O-Bis(trimethylsilyl) trifluoroacetamide (BSTFA) reagent purchased from Sigma Aldrich. Stearic and palmitic acids were also calibrated for on the GC and derivatized using BSTFA. Calibration standards were prepared by adding 500 µL of each calibration stock to a GC vial with 100 µL of internal standard stock solution and 100 µL BSTFA reagent (BSTFA for diterpenoids only). The mixture was briefly purged with nitrogen and heated to 75°C for 1 h prior to analysis by GC/FID. To validate the calibrations, 1 mg of four different samples of slash pine oleoresin were dissolved in 2 mL of the hexane/acetone solvent. Fast-GC analysis was performed

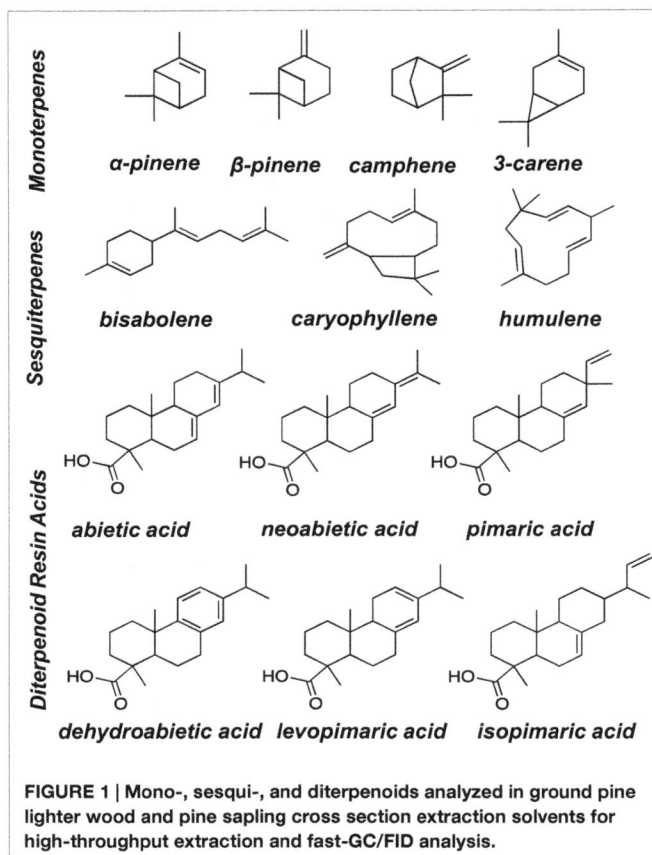

FIGURE 1 | Mono-, sesqui-, and diterpenoids analyzed in ground pine lighter wood and pine sapling cross section extraction solvents for high-throughput extraction and fast-GC/FID analysis.

on each oleoresin sample using a 500 μL underivatized aliquot with 50 μL of hexadecane internal standard stock solution and a 500 μL aliquot that was derivatized by adding to 50 μL internal standard, and 100 μL BSTFA reagent with brief nitrogen purging and heating to 75°C for 1 h prior to analysis by GC/FID. Trace terpenoids for which standards do not exist were not included in the calibration. However, they could be included when standards become available.

Gas Chromotography

An Agilent 6890 GC/FID was used to analyze the terpenoid calibration standards, derivatized pine resin, and the extracted samples from pine lighter wood (PLW) and loblolly pine sapling samples. The GC was equipped with an Agilent 10 m × 0.10 mm × 0.10 μm DB-5 column LTM module used in a Gerstel MACH unit. The LTM column was connected to the inlet and the detector by 0.5 m of 0.10 mm fused silica deactivated transfer lines. The inlet was run in splitless mode at 250°C and the column flow was 0.6 mL/min. The GC oven was isothermal at 250°C and the detector temperature was 250°C. The LTM oven program began at 60°C for 0.75 min and was ramped to 325°C at a rate of 150°C/min and held at the final temperature for 1 min for a total run time of approximately 3.5 min. The rapid nature of this method does not differentiate isomers, for example (+) and (−) α-pinene, and developed for screening large sample sets of pine for total terpenoid content.

Extraction of Terpenoids from Pine Lighter Wood

Pine lighter wood was used to optimize extraction conditions because of its high terpenoid content. PLW samples were air dried to 5 wt% water and Wiley milled to 20 mesh particle size. Four extraction solvent systems were tested to determine the solvent that yielded the maximum amount of terpenes extracted from the PLW: hexane (H), hexane:acetone (v/v, 1:1), hexane:diethyl ether (v/v, 1:1), and hexane:ethyl acetate (v/v, 1:1). Approximately 10 mg of ground PLW was added to a 4 mL borosilicate glass vial and 2 mL of extraction solvent. Vials were then stored at temperatures of either −22, 2, or 22°C for times of either 0.25, 1, 5, or 24 h to evaluate the influence of time and temperature on each solvent extraction system. The solvent was transferred to another vial using a pipette and the biomass was dried in a vacuum oven at 40°C overnight. Next, 4 mg of the dried biomass was added to Frontier pyrolyzer cups with Type A/D glass fiber filters for py-MBMS analysis. Unextracted PLW samples were also prepared for py-MBMS analysis.

The extract solvent containing the terpenoid components was then aliquoted for separate analysis of mono-/sesquiterpenoids and resin acids. Mono- and sesquiterpenoids were analyzed by combining neat solvent extract with hexadecane internal standard and analyzed using the GC method described in the previous section. To analyze for resin acids, approximately 1 mL of the extract solvent was combined with 0.25 mL of 0.1 M aqueous ammonium carbonate. The aqueous layer was removed with a pipette and the organic layer was then dried over 100 mg of 3Å molecular sieves for 1 h. Then, 400 μL of the dried organic layer was added to a GC vial with 50 μL of internal standard and 100 μL of BSTFA, heated to 75°C for 1 h and analyzed by GC using the method described in herein. A flow chart of the extraction process and sample preparation is shown in **Figure 2**.

The extraction and derivatization method was validated for recovery of total terpenoids by a standard addition method. The extract solvent from ground PLW, as well as the extract solvent from freshly cut pine sapling cross sections, was spiked using diterpenoid standards. Terpenoids were extracted as described previously and the extract solvent was spiked with two concentrations of abietic acid and neoabietic acid (40–80 μg/400 μL aliquot extract) prior to sample analysis. For these experiments, spiked extraction solvent was washed with ammonium carbonate, dried with molecular sieves, and added to a GC vial containing 100 μL of internal standard stock solution and 100 μL of BSTFA or Methyl-8 reagent (Thermo Scientific). For BSTFA, the vials were heated to 75°C for 1 h; and for the Methyl-8 derivatives, the vials were heated to 85°C for 30 min prior to fast-GC analysis. Standards were also derivatized with Methyl-8 for calibration to test derivatization efficiency.

Extraction Efficiency Evaluation by Pyrolysis-Molecular Beam Mass Spectrometry

Pine samples before and after extractions were pyrolyzed using a Frontier PY-2020 iD autosampler pyrolysis unit and pyrolysis vapors were analyzed with an Extrel Model Max1000 MBMS. (Sykes et al., 2009) Pyrolysis was conducted under a He flow

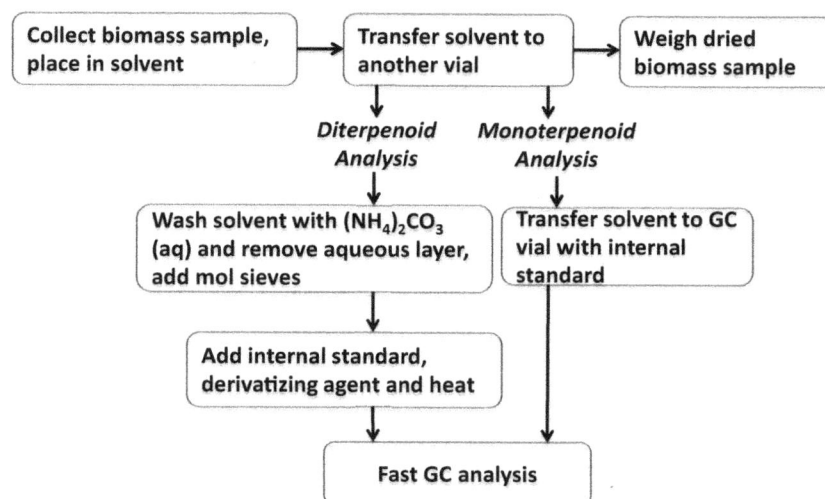

FIGURE 2 | Terpenoid extraction flow chart.

rate of 0.9 L/min (STP) and furnace temperature of 500°C for a 1.5 min acquisition time per sample, although pyrolysis was complete in <30 s. The interface and transfer lines were maintained at 350°C. Calibration standards for py-MBMS reference spectra of terpenoids were prepared in acetone. Abietic acid was used as a reference for diterpenoids ($m/z = 302, 285, 239$), caryophyllene was used to reference sesquiterpenoids ($m/z = 204$), and α-pinene was used to reference monoterpenes ($m/z = 93$). Solutions of each terpenoid (40 μL of 5 mg/mL) were added to empty pyrolyzer cups containing glass fiber filter disks. The solvent was allowed to evaporate prior to py-MBMS analysis.

Extraction of Terpenoids from Pine Saplings

Greenhouse-grown loblolly pine saplings (16-month-old) were obtained from ArborGen and grown in a greenhouse at the National Renewable Energy Laboratory prior to sampling. The plants were trimmed 1 cm above the soil and five cross sections, each 2″ apart, were taken from the tree, while the bark was removed immediately prior to taking each cross section sample. Each cross section sample, approximately 50 mg dry weight (100 mg wet weight), was immediately added to a vial containing 2 mL of hexane/acetone and stored at room temperature for 1 h prior to sample analysis by fast-GC/FID. After the biomass was extracted, it was dried in a vacuum oven at 40°C overnight and then 4 mg of the dried biomass was added to 80μl pyrolyzer cups with glass fiber filters for py-MBMS analysis. Unextracted cross section samples were also prepared for py-MBMS analysis.

RESULTS

Extraction Optimization and py-MBMS Analysis

Pine lighter wood was chosen as a feedstock to optimize the extraction conditions, since it is known to have high oleoresin content (Beglinger, 1958). Py-MBMS of the PLW before and after the extractions was used to quantify the extraction efficiency of terpenoid removal. The intensity of the ions corresponding to the presence of terpenoids in unextracted PLW was considered the starting amount of terpenoids (100%) present in the samples. **Figure 3A** shows the comparison of intensities of the diterpenoid-based ion, $m/z = 302$, before and after extractions where the PLW sample was in solvent for 1 h at 22°C (room temperature) using the four different solvent systems. The percent values reported for each extraction solvent correspond to the percentage intensity of each ion lost from the unextracted material, reflecting the approximate extraction efficiency. Ions corresponding to the presence of mono- and sesquiterpenoids were also monitored, but their presence and intensity can also result from the fragmentation of the diterpenoids present in the samples. Hence, these ions were only semi-quantitative and were not used for comparison. Py-MBMS results indicated that the solvent systems that incorporated the polar solvents extracted more terpenoids than the hexane-only extractions. It was also found that the extractions performed at colder temperatures and over longer periods of time (beyond 1 h) did not influence the amount of terpenoids extracted. Error bars in **Figure 3A** are SD for triplicate extractions. **Figure 3B** shows that the extraction is also efficient for sapling cross section samples. (The data shown in **Figure 3B** correspond to the extraction from a single sapling's cross sections; hence, there are no replicates or error bars, which incorporates the biological variability, but not the variability associated with the method). Therefore, the optimal extraction conditions were determined to be the hexane/acetone solvent system for 1 h at 22°C. Additional research to determine the variability of terpenoids within a tree and across trees is currently under way, but was not the focus of this method development work.

The amount of PLW used to determine the terpenoid content (approximately 10 mg in 2 mL solvent) was chosen as

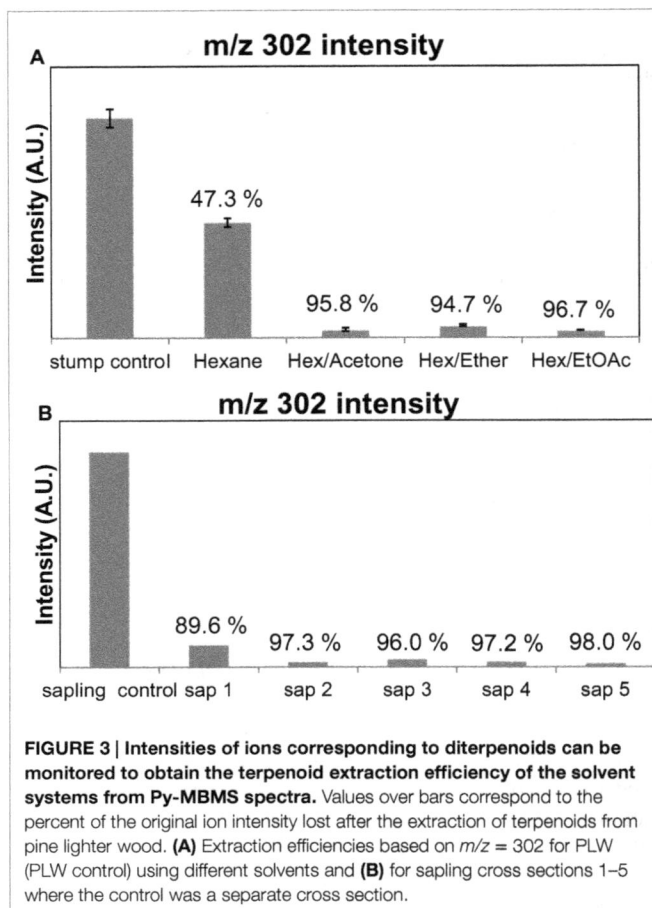

FIGURE 3 | Intensities of ions corresponding to diterpenoids can be monitored to obtain the terpenoid extraction efficiency of the solvent systems from Py-MBMS spectra. Values over bars correspond to the percent of the original ion intensity lost after the extraction of terpenoids from pine lighter wood. **(A)** Extraction efficiencies based on m/z = 302 for PLW (PLW control) using different solvents and **(B)** for sapling cross sections 1–5 where the control was a separate cross section.

the concentration of the analytes detected from this amount of sample was within the dynamic range of the calibrations for each analyte. Fresh pine cross section or core samples containing high moisture and lower terpenoid content need to be used in larger quantities, ranging from 20 to 150 mg of dry weight material for 2 mL of extract solvent. While each terpenoid analyte has its own detection limit, the limit of quantification can vary from detector to detector, this method allows for determination of most analytes as low as 0.01 dry wt% of the biomass and some analytes as low as 0.002 dry wt%. The speed of preparation of the samples can vary by individual, but combined with the GC analysis of 3.5 min, the extraction and total terpenoid analysis of 200 separate pine samples per week (800 injections total, counting duplicates) are possible using this extraction technique. By contrast, typical GC methods are on the order of 20–40 min, making the GC analysis of this method approximately 10 times faster, not counting the increased cooling speed between injections of the LTM MACH.

Validation of Derivatization of the Terpenoid Extracts for GC Analysis

Table 1 shows the terpenoid content determined in PLW and PS from fast-GC analysis using the hexane/acetone solvent system for 1 h at 22°C using BSTFA as a derivatizing reagent. The BSTFA was chosen as the optimal derivatizing agent based on the ability

to recover known amount of resin acid spiked in extract solvent (where the concentration of resin acids in the extract was previously determined based on external calibrations). Extraction solvent from fresh-cut PS and extraction solvent from ground PLW were spiked with abietic and neoabietic acid to test for the recovery of these resin acids using the selected extraction method. The 1 h room temperature hexane/acetone extraction solvents from PLW samples and PS samples were prepared according to the procedure described in the experimental section and analyzed for total terpenoid content. The extraction solvent was also aliquoted and spiked with abietic and neoabietic acid prior to sample analysis (ammonium carbonate washing and molecular sieve drying) and analyzed for terpene content using the two different derivatizing agents. **Table 2** shows the recovery of the total resin acids from the pine sapling cross section samples. The results suggest that the Methyl-8 reagent is not a favorable derivatizing reagent due to its inability to recover all of the resin acids known to be present in the extraction solvent from the fresh-cut pine sapling samples. The Methyl-8 is likely deactivated by the presence of any water or interfering metabolites that prevent methylation of the resin acids. Methyl-8 was capable of recovering more of the resin acids from the lighter wood when not washing or drying the extract solvent, indicating that the fresh samples may contain water or other metabolites that inhibit the methylation of the resin acids. In addition to inefficient derivatization of the resin acids, the Methyl-8 reagent also showed high standard deviations from replicated experiments using the same extract solvent from a particular pine sample.

The majority of terpenoids extracted from pine samples can be successfully resolved using a 3.5 min ballistic GC method as shown in the chromatograms in **Figure 4**. Monoterpenes that were monitored included α- and β- pinene, carene, and camphene. Sesquiterpenes that were accounted for included humulene, caryophyllene, and bisabolene. Diterpenoid resin acids that were quantified based on the presence of their trimethylsilyl derivatives included pimaric, levopimaric, isopimaric, palustric, dehydroabietic, abietic, and neoabietic acids. This method was unable to resolve isopimaric from palustric acid and levopimaric from dehydroabietic acid so the response from these compounds was averaged for their content determination. Other terpenoid compounds can also be calibrated and accounted for using this method as long as standards are available. While other terpenoids may be present in conifer or pine oleoresin that are unaccounted for using limited calibrated standards or could be co-eluting (particularly the diterpenoids), the total terpenoid content determination may still be accurately reflected using representative standards and responses for each compound, as demonstrated in the spiking experiment (results in **Table 2**) and additionally discussed below.

To validate the calibration and wt% recovery of the terpenoid components based on this analysis procedure, 1 mg of four resin samples secreted from slash pine were extracted in 2 mL of hexane/acetone using the same derivatization method as the calibration standards and the pine extract solvents. The fast-GC analysis of the resins, assumed to be almost entirely terpenoid components, accounted for an average of 97% (±6%) of the mass of the resins based on the presence of the calibrated compounds.

TABLE 1 | Terpenoid content (dry wt%) of pine lighter wood (PLW) and pine sapling cross sections (PS).

	α-pinene	β-pinene	Pimaric acid	Isopimaric/ palustric acid	Dehydroabietic/ levopimaric acid	Abietic acid	Neoabietic acid	Total
PLW[a]								
	0.26 (±0.08)	0.03 (±0.03)	0.95 (±0.11)	2.09 (±0.08)	2.48 (±0.11)	13.38 (±0.70)	2.39 (±0.17)	21.57 (±0.91)
PS[b]								
1	0.43 (±0.00)	0.10 (±0.00)	0.01 (±0.02)	0.14 (±0.02)	0.28 (±0.01)	0.16 (±0.01)	0.04 (±0.00)	1.16 (±0.02)
2	0.69 (±0.00)	0.17 (±0.00)	0.00 (±0.00)	0.03 (±0.01)	0.02 (±0.01)	0.10 (±0.00)	0.17 (±0.02)	1.18 (±0.02)
3	0.46 (±0.01)	0.11 (±0.00)	0.04 (±0.00)	0.16 (±0.00)	0.38 (±0.01)	0.59 (±0.03)	0.08 (±0.00)	1.81 (±0.04)
4	0.57 (±0.00)	0.14 (±0.00)	0.03 (±0.00)	0.13 (±0.01)	0.28 (±0.01)	0.25 (±0.01)	0.05 (±0.00)	1.46 (±0.02)
5	0.78 (±0.01)	0.13 (±0.00)	0.05 (±0.00)	0.23 (±0.02)	0.58 (±0.02)	0.40 (±0.02)	0.08 (±0.00)	2.25 (±0.02)

[a]Averaged from analysis of three extractions.
[b]Averaged from duplicate GC analysis of each cross section extraction where 1 corresponds to the cross section at the bottom of the sapling and 5 is the top.
Dry Wt% = 100 × (mass of terpenoid recovered in 2 mL)/(Mass dried, extracted biomass + mass of total terpenoids recovered).

TABLE 2 | Resin acid recoveries after spiking extract solvents from pine sapling cross sections and pine lighter wood.

Spiking experiment	Sapling extract (total terpenoid recovery) BSTFA	PLW extract (total terpenoid recovery) BSTFA	Sapling extract (total terpenoid recovery) Methyl-8	PLW extract (total terpenoid recovery) Methyl-8
Abietic acid (level 1)	86% (±6%)	95 (±2%)	39% (±28%)	44% (±27%)
Abietic acid (level 2)	97% (±12%)	100 (±6%)	18% (±10%)	37% (±19%)
Neoabietic acid (level 1)	94% (±12%)	94 (±6%)	34% (±6)	60% (±37%)
Neoabietic acid (level 2)	90% (±12%)	100 (±5%)	25% (±6%)	19% (±6%)

Extract solvents from the samples were spiked, the solvents were worked-up according to the procedure and derivatized using two different reagents, BSTFA and Methyl-8. Level 1 is the 40 μg/400 μL extract concentration spiking and level 2 is 80 μg/400 μL extract concentration spiking.

Hence, the derivatization efficiency of the pine resin was similar to that of the standards and the solvent-extracted oleoresin from the pine samples can be accurately quantified relative to these standards.

Terpenoid Content of Pine Lighter Wood

As shown in **Table 1**, the PLW sample, having been ground and dried, contained very few volatile monoterpenes, primarily α- and β-pinene. Carene and camphene were not detected in the PLW extract solvent. No calibrated sesquiterpenes were identified in the PLW using this method. Trimethylsilyl diterpenoid resin acids (pimaric, levopimaric, isopimaric, palustric, dehydroabietic, abietic, and neoabietic acids) were detected in the PLW extract solvent. Several types of terpenoids detected in PLW in this study agree with reported terpenoids found in other pine sources (Nerg et al., 1994; Manninen et al., 2002; Bojovic et al., 2005; Ormeño et al., 2007; Bohlmann and Keeling, 2008; Rodrigues-Corrêa et al., 2012; Achotegui-Castells et al., 2013) and show some similarities with terpenoids present in spruce (Martin et al., 2002; Zhao et al., 2010). For example, the most abundant monoterpenes extracted from PLW were in agreement with the

most abundant monoterpenes present in pine wood and foliage, being α- and β-pinene. Other monoterpenes and sesquiterpenes have been detected in the foliage of pine but may have not been present initially in the PLW or could have been lost during the drying of the PLW in this study.

Diterpenoids extracted from the PLW are representative of typical oleoresin secretions from conifer trees. The most abundant diterpenoids present in PLW included abietic acid (> 13 dry wt%) and neoabietic acid (> 2 dry wt%), with smaller quantities of isopimaric/palustric acid and dehydroabietic/levopimaric acid (being approximately 2 dry wt%). These diterpenoids are also similar to those identified from spruce (Martin et al., 2002). PLW, the resinous portion of felled trees or stumps, is known to have higher oleoresin content than the wood, foliage, or bark from the live tree (Beglinger, 1958). The PLW in this study was shown to have total terpenoid content in excess of 20 wt% (dry basis) if referring to the maximum amount of terpenoids determined from the hexane/acetone extractions. The higher terpenoid composition from PLW in this study is consistent in comparison to the total terpenoid content, being <5 wt%, from the greenhouse-grown loblolly saplings in this study and that reported for pine and spruce in the literature (Martin et al., 2002; Thompson et al., 2006; Ormeño et al., 2007; Zhao et al., 2010; Achotegui-Castells et al., 2013). The SD of the terpenoid content determined for the PLW as reported in **Table 1** reflects the reproducibility of the method as each sample of PLW is considered to be homogeneous and should yield statistically similar results for each analysis.

Terpenoid Content of Pine Sapling Cross Sections

Chromatograms of the extract obtained from a greenhouse-grown loblolly pine sapling cross section (PS) are shown in **Figure 4**. Mono- and sesquiterpenoids elute before 2.1 min, fatty acids elute around 2.5 min, and trimethylsilyl derivatives of resin acids elute after 2.6 min. PS samples were measured for total terpenoid content using the same procedure and GC analysis of the PLW. As shown in **Figure 5** and **Table 1**, the PS had an overall lower terpenoid content, being 1–3 wt%, than the PLW that had greater than 20 wt% terpenoids. The PS samples, extracted fresh, contained higher monoterpene content, being mostly α-pinene and some β-pinene, than the PLW in each of the cross sections. No

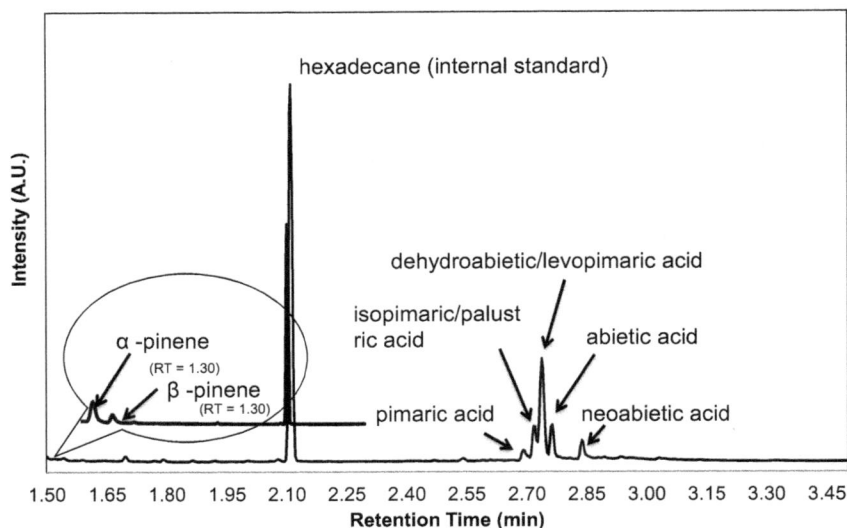

FIGURE 4 | A chromatogram obtained from an underivatized extract from a greenhouse-grown loblolly pine sapling cross section (below 2.05 min) shows mono- and sesquiterpenes. BSTFA derivatized loblolly extract, the chromatogram beyond 2.05 min, is obtained to measure the resin acid diterpenoid content.

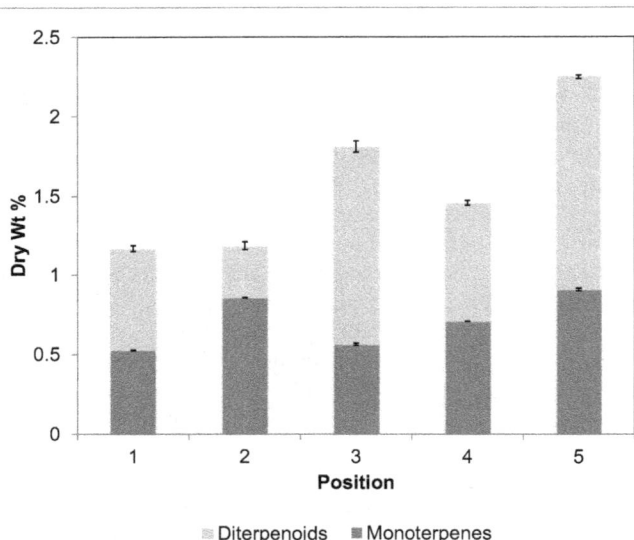

FIGURE 5 | Terpenoid content by dry wt%, separated into mono- and diterpenoids, for different cross sections of a fresh-cut, greenhouse-grown, loblolly pine sapling. Position 1 corresponds to the bottom of the sapling trimmed 1 cm above the base of the soil and each consecutive number is separated by 2″, where position 5 is the top cross section of the sapling, approximately 10 in above the soil. Error bars are for duplicate analysis by fast-GC. 4.

acids were the most abundant diterpenoids present in the sapling, being 0.02–0.6 dry wt% of the cross section samples. Abietic acid was also present in high abundance, being 0.1–0.6 dry wt% of the sapling samples. Neoabietic acid and isopimaric/palustric acids were present in lower abundances, being around 0.02–0.2 wt%, and pimaric acid was the lowest, being 0–0.05 dry wt%. Py-MBMS analysis of cross sections before and after extractions showed the 1 h room temperature hexane/acetone extractions to be 87.5 ± 3.6% efficient (the average from five extracted cross sections based on all diterpenoid ion intensities monitored). The SD of each sapling cross section analyzed (reported in **Table 1**) reflects the reproducibility of the GC analysis. As a single cross section is unique in terpenoid content, it was not possible to extrapolate the reproducibility of the method by comparing terpenoid content from different cross sections. Hence, method reproducibility is inferred from the PLW experiments and is <5% for a homogeneous sample.

DISCUSSION

A variety of solvent systems were used to extract terpenoids from milled PLW and greenhouse-grown loblolly pine sapling cross sections. Different extraction times and temperatures were tested in order to determine the optimal conditions for extracting the maximum amount of total terpenoids from pine wood samples. Py-MBMS analysis of the biomass samples before and after extraction indicated that the single step extraction using hexane/acetone extractions were efficient at extracting the maximum amount of terpenoid components and that the extractions should be performed over 1 h at 22°C. The extraction solvents from the ground PLW and pine sapling cross sections were analyzed by fast-GC/FID, allowing for rapid analysis of the extracted samples to determine their terpenoid content (based on

calibrated sesquiterpenes were found to be present in the PS. Both the mono- and diterpenoid content of the cross sections varied along the length of the sapling where the top of the plant (position 5) contained more terpenoids than the lower (position 1) and middle sections (positions 2–4). Dehydroabietic/levopimaric dry wt%). Ground PLW contained more than 20 wt% terpenoids, whereas, the pine sapling cross sections had varying terpenoid content along the length of the sapling, ranging from 1 to 2.3% dry wt total terpenoids. Model compound spiking studies using two types of resin acids indicated the preparation of the samples allowed for complete recovery of the terpenoids into the extract solvent and derivatization of extruded pine resin showed that 97% of terpenoid content could be accounted for using the suggested calibration standards and the suggested GC method. In order to accurately quantify all diterpenoids present, it is essential that all water be removed from the aqueous fraction and freshly activated molecular sieves be used to dry the organic layer. It was found that the average ratio of monoterpenes to diterpenoids in fresh samples should fall within the range of approximately 0.2–0.6 for fresh samples as anything outside of this indicates that there could be inefficient extraction, monoterpene loss, or lack of derivatization of the diterpenoid. This issue was amplified with the fresh wood and it is possible that water in the biomass could be complicating the analysis.

This method utilizes a single extraction step using a mixture of a polar and non-polar solvent to extract all types of terpenoids at once and does not require the use of specialized equipment, such as a shaker table or Soxhlet apparatus. The method also does not need the samples to be ground as cross sections of saplings could be extracted efficiently. The preparation and analysis of 200 pine samples/week (counting duplicate analysis of both monoterpene and diterpenoid aliquoted extracts, or 800 injections/week) is possible using this method coupled with fast-GC analysis. Overall, this method is capable of rapidly determining terpenoid content in pine biomass samples, while minimizing solvent extraction steps, equipment usage, sample preparation and handling and increasing GC analysis throughput.

AUTHOR CONTRIBUTIONS

AH-W is a postdoctoral researcher. She performed the extraction experiments, method development, GC analysis, wrote, and edited the manuscript. RS is a PI on the project and did py-MBMS analysis and helped with method development. GP and MD are PIs on the project and helped write/review/edit the manuscript. GP provided biomass samples, as well as method development, sampling and extraction insight, and guidance.

ACKNOWLEDGMENTS

The authors would like to acknowledge Dr. Gabriella Papa and Dr. James Kirby at the University of California, Berkeley, for purified bisabolene and collaborations with this project.

FUNDING

This work was supported by the U.S. Department of Energy (DOE) Advanced Research Projects Agency – Energy (ARPA-E) under award No. DE-AR0000209. This work was also supported by the U.S. Department of Energy under Contract No. DE-AC36-08-GO28308 with the National Renewable Energy Laboratory.

REFERENCES

Achotegui-Castells, A., Llusià, J., Hódar, J., and Peñuelas, J. (2013). Needle terpene concentrations and emissions of two coexisting subspecies of Scots pine attacked by the pine processionary moth (*Thaumetopoea pityocampa*). *Acta Physiol. Plant* 35, 3047–3058. doi:10.1007/s11738-013-1337-3

Beglinger, E. (1958). *Distillation of Resinous Wood*. Madison, WI: USDA Forest Service, 496.

Bohlmann, J., and Keeling, C. I. (2008). Terpenoid biomaterials. *Plant J.* 54, 656–669. doi:10.1111/j.1365-313X.2008.03449.x

Bojovic, S., Jurc, M., Drazic, D., Pavlovic, P., Mitrovic, M., Djurdjevic, L., et al. (2005). Origin identification of *Pinus nigra* populations in southwestern Europe using terpene composition variations. *Trees* 19, 531–538. doi:10.1007/s00468-005-0411-x

Evans, R. J., and Milne, T. A. (1987). Molecular characterization of the pyrolysis of biomass. *Energy Fuels* 1, 123–137. doi:10.1021/ef00002a001

Firor, R. L. (2011). *Analysis of Natural Oils and Extracts Using the Low Thermal Mass LTM Series II System*. Wilmington, DE: Agilent Technologies.

Fojtová, J., Lojková, L., and Kubáň, V. (2008). GC/MS of terpenes in walnut-tree leaves after accelerated solvent extraction. *J. Sep. Sci.* 31, 162–168. doi:10.1002/jssc.200700371

Harvey, B. G., Wright, M. E., and Quintana, R. L. (2009). High-density renewable fuels based on the selective dimerization of pinenes. *Energy Fuels* 24, 267–273. doi:10.1021/ef900799c

Hellier, P., Al-Haj, L., Talibi, M., Purton, S., and Ladommatos, N. (2013). Combustion and emissions characterization of terpenes with a view to their biological production in cyanobacteria. *Fuel* 111, 670–688. doi:10.1016/j.fuel.2013.04.042

Keeling, C. I., and Bohlmann, J. (2006). Diterpene resin acids in conifers. *Phytochemistry* 67, 2415–2423. doi:10.1016/j.phytochem.2006.08.019

Lewinsohn, E., Savage, T. J., Gijzen, M., and Croteau, R. (1993). Simultaneous analysis of monoterpenes and diterpenoids of conifer oleoresin. *Phytochem. Anal.* 4, 220–225. doi:10.1002/pca.2800040506

Luan, W., and Szelewski, M. (2008). *Ultra-Fast Total Petroleum Hydrocarbons (TPH) Analysis with Agilent Low Thermal Mass (LTM) GC and Simultaneous Dual-Tower Injection*. Wilmington, DE: Agilent Technologies.

Manninen, A.-M., Tarhanen, S., Vuorinen, M., and Kainulainen, P. (2002). Comparing the variation of needle and wood terpenoids in Scots pine provinces. *J. Chem. Ecol.* 28, 211–227. doi:10.1023/A:1013579222600

Martin, D., Tholl, D., Gershenzon, J., and Bohlmann, J. (2002). Methyl jasmonate induces traumatic resin ducts, terpenoid resin biosynthesis, and terpenoid accumulation in developing xylem of Norway spruce stems. *Plant Physiol.* 129, 1003–1018. doi:10.1104/pp.011001

Meylemans, H. A., Quintana, R. L., and Harvey, B. G. (2012). Efficient conversion of pure and mixed terpene feedstocks to high density fuels. *Fuel* 97, 560–568. doi:10.1016/j.fuel.2012.01.062

Monteiro, J., and Veloso, C. (2004). Catalytic conversion of terpenes into fine chemicals. *Top. Catal.* 27, 169–180. doi:10.1023/B:TOCA.0000013551.99872.8d

Nerg, A., Kainulainen, P., Vuorinen, M., Hanso, M., Holopainen, J. K., and Kurkela, T. (1994). Seasonal and geographical variation of terpenes, resin acids and total phenolics in nursery grown seedlings of Scots pine (*Pinus sylvestris* L.). *New Phytol.* 128, 703–713. doi:10.1111/j.1469-8137.1994.tb04034.x

Ormeño, E., Fernandez, C., and Mévy, J.-P. (2007). Plant coexistence alters terpene emission and content of Mediterranean species. *Phytochemistry* 68, 840–852. doi:10.1016/j.phytochem.2006.11.033

Ormeño, E., Gentner, D. R., Fares, S., Karlik, J., Park, J. H., and Goldstein, A. H. (2010). Sesquiterpenoid emissions from agricultural crops: correlations to monoterpenoid emissions and leaf terpene content. *Environ. Sci. Technol.* 44, 3758–3764. doi:10.1021/es903674m

Palmer, R. C. (1943). Solvents from pine. *Ind. Eng. Chem.* 35, 1023–1025. doi:10.1021/ie50406a003

Peralta-Yahya, P. P., Ouellet, M., Chan, R., Mukhopadhyay, A., Keasling, J. D., and Lee, T. S. (2011). Identification and microbial production of a terpene-based advanced biofuel. *Nat. Commun.* 2, 1–8. doi:10.1038/ncomms1494

Renninger, N. S., Ryder, J. A., and Fisher, K. J. (2011). *Jet Fuel Compositions and Methods of Making and Using Same*. Emeryville, CA: Google Patents.

Rodrigues-Corrêa, K. C. D. S., de Lima, J. C., and Fett-Neto, A. G. (2012). Pine oleo-resin: tapping green chemicals, biofuels, food protection, and carbon sequestration from multipurpose trees. *Food Energy Secur.* 1, 81–93. doi:10.1002/fes3.13

Schmidt, A., Nagel, R., Krekling, T., Christiansen, E., Gershenzon, J., and Krokene, P. (2011). Induction of isoprenyl diphosphate synthases, plant hormones and defense signalling genese correlates with traumatic resin duct formation in Norway spruce (*Picea abies*). *Plant Mol. Biol.* 77, 577–590. doi:10.1007/s11103-011-9832-7

Susaeta, A., Peter, G. F., Hodges, A. W., and Carter, D. R. (2014). Oleoresin tapping of planted slash pine (*Pinus elliottii* Engelm. var. elliottii) adds value and management flexibility to landowners in the southern United States. *Biomass Bioenergy* 68, 55–61. doi:10.1016/j.biombioe.2014.06.003

Sykes, R., Yung, M., Novaes, E., Kirst, M., Peter, G., and Davis, M. (2009). "High-throughput screening of plant cell-wall composition using pyrolysis molecular beam mass spectroscopy," in *Biofuels*, ed. Mielenz J. R. (Golden, CO: Humana Press), 169–183.

Thompson, A., Cooper, J., and Ingram, L. L. Jr. (2006). Distribution of terpenes in heartwood and sapwood of loblolly pine. *For. Prod. J.* 56, 46–48.

Vallinayagam, R., Vedharaj, S., Yang, W. M., Lee, P. S., Chua, K. J. E., and Chou, S. K. (2014). Pine oil–biodiesel blends: a double biofuel strategy to completely eliminate the use of diesel in a diesel engine. *Appl. Energy* 130, 466–473. doi:10.1016/j.apenergy.2013.11.025

Varming, C., Andersen, M. L., and Poll, L. (2006). Volatile monoterpenes in black currant (*Ribes nigrum L.*) juice: effects of heating and enzymatic treatment by β-glucosidase. *J. Agric. Food Chem.* 54, 2298–2302. doi:10.1021/jf051938k

Zhao, T., Krokene, P., Björklund, N., Långström, B., Solheim, H., Christiansen, E., et al. (2010). The influence of *Ceratocystis polonica* inoculation and methyl jasmonate application on terpene chemistry of Norway spruce, *Picea abies*. *Phytochemistry* 71, 1332–1341. doi:10.1016/j.phytochem.2010.05.017

5

Performance of Separation Processes for Precipitated Calcium Carbonate Produced with an Innovative Method from Steelmaking Slag and Carbon Dioxide

Sebastian Teir[1]*, Toni Auvinen[2], Arshe Said[3], Tuukka Kotiranta[4] and Heljä Peltola[4]

[1] VTT Technical Research Centre of Finland Ltd., Espoo, Finland, [2] Outotec Dewatering Technology Center, Lappeenranta, Finland, [3] Department of Energy Technology, School of Engineering, Aalto University, Espoo, Finland, [4] Outotec Research Center, Pori, Finland

Edited by:
Renato Baciocchi,
University of Rome
Tor Vergata, Italy

Reviewed by:
Rafael Mattos Dos Santos,
Sheridan College Institute of
Technology and Advanced
Learning, Canada
Giulia Costa,
University of Rome Tor Vergata, Italy

***Correspondence:**
Sebastian Teir
sebastian.teir@vtt.fi

Specialty section:
This article was submitted to Carbon
Capture, Storage, and Utilization,
a section of the journal
Frontiers in Energy Research

Citation:
Teir S, Auvinen T, Said A, Kotiranta T
and Peltola H (2016) Performance of
Separation Processes for Precipitated
Calcium Carbonate Produced with an
Innovative Method from Steelmaking
Slag and Carbon Dioxide.
Front. Energy Res. 4:6.

In this work, experiments were performed to determine the filterability of calcium carbonate produced with an alternative calcium carbonate production concept. The concept uses steelmaking slag as raw material and has potential to fix CO_2 emissions and utilize steelmaking slag, simultaneously. As calcium carbonate is precipitated in a solution containing ammonium chloride, calcium chloride, and ammonia, the product needs to be washed and hence filtered. In this work, different separation processes, including washing, filtering, and drying, were tested on two calcium carbonate slurries produced from steel converter slag and CO_2 by a laboratory-scale pilot facility, with the aim of obtaining a solid product with a low chloride content using a minimum amount of washing water. The order of maximum filtration rates achievable of the calcium carbonate slurries was determined by experimental work. The tests included pressure filtration and vacuum filtration and the test series contained altogether 21 different filtration cycles with varying combinations of filtering, washing, and drying steps. The filtered cakes were analyzed by their residual moisture content, chloride content, and conductivity, and the filtrates by their residual solids content, chloride content, and conductivity. Pressure filtration gave a high capacity (400–460 kg/m²h) and a low cake residual moisture content (12–14 wt-%). Vacuum filtration gave slightly higher filtration rates (500–610 kg/m²h at the lowest residual chloride contents of the cakes), but the cake residual moisture also stayed higher (25–26 wt-%). As the vacuum filtration tests used a filter cloth with higher permeability than that of the pressure filtration tests, a slightly higher filtration rate was expected. However, both filtration technologies seem suitable for filtering and washing calcium carbonate prepared with the studied method as a residual chloride content as low as 10 ppm of the filtered solids can be achieved with quite a small amount of washing water and the filtration rate is fast.

Keywords: mineralization, carbon dioxide, CCU, utilization, filtration, PCC, GCC

INTRODUCTION

Calcium carbonate is the most commonly used filler material in paper making (Naydowski et al., 2001). Ground calcium carbonate (GCC) is manufactured by grinding high quality limestone to very small sizes and is mostly used as a pigment, included as an externally applied coating in coated papers. Therefore, GCC has a broad distribution of shapes and sizes, which reduces their optical performance (Clark, 1992). Precipitated calcium carbonate (PCC) is also used in papermaking as a coating and filler material. In conventional production of synthetic (i.e., precipitated) calcium carbonate, flue gas containing CO_2 is bubbled though a hydrated lime slurry (calcium hydroxide), from which calcium carbonate precipitates. By adjusting the precipitation process parameters, the shape and size of the crystals produced can be controlled to optimize their optical properties for use in paper making. PCC is normally also brighter than GCC, since organic impurities and some metal oxides can be separated during the PCC production process (Naydowski et al., 2001). The production of PCC binds CO_2 as carbonate, but CO_2 is also emitted when limestone is calcined for providing lime for the process. PCC production requires also relatively pure limestone. To minimize transportation costs, the limestone is calcined before transportation to the PCC production facility, which is normally located at the paper mill site.

An alternative production concept for calcium carbonate production is being developed that omits the need for fresh limestone and its calcination (Eloneva et al., 2009). This concept has the potential to reduce CO_2 emissions and simultaneously utilize steelmaking slag. In this process, calcium is selectively extracted from steelmaking slag using ammonium salt solutions in the extraction step:

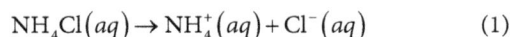

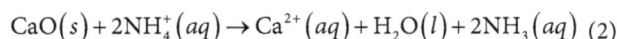

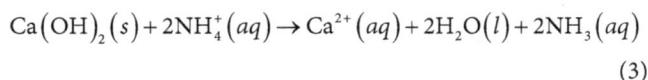

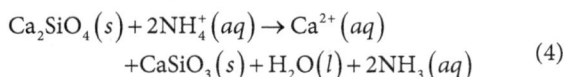

$$NH_4Cl(aq) \rightarrow NH_4^+(aq) + Cl^-(aq) \tag{1}$$

$$CaO(s) + 2NH_4^+(aq) \rightarrow Ca^{2+}(aq) + H_2O(l) + 2NH_3(aq) \tag{2}$$

$$Ca(OH)_2(s) + 2NH_4^+(aq) \rightarrow Ca^{2+}(aq) + 2H_2O(l) + 2NH_3(aq) \tag{3}$$

$$Ca_2SiO_4(s) + 2NH_4^+(aq) \rightarrow Ca^{2+}(aq) + CaSiO_3(s) + H_2O(l) + 2NH_3(aq) \tag{4}$$

Calcium carbonate is subsequently precipitated in the carbonator by bubbling CO_2-containing flue gas through the solution:

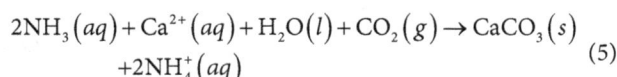

$$2NH_3(aq) + Ca^{2+}(aq) + H_2O(l) + CO_2(g) \rightarrow CaCO_3(s) + 2NH_4^+(aq) \tag{5}$$

The calcium carbonate is separated by filtration, and the solvent is returned to the extractor, as the ammonium salt is regenerated in the carbonator (**Figure 1**). Both the calcium carbonate and the unreacted part of the slag from the extractor (slag residue) are washed in order to remove solvent remnants (mainly ammonium salt). Laboratory experiments have shown that calcium carbonate

purity up to 99.8 wt-% can be achieved (Eloneva et al., 2009). Similarly, the particle shape can be adjusted – both rhombohedral and scalenohedral calcite can be produced, as well as aragonite. A pilot facility of the concept was recently completed at Aalto University, enabling production of 5–10 kg batches of calcium carbonate (Said and Järvinen, 2015). Although the particles that have been produced with the pilot so far are coarser than those required for paper applications (typically $P_{50} < \sim 1$–2 μm for filling applications), the calcium carbonate could be subsequently ground and possibly be an alternative to GCC in coating applications (Teir et al., 2015). As grinding is expected to result in a broad particle size distribution, the calcium carbonate is not necessarily suitable as an alternative to PCC in filler applications, where a narrow particle size distribution is needed.

One challenge of the concept presented in **Figure 1** is that the produced calcium carbonate is precipitated in a solution of ammonia and ammonium chloride, which raises concern that chloride may be embedded in the product and hinders its commercial use. However, the maximum chloride content that can be allowed in a PCC product is not clear as there does not seem to be any standard or guideline available for this. For comparison, commercial PCC is also produced as a by-product of ammonia production in the Solvay process (Ciullo, 1996). As the calcium carbonate is produced in presence of sodium chloride, the Solvay PCC product contains 0.10% NaCl (**Table 1**), which has contributed to the Solvay PCC process being less used than the conventional carbonation process (Mattila and Zevenhoven, 2014a). Also ESAPA (2004) reports difficulties with commercializing calcium carbonate containing chloride impurities. According to ESAPA, attempts were made to recover $CaCO_3$ produced as a by-product from brine purification for a Solvay process, but

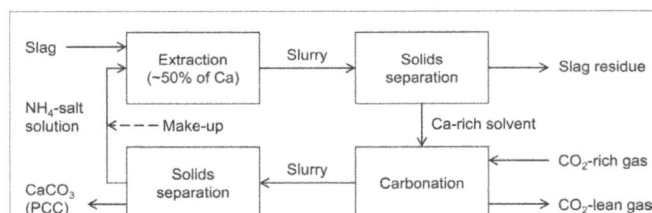

FIGURE 1 | The studied concept for producing calcium carbonate from CO_2-rich gas and slag.

TABLE 1 | Chemical composition of PCC products by process (unit: %, Statton, 2012).

	PCC (carbonation)	PCC (Solvay)
$CaCO_3$	98.36	98.62
$CaSO_4$	0.08	0.63
$MgCO_3$	0.7	0.21
Al_2O_3	0.09	0.01
Fe_2O_3	0.07	0.01
SiO_2	0.1	0.02
NaCl	–	0.10
% H_2O loss[a]	0.6	0.30
pH[b]	9.4	8.5

[a]At 110°C.
[b]Saturated solution.

the chloride content in the effluent to be treated and impurities remaining after treatment were found to be problematic. Due to these impurities, the product could not compete with more pure products available on the market. In the conventional PCC production process, PCC is precipitated in water and requires no further purification.

Although no detailed product quality assessment has yet been performed on the calcium carbonate produced by the studied concept, the analyses performed so far indicate that the chloride can be removed by washing and does not end up in the product (Mattila et al., 2012). According to the best knowledge of the authors, the only published data on filtration and washing performance is that of Hudd (2014), who performed washing experiments of PCC (commercial rhombohedral calcite) and ammonia chloride mixtures using both a vacuum filter and pressure filter. Hudd studied a staged filtration, with both countercurrent and crosscurrent flow of washing water, so the experimental parameters for each filtration stage were kept constant. Hudd assumed that the filtered cake needs to be separated from the filter cloth and mixed with washing water after which the mixture is filtered again, and the procedure is repeated using either fresh or recycled washing water a few times. Although a low chloride content in the calcium carbonate can be achieved with this method, it requires a relatively large amount of washing water.

The objective of the work presented here was to select suitable filtration technologies and determine the maximum filtration rates achievable for the calcium carbonate slurries produced with the new concept. The focus was on minimizing the amount of residual chloride (i.e., solvent remnants of the ammonium chloride solution) on the carbonate product by varying the parameters for filtration and washing while keeping the filtration rate high. In order to minimize the use of washing water, a different washing approach than that used by Hudd was tested: washing water was added directly on top of the filtered cake, after which the water was pressed through the cake. The concentration of possible metals leached from steel slag was not studied here as this can be minimized by managing process parameters (Mattila and Zevenhoven, 2014b). The tests included pressure filtration and vacuum filtration, and the test series contained altogether 21 different filtration cycles with varying combinations of filtering, washing, and drying steps. The filtered cakes were analyzed by their residual moisture content, chloride content,

and conductivity, and the filtrates by their residual solids content, chloride content, and conductivity.

MATERIALS AND METHODS

Production of Calcium Carbonate for Filtration Tests

The steelmaking slag used in the experiments was steel converter slag from Raahe steel plant in Finland. The steel converter slag had been ground to a particle size <250 μm. The chemical composition of the slag was analyzed by X-ray fluorescence (XRF) spectroscopy (**Table 2**).

The pilot facility at Aalto University was used for preparing two batches of calcium carbonate slurry for subsequent filtration tests. The pilot facility consists of three reactors of 200 l volume each, connected to two stages of filtration (a full description of the test facility can be found in Said and Järvinen, 2015). So far, both rhombohedral calcite and aragonite has been successfully produced with the pilot plant (Said and Järvinen, 2015). Therefore, one batch of each calcium carbonate type was produced. The parameters are summarized in **Table 3** and the procedure was as follows: first, a batch of 17 kg of steel slag was mixed with 170 l, 1M NH_4Cl (aq) solvent to selectively dissolve calcium from the slag at room temperature (~20°C). The mixture was agitated for 1 h at 200 rpm, after which the Ca-rich solution was separated from the residual slag through 1-μm filter bags. To maximize the removal of solid particles, the filtrate was further pumped through a series of filters consisting of two 1-μm filters and one 0.45-μm filter. After filtration, 150 l of the Ca-rich solution was pumped to a carbonation reactor, where it was heated up to the desired temperature. After reaching the targeted temperature, pure CO_2 gas was bubbled through the solution, forming a slurry containing calcium carbonate precipitate. After carbonation, the slurry from the reactor was preliminarily filtered using an identical filtration system as described above for separating the PCC from the liquid. Both slurries were produced with identical parameters except for the carbonation reactor temperature, which was set to 20°C for precipitating rhombohedral calcite and 55°C for producing aragonite (**Table 3**). After preliminary filtration, the moist calcium carbonate solids and 100 l additional filtrate were transported in separate containers to Outotec Dewatering Technology Center for further filtration tests. The particle size

TABLE 2 | Summary of the XRF analysis of steel converter slag (only compounds found in amounts ≥0.1 wt-% listed).

CaO	FeO	SiO_2	MnO	Al_2O_3	MgO	V_2O_3	Ti	P	Cr	Na_2O
51.40	14.60	13.70	1.80	1.60	1.50	2.05	0.55	0.45	0.25	0.1

TABLE 3 | Summary of calcium carbonate batch production parameters.

| Batch | Extraction | | | Carbonation | | |
	Solvent (L)	Concentration of NH_4Cl (M)	Slag (kg)	T (°C)	Solution (L)	CO_2 (L/min)
Calcite	170	1	17	20	150	13
Aragonite	170	1	17	55	150	13

distributions of the produced calcium carbonates were measured using a laser diffraction particle size analyzer.

Test Methods for Filtration and Washing

When choosing a process suitable for solid–liquid separation of calcium carbonate, for this concept, the requirements for efficient washing must be taken into consideration. In addition, the liquid content of the product should also be as low as possible to maximize the recovery of ammonium chloride and minimize the need for make-up. Solid–liquid separation processes considered in this work were processes that allow large-scale (~100,000 tpa) separation of calcium carbonate at the lowest possible cost considering the requirements: efficient filtration for particle sizes <100 μm, minimum amount of liquid in product, and efficient washing of the product. A comparative performance of the main types of commercial solid–liquid separation processes (**Table 4**) is given by Couper et al. (2010). Sedimentation, centrifugation, liquid cyclones, screens, and ultrafiltration have both poor wash possibilities and poor removal of liquid from the solid product. In contrast, vacuum drum filters, horizontal pressure filters, and basket centrifuges are good at removing liquids from the solid product and have excellent wash possibilities. For the experimental work, vacuum filtration and pressure filtration were selected, since both are well suited for large quantities of solids. However, as the pilot facility at Aalto was only able to produce about 5 kg PCC per day (one batch), only small, laboratory-size equipment was used for the test series (0.5–2.7 l of slurry filtered per test). The pressure filtration and vacuum filtration were separately optimized for minimizing the chloride content, while keeping the filtration rate as high as possible. In order to minimize the use of washing water, a different washing approach than Hudd (2014) was tested: washing water was added directly on top of the filtered cake, after which the water was pressed through the cake.

Pressure Filtration Tests

The pressure filtration test cycle is schematically presented in **Figure 2**. In pressure filtration, the slurry is fed into a filter chamber by pumping. Part of the filtrate passes through a filter cloth and exits beneath the chamber. Pressure is produced by pumping water or air over a diaphragm that expands and presses the slurry, removing more liquid out of the slurry. This is also referred to as the first pressing. After the first pressing, solids can be washed if needed. In washing of the solids, the washing liquid is fed into the empty space above the cake that has been formed inside the chamber as a result from the first pressing. Then the washing liquid is pressed through the cake (the second pressing). After the cake has been pressed, it is air dried with pressurized air, after which the cake is mechanically removed from the filter cloth ("solids discharge" in **Figure 2**). The cake, i.e., filtered solids, is then ready for further processing. For the experiments, an Outotec Labox 100 test unit was used with an AINO K11 filter cloth having the filtration area of 0.01 m². The filter cloth has a permeability of 0.08 m³/m² min (air permeability measured at 200 Pa). While industrial-scale filter units normally consist of several filter chambers run in parallel, the test unit had only one chamber (**Figure 3**). Parameters that were varied were chamber depth, temperature, wash liquid (water) volume, pressure, and

the quantity of slurry used (**Table 5**). Air was employed as pressing media. The produced cake was further analyzed with various analytical methods (see Analysis Methods).

Vacuum Filtration

In vacuum filtration, liquid is drawn out from the slurry through a filter cloth by forming a vacuum on the opposite side of the filter cloth. The slurry sample is poured into a cylinder on top of the filter cloth, and then vacuum is applied to the slurry underneath it. The solid is trapped by the filter, and the liquid is drawn by the vacuum through the filter into the flask below. The time period that there is excess water on the cake is called the filtration time. When the excess water has been sucked out of the cake, the vacuum is turned off, the mother liquid is collected, and then the cake can be washed. Washing is done by turning the vacuum back on and by pouring the washing liquid on the cake carefully to ensure its equal distribution on the cake. When there is no more washing liquid on the cake surface, the cake drying begins. Washing can be performed in multiple stages and also in a co-current or a counter current mode. In the experiments, an Outotec Larox® Büchner (BVB) test unit was used with a filtration area of 0.01 m². An ARTO S11 filter cloth, which is the tightest filter cloth available for the test equipment, was used. The filter cloth has a permeability of 0.3 m³/m² min (air permeability at 200 Pa). During filtration, the vacuum was kept at 0.5 bar for all experiments. The Büchner test unit setup is illustrated in **Figure 4**. Parameters that were varied were slurry volume, temperature, and wash liquid (water) volume (**Table 6**).

Analysis Methods

The moisture contents of the filtered cakes were analyzed from ~50 g of filtered samples (wet samples), which were dried in a laboratory oven at 60°C in air. The drying time was 38 h 15 min. The moisture contents were calculated from the mass losses of the samples. The filtrate solids contents were measured by pouring 100 ml of filtrate through a filter paper. The paper with solids on was then dried in the oven overnight at 105°C and weighed.

The chloride contents were determined as follows. A preweighed sample (~10 g) of the dry cake was mixed with 100 ml ion exchanged water. The suspension was stirred vigorously with a magnetic stirrer for 60 min. The suspension was filtered and the chloride content of the clear solution was analyzed by an ion chromatograph (IC). In the case of the lowest chloride contents, potentiometric titration with $AgNO_3$ was used instead of ion chromatography. In these analyses, increased sample weights (~25 g) were also used. The chloride contents corresponding to both wet cake and dry cake were calculated from the analyzed chloride content of the solution. The assumption for measuring the chloride contents by this manner was that all the chlorides in the filtered cake are water soluble. It was also assumed that the chlorides were situated on the particle surfaces, and not locked inside the particles, so they were able to dissolve into water during stirring of the suspension. Two stirring times (30 and 60 min) were tested before starting the analysis series, and they were found to give almost identical results. For the analysis series, the longer stirring time (60 min) was adopted.

TABLE 4 | Comparative performance of solid–liquid separation equipment[a] (Couper et al., 2010).

	Product parameters			Feed conditions favouring use			Equipment characteristics		Direct costs		
	Solids in liquid product	Liquid in solid product	Wash possibilities[b]	Solids concentration	Solids density	Particle size	Power	Space and holdup	Initial	Operating	Maintenance
Filtration											
Vacuum drum filter	F	G	E[c]	High to medium	–	Medium	High	Medium	High	High	Medium
Disc filters	F	G	P to F	Medium	–	Fine	High	Medium	Medium to high	High	Medium
Horizontal filters	F	G	G to E[c]	high to medium	–	Coarse	High	Medium	Medium	High	Medium
Precoat filter	E	P[d]	P to F[d]	Very low	–	Slimy	High to medium	Medium	High	Very high	Medium
Leaf (Kelly) filter	G to E[c]	F	F to G	Low	–	Fine, slimy	Medium to low	Medium	Medium	Very high	Medium
Sedimentation											
Thickener	G to E	P	P	Medium	Dense	Medium	Low	Very high	Medium to low	Low	Very low
Clarifier	G	P	very P	Low	Medium dense	Fine	Very low	Very high	Medium to low	Low	Very low
Classifier	P	P	P to F	Medium	Dense	coarse	Low	High	Medium to low	Low	Very low
Centrifugation											
Disc	F to G	P	P	Low to medium	Medium	Fine	High	Low	High	High	High
Solid bowl	P	F	P to F	Medium to high	Medium	Medium to fine	High	Low	Medium to high	High	High
Basket	P to F	E	E[c]	Medium to high	–	Coarse	High	Low	Medium	High	High
Liquid cyclones											
Large	P	P to F	P	Low to medium	High	Medium	Medium to low	Low	Very low	Medium	High
Small multiple	P to F	P	Very P	Low	Medium to high	Fine	Medium to low	Low	Low	Medium	Medium
Screens	P	P to F	P	Medium to high	–	Coarse to medium	Low	Very low	Very low	Medium	Medium to high
Ultrafiltration	E	P to F	P	Low	–	Very fine	Medium to high	High	High	High	Very high

[a]P, poor; F, fair; G, good; E, excellent.
[b]Decantation wash always possible.
[c]Displacement wash feasible.
[d]Solids product contaminated by precoat material.

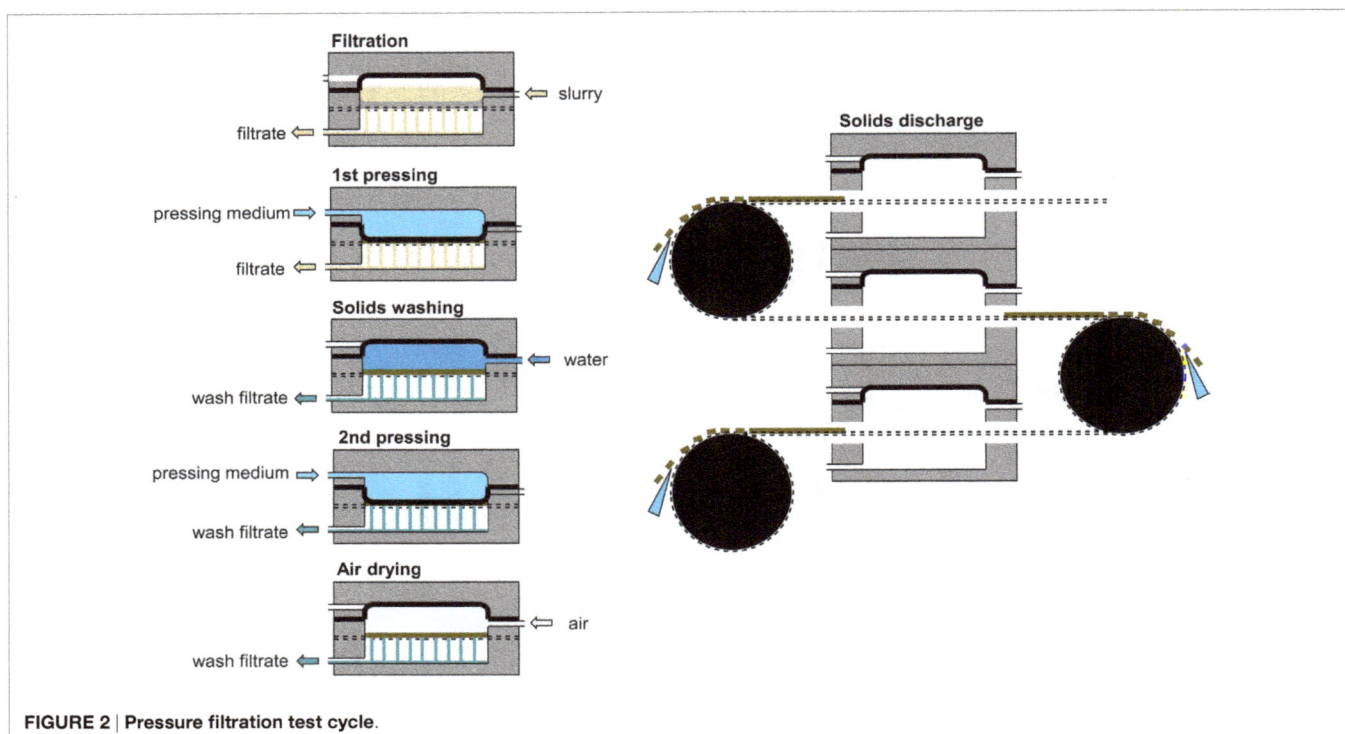

FIGURE 2 | Pressure filtration test cycle.

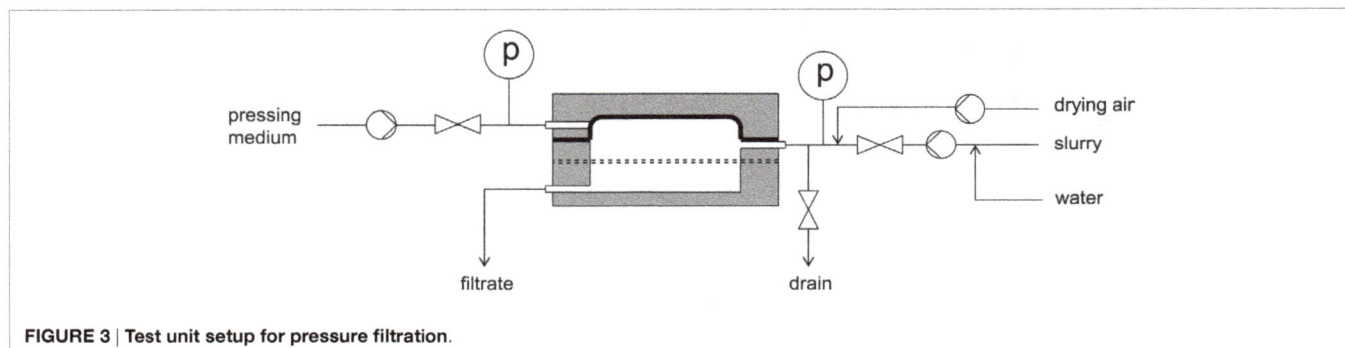

FIGURE 3 | Test unit setup for pressure filtration.

As the chloride analyses were performed at another location, Outotec Research Center, at the end of the filtration test series, conductivities of the filtrates were measured online for getting instant indications of the washing results during the tests. The conductivities were measured with a HANNA instruments conductivity meter. The cake conductivity was measured after mixing dry cake in the ratio of 1:1 with distilled water.

Solid samples of the washed and dried cakes from the pressure filtration tests were analyzed using a scanning electron microscope (SEM) equipped with energy-dispersive x-ray spectroscopy (EDS), as well as with X-ray diffraction (XRD).

RESULTS

The moist carbonate samples and the filtrate were transported by road from the Aalto pilot plant to the Outotec premises for the filtration tests. The results from the particle size analysis of the produced calcium carbonate showed a wide size distribution of particles, with P_{50} <41 µm and P_{80} <84 µm for the calcite batch (**Figure 5**) and P_{50} <27 µm and P_{80} <50 µm for the aragonite batch (**Figure 6**). These samples had already been pre-filtered at Aalto, so the solid contents needed to be altered to simulate the solid contents of the slurry at a full-scale facility before filtration. It is very likely that a thickener would be used for raising the solid contents of the slurry prior to filtration to about 20–40 wt-%. However, the performance of a thickener was not included in these tests, so different solid concentrations were chosen for the two slurries. Some of the liquid had leaked from the calcite containers, resulting in a solids content of 60 wt-% for the calcite slurry. The calcite sample was therefore diluted to the solids content of 25 wt-% for the filtration experiments with its own NH_4Cl filtrate that was delivered with the samples. The moisture content of the aragonite batch was at 40 wt-% solids, so this batch was used without dilution. The pressure filtration tests were performed with the filter cloth AINO K11 while the vacuum filtration tests were performed with the filter cloth ARTO S11 (the tightest filter cloth

TABLE 5 | Parameters used in the pressure filtration tests.

Test number	1	2	5	6	7	12	13	14	15
Slurry	Calcite	Calcite	Calcite	Calcite	Calcite	Aragonite	Aragonite	Aragonite	Aragonite
Chamber depth (mm)	33	60	60	60	60	33	60	60	60
Solids in slurry (wt-%)	19.7	17.7	18.2	20	19.9	37.5	38.2	42	37.3
Temperature (°C)									
Slurry	17	17	18	18	18	55	55	55	53
Wash liquid	–	–	24	24	24	–	–	24	50
Duration of (min)									
Pumping	1.5	1	0.75	1	1	1	1	1	1
First pressing	–	1.5	0.5	0.5	–	1.5	1.5	–	0.5
Washing	–	–	0.5	1.33	1.5	–	–	5	1.5
Second pressing	–	–	1.5	1	1	–	–	1.5	1.5
Air drying	–	1	1	1	1	1	1	1	1
Technical time	4	4	4	4	4	4	4	4	4
Total cycle time (min)	5.5	7.5	8.25	8.83	8.5	7.5	7.5	12.5	9.5
Pressure of (bar)									
Slurry feed	6	4	4	4	4	6	4	4	4
First pressing	–	16	6	8	–	16	12	–	8
Wash liquid	–	–	5	2	–	–	–	6	6
Second pressing	–	–	16	12	12	–	–	12	12
Quantity of slurry (l)	1.8	2.7	2	2.3	2.5	0.7	1.3	1.2	1.5
Quantity of filtrate during									
Pumping (kg)	1.51	2.12	1.68	2.04	2.03	0.276	0.605	0.663	0.79
First pressing (kg)	–	0.12	0.017	0.01	–	0.108	0.136	–	0.078
Washing (l)	–	–	0.729	0.607	0.709	–	–	0.037	0.078
Second pressing (kg)	–	–	0.185	–	0.106	–	–	0.137	0.111
Air drying (kg)	–	0.12	0.083	–	0.151	0.129	0.179	0.19	0.23
Total (w/o wash filtrate)	1.51	2.36	1.784	2.053	2.181	0.513	0.92	0.853	1.102
Consumption of wash liquid (l)	–	–	–	1	1	–	–	0.04	1
Air flow (l/min)	–	30	35	35	25	12.5	10	15	<10
Air pressure (bar)	–	4	3	3	3	6	6	6	6
Cake									
Moisture (wt-%)	29.3	21.3	19	12	11.9	11.7	13.5	11	13.7
Thickness, average (mm)	36	48	33	39	42	23	43	45	52
Dry weight (kg)	0.42	0.55	0.42	0.54	0.56	0.34	0.64	0.69	0.73
Filtration rate (dry solids, kg/m²h)	452.7	436.1	307.2	363.6	398.8	270.7	508.9	329.5	463.1
Filtration rate (filtrate, l/m² h)	1647	1888	1297	1395	1540	410	736	409	696
Wash ratio (m³/ton DS)	–	–	2.4	1.9	1.8	–	–	0.1	1.4
Washing results									
Cake conductivity (μS/m)	–	7000	70	57	52	9100	9700	8500	110
Wash filtrate conductivity (μS/m)	–	–	1285	–	330	–	–	–	–
Cake chloride content (ppm)	–	10022		18	10	5657	5812	4655	109

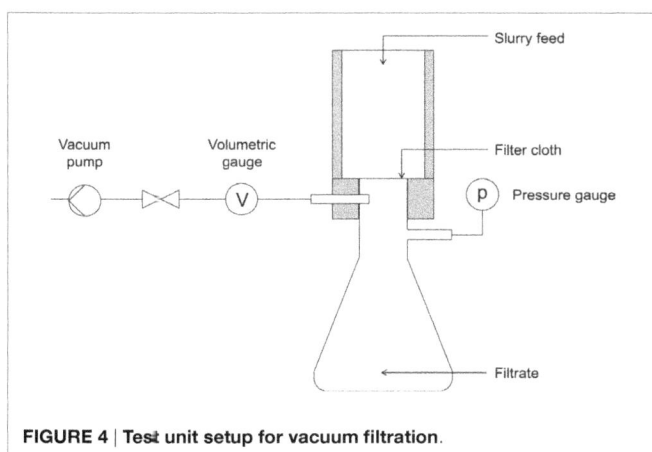

FIGURE 4 | Test unit setup for vacuum filtration.

for Büchner at the moment) and a vacuum of 0.5 bar. Both filter cloths produced a very clear filtrate for both the calcite and the aragonite slurry. AINO K11 gave filtrates with <10 mg/l solids, while ARTO S11 gave filtrates with ~10–15 mg/l solids. No cake cracking was observed in any of the vacuum filtration tests.

Pressure Filtration
A detailed listing of the parameters used in each pressure filtration experiment is given in **Table 5**.

Tests with Calcite Slurry
The first tests were performed using the calcite slurry. The initial test run with a 33-mm chamber showed that the slurry filtered very easily (test number 1, **Table 5**). The chamber was full after ~40 s of pumping, and the cake solids content was above 70 wt-%

TABLE 6 | Parameters used in the vacuum filtration tests.

Test number	3	4	8	9	10	11	16	17	18	19	20	21
Slurry[a]	Calc	Calc	Calc	Calc	Calc	Calc	Arag	Arag	Arag	Arag	Arag	Arag
Pressure drop over wet filtercloth (bar)	0.15	0.2	0.2	0.2	0.25	0.25	0.15	0.25	0.25	0.25	0.25	0.25
Filtration												
Quantity of slurry (ml)	500	1000	750	1000	1000	1000	500	750	1000	1000	1000	1000
Settling time (s)	5	5	5	5	5	5	5	5	5	5	5	5
Separation time (s)	15	44	28	44	50	52	15	31	52	61	54	52
Drying time (s)	60	60	60	10	10	10	60	60	60	10	10	10
Mother liquor (ml)	420	840	625	800	785	780	290	440	580	530	515	520
Washing												
Temperature (°C)	–	–	–	20	20	22	–	–	–	50	50	55
Volume in (ml)	–	–	–	225	450	1125	–	–	–	270	540	1080
Volume out (ml)	–	–	–	250	480	1160	–	–	–	340	590	1135
Wash time (s)	–	–	–	20	42	111	–	–	–	49	96	193
Drying												
Air flow (l/min)	15	15	15	15	15	15	15	10	10	10	10	10
Vacuum (bar)	0.4	0.4	0.35	0.4	0.4	0.45	0.4	0.45	0.45	0.47	0.47	0.47
Drying time (s)	–	–	–	60	60	60	–	–	–	60	60	60
Cake												
Thickness (mm)	10	20	16	21	21	21	24	37	49	49	48	48
Moisture (wt-%)	21.5	24.7	23.7	21.1	25.0	23.7	25.6	26.0	28.0	25.6	28.4	26
Dry weight (g)	109.9	227.1	181.6	242.7	229.6	232.8	266.1	402.8	534.6	561	523	543.2
Conductivity (µS/m)	13500	–	–	103	106	73	21100	23000	22700	7400	895	174
Chloride content (ppm)	8553	8771	15133	252	3	48	13464	13993	10826	4455	598	127
Total time (s)	80	109	93	139	167	238	80	96	117	185	225	320
Wash ratio (l/kg DS)	–	–	–	0.9	2	4.8	–	–	–	0.5	1	2
Filtration rate (dry solids, kg/m² h)	494	750	703	629	494	352	1197	1151	1645	1092	837	611
Filtration rate (filtrate, l/m² h)	2250	3303	2903	2590	2156	1513	2250	2813	3077	1946	1600	1125

[a]Calc, calcite; Arag, aragonite.

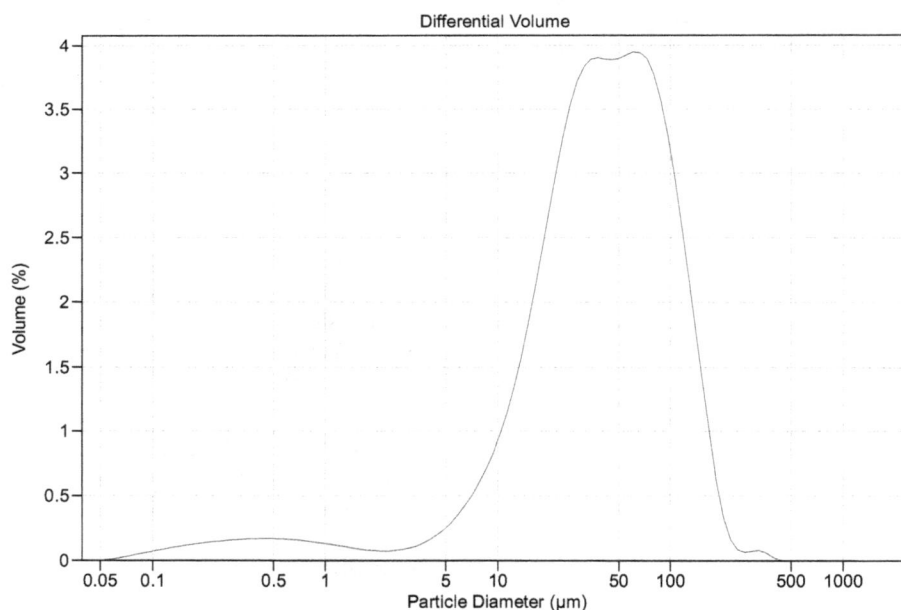

FIGURE 5 | Particle size distribution of the solids in the calcite slurry.

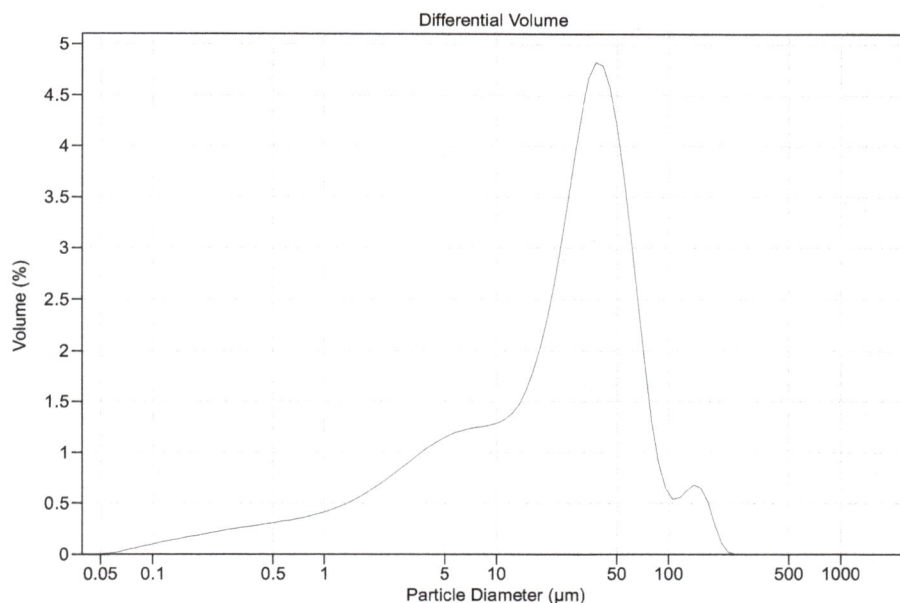

FIGURE 6 | Particle size distribution of the solids in the aragonite slurry.

even without pressing and drying. Pressing or drying was not performed due to a leakage of pressing air in the test unit caused by a mishap in the unit assembly.

As the slurry was found to be easily filtered, test number 2 was performed with a 60-mm filter chamber and the pumping pressure was reduced to 4 bar. Pressing was then performed with 16-bar pressure, followed by 1 min air drying with 4-bar pressure, which resulted in a cake conductivity of 7000 µS/m indicating a high salt concentration (the analyses carried out later verified a high cake chloride content of 10022 ppm) and that subsequent washing is needed.

Test runs number 5–7 were performed with solids washing using various pressures and pressure schemes: test number 6 was performed with a higher pressure in the first pressing (8 vs. 5 bar in test number 5) and a lower washing liquid pressure (2 vs. 5 bar in test number 5). Test run number 7 was performed without the first pressing, and the wash liquid was introduced right after the slurry feed. Washing filtrate samples were taken at the beginning of washing, after 45 s of washing and during pressing. The conductivity measurements of these samples are shown in **Figure 7**. Test number 7 produced the highest filtration rate with washing: 399 kg/m²h. The cake produced with this test had the lowest chloride content (10 ppm) and a low moisture content (12 wt-%).

Tests with Aragonite Slurry

The first test run with aragonite (test number 12, **Table 5**) showed that the aragonite slurry was about as easy to filter as the calcite slurry. Further tests were therefore performed using the higher 60-mm chamber instead of the 33-mm chamber used in test number 12. Pressures were also lowered in the slurry feed to 4 bar (6 bar in test number 12) and in the last pressing to 12 bar (16 bar in test number 12). Test run number 13 was also performed

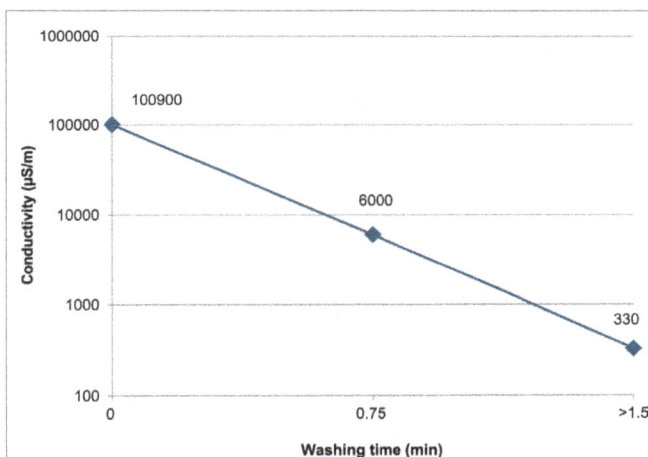

FIGURE 7 | Wash filtrate conductivity during solids washing in Test run #7 (>1.5 min means that this sample was taken during pressing, after a washing time of 1.5 min).

without solids washing while the filtration cycle was the same as in test number 12. Test number 12 resulted in a cake conductivity of 9100 µS/m (which corresponded to a cake chloride content of 5700 ppm), which indicated that subsequent washing is needed.

Test runs number 14 and 15 were performed with solids washing using room temperature water (**Table 5**). Washing was performed right after the slurry feed, without the first pressing. It was observed that the washing liquid could not penetrate the cake. After 5 min of washing, only ~40 ml of washing liquid had been introduced to the cake although a relatively high pressure of 6 bar had been used. Test run number 15 was performed with solids washing after the first pressing. The temperature of the wash liquid was also raised to ~50°C, in order to simulate real process conditions (aragonite is produced at a temperature of 50°C or more). A slurry feed of 1 min was followed by 0.5 min first pressing with 8-bar pressure. Then washing liquid was introduced to the cake with 6-bar pressure. After 1.5 min, 1 l of wash liquid had been introduced to the cake, and the cake conductivity was measured to be 110 µS after washing. Test run number 15 gave a filtration rate of 463 kg/m²h with the final chloride content of 109 ppm and moisture content of 14 wt-% in the cake.

Vacuum Filtration

A detailed listing of the parameters used in each pressure filtration experiment is given in **Table 6**.

Tests with Calcite Slurry

Büchner vacuum filtration tests were included as it was observed in the first pressure filter tests that the calcite slurry was very easy to filter. Test runs number 3, 4, and 8 were performed without solids washing and varying the slurry sample volume with 500–1000 ml for finding the volume giving the maximal filtration rate (**Table 6**). Larger slurry samples gave higher rates, and therefore 1000 ml was selected for subsequent washing tests.

The optimal wash ratio was tested by varying the washing liquid volume in test run numbers 9, 10, and 11 [0.9, 2.0, and 4.8 l/kg DS (dry solids), respectively]. The filtration cycle was otherwise equal in all three runs: after filtration there was 10 s intermediate drying, followed by solids washing and 60 s final drying (**Table 6**). The washing liquid temperature was ~20°C in all tests. The cake conductivity dropped from the original 13,500 µS/m without washing to ~100 µS/m with 1 l/kg DS wash ratio. Higher wash ratios did not affect the cake conductivity much (**Figure 8**). The chloride analyses varied somewhat, giving the lowest chloride content in the cake with the wash ratio of 2.0 l/kg DS. The filtration rate dropped from 629 kg/m²h for 0.9 l/kg DS wash ratio to 352 kg/m²h for 4.8 l/kg DS (**Figure 9**). The moisture contents in the cakes were 21–25 wt-%.

Tests with Aragonite Slurry

Test runs number 16–21 were performed using the aragonite slurry preheated to a temperature of ~55°C (**Table 6**). The slurry sample volume was varied in test runs number 16–18 without solids washing between 500 and 1000 ml, respectively, in order to find the volume giving the maximal filtration rate. Again, larger slurry samples gave higher rates, so 1000 ml was selected for the subsequent washing tests.

Different solids washing ratios were tested in test run numbers 19, 20, and 21 (0.5, 1.0, and 2.0 l/kg DS, respectively) with a washing water temperature of ~50°C (**Table 6**). Before solids washing, the separation phase was followed by 10 s intermediate drying. After washing, 60 s of drying was performed. **Figure 10** shows how the cake conductivity and chloride content dropped with higher wash ratios, with the lowest content of 127 ppm Cl for the 2 l/kg DS wash ratio. The filtration rate dropped from 1092 kg/m²h for 0.5 l/kg DS wash ratio to 611 kg/m²h for 2.0 l/kg DS (**Figure 9**). The moisture contents in the cakes were 26–28 wt-%.

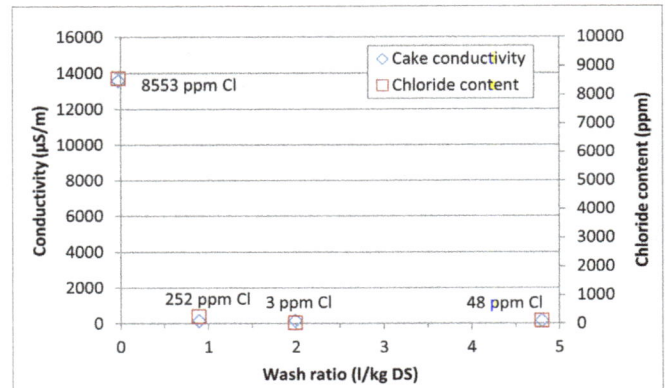

FIGURE 8 | Conductivity and final chloride content of the calcite cake with vacuum filtration using different wash ratios.

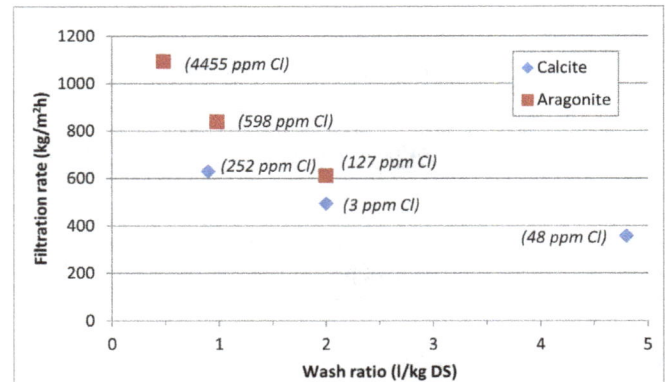

FIGURE 9 | Filtration rate of the washing tests of calcium carbonate cakes with vacuum filtration. The final chloride content of the cakes added as text next to the corresponding samples.

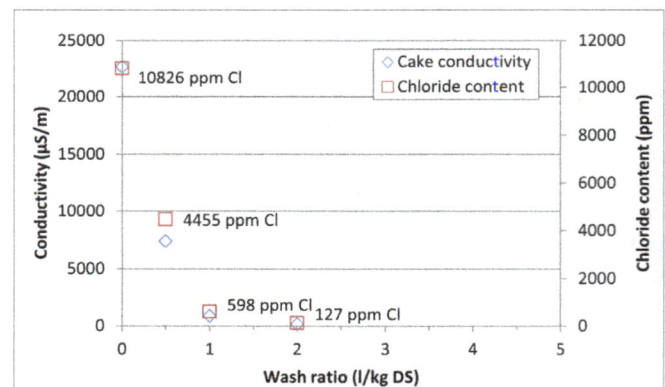

FIGURE 10 | Conductivity and final chloride content of the aragonite cake with vacuum filtration using different wash ratios.

Since the aragonite slurry required much more washing time than calcite slurry, a washing ratio of 5 l/kg was not tested as it would have significantly increased the washing time thus lowering the filtration rate.

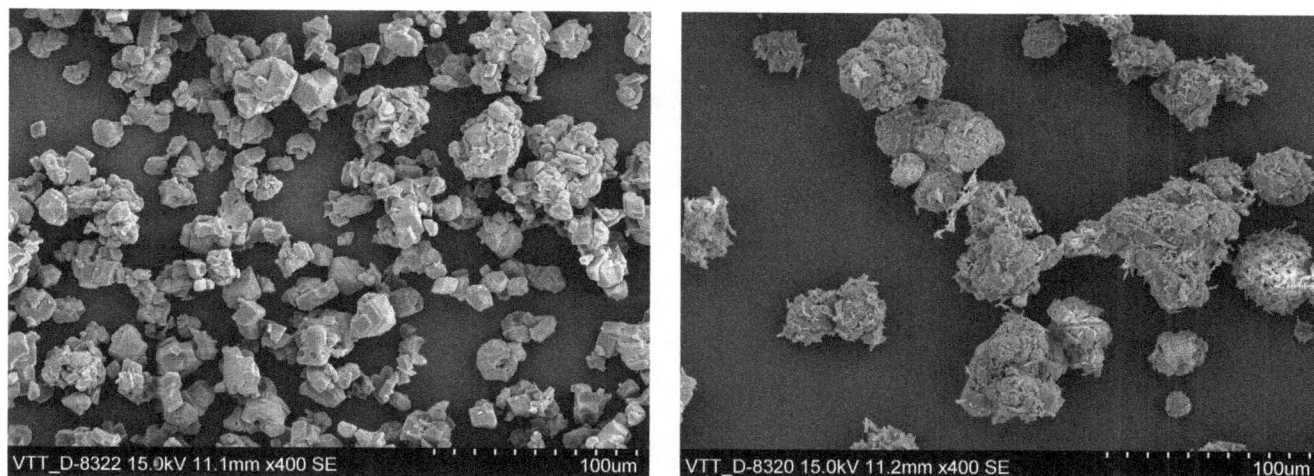

FIGURE 11 | SEM images of solids separated by pressure filtration from the calcite slurry (test number 5, left) and from the aragonite slurry (test number 15, right).

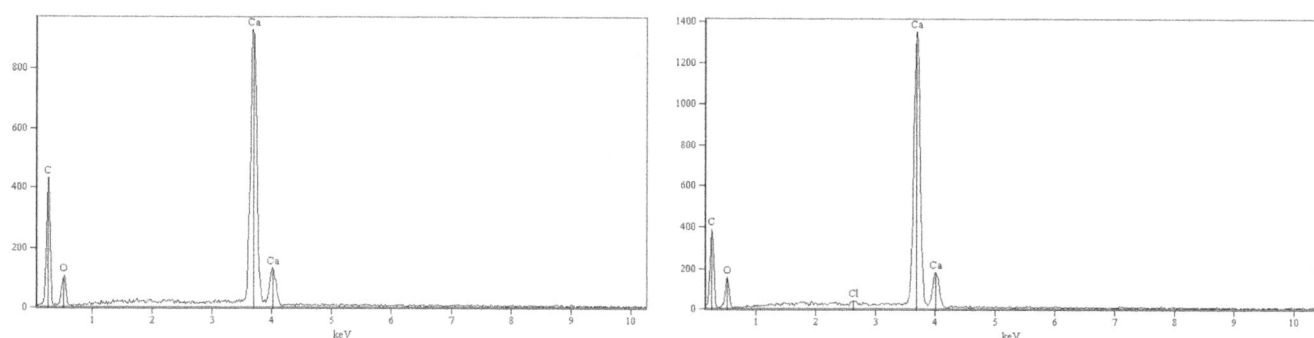

FIGURE 12 | EDS spectra of solids separated by pressure filtration from the calcite slurry (left) and from the aragonite slurry (right).

Mineralogy and Morphology of the Filtered and Washed Cakes

The SEM image of solids from the filtered, washed, and dried calcium carbonate cakes show that the calcite batch contained mostly rhombohedral calcite, somewhat agglomerated (**Figure 11**, left-hand image). The only identified crystal structure in the XRD analyses was calcite (Figure S1 in Supplementary Material). However, the XRD analyses of the aragonite sample identified both aragonite and calcite phases, with calcite being the dominating phase[1] (Figure S2 in Supplementary Material). The SEM image of the aragonite sample shows needle-shape crystals (typical for aragonite) embedded in large blocks of (probably calcite) agglomerates (**Figure 11**, right-hand image). While the EDS analysis for the calcite solids identifies only the expected elements contained in calcite (**Figure 12**, left-hand image), the

EDS spectrum from the aragonite solids also identifies traces of chloride (**Figure 12**, right-hand image).

DISCUSSIONS AND CONCLUSION

The experimental results show that both the calcium carbonate slurries produced with the studied concept are easy to filter with pressure filtration and with vacuum filtration. Pressure filtration gave a high capacity (400–460 kg/m²h) and a low cake residual moisture content (12–14 wt-%). Vacuum filtration gave slightly higher filtration rates (500–610 kg/m²h at the lowest chloride contents of the cakes), but the cake residual moisture also stayed higher (25–26 wt-%). However, vacuum filtration used a filter cloth with slightly higher permeability, which is expected to give a higher filtration rate. The cake thickness achieved (up to 5 cm) is similar to that of industrial scale filters.

For calcite, the washing liquid penetrated the cake easily, even with a low wash water pressure. The results show that washing of the calcite slurry can be performed either directly after the slurry

---

[1] For the sake of differentiating between the two samples tested, this is still referenced to as the "aragonite" batch in the text.

feed, or after an intermediate (or first) pressing. The chloride content of the calcite cake can be dropped from 10000 ppm to ~10 ppm using a 2 l/kg DS wash ratio. This is an important result when considering industrial scale-up of the process. Although Hudd (2014) assumed that multiple filtration stages would be needed, the results presented here show that the residual chloride content of the calcium carbonate can be minimized using filtration and washing in a single stage, without the need for multiple filtration and washing stages. However, the acceptable level of chloride in calcium carbonate for paper applications is yet to be determined.

High capacities were achieved also for the aragonite slurry, with both filtration technologies carrying out the washing treatment at 55°C. However, SEM images revealed that the aragonite batch was not the best representative of aragonite. Previous experiences at similar parameters with the Aalto pilot facility had yielded typical aragonite needle-shaped particles (Said and Järvinen, 2015). However, the previously reported aragonite precipitation had been performed for a shorter time (40 min) and at a slightly higher temperature (58°C) than the precipitation of the materials in this study. The longer reaction time with CO_2 may have caused a lower final solution pH, causing calcium carbonate to dissolve and recrystallize as calcite. Previous (unreported) experiments for producing aragonite with the pilot facility have indicated that aragonite formation is sensitive to the volume flow of CO_2. This may be partly due to the temperature of the incoming gas as it is not preheated before entering the reactor. It is also possible that the slurry underwent Ostwald ripening[2] during transportation. However, due to the large agglomerations visible in the SEM images the more likely explanation is that the reheating of the aragonite slurry for the filtration tests had caused more calcium carbonate to precipitate as calcite, since it is well known that calcium carbonate is less soluble in water at higher temperatures. The precipitation of calcium carbonate during reheating had possibly formed bigger agglomerates that could have enclosed some of the solution. If this is the case, the chloride in the filter cake, which was detected both with EDS and the pulping method, comes from this enclosed solution, because filtration is not very efficient at removing solution enclosed by particle agglomerates. As we cannot rule out the possibility that our aragonite sample had been altered in the filtration testing, it is something that needs more attention in future experimental work.

The aragonite slurry was also more difficult to wash, and the test parameters that were successfully used for calcite filtration were not suitable for filtration of the aragonite batch. The difficulty in aragonite filtration could partly also be due to a thicker cake in the aragonite experiments. Using a temperature of 50°C for the washing water and higher pressure/vacuum, washing of the aragonite cake was also performed successfully with both filtration technologies, although the residual chloride content remained higher than that for the calcite cake. The cake residual chloride content with pressure filtration technology dropped from ~5700 to 109 ppm with 1.4 l/kg DS wash ratio. With vacuum

filtration, the chloride content dropped from >13,000 to 127 ppm with 2.0 l/kg DS wash ratio.

The final chloride content in the cake produced from the aragonite slurry was higher (around 100 ppm) than that produced from the calcite slurry (around 10 ppm). This is probably due to the differences in particle size distribution of the solids in the two slurries, but also the different crystal morphologies. Although varying the wash ratio with 1–5 l/kg DS had little effect on the final chloride content of the calcite cake (**Figure 8**), it seems to have a larger effect on that of the aragonite cake (**Figure 10**). Sadly, either the filtration or the precipitation procedure had resulted in an unrepresentative aragonite sample, consisting both of calcite and aragonite. This became evident first weeks after the filtration experiments when SEM images were taken of the filtered batches. However, it does imply that the aragonite filtration needs to be performed immediately after precipitation, and that the temperature of the process needs to be controlled and monitored. Further testing with a representative aragonite sample is therefore required. Also further tests with calcite are recommended for evaluating the washing liquid consumption more accurately and the effect of particle size on the washing result.

In summary, the results showed that a low chloride content of the cake can be achieved with quite a small amount of washing water and the filtration rate is fast with both filtration methods tested. More work with larger filter units is needed for optimizing the filtration for a full-scale commercial filtration plant. It is very likely that a thickener would be an effective option for concentrating the slurry before the filter, due to the relatively large-sized particles. The selection of the filtration method depends also on possible additional requirements by the subsequent milling of calcium carbonate, required for grinding the produced calcium carbonate down to the particle sizes required by various paper applications. Possible milling options for the produced calcium carbonate include jet milling and vertical fine grinding mills. In jet milling, it is better to have a low moisture content in the cake as the moisture will evaporate during milling. In this case, pressure filtration would be preferred as the washing liquid penetrates the cakes very easily giving a lower residual moisture content in the cake than vacuum filtration. Further tests with pressure filtration could for instance be carried out with an Outotec PF 0.1 filtration unit as it allows easier control of the washing liquid volumes used. For vertical fine grinding mills both filtration options should be suitable as this mill type is less sensitive to cake moisture content. Therefore, further tests for both vacuum and pressure filtration are recommended to be performed with bigger test units. In addition, milling of the produced calcium carbonate should also be tested.

AUTHOR CONTRIBUTIONS

ST wrote the main part of the final manuscript (based on TA's report) and assisted in the analysis of the filtration results. TA performed the filtration experiments, participated in the analyses, and wrote the main part of the report of the filtration results. TK assisted with both the report and the manuscript and assisted in analysis of the results. HP coordinated the analysis work and participated in writing both the report and the final

---

[2]Ostwald ripening is the name of the phenomenon when small crystals or sol particles dissolve over time and redeposit onto larger crystals or sol particles.

manuscript. AS made the calcium carbonate batches and provided the analyses for the steel slag, and assisted in the analysis of the results.

ACKNOWLEDGMENTS

This work was carried out in the Carbon Capture and Storage Program (CCSP) research program coordinated by CLEEN Ltd. with funding and support from Outotec Oyj and the Finnish Funding Agency for Technology and Innovation, Tekes. SSAB Europe Oy kindly provided the steelmaking slag for the experiments. Tom E. Gustafsson at VTT Expert Services is acknowledged for SEM imaging and EDS analyses.

REFERENCES

Ciullo, P. A. (1996). *Industrial Minerals and Their Uses: A Handbook and Formulary*. Westwood, NJ: Noyes Publications.

Clark, E. B. (1992). "Lime Used for Precipitated Calcium Carbonate (PCC) for the Paper Industry," in *Innovations and Uses for Lime*, Vol. 1135, eds D. D. Walker, T. B. Hardy, D. C. Hoffman, and D. D. Stanley (Philadelphia, PA: American Society for Testing and Materials, ASTM STP), 1–7.

Couper, J. R., Penner, W. R., Fair, J. R., and Walas, S. M. (2010). *Chemical Process Equipment – Selection and Design. Revised 2nd edition*. Burlington, MA: Elsevier.

Eloneva, S., Teir, S., Revitzer, H., Salminen, J., Said, A., Fogelholm, C.-J., et al. (2009). Reduction of CO_2 emissions from steel plants by using steelmaking slags for production of marketable calcium carbonate. *Steel Res. Int.* 80, 415–421. doi:10.2374/SRI09SP028

ESAPA. (2004). *IPPC BAT Reference Document – Large Volume Solid Inorganic Chemicals Family: Process BREF for Soda Ash*. ESAPA – European Soda Ash Producers Association. Available at: http://www.cefic.org/Documents/Industry%20sectors/ESAPA_Soda_Ash_Process_BREF3.pdf

Hudd, H. (2014). *Post-Treatment of Precipitated Calcium Carbonate (PCC) Produced from Steel Converter Slag*. M.Sc. Thesis, Åbo Akademi University, Turku.

Mattila, H.-P., Grigaliūnaitė, I., and Zevenhoven, R. (2012). Chemical kinetics modeling and process parameter sensitivity for precipitated calcium carbonate production from steelmaking slags. *Chem. Eng. J.* 192, 77–89. doi:10.1016/j.cej.2012.03.068

Mattila, H.-P., and Zevenhoven, R. (2014a). "Production of precipitated calcium carbonate from steel converter slag and other calcium-containing industrial wastes and residues," in *Advances in Inorganic Chemistry*, Vol. 66, Chap. 10. eds M. Aresta and R. van Eldik (Waltham, MA: Academic Press), 347–384. doi:10.1016/B978-0-12-420221-4.00010-X

Mattila, H.-P., and Zevenhoven, R. (2014b). Design of a continuous process setup for precipitated calcium carbonate production from steel converter slag. *ChemSusChem* 7, 903–913. doi:10.1002/cssc.201300516

Naydowski, C., Hess, P., Strauch, D., Kuhlmann, R., and Rohleder, J. (2001). "Calcium carbonate and its industrial application," in *Calcium Carbonate: From the Cretaceous Period into the 21st Century*, eds F. W. Tegethoff, J. Rohleder, and E. Kroker (Basel: Birkhäuser Verlag), 197–311.

Said, A., and Järvinen, M. (2015). "Demonstration pilot plant for the production of precipitated calcium carbonate (PCC) from steelmaking slag and carbon dioxide," in *Proceedings from the 5th International Conference on Accelerated Carbonation for Environmental and Material Engineering (ACEME)* (New York City). Available at: http://www3.aiche.org/proceedings/Conference.aspx?ConfID=ACEME-2015

Statton, P. (2012). "An overview of the North American calcium carbonate market," in *Presentation at the 23rd Annual Canadian Conference on Markets for Industrial Minerals* (Quebec). Available at: http://roskill.com/wp/wp-content/uploads/2014/11/download-roskills-paper-on-the-north-american-calcium-carbonate-market.attachment1.pdf

Teir, S., Kotiranta, T., Parviainen, T., and Mattila, H.-P. (2015). "Case study for utilisation of CO_2 from flue gases in PCC production from steelmaking slag," in *Proceedings from the 5th International Conference on Accelerated Carbonation for Environmental and Material Engineering (ACEME)* (New York City). Available at: http://www3.aiche.org/proceedings/Conference.aspx?ConfID=ACEME-2015

Conflict of Interest Statement: Author ST has filed a patent application related to the concept studied in this paper: Teir S, Eloneva S, Revitzer H, Zevenhoven R, Salminen J, Fogelholm C-J, Pöyliö E (2009) Method of producing calcium carbonate from waste and by-products. International application published under the patent cooperation treaty (PCT), WO 2009/144382, PCT/FI2009/050455.

The remaining co-authors declare that the research was conducted in the absence of any commercial or financial relationships that could be construed as a potential conflict of interest.

6

Non-invasive rapid harvest time determination of oil-producing microalgae cultivations for biodiesel production by using chlorophyll fluorescence

Yaqin Qiao[1,2], Junfeng Rong[3], Hui Chen[1], Chenliu He[1] and Qiang Wang[1]*

[1] Key Laboratory of Algal Biology, Institute of Hydrobiology, Chinese Academy of Sciences, Wuhan, China, [2] University of Chinese Academy of Sciences, Beijing, China, [3] SINOPEC Research Institute of Petroleum Processing, Beijing, China

Edited by:
P. C. Abhilash,
Banaras Hindu University, India

Reviewed by:
Jason Ryan Hattrick-Simpers,
University of South Carolina, USA
Shanmugaprakash Muthusamy,
Kumaraguru College of
Technology, India

*Correspondence:
Qiang Wang,
Key Laboratory of Algal Biology,
Institute of Hydrobiology, Chinese
Academy of Sciences, 7 South
Donghu Road, Hubei Province,
Wuhan 430072, China
wangqiang@ihb.ac.cn

Specialty section:
This article was submitted to
Bioenergy and Biofuels,
a section of the
journal Frontiers in Energy Research

Citation:
Qiao Y, Rong J, Chen H, He C and
Wang Q (2015) Non-invasive rapid
harvest time determination of
oil-producing microalgae cultivations
for biodiesel production by using
chlorophyll fluorescence.
Front. Energy Res. 3:44.

For the large-scale cultivation of microalgae for biodiesel production, one of the key problems is the determination of the optimum time for algal harvest when algae cells are saturated with neutral lipids. In this study, a method to determine the optimum harvest time in oil-producing microalgal cultivations by measuring the maximum photochemical efficiency of photosystem II, also called Fv/Fm, was established. When oil-producing *Chlorella* strains were cultivated and then treated with nitrogen starvation, it not only stimulated neutral lipid accumulation, but also affected the photosynthesis system, with the neutral lipid contents in all four algae strains – *Chlorella sorokiniana* C1, *Chlorella* sp. C2, *C. sorokiniana* C3, and *C. sorokiniana* C7 – correlating negatively with the Fv/Fm values. Thus, for the given oil-producing algae, in which a significant relationship between the neutral lipid content and Fv/Fm value under nutrient stress can be established, the optimum harvest time can be determined by measuring the value of Fv/Fm. It is hoped that this method can provide an efficient way to determine the harvest time rapidly and expediently in large-scale oil-producing microalgae cultivations for biodiesel production.

Keywords: chlorophyll fluorescence, harvest time, oil-producing microalgae, neutral lipid, photosynthesis

Introduction

The limiting supply, increasing cost, and ever increasing environmental pollution and health problems from conventional fossil fuels is placing an increasing demand on biodiesel to replace some fossil fuel usage (Hill et al., 2006; Hu et al., 2006; Hu et al., 2008; Amaro et al., 2011). With this in mind, microalgae have attracted considerable attention in recent years as potential sources of renewable fuel, due to their fast growth rate, high adaptability to environment conditions, and their no competition with crops for arable land and potable water (Santos et al., 2011). Microalgae have been promising feedstock for biodiesel production (Zhang et al., 2014b). The green microalgae genus *Chlorella* (*Chlorophyta*), which are capable of photoautotrophic, mixotrophic, and heterotrophic growth with high biomass accumulation, appears to contain good candidate strains for biodiesel production (Petkov and Garcia, 2007).

The cell metabolism of microalgae varies with the change in growth conditions. Nitrogen (N) starvation is one of the most effective environmental stresses and stimulates the accumulation of lipids in many microalgae (Illman et al., 2000). The general principle is that the lack of N limits protein biosynthesis, and the fixed carbon from photosynthesis will then channeled into high energy density compounds such as triglycerides and/or starch (Scott et al., 2010). With limited N supply, *Chlorella* could accumulate doubled or even tripled amount of lipid (Converti et al., 2009; Widjaja et al., 2009). However, even though cultivation with N limiting media results in an increase in the lipid content on a per cell weight basis, it also lowers the total biomass productivity significantly, resulting in decreased lipid productivity. A two-stage cultivation strategy has then been proposed by Zhu et al. (2014) to avoid this problem, in which a full-strength medium was used in the first stage to promote algal cell growth and to accumulate biomass, then the cells were subjected to N-starvation conditions in the second stage to trigger the target product accumulation (Zhu et al., 2014).

For the application of microalgae in biodiesel production, there remain many problems that need to be solved, and development of an effective method to ascertain the optimum time for algal harvest to obtain high neutral lipid productivity is one of the most important challenges. The chlorophyll-fluorescence measurement has been utilized for decades for non-invasive analyses of stress-induced perturbations to photosynthesis (Schreiber, 1973; Conroy et al., 1986). The maximum quantum efficiency of photosystem II (PSII, Fv/Fm) measured under dark-adapted conditions represents the theoretical capacity for light energy absorbed by PSII to be utilized in photosynthesis. The Fv/Fm value, which remains relatively constant under normal conditions, reflects the potential quantum efficiency of PSII and is used as one of the most sensitive indicators of the photosynthetic performance, which decreases when exposed to stress (Maxwell and Johnson, 2000; Oxborough, 2004). The Fv/Fm value was got with no extra agent add to the algae suspension and no harm to the algae influencing the downstream apply of microalgae. And the measurement of Fv/Fm was performed in action light that is the optimal light condition for algae growth for a few minutes and only no more one second saturation pulse. So Fv/Fm value was rapid and no-invasive parameter to characterize microalgae.

In this study, it was found that the lipid content increased with the decreasing Fv/Fm in the green algae *Chlorella* sp. C2 treated by N starvation, which is consistent with the reports in our previous study (Zhang et al., 2013). On this basis, four oil-producing green algae strains – *Chlorella sorokiniana* C1, *Chlorella* sp. C2, *C. sorokiniana* C3, and *C. sorokiniana* C7 – were cultivated and treated under N-starvation conditions to investigate the relationship between neutral lipid accumulation and the Fv/Fm values. A significant negative correlation between the neutral lipid contents and the Fv/Fm values was found for all four algal strains. Then, a method to determine the optimum harvest time in oil-producing microalgae cultivation by measuring Fv/Fm only was established and validated in lab-scale cultivation. This could provide a rapid and inexpensive way to determine the harvest time in large-scale oil-producing microalgae cultivation for biodiesel production.

Materials and Methods

Growth Conditions and N-Starvation Treatment

The N-sufficient medium (N+) used was full-strength BG11 medium (Tran et al., 2014). The N-deficient medium (N−) was BG11 without NaNO3. Alga strains *C. sorokiniana* C1, *Chlorella* sp. C2, *C. sorokiniana* C3 and *C. sorokiniana* C7 in the exponential phase were each inoculated with an initial OD_{700} of 0.05 into a 1 l flask containing 500 ml BG11 medium at controlled temperature of 25°C, under continuous illumination of white fluorescence light at 70 μmol m^{-2} s^{-1}, and bubbled continuously with filtered air. For the N-starvation treatment, the cells were harvested during the middle exponential growth phase (OD_{700} approximately 0.8, about 1.1×10^7 cells ml^{-1}) and centrifuged 3 min at 3,000 g at room temperature (about 25°C). The pellet was then washed and resuspended in the N-medium to OD_{700} 0.8. Then, cultivation of the cultures was continued with the same light, temperature, and air supply as the N-sufficient medium.

Fluorescence Microscope Analysis

Microscopic visualization of algae cells were carried out using a confocal scanner (Zeiss LSM 710 NLO) and fluorescence microscope (OLYMPUS system microscope BX53, Japan) as described previously (Zhang et al., 2013). The lipid bodies were stained with the fluorescent dye, Bodipy 505/515 (Invitrogen Molecular Probes, Carlsbad, CA, USA) with a final labeling concentration of 1 mM, according to Cooper et al. (2010). Confocal laser scanning microscopy (CLSM) analysis was carried out as previously described in Zhang et al. (2013, 2014a). Fluorescence microscope analysis of the green bodipy fluorescence was excited at 488 nm and detected at 505–515 nm. The red auto-fluorescence of the chloroplasts was simultaneously detected at 650–700 nm.

Thin Layer Chromatography Analysis of Lipid

The cell lipids were extracted and thin layer chromatography (TLC) analyzed according to Reiser and Somerville (1997) with minor modifications as described by Zhang et al. (2013).

Flow Cytometry Analysis

Flow cytometry (FCM) analysis was used to visualize the florescence of lipid in large number cells level. Five hundred microliters of cells (1.1×107 cells ml^{-1}) were stained with Bodipy 505/515 at 37°C for 30 min under dark condition and then analyzed by FACS Aria Flow Cytometer (Becton Dickinson, San Jose, CA, USA) equipped with a laser emitting at 488 nm and an optical filter FL1 (530/30 nm). And Flow Jo software (Tree Star, San Carlos, CA, USA) was used to analysis the collected data from Flow Cytometer.

Pigments Quantification

Total pigments were extracted with 100% methanol. One milliliter of cells (1.1×10^7 cells ml^{-1}) was collected and centrifuged at 3,000 g for 3 min at room temperature. The pellet was resuspended in 1 ml of 100% methanol for 12 h. And the pigment concentrations were determined spectrophotometrically and calculated according to the formula developed by Lichtenthaler (1987) as chlorophyll a (Chl a) (microgram per milliliter) = 16.72

$A_{665.2} - 29.16$ $A_{652.4}$, chlorophyll b (Chl *b*) (microgram per milliliter) = 34.09 $A_{652.4} - 15.28$ $A_{665.2}$, total chlorophylls (Chl $a + b$) (microgram per milliliter) = 1.44 $A_{665.2} + 24.93$ $A_{652.4}$, total carotenoids (Car) (microgram per milliliter) = $(1000$ $A_{470} - 21.63$ Chl a $- 104.96$ Chl b)/221.

Chl Fluorescence Analysis

A Dual-PAM-100 Chl fluorometer (Walz, Germany) was used for all Chl fluorescence measurements. Cells were fully dark-adapted for 15 min before the measurement of Initial (Fo) and maximum Chl fluorescence level. The maximum quantum yields of PSII electron transport was calculated as Fv/Fm = (Fm − Fo)/Fm according to Genty et al. (1989).

Photosynthetic Steady-State Oxygen Evolution and Dark Respiration Rates Measurement

Photosynthetic oxygen evolution and dark respiration rates were measured using a Clark-type oxygen electrode (Oxylab 2, Hansatech, UK) at 20°C as described by Zhang et al. (2013). The dark respiration rates were measured with 10 μM NaHCO3 in cell suspensions (2 ml) under dark condition and the oxygen evolution rates were measured with cell suspensions (2 ml) illuminated at a quantum flux density of 600 μmol m^{-2} s^{-1}. The collected data were the rate of oxygen changes (μmol O$_2$ min^{-1}). The chlorophyll (Chl a) contents were determined as described in above 2.5 pigments quantification. And the result of oxygen evolution and dark respiration rates were expressed with the rate of oxygen released (oxygen evolution μmol O$_2$ mg^{-1} Chla h^{-1}) and consumption (respiration rate μmol O$_2$ mg^{-1}Chla h^{-1}).

Assessment of Lipid Peroxidation and ROS Scavenging Enzyme Activity Assays

The malondialdehyde (MDA) level, and the enzyme activities of catalase (CAT), peroxidase (POD), and superoxide dismutase (SOD) were measured according to Shi et al. (2009) as described by Zhang et al. (2013).

Lab-Scale (3L) Cultivation

The lab-scale cultivation was performed in a 3 l photobioreactor (Zhang et al., 2014a). The four algal strains were inoculated with an initial OD$_{700}$ of 0.05 into the photobioreactor flask containing 2 l BG11 medium at controlled temperature of 25°C, under continuous illumination of white fluorescence light at 70 μmol m^{-2} s^{-1}, and bubbled continuously with filtered air. Cells reached their mid logarithmic growth phase (about 0.8 at OD$_{700}$) were harvested by centrifugation for 3 min at 3,000 *g* at room temperature. The algae pellets were then washed and resuspended in N-medium to OD$_{700}$ 0.8. Then, cultivation of the cultures continued under the same growth conditions.

Statistical Analysis

Each result shown represents at least the mean of three independent biological replicates. The statistical analysis of the collected data was made using SPSS-13. The correlation between the variables was analyzed with Pearson's correlation.

Results

N Starvation Stimulates Neutral Lipid Accumulation in *Chlorella* sp. C2

To understand the regulation of neutral lipid accumulation in *Chlorella* sp. C2, the cell neutral lipid levels at various time intervals of N depletion were extracted and examined using TLC. As shown in **Figure 1A**, neutral lipid accumulation in both N− (**Figure 1A**, lanes 1, 2) and N+ medium (**Figure 1A**, lanes 5, 6) were undetectable during the first 2 days of cultivation, which came into sight after 2 days in the cells cultured in N-medium (**Figure 1A**, lane 3), and significant accumulation could be observed after 8 days of N starvation (**Figure 1A**, lane 4). While as contrast, only a trace amount of neutral lipids could be detected in N repletion medium grown cells (**Figure 1A**, lane 8) even after 8 days.

A CLSM analysis was used to visualize the neutral lipid accumulation of the algal cells cultured in N-medium at different stages. In accordance with the TLC (**Figure 1A**) analysis, **Figure 1B** shows that no green Bodipy 505/515 fluorescence could be detected at 0 day (**Figure 1B**) and 0.5 day (**Figure 1B**) after N-treatment, while a weak green fluorescence was first detected after 2 days (**Figure 1B**) and a strong green fluorescence signal was observed after 8 days of treatment (**Figure 1B**). Moreover, to further characterize the neutral lipid accumulation during N starvation at statistical level, a group of Chlorella sp. C2 cells (>10,000) at different stages were analyzed using FCM. Compared with day 0, the Bodipy fluorescence intensity of the

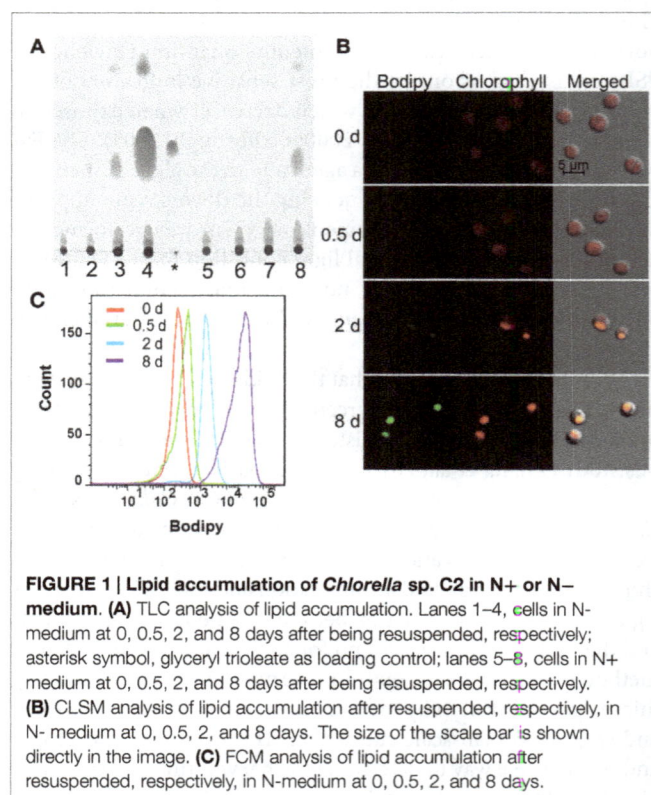

FIGURE 1 | Lipid accumulation of *Chlorella* sp. C2 in N+ or N− medium. (A) TLC analysis of lipid accumulation. Lanes 1–4, cells in N-medium at 0, 0.5, 2, and 8 days after being resuspended, respectively; asterisk symbol, glyceryl trioleate as loading control; lanes 5–8, cells in N+ medium at 0, 0.5, 2, and 8 days after being resuspended, respectively. **(B)** CLSM analysis of lipid accumulation after resuspended, respectively, in N- medium at 0, 0.5, 2, and 8 days. The size of the scale bar is shown directly in the image. **(C)** FCM analysis of lipid accumulation after resuspended, respectively, in N-medium at 0, 0.5, 2, and 8 days.

cell populations at days 0.5, 2, and 8 of N starvation showing a constant increase, indicating constantly increasing levels of neutral lipid (**Figure 1C**). Thus, as in *Chlorella* sp. C3 (Zhang et al., 2013), days 0, 0.5, 2, and 8 of N- treatment could be defined as the control stage (Cs), pre-oil droplet formation stage (PDFs), oil droplet formation stage (ODFs), and late-oil droplet formation stage (LDFs), respectively. These were also the key stages in the further tests.

N starvation Damages Photosynthesis System

Nitrate is one of the most important elements contributing to algae growth, so its depletion may dramatically change the physiological activity. As a stressor, the N starvation will not only induce lipid accumulation but also lead to the depression of photosynthesis. As shown in **Figure 2A**, all the pigment contents, including Chl *a*, Chl *b*, Chl *a + b*, and carotenoid (Car), had a significant decrease before the ODFs and then decreased continuously during neutral lipid accumulation in N-medium.

In photosynthetic organisms, nutrient stresses could be generally detected with a decrease in the Fv/Fm value, as the Fv/Fm value in healthy cells has been reported to be constant (Conroy et al., 1986; Genty et al., 1989; Maxwell and Johnson, 2000), but decreases under various stresses (Conroy et al., 1986; Lu et al., 1999; Lee et al., 2013; Zhang et al., 2013). **Figure 2B** shows the Fv/Fm value declined linearly during

oil droplet formation, suggesting that the algae cell was under environmental stress.

Rates of steady-state photosynthetic oxygen evolution and dark respiration were examined to further understand the variation in photosynthesis of *Chlorella* sp. C2 during N-induced oil droplet formation. Compared with the untreated cells, both the photosynthetic oxygen evolution rate and the respiration rate of the N starved cells reduced significantly during oil droplet formation (**Figure 2C**), indicating that severe damage to the photosynthetic apparatus, as well as to the respiratory apparatus, had occurred.

Most stresses in oxygenic photosynthetic organisms would ultimately lead to oxidative stress (Elstner, 1991). Lipid peroxidation level, the most commonly accepted indicator of oxidative stress (Apel and Hirt, 2004; Zhang et al., 2013), could be estimated by measuring the formation of MDA in cells. During oxidative stress, ROS, including $^{\bullet}O_2^-$, $^1O_2$, and H_2O_2, levels increase (Apel and Hirt, 2004), and the antioxidant enzyme, including SOD, POD, and CAT, will be induced or activated in cells (Ali et al., 2005). As shown in **Figure 3**, N starvation induced an increase in the MDA level in the cells. The relative activities of SOD and CAT dramatically increased at the PDFs and dropped gradually at the ODFs and LDFs. By contrast, the relative activities of POD declined slightly at the PDFs, and then increased at the ODFs and LDFs. Therefore, this suggests that POD, CAT, and SOD

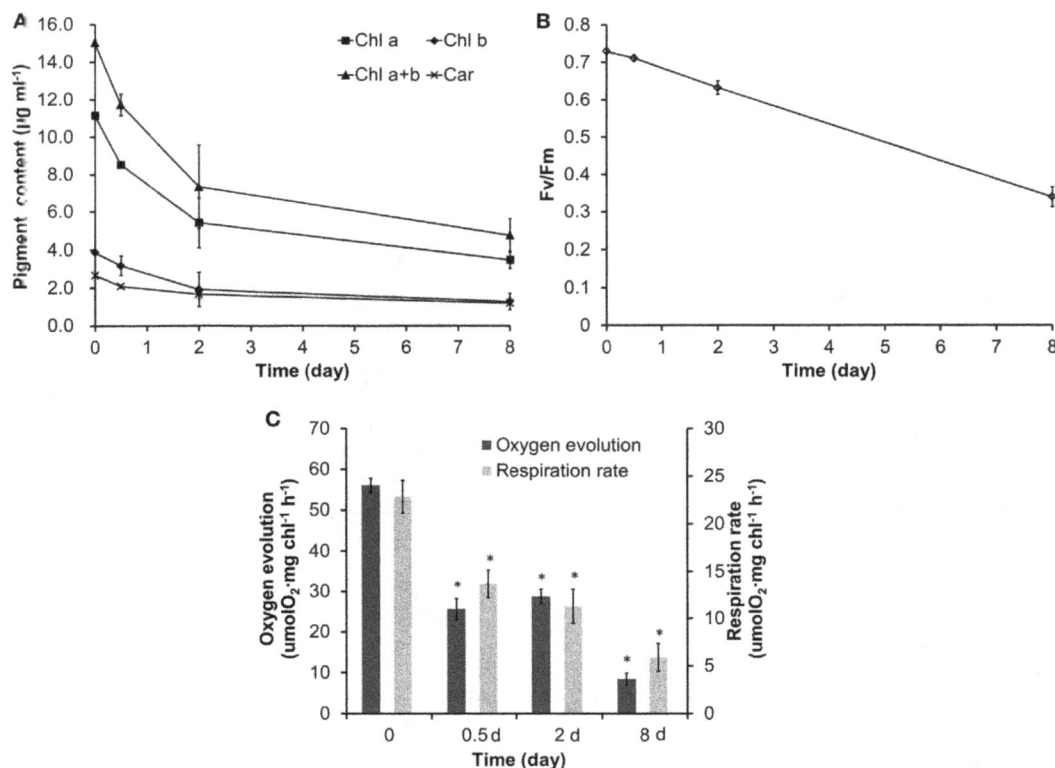

FIGURE 2 | The variation of photosynthetic physiology in *Chlorella* sp. C2 during lipid droplet formation. (A) Pigment content. (B) The maximum photochemical efficiency of PSII (Fv/Fm). (C) Steady-state oxygen evolution and in dark respiration. All data points in the current and following figures represent the means of three replicated studies in each independent culture, with the SD of the means (*$p < 0.05$; **$p < 0.01$).

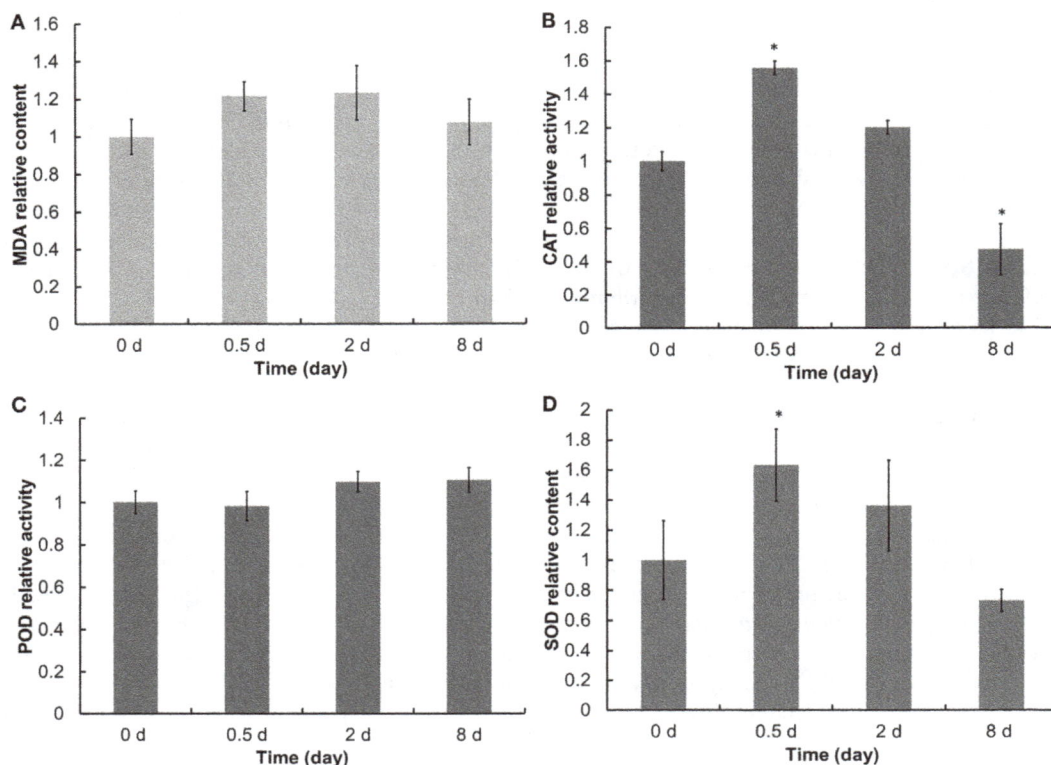

FIGURE 3 | Lipid peroxidation level and antioxidant enzymes activities of *Chlorella* **sp. C2 during OD formation. (A–D)** represents the MDA content, CAT, POD, and SOD activities, respectively.

play important roles in scavenging ROS and reducing MDA in *Chlorella* sp. C2 cells under N starvation at different stages (**Figure 3**).

The above results suggest that both the neutral lipid accumulation and the damage to the photosynthetic system might be caused by N starvation-induced oxidative stress. Furthermore, as the most sensitive indicator for stress response under unfavorable conditions, the Fv/Fm value decreased linearly (**Figure 2B**) along with the increasing triacylglycerol (TAG) accumulation (**Figure 1**) under N starvation. So it is predicted that the Fv/Fm value may be an excellent indicator for identifying the level of N-induced TAG accumulation and could provide a non-invasive and quick tools of determining the timing for microalgae harvest if a significant relationship existed between the Fv/Fm value and the TAG content.

Fv/Fm Value Can be Used to Estimate Neutral Lipid Levels in the Four Oil-Production Algae

Besides *Chlorella* sp. C2, three further oil-producing green algae strains – *C. sorokiniana* C1, *C. sorokiniana* C3, and *C. sorokiniana* C7 – were selected to identify the relationship between the Fv/Fm value and the neutral lipid level. During N-starvation treatment, the neutral lipid levels in cells of all four strains rose gradually with the prolongation of N-depletion stress (**Figure 4**, green fluorescence). However, Chl auto-fluorescence (**Figure 4** red

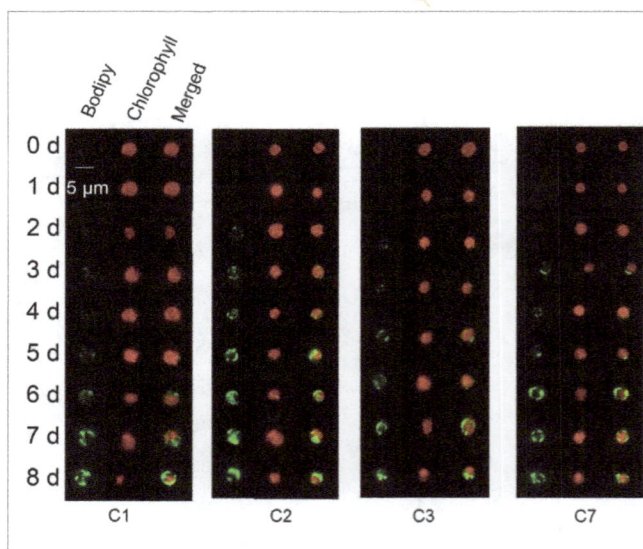

FIGURE 4 | Representative fluorescence microscope analysis of *Chlorella sorokiniana* **C1 (C1),** *Chlorella* **sp. C2 (C2),** *Chlorella sorokiniana* **C3 (C3), and** *Chlorella sorokiniana* **C7 (C7) labeled** *in vivo* **with Bodipy 505/515.** Bodipy 505/515 (green) was excited with an argon laser at 488 nm and detected at 505–515 nm. Chl autofluorescence (red) was detected simultaneously at 650–700 nm. The cells were stained after resuspended, respectively, in N-medium for 0–8 days. The size of the scale bar is shown directly in the image.

fluorescence) signals became more heterogeneous as the chloroplast shapes became abnormal. Moreover, in accordance with the fluorescence microscope results (**Figure 4**), the TLC results also showed that increasing neutral lipid levels were detected with time (**Figure 5A**). The relative contents of neutral lipid were measured using Image J (v1.41, NIH), and it was seen that the contents had significantly positive association with time under N starvation in all four algae (**Figure 5B**, $r > 0.9$, $p < 0.01$, correlation test by SPSS-13). As show in **Figure 6**, the values of Fv/Fm in all four algae declined linearly during N starvation, and

the Fv/Fm values were significant negatively correlated with time (correlation test, $r < -0.9$, $p < 0.01$). Notably, during N starvation, the significant negative linear correlations appeared between the Fv/Fm values and relative lipid content in all four algae (**Figure 7**, correlate test, $r > 0.9$, $p < 0.01$). Therefore, for these four oil-rich algae, in which we have established the relationship between the neutral lipid content and the Fv/Fm values, the neutral lipid level under N starvation can be got indirectly by measuring the Fv/Fm value. This can therefore be used to determine the optimal harvest time of microalgae for lipid production.

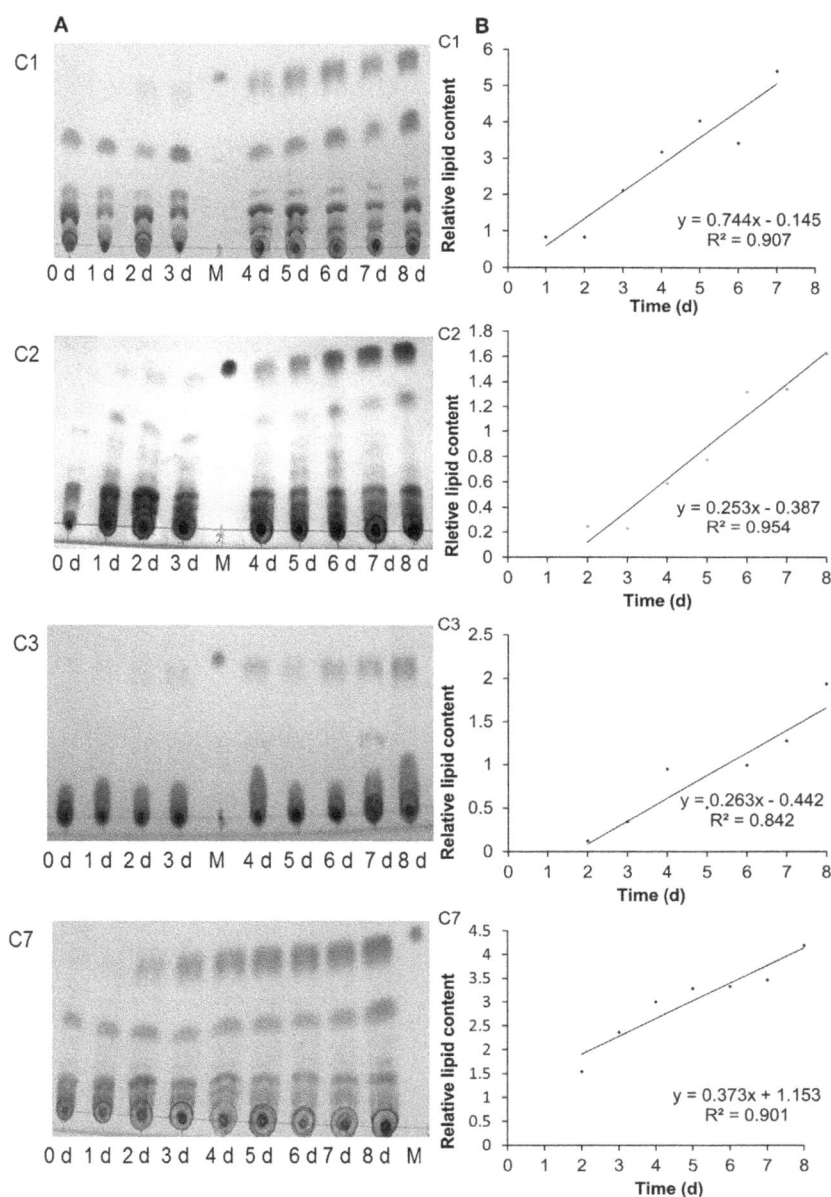

FIGURE 5 | TLC analysis of the neutral lipid accumulation in *Chlorella sorokiniana* C1 (C1), *Chlorella* sp. C2 (C2), *Chlorella sorokiniana* C3 (C3), and *Chlorella sorokiniana* C7 (C7) cells under N starvation. The neutral lipid accumulation was detected at 0–8 days after N starvation. M symbol, glyceryl trioleate as loading control **(A)**. The variation of the relative neutral lipid content with the prolog of N-starvation time **(B)**.

Lab-Scale Experiments Demonstrated the Application of the Fv/Fm Index for Determining the Harvest Time in a Given Oil-Producing Algae Cultivation

To further test the practicability of using the Fv/Fm value for determining the harvest time in a given oil-producing algae,

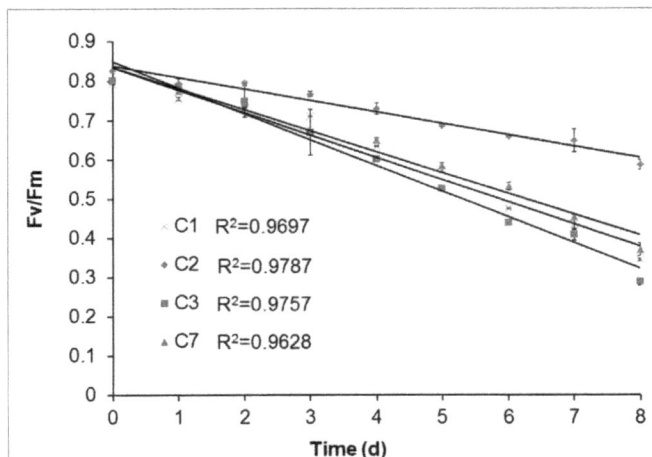

lab-scale experiments using four algal strains were performed in 3 l bioreactors. During the N-starvation time, the algae were collected at random time points to detect the neutral lipid accumulation and the Fv/Fm values. Similar to our previous results, the accumulation of neutral lipids increased with the decreasing Fv/Fm values in all four algae in the 3 l bioreactors (**Figure 8**). This indicated that the Fv/Fm value could be an indicator for determining the neutral lipid level in a given oil-producing algae cultivated in a 3 l or even larger scale bioreactor. Furthermore, for a given oil-rich microalgae, this indicates that if the relationship between the neutral lipid content and the Fv/Fm value under nutrient stress can be established, the optimum harvest time for lipid production in cultivations under nutrient stress can be determined by measuring the Fv/Fm value.

Discussion

One of the challenges of microalgae biodiesel production is the measurement of lipid content to determine the optimum harvest time obtaining high levels of lipids. In order to determine the lipid accumulation level in microalgae cells, lipid extraction and detection are required using traditional technology. TLC and the gravimetric method are the two most commonly used methods for lipid-content detection (Bligh and Dyer, 1959; Reiser and Somerville, 1997). However, the lipid extraction and detection steps used are complicated, need many organic reagents, resulting in environmental toxicity, and are laboring intensive and expensive. Recently, fluorescent probes such as the lipophilic probes Nile red (Cooksey et al., 1987) and BODIPY 505/515

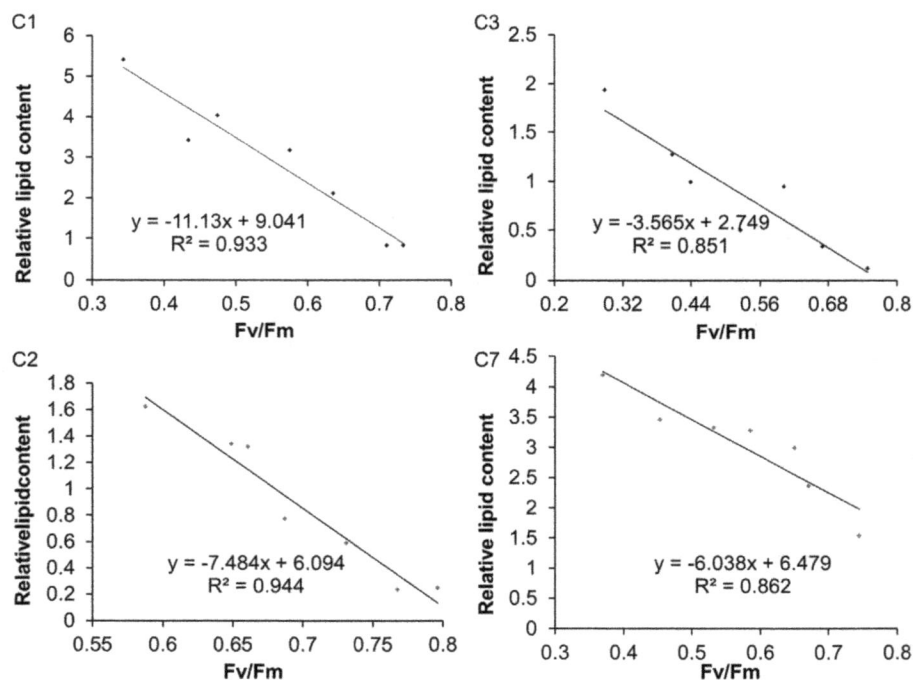

FIGURE 6 | The variation of maximum photochemical efficiency of PSII (Fv/Fm) of *Chlorella sorokiniana* C1 (C1), *Chlorella* sp. C2 (C2), *Chlorella sorokiniana* C3 (C3), *Chlorella sorokiniana* C7 (C7) with the prolog of N-starvation time.

FIGURE 7 | The correlation between Fv/Fm and relative neutral lipid content in *Chlorella sorokiniana* C1 (C1), *Chlorella* sp. C2 (C2), *Chlorella sorokiniana* C3 (C3), and *Chlorella sorokiniana* C7 (C7) under N starvation.

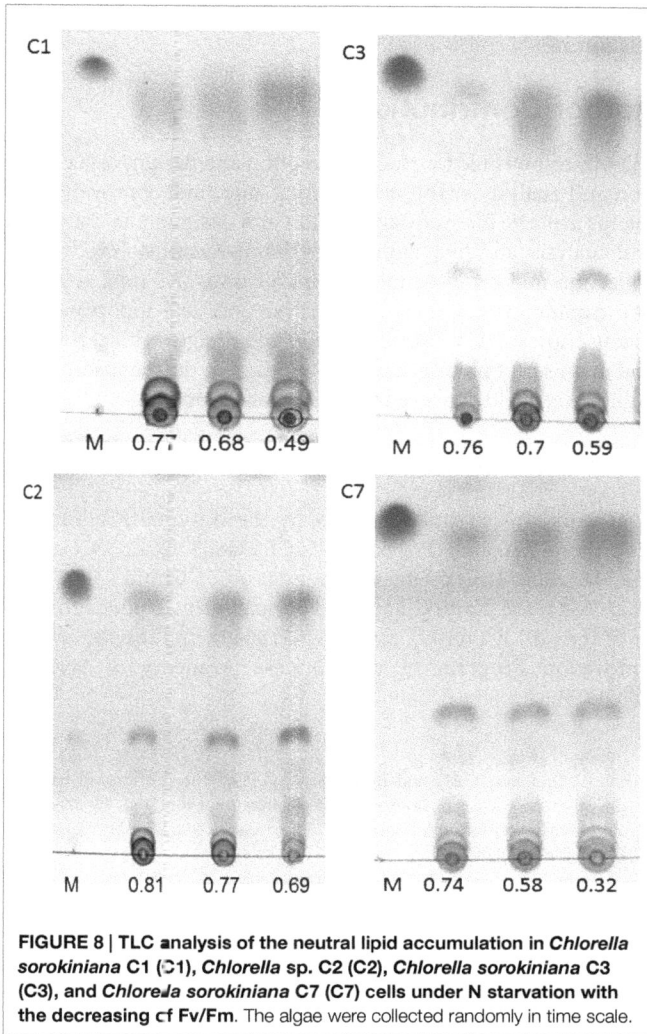

FIGURE 8 | TLC analysis of the neutral lipid accumulation in *Chlorella sorokiniana* C1 (C1), *Chlorella* sp. C2 (C2), *Chlorella sorokiniana* C3 (C3), and *Chlorella sorokiniana* C7 (C7) cells under N starvation with the decreasing of Fv/Fm. The algae were collected randomly in time scale.

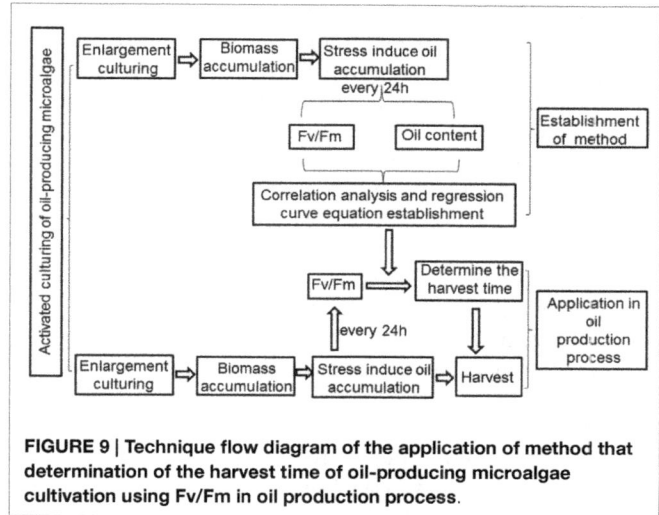

FIGURE 9 | Technique flow diagram of the application of method that determination of the harvest time of oil-producing microalgae cultivation using Fv/Fm in oil production process.

(Cooper et al., 2010), which can measure the neutral lipid level in intact cells without lipid extraction, have been used to estimate lipid accumulation. However, there are also some disadvantages of these florescent probes, namely, the relatively high cost of time and money for staining and detecting the fluorescent probe, and the potential errors caused by the different permeability of the fluorescent probe into diverse microalgae cells. Therefore, a simple and low-cost method is needed to establish the lipid level to then determine the optimum harvest time rapidly and expediently in oil-producing microalgae cultivations. In our study, we established a new method by measuring the chlorophyll-fluorescence parameter Fv/Fm.

Under N-starvation condition, microalgae cells preferentially degraded nitrate containing macromolecules, resulting in a decrease of total nitrogen content as well as the accumulation of excess carbon in the form of lipids (Dawes, 1976). In this article, the lipid content was linearly increased after 2 days of N-starvation treatment (**Figures 1** and **5**). N depletion also induced the decrease of Fv/Fm values, which may be a consequence of reduced

photosynthetic pigment (**Figures 2** and **4**). Previous studies also reported the decrease in Fv/Fm values under osmotic, light, and nutrient stress (Lu et al., 1999; Beardall et al., 2001). Under N-starvation condition, the neutral lipid accumulation increasing and Fv/Fm decreasing occurs simultaneously. So we predicted there may be some relationship between lipid increasing and Fv/Fm decreasing. The result shows that the lipid increasing was negatively correlative with Fv/Fm decreasing. Therefore, Fv/Fm may be a tool to determine the lipid accumulation for microalgae under N starvation. A previous research reported the use of PAM fluorometry to measure the biosynthesis of neutral lipid under nutrient stress (White et al., 2011). So the measurement of Fv/Fm for the determination of harvest time is a feasible method. The Fv/Fm value was obtained from specimens in the dark-adapted state, and the measurement could be completed within a few seconds using a single saturating pulse. So this procedure is simple, rapid, non-invasive, low-cost, and highly appropriate for large-scale application.

Even though Fv/Fm can be a parameter to determine lipid content, there were some interesting reports. A study using the Fv/Fm value as a screening tool for oil-rich mutant microalgae reported that under stress conditions, for different algae, algae with high remaining Fv/Fm values also have a high total lipid content, thus presenting a positive correlation ($R^2 = 0.906$) between the lipid content and the Fv/Fm value (Huangfu et al., 2013). Similar phenomenon was also observed in naturally occurring microalgae strains (Pan et al., 2011). In our study, the Fv/Fm values were significantly negatively correlated with the relative lipid content (**Figure 7**, $r < -0.9$, $p < 0.01$), and the result was verified in a 3 l scale culture by testing samples at random time points (**Figure 8**). The difference of our negative correlation and the previous positive correlation report may be because that the previous report focused on the comparison of different algae, while we concentrate on the comparison of the given algae at different stress time. Under stress condition, the Fv/Fm value decreased and the lipid content increased. However, for different algae or mutants under same stress, the alga with higher remaining

Fv/Fm value behaved a higher toleration to the stress and has a higher potential photosynthesis for fixing more carbon and supplying more energy and carbon source for lipid production. So the remaining Fv/Fm was positive correlation with lipid content. As for our study, Fv/Fm value is the most sensitive indicator for stress response under unfavorable conditions, and neutral lipid accumulation coupled with N starvation-induced oxidative stress (Zhang et al., 2013). With the decreasing of Fv/Fm under N starvation, the lipid content continuously accumulated to saturated point with a slower and slower rate. Therefore, for the given algae, comparison with the stress time, the Fv/Fm value was negative correlation with lipid content. Thus, we predicted that for a given oil-producing algae under stress conditions, there is a negative relationship between the Fv/Fm value and the TAG content that can be used to estimate the TAG content and thus determine the optimum harvest time for lipid production.

In summary, we have established a simple and convenient method to determine the harvest time of microalgae under stress conditions for lipid production. For the technological process shown in **Figure 9**, for a given oil-rich microalgae, we can establish the relationship between the lipid content and the Fv/Fm value, thus obtaining a range for the Fv/Fm value around the lipid saturation point. When cultivating algae on a large scale for oil production, we can determine the optimum harvest time by measuring the Fv/Fm value and referring it to the established relationship.

Author Contributions

YQ was responsible for study conception and design, data collection and analysis, manuscript writing, and final approval of the manuscript; JR for study conception and design, data collection and analysis, and final approval of the manuscript; HC for data collection and analysis, manuscript writing, and final approval of the manuscript; CH for data analysis and final approval of the manuscript; and QW for conception and design, critical revision and manuscript writing, and final approval of the manuscript. All authors read and approved the final manuscript.

Acknowledgments

This work was supported jointly by the National Program on Key Basic Research Project (2012CB224803, 2011CB200902), the National Natural Science Foundation of China (31300030, 31270094), the Natural Science Foundation of Hubei Province of China (2013CFA109), Sinopec (S213049), and the Knowledge Innovation Program of the Chinese Academy of Sciences (Y35E05).

References

Ali, M. B., Yu, K. W., Hahn, E. J., and Paek, K. Y. (2005). Differential responses of anti-oxidants enzymes, lipoxygenase activity, ascorbate content and the production of saponins in tissue cultured root of mountain *Panax ginseng* C.A. Mayer and *Panax quinquefolium* L. in bioreactor subjected to methyl jasmonate stress. *Plant Sci.* 169, 83–92. doi:10.1016/j.plantsci.2005.02.027

Amaro, H. M., Guedes, A. C., and Malcata, F. X. (2011). Advances and perspectives in using microalgae to produce biodiesel. *Appl. Energy* 88, 3402–3410. doi:10.1016/j.apenergy.2010.12.014

Apel, K., and Hirt, H. (2004). Reactive oxygen species: metabolism, oxidative stress, and signal transduction. *Annu. Rev. Plant Biol.* 55, 373–399. doi:10.1146/annurev.arplant.55.031903.141701

Beardall, J., Young, E., and Roberts, S. (2001). Approaches for determining phytoplankton nutrient limitation. *Aquat. Sci.* 63, 44–69. doi:10.1007/PL00001344

Bligh, E. G., and Dyer, W. J. (1959). A rapid method of total lipid extraction and purification. *Can. J. Biochem. Physiol.* 37, 911–917. doi:10.1139/o59-099

Conroy, J. P., Smillie, R. M., Kuppers, M., Bevege, D. I., and Barlow, E. W. (1986). Chlorophyll a fluorescence and photosynthetic and growth responses of *Pinus radiata* to phosphorus deficiency, drought stress, and high CO2. *Plant Physiol.* 81, 423–429. doi:10.1104/pp.81.2.423

Converti, A., Casazza, A. A., Ortiz, E. Y., Perego, P., and Del Borghi, M. (2009). Effect of temperature and nitrogen concentration on the growth and lipid content of *Nannochloropsis oculata* and *Chlorella vulgaris* for biodiesel production. *J. Chem. Eng. Process.* 48, 1146–1151. doi:10.1016/j.cep.2009.03.006

Cooksey, K. E., Guckert, J. B., Williams, S. A., and Callis, P. R. (1987). Fluorometric-determination of the neutral lipid-content of microalgal cells using nile red. *J. Microbiol. Methods* 6, 333–345. doi:10.1016/0167-7012(87)90019-4

Cooper, M. S., Hardin, W. R., Petersen, T. W., and Cattolico, R. A. (2010). Visualizing "green oil" in live algal cells. *J. Biosci. Bioeng.* 109, 198–201. doi:10.1016/j.jbiosc.2009.08.004

Dawes, E. (1976). "Endogenous metabolism and the survival of starved prokaryotes," in *The Survival of Vegetative Microbes*, eds Gray T.R.G. and Postgate J.R. (Cambridge: Cambridge University Press), 19–53.

Elstner, E. F. (1991). "Mechanisms of oxygen activation in different compartments of plant cells," in *Active Oxygen/Oxidative Stress in Plant Metabolism*, eds Pell E. J. and Steffen K. L. (Rockville, MD: American Society of Plant Physiologists), 13–25.

Genty, B., Briantais, J. M., and Baker, N. R. (1989). The relationship between the quantum yield of photosynthetic electron-transport and quenching of chlorophyll fluorescence. *Biochim. Biophys. Acta* 990, 87–92. doi:10.1016/S0304-4165(89)80016-9

Hill, J., Nelson, E., Tilman, D., Polasky, S., and Tiffany, D. (2006). Environmental, economic, and energetic costs and benefits of biodiesel and ethanol biofuels. *Proc. Natl. Acad. Sci. U.S.A.* 103, 11206–11210. doi:10.1073/pnas.0604600103

Hu, Q., Sommerfeld, M., Jarvis, E., Ghirardi, M., Posewitz, M., Seibert, M., et al. (2008). Microalgal triacylglycerols as feedstocks for biofuel production: perspectives and advances. *Plant J.* 54, 621–639. doi:10.1111/j.1365-313X.2008.03492.x

Hu, Q., Zhang, C., and Sommerfeld, M. (2006). Biodiesel from algae: lessons learned over the past 60 years and future perspectives. *PSA Abstracts. J. Phycol.* 42, 12–12. doi:10.1111/j.1529-8817.2006.20064201.x

Huangfu, J. Q., Liu, J., Sun, Z., Wang, M. F., Jiang, Y., Chen, Z. Y., et al. (2013). Antiaging effects of astaxanthin-rich alga *Haematococcus pluvialis* on fruit flies under oxidative stress. *J. Agric. Food Chem.* 61, 7800–7804. doi:10.1021/jf402224w

Illman, A. M., Scragg, A. H., and Shales, S. W. (2000). Increase in *Chlorella* strains calorific values when grown in low nitrogen medium. *Enzyme Microb. Technol.* 27, 631–635. doi:10.1016/S0141-0229(00)00266-0

Lee, C. W., Lim, J. H., and Heng, P. L. (2013). Investigating the spatial distribution of phototrophic picoplankton in a tropical estuary. *Environ. Monit. Assess.* 185, 9697–9704. doi:10.1007/s10661-013-3283-3

Lichtenthaler, H. K. (1987). "Chlorophylls and carotenoids: pigments of photosynthetic biomembranes," in *Methods in Enzymology*, ed. Lester Packer R. D. (San Diego: Academic Press), 350–382.

Lu, C. M., Torzillo, G., and Vonshak, A. (1999). Kinetic response of photosystem II photochemistry in the cyanobacterium Spirulina platensis to high salinity is characterized by two distinct phases. *Aust. J. Plant Physiol.* 26, 283–292. doi:10.1071/PP98119

Maxwell, K., and Johnson, G. N. (2000). Chlorophyll fluorescence-a practical guide. *J. Exp. Bot.* 51, 659–668. doi:10.1093/jexbot/51.345.659

Oxborough, K. (2004). Imaging of chlorophyll a fluorescence: theoretical and practical aspects of an emerging technique for the monitoring of photosynthetic performance. *J. Exp. Bot.* 55, 1195–1205. doi:10.1093/Jxb/Erh145

Pan, Y. Y., Wang, S. T., Chuang, L. T., Chang, Y. W., and Chen, C. N. (2011). Isolation of thermo-tolerant and high lipid content green microalgae: oil accumulation is predominantly controlled by photosystem efficiency during stress treatments in *Desmodesmus*. *Bioresour. Technol.* 102, 10510–10517. doi:10.1016/j.biortech.2011.08.091

Petkov, G., and Garcia, G. (2007). Which are fatty acids of the green alga *Chlorella*? *Biochem. Syst. Ecol.* 35, 281–285. doi:10.1016/j.bse.2006.10.017

Reiser, S., and Somerville, C. (1997). Isolation of mutants of *Acinetobacter calcoaceticus* deficient in wax ester synthesis and complementation of one mutation with a gene encoding a fatty acyl coenzyme a reductase. *J. Bacteriol.* 179, 2969–2975.

Santos, C. A., Ferreira, M. E., da Silva, T. L., Gouveia, L., Novais, J. M., and Reis, A. (2011). A symbiotic gas exchange between bioreactors enhances microalgal biomass and lipid productivities: taking advantage of complementary nutritional modes. *J. Ind. Microbiol. Biotechnol.* 38, 909–917. doi:10.1007/s10295-010-0860-0

Schreiber, U. (1978) Chlorophyll fluorescence assay for ozone injury in intact plants. *Plant Physiol.* 61, 80–84. doi:10.1104/pp.61.1.80

Scott, S. A., Davey, M. P., Dennis, J. S., Horst, I., Howe, C. J., Lea-Smith, D. J., et al. (2010). Biodiesel from algae: challenges and prospects. *Curr. Opin. Biotechnol.* 21, 277–286. doi:10.1016/j.copbio.2010.03.005

Shi, S., Tang, D., and Liu, Y. (2009). Effects of an algicidal bacterium *Pseudomonas mendocina* on the growth and antioxidant system of *Aphanizomenon flos-aquae*. *Curr. Microbiol.* 59, 107–112. doi:10.1007/s00284-009-9404-0

Tran, T. H., Govin, A., Guyonnet, R., Grosseau, P., Lors, C., Damidot, D., et al. (2014). Influence of the intrinsic characteristics of mortars on their biofouling by pigmented organisms: comparison between laboratory and field-scale experiments. *Int. Biodeterior. Biodegradation* 86, 334–342. doi:10.1016/j.ibiod.2013.10.005

White, S., Anandraj, A., and Bux, F. (2011). PAM fluorometry as a tool to assess microalgal nutrient stress and monitor cellular neutral lipids. *Bioresour. Technol.* 102, 1675–1682. doi:10.1016/j.biortech.2010.09.097

Widjaja, A., Chien, C.-C., and Ju, Y.-H. (2009). Study of increasing lipid production from fresh water microalgae *Chlorella vulgaris*. *J. Taiwan Inst. Chem. Eng.* 40, 13–20. doi:10.1016/j.jtice.2008.07.007

Zhang, X., Chen, H., Chen, W., Qiao, Y., He, C., and Wang, Q. (2014a). Evaluation of an oil-producing green alga *Chlorella* sp. C2 for biological DeNOx of industrial flue gases. *Environ. Sci. Technol.* 48, 10497–10504. doi:10.1021/es5013824

Zhang, X., Rong, J., Chen, H., He, C., and Wang, Q. (2014b). Current status and outlook in the application of microalgae in biodiesel production and environmental protection. *Front. Energy Res.* 2:32. doi:10.3389/fenrg.2014.00032

Zhang, Y. M., Chen, H., He, C. L., and Wang, Q. (2013). Nitrogen starvation induced oxidative stress in an oil-producing green alga *Chlorella sorokiniana* C3. *PLoS ONE* 8:e69225. doi:10.1371/journal.pone.0069225

Zhu, S., Wang, Y., Huang, W., Xu, J., Wang, Z., Xu, J., et al. (2014). Enhanced accumulation of carbohydrate and starch in *Chlorella zofingiensis* induced by nitrogen starvation. *Appl. Biochem. Biotechnol.* 174, 2435–2445. doi:10.1007/s12010-014-1183-9

Conflict of Interest Statement: The authors declare that the research was conducted in the absence of any commercial or financial relationships that could be construed as a potential conflict of interest.

7

The Enzymatic Conversion of Major Algal and Cyanobacterial Carbohydrates to Bioethanol

Qusai Al Abdallah[1]*, B. Tracy Nixon[2] and Jarrod R. Fortwendel[1]

[1] *Department of Clinical Pharmacy, University of Tennessee Health Science Center, Memphis, TN, USA,* [2] *Department of Biochemistry and Molecular Biology, The Pennsylvania State University, University Park, PA, USA*

Edited by:
Arumugam Muthu,
Council of Scientific and
Industrial Research, India

Reviewed by:
Wenjie Liao,
Sichuan University, China
Sachin Kumar,
Sardar Swaran Singh National
Institute of Renewable Energy, India

*Correspondence:
Qusai Al Abdallah
qalabdal@uthsc.edu

Specialty section:
This article was submitted
to Bioenergy and Biofuels,
a section of the journal
Frontiers in Energy Research

Citation:
Al Abdallah Q, Nixon BT and
Fortwendel JR (2016) The Enzymatic
Conversion of Major Algal and
Cyanobacterial Carbohydrates to
Bioethanol.
Front. Energy Res. 4:36.

The production of fuels from biomass is categorized as first-, second-, or third-generation depending upon the source of raw materials, either food crops, lignocellulosic material, or algal biomass, respectively. Thus far, the emphasis has been on using food crops creating several environmental problems. To overcome these problems, there is a shift toward bioenergy production from non-food sources. Algae, which store high amounts of carbohydrates, are a potential producer of raw materials for sustainable production of bioethanol. Algae store their carbohydrates in the form of food storage sugars and structural material. In general, algal food storage polysaccharides are composed of glucose subunits; however, they vary in the glycosidic bond that links the glucose molecules. In starch-type polysaccharides (starch, floridean starch, and glycogen), the glucose subunits are linked together by α-(1→4) and α-(1→6) glycosidic bonds. Laminarin-type polysaccharides (laminarin, chrysolaminarin, and paramylon) are made of glucose subunits that are linked together by β-(1→3) and β-(1→6) glycosidic bonds. In contrast to food storage polysaccharides, structural polysaccharides vary in composition and glycosidic bond. The industrial production of bioethanol from algae requires efficient hydrolysis and fermentation of different algal sugars. However, the hydrolysis of algal polysaccharides employs more enzymatic mixes in comparison to terrestrial plants. Similarly, algal fermentable sugars display more diversity than plants, and therefore more metabolic pathways are required to produce ethanol from these sugars. In general, the fermentation of glucose, galactose, and glucose isomers is carried out by wild-type strains of *Saccharomyces cerevisiae* and *Zymomonas mobilis*. In these strains, glucose enters glycolysis, where is it converted to pyruvate through either Embden–Meyerhof–Parnas pathway or Entner–Doudoroff pathway. Other monosaccharides must be converted to fermentable sugars before entering glycolysis. In contrast, microbial wild-type strains are not capable of producing ethanol from alginate, and therefore the production of bioethanol from alginate was achieved by using genetically engineered microbial strains, which can simultaneously hydrolyze and ferment alginate to ethanol. In this review, we emphasize the enzymatic hydrolysis processes of different algal polysaccharides. Additionally, we highlight the major metabolic pathways that are employed to ferment different algal monosaccharides to ethanol.

Keywords: bioethanol, algal carbohydrates, food reserves, structural polysaccharides, enzymatic hydrolysis, fermentation

INTRODUCTION

During the last decade, growing concerns over depleting fossil oil reserves and increasing greenhouse gas emissions have led to the use of food crops as biomass for making what is called first-generation biofuel. Nevertheless, expansion in biofuel production from food crops has drawn attention to several environmental impacts, such as the conversion of agricultural land from producing food crops to producing biofuel crops and the deforestation of hundreds of thousands of acres (Groom et al., 2008; Searchinger et al., 2008; Naik et al., 2010).

To overcome these environmental problems, there is a shift toward the production of biofuel from non-food biomass sources, such as lignocellulosic and algal biomass sources, which are also known as the second- and third-generation biofuel crops, respectively (Badger, 2002; Zheng et al., 2009; Brennan and Owende, 2010). However, commercial production of second-generation biofuels is only limited to countries with large agricultural and forestry lands. Therefore, algal biomass is an emerging alternative for the production of biofuels.

The production of biofuel from algae has several advantages over the first- and second-generation of biomass sources [discussed by John et al. (2011)]. First, algae serve as non-food feedstock, which does not compromise our food security. Second, algae grow in aquatic habitats and thereby do not compete with food crops on agricultural land, or cause deforestation. Third, algal biomass can be used to produce two types of biofuel (bioethanol and biodiesel) since they accumulate high amounts of carbohydrates and lipids. Finally, the fresh water requirement for algal growth is significantly lower than plant demands to produce the same volume of biofuel. Nevertheless, there are several constraints that restrict the production of biofuel from algae [discussed by Hannon et al. (2010), Singh et al. (2011), and Behera et al. (2015)].

The hydrolysis of algal carbohydrates to basic sugars is primarily carried out using chemical and enzymatic methods. Although the chemical method yields high concentrations of fermentable sugars in a short time, this method requires harsh reaction conditions producing byproducts, which might inhibit the fermentation process and require costly disposal processes. In contrast, enzymatic hydrolysis produces high amounts of fermentable sugars under mild conditions without producing inhibitory byproducts (Chen et al., 2013).

Algae produce a wide spectrum of polysaccharides that are specific to an algal group, family, or species. The enzymatic hydrolysis of algal polysaccharides requires a wider range of enzymatic mixtures, compared to plants. This review focuses on the enzymatic hydrolysis steps of the major algal carbohydrates and their fermentation process to ethanol. Since the scope of this topic is broad, only the fundamental concepts of the field are addressed in this review. Nevertheless, we will refer the reader to other reviews that are complementary to this topic.

THIRD-GENERATION BIOFUELS FROM ALGAL BIOMASS

Algae are photosynthetic eukaryotes that are distinguishable from cyanobacteria, which are photosynthetic prokaryotes (Brodie and Lewis, 2007). Because of their importance for biofuel production, this review will cover cyanobacteria as well.

Algae vary dramatically in size and morphology from microscopic unicellular phytoplanktons to 50-m long seaweeds. Based on their morphology and size, algae are classified into microalgae and macroalgae. Currently, microalgae are the major source for third-generation biofuels. In contrast, only small amount of cyanobacterial biomass are utilized for bioethanol production. Additionally, development of methods that overcome obstacles in using macroalgae would greatly improve harvesting bioethanol from natural, renewable biomaterials. The advantages and disadvantages of relevant algal sources are summarized in **Table 1**.

Microalgae Are the Current Source for Third-generation Biofuels

Microalgae are microscopic in size (measured in micrometers) and exist as single cells; or unspecialized multicellular filaments and colonies (Satyanarayana et al., 2011). They are highly diverse including 40,000 species that belong to nearly all major algal groups with the exception of brown algae [reviewed by Metting (1996), Dahiya (2015), and Kim (2015)].

Microalgae exhibit several features that favor using them for industrial production of biofuel. First, they lack specialized tissues and structures, which simplify the cultivation and harvesting processes. In addition, microalgae exhibit high rates of asexual growth and yield huge amount of biomass from low inoculum (Packer, 2009; Chen et al., 2010). Furthermore, microalgae accumulate large amounts of polysaccharides and triacylglycerols – storage lipids and energy sources, and thereby they are suitable for simultaneous production of bioethanol and biodiesel (Mata et al., 2010; Singh et al., 2011; Suutari et al., 2015).

The commercial production of microalgal biomass is obtained from cultivating the freshwater algae *Chlorella* and *Haematococcus*, and marine algae, such as *Dunaliella*, *Phaeodactylum*, and *Tetraselmis* (Lee, 1997; Wikfors and Ohno, 2001; Carlsson et al., 2007; Benemann, 2013; Borowitzka, 2013). Additionally, other microalgae have been shown to be a potential source for third-generation biofuels due to their high oil and carbohydrates contents (Singh et al., 2011).

One of the challenges for commercial cultivation of microalgae is the economic feasibility. In their marine natural habitats, the productivity of microalgae is very low, not exceeding 10% of that for macroalgae under the same conditions. Such low yield of microalgal biomass is not sufficient for the industrial production of bioethanol. To improve the yield, microalgae should be cultivated in artificial systems (Lüning and Pang, 2003). The most common two methods for the cultivation of microalgae are the outdoor open pond system and the closed photobioreactor [for reviews, refer to Brennan and Owende (2010) and Benemann (2013)]. The photobioreactor system, which produces high biomass under controlled growth conditions, requires high capital and operating costs (Pruvost et al., 2016). In contrast, cultivation of microalgae in open ponds involves lower capital and operating costs but offers low productivity. Additionally, microalgal cultures growing in open ponds are exposed to contaminants and affected by seasonal variations (Chisti, 2016). In both systems, microalgal density must be controlled to maintain a viable culture (Wang et al., 2009). Other challenges associated with biofuel from

TABLE 1 | Comparison between relevant algal sources and the advantages and disadvantages of employing each for the production of third-generation biofuels.

Algal source	Characteristics	Current industrial applications	Advantages and disadvantages
Microalgae	(1) Eukaryote (2) Microscopic	(1) The major source for third-generation biofuels. (2) A major source for several nutritional and pharmaceutical products.	Advantages: (1) Easy to cultivate and harvest. (2) High growth rates in artificial growth systems. (3) Suitable for simultaneous production of bioethanol and biodiesel. Disadvantages: (1) Several microalgae that are employed for biofuel production require freshwater for growth. (2) Low productivity levels in their marine natural habitats, therefore microalgae should be cultivated in artificial systems, which involve capital and operating costs. (3) Microalgae that are cultivated in open ponds are normally cultured outdoors, and thus biomass production is heavily affected by contamination with epiphytes, microbial infections, and seasonal variations.
Cyanobacteria	(1) Prokaryote (2) Microscopic	(1) A minor source for third-generation biofuels. (2) A major source for several nutritional and pharmaceutical products.	Advantages: (1) Easy to cultivate and harvest. (2) The simple nutrient requirements of cyanobacteria make its cultivation and harvesting simple and inexpensive. (3) Higher photosynthetic levels and growth rates than algae and plants (only if light is provided in saturating amounts). Disadvantages: (1) Cyanobacteria are not suitable for biodiesel production. (2) Accumulate significantly lower amounts of carbohydrates (% of dry weight) than microalgae. (3) Saturating amounts of light must be provided to reach highest photosynthesis rates from cyanobacteria.
Macroalgae	(1) Eukaryote (2) Macroscopic	(1) A major source for several nutritional and pharmaceutical products.	Advantages: (1) Macroalgae produce more biomass in their marine natural habitats and therefore do not require cultivation in costly artificial systems. (2) Does not require freshwater for cultivation. (3) Significantly higher annual biomass production for non-biofuel industrial purposes. This carbohydrate-rich biomass can be employed for biofuel production purposes as well. Disadvantages: (1) Macroalgae are not suitable for the production of biodiesel. (2) The production is heavily affected by contaminants and seasonal variations. (3) To achieve the commercial levels of bioethanol from macroalgae, glucose-, and non-glucose-based sugars must be fermented.

microalgae have been discussed in detail elsewhere (Hannon et al., 2010; Benemann, 2013; Scaife et al., 2015).

Macroalgae Are an Unexploited Potential Source for Bioethanol

Macroalgae refer to the macroscopic seaweeds. They are characterized by forming multicellular specialized tissues and defined structures that are comparable to plant leaves and roots (John and Anisha, 2012; Murphy et al., 2013). Macroalgae are less versatile than microalgae and are distributed primarily over green, red, and brown algae (Jung et al., 2013). In comparison to terrestrial plants, macroalgae grow faster and produce more biomass per area due to their high photosynthetic efficiency (Murphy et al., 2013; Yanagisawa et al., 2013).

Although commercial third-generation biofuels are derived from microalgal biomass, seaweeds (specifically red and brown macroalgae) serve as an unexploited potential source for bioethanol production due to two facts. First, macroalgae combine high biomass productivity with low capital and operating costs owing to the fact that macroalgae are harvested from naturally

occurring stocks or aquacultured sea farms. Such cultivation systems require capital and operating costs that are significantly lower than the microalgal open ponds, nevertheless they provide high biomass productivity (Carlsson et al., 2007; Bruton et al., 2009; Jung et al., 2013; Yanagisawa et al., 2013; Rajkumar et al., 2014). Second, macroalgae are cultivated worldwide on a large scale for non-biofuel purposes. The remainder of the biomass, which is rich in carbohydrates, can be hydrolyzed to produce ethanol. In fact, the worldwide biomass production from macroalgae greatly surpasses that of microalgae. For example, in 2010, approximately 9 million and 6.75 million wet metric tons (WMT) were harvested from red and brown macroalgae, respectively (Jung et al., 2013). In comparison, a total of only 6.2 thousand dry metric tons (DMT) were produced by major microalgal species, such as *Chlorella* sp., *Dunaliella salina*, and *Haematococcus pluvialis*, in the same year [refer to Table 2 in Jung et al. (2013)]. Additionally, 93% of the worldwide cultivated macroalgal biomass is produced from only four genera that belong to the brown algae (*Laminaria* 65.8% and *Undaria* 9.8%) and red algae (*Porphyra* 12.6% and *Gracilaria* 4.8%) (Zemke-White and

Ohno, 1999). The potential application of macroalgae for biofuel production has been reviewed by others (Murphy et al., 2013).

The production of biofuels from macroalgae has several environmental advantages [discussed by Hughes et al. (2012)] and is challenged by several obstacles, which will be discussed here. First, in contrast to microalgal feedstocks, which are used for simultaneous production of bioethanol and biodiesel, macroalgae accumulate considerable amounts of carbohydrates, and thus can be used to produce bioethanol only [see Table 1 in Singh et al. (2011); Tables 1 and 2 in Suutari et al. (2015)]. Second, macroalgae are normally cultured outdoors, and thus biomass production is heavily affected by contamination with epiphytes (Lüning and Pang, 2003) and microbial infections (Ramaiah, 2006). Third, the macroalgal carbohydrates content varies depending on the alga growth stage and seasonal variations (Suutari et al., 2015). Fourth, macroalgae accumulate lower amounts of glucan food reserves (i.e., glucose-based polysaccharides) in comparison to microalgae, while producing high amounts of non-glucose-based sugars, such as mannitol and cell wall polysaccharides. Therefore, the industrial production of bioethanol from macroalgae requires fermentation of both glucose- and non-glucose-based sugars (Yanagisawa et al., 2013).

Cyanobacteria Serve as a Minor Source for Third-generation Bioethanol

Spirulina sp. is the most commonly grown cyanobacterium for commercial use. Its biomass is used primarily for human and animal consumption; however, only a small portion is directed toward biofuel production (Ciferri, 1983; Wikfors and Ohno, 2001; Habib et al., 2008; Benemann, 2013). Additionally, several cyanobacterial strains of *Synechococcus* species have been genetically modified for enhanced commercial production of bioethanol [reviewed by Dexter et al. (2015)].

The production of biofuel from cyanobacteria has several advantages [discussed by Quintana et al. (2011) and Sarsekeyeva et al. (2015)]. Among these advantages is the fact that many cyanobacteria, e.g., *Spirulina* and *Synechococcus*, accumulate high amounts of glycogen, which can be easily extracted and fermented to ethanol (See Conversion of Glycogen to Glucose). However, there are several disadvantages of using cyanobacterial biomass for biofuel production. For example, in contrast to microalgae, which store high amounts of lipids and carbohydrates, cyanobacteria do not accumulate significant amounts of lipids, and therefore they are not suitable for biodiesel production (Quintana et al., 2011). Other challenges that constrain bioethanol production from cyanobacteria have been discussed by other reports (Nozzi et al., 2013).

THE MAJOR ALGAL POLYSACCHARIDES

Similar to plants, photosynthesis in algae is divided into two steps: the light-dependent reactions and Calvin cycle. In the light-dependent reactions, light energy is absorbed at the thylakoid membranes in the chloroplasts, where it is converted into adenosine triphosphate (ATP) and the reduced form of nicotinamide–adenine dinucleotide phosphate (NADPH). The energy molecules NADPH and ATP are then employed by Calvin cycle to metabolize carbon dioxide and water into sugar [see Figure 1 in Moroney and Ynalvez (2001)]. For reviews on algal photosynthesis, we recommend the reader to refer to Moroney and Ynalvez (2001).

Algae produce a wide range of polysaccharides depending on the algal species (**Table 2**). This diverse collection of polysaccharides functions primarily as food reserves or structural material (**Figure 1**). Here, we describe the most economically important algal sugars, which have received considerable amount of research interest. The advantages and disadvantages of employing these sugars for bioethanol production are highlighted in **Table 3**. For more information about algal polysaccharides, we refer the reader to previously published reviews (Peat and Turvey, 1965; Percival, 1970, 1979; Avigad and Dey, 1997; Grant Reid, 1997; Synytsya et al., 2015).

Food reserves are easily fermented into ethanol and thus are the primary source for industrial third-generation bioethanol. In contrast, the hydrolysis of structural carbohydrates is challenging due to their rigidity. Therefore, optimization of the hydrolysis process of structural carbohydrates carries the promise of maximizing ethanol yield. In this section, we will first review the major algal food reserves. Additionally, we will discuss major algal structural polysaccharides because of their potential in enhancing the yield of bioethanol from algal feedstock.

Food Reserves

The majority of algae store their food reserves in the form of starch-type polysaccharides (such as starch, floridean starch, and glycogen) (Viola et al., 2001; Deschamps et al., 2008; Quintana et al., 2011) or laminarin-type polysaccharides (such as chrysolaminarin and laminarin) (Michel et al., 2010). Additionally, brown algae accumulate large amounts of mannitol, which functions as an antioxidant and regulator of cell osmolarity (Davis et al., 2003; Iwamoto and Shiraiwa, 2005).

Starch is a homopolysaccharide of glucose units that are linked by α-(1→4) glycosidic bonds to form a linear amylose; and *via* α-(1→6) bonds to form amylopectin (**Table 2**) (Fengel and Gert, 1989; Busi et al., 2014). In contrast to plants, which store starch granules in the amyloplast, most algae lack the amyloplast, and therefore store starch grains in the chloroplast (Busi et al., 2014). Exceptions to this are the red algae, Dinophyta (Dinoflagellates), and Glaucophyta, which store their food reserves in the cytosol (Radakovits et al., 2010). Several algal species have been reported to store relatively high concentrations of starch, reaching in some species to about 50% of the dry weight [see Table 1 in John et al. (2011)].

Floridean starch is another main food reserve polysaccharide (**Table 2**). It is a starch derivative that is synthesized by red algae (Rhodophyta). Its granules differ from starch by lacking amylose and thereby are composed completely of amylopectin (Viola et al., 2001). The red alga *Seirospora griffithsiana* stores up to 80% of its cell volume as floridean starch. The granules are similar in structure to plant starch but more variable in size (diameter: 0.3–1.7 µm) and shape (Sheath et al., 1981).

Laminarin and chrysolaminarin are the third major food reserves. They are linear polymers of β-(1→3) glucan repeating units with β-(1→6) branches in the ratio of 15:1 for laminarin

TABLE 2 | Chemical structure and distribution of food reserves and structural polysaccharides among different groups of algae.

Major algal carbohydrate	Composition	Glycosidic bonds	Chemical structure	Algal source
Starch -Amylose	Glucose_n	α-(1→4)		Chlorophyta Dinophyta Cryptophyta Glaucocystophyta
-Amylopectin		α-(1→4) and α-(1→6)		
Floridean starch	Glucose_n	α-(1→4) and α-(1→6)	See amylopectin	Rhodophyta
Glycogen	Glucose_n	α-(1→4) and α-(1→6)	See amylopectin	Cyanophyta
Laminarin	Glucose_n	β-(1→3) and β-(1→6)		Phaeophyceae
Chrysolaminarin	Glucose_n	β-(1→3) and β-(1→6)	See laminarin	Bacillariophyceae Xanthophyceae Chrysophyceae Haptophyta Chlorarachniophyta
Paramylon	Glucose_n	β-(1→3)		Euglenophyta
Cellulose	Glucose_n	β-(1→4)		All algal groups expect diatoms
Agarose	[Galactose and anhydro-L-galactose]_n	β-(1→4) and α-(1→3)		Rhodophyta
Carrageenan	[Galactose and anhydro-D-galactose]_n	β-(1→4) and α-(1→3)	See agarose	Rhodophyta
Alginate	α-L-guluronate_n and β-D-mannuronate_n	β-(1→4) and α-(1→4)		Phaeophyceae
Ulvan	Sulfated rhamnose, glucuronic acid, iduronic acid, xylose, and sulfated xylose	β-(1→4) α-(1→4) β-(1→4) β-(1→4)	[Glucuronic acid – sulfated rhamnose] [Iduronic acid – sulfated rhamnose] [Xylose – sulfated rhamnose] [Sulfated xylose – sulfated rhamnose]	*Ulva* and *Enteromorpha* sp. (green algae)
Fucoidan	Sulfated L-fucose_n	Predominantly α-(1→2)		Family Laminariaceae (brown algae)

n is the number of repeating units of the molecule.

M, β-D-mannuronate.

G, α-L-guluronaten.

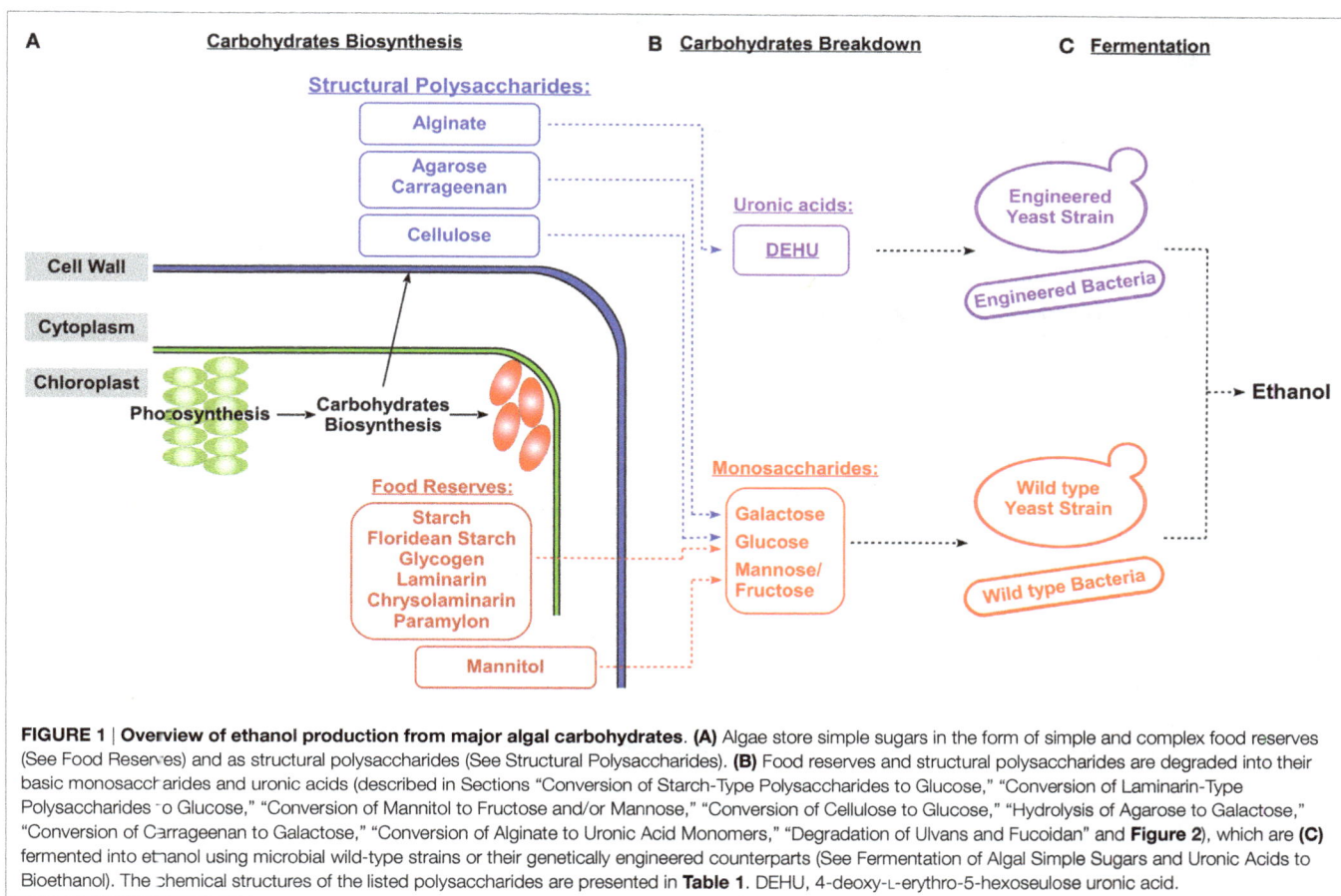

FIGURE 1 | Overview of ethanol production from major algal carbohydrates. (A) Algae store simple sugars in the form of simple and complex food reserves (See Food Reserves) and as structural polysaccharides (See Structural Polysaccharides). **(B)** Food reserves and structural polysaccharides are degraded into their basic monosaccharides and uronic acids (described in Sections "Conversion of Starch-Type Polysaccharides to Glucose," "Conversion of Laminarin-Type Polysaccharides to Glucose," "Conversion of Mannitol to Fructose and/or Mannose," "Conversion of Cellulose to Glucose," "Hydrolysis of Agarose to Galactose," "Conversion of Carrageenan to Galactose," "Conversion of Alginate to Uronic Acid Monomers," "Degradation of Ulvans and Fucoidan" and **Figure 2**), which are **(C)** fermented into ethanol using microbial wild-type strains or their genetically engineered counterparts (See Fermentation of Algal Simple Sugars and Uronic Acids to Bioethanol). The chemical structures of the listed polysaccharides are presented in **Table 1**. DEHU, 4-deoxy-L-erythro-5-hexoseulose uronic acid.

and 11:1 for chrysolaminarin (**Table 2**) (Beattie et al., 1961; Michel et al., 2010). Laminarin is synthesized by brown algae, and it forms either a G-chain – with glucose molecule at the reducing end – or an M-chain – with mannitol at the reducing end (Kadam et al., 2015; Motone et al., 2016). Accumulation of high amounts of laminarin in algae has been reported in several seaweeds, comprising up to 32–35% of dry weight [refer to Table 1 in Kadam et al. (2015)]. Chrysolaminarin is the food reserve polysaccharide in diatoms, and it is comprised only of glucose molecules (G-chains) at the reducing end (Beattie et al., 1961; Michel et al., 2010).

Glycogen is the food reserve form in cyanobacteria. Glycogen is made of glucose subunits that are linked together by α-(1→4) and α-(1→6) glycosidic bonds. Glycogen and amylopectin (one of starch granule constituents) are similar in structure; however, glycogen is more branched and forms smaller granules (diameter is 42 nm) in comparison to starch granules (diameter 100–100,000 nm) (Ball et al., 2011).

In addition to the previously described major polysaccharide forms, other granule forms exist among algae but to a lesser degree. For instance, algae of the class Euglenophyta store their food reserves in the cytoplasm as paramylon. The granules of paramylon (or paramylum) are membrane bound and are composed of linear β-(1→3) glucan repeating units (**Table 2**). Its chemical structure is similar to that of laminarin

but does not form β-(1→6) branches (Barsanti et al., 2011; Monfils et al., 2011).

Mannitol is a sugar alcohol of the aldohexose D-mannose. In brown algae, it serves as a storage sugar, and an antioxidant, and protects against osmotic stress. It accumulates in the cell as a monosaccharide (i.e., free mannitol sugar) and as part of the laminarin polysaccharide – forming the laminarin M-chain, where mannitol is at the reducing end (see laminarin above) (Davis et al., 2003). Mannitol is one of the major storage carbohydrates in brown algae, and some brown algae accumulate high amounts of mannitol reaching approximately 25% of *Laminaria hyperborea*, 30% of *Laminaria japonica*, and 33% of several *Sargassum* and *Turbinaria* dry weight (Ota et al., 2013). Mannitol is produced in brown algae from fructose-6-phosphate, which is reduced to mannitol-1-phosphate *via* Mannitol-1-phosphate 5-dehydrogenase (EC1.1.1.17). In the second step, mannitol-1-phosphate is converted to D-mannitol by mannitol-1-phosphatase (EC3.1.3.22) [see Figure 1B (b) in Iwamoto and Shiraiwa (2005)].

Structural Polysaccharides

Structural polysaccharides are another putative source to increase bioethanol yield from algae. Their main function is to confer rigidity to the algal cell wall. In contrast to plants, which usually have a lignocellulosic cell wall, the composition of algal cell wall

TABLE 3 | The advantages and disadvantages of employing different algal sugars for the production of third-generation biofuels.

Algal sugar	Advantages and disadvantages
Starch	Advantages: (1) Abundant polysaccharide that is synthesized by green algae and several plants. (2) There are abundant sources for amylolytic enzymes. (3) The availability of genetically engineered microbes that can simultaneously hydrolyze starch and ferment it to ethanol. Disadvantages: (1) Starch annual production from green algae is significantly lower than floridean starch from red algae and laminarin from brown algae.
Floridean starch	Advantages: (1) The annual production levels of floridean starch from red macroalgae are hundreds of times more than starch from green micro- and macroalgae. (2) The chemical structure is similar to starch, therefore can be degraded by the amylolytic enzymes, and fermented by microbial strains that ferment starch. Disadvantages: (1) There are only few scientific studies that investigated bioethanol production from floridean starch.
Glycogen	Advantages: (1) Can be easily extracted from cyanobacteria. (2) Similar to starch, therefore can be degraded by the amylolytic enzymes, and fermented by microbial strains that ferment starch. Disadvantages: (1) Glycogen annual production is significantly lower than floridean starch and laminarin.
Laminarin, chrysolaminarin, and paramylon	Advantages: (1) The annual production levels of laminarin-type polysaccharides are hundreds of times more than starch from green algae. (2) The chemical structure is similar to fungal cell wall β-glucans, and thus lytic enzymes that hydrolyze fungal cell wall can be utilized to degrade laminarin-type polysaccharides. Disadvantages: (1) There are only few scientific studies that investigated bioethanol production from laminarin-type polysaccharides.
Mannitol	Advantages: (1) Soluble simple sugar that does not require hydrolysis process. (2) Accumulate in brown algae in high concentrations. Disadvantages: (1) Must be converted into fructose-6-phosphate before fermentation.
Cellulose	Advantages (1) Abundant polysaccharide that is synthesized by most of algae and all plants. (2) There are abundant sources for cellulolytic enzymes. Disadvantages: (1) Rigid polysaccharide that requires a pretreatment step before hydrolysis.
Agarose	Advantages: (1) High annual production from red algae. Disadvantages: (1) Low solubility and therefore requires pretreatment before enzymatic hydrolysis. (2) The enzymatic hydrolysis of agarose produces a non-fermentable sugar, 3,6-anhydro-L-galactose, which must be converted into fermentable sugar using agar-degrading microorganisms.
Carrageenan	Advantages: (1) High annual production from red algae. Disadvantage: (1) Carrageenolytic enzymes are not common among microbes. (2) The enzymatic hydrolysis of carrageenan produces a non-fermentable sugar, 3,6-anhydro-D-galactose, which must be converted into fermentable sugar using carrageenan-degrading microorganisms.
Alginate	Advantages: (1) High annual production from brown algae. Disadvantages: (1) Fermentation of alginate requires genetically engineered strains.

varies among algal groups. Cellulose is the major algal cell wall polysaccharide, and it is present in most algal groups. In addition to cellulose, algae incorporate significant amounts of other polysaccharides into their cell wall, which can be converted into ethanol. Such polysaccharides can be specific to an algal group, such as the red algae, which contain agarose and carrageenan; and the brown algae, which are rich in alginate (Vreeland and Kloareg, 2000; Murphy et al., 2013). Variations in algal cell wall contents can also be found within families and genera of the same group. For example, the cell wall of the green seaweeds *Ulva* and *Enteromorpha* sp. contains high amounts of ulvan, while fucoidan is found in the members of family Laminariaceae of the brown algae (Jiao et al., 2011; Ale and Meyer, 2013).

Cellulose is a homopolysaccharide made of a linear chain of D-glucose units, which are connected together by β-(1→4) glycosidic bonds (**Table 2**). Cellulose chains aggregate together by intra- and inter-molecular hydrogen bonds to form cellulose microfibrils. Microfibrils are packed together to form fibrils, which in turn aggregate to form cell wall fibers (Brown and Saxena, 2000; Pu et al., 2013). With exception to diatoms, cellulose is found in the majority of algal cell walls.

Agarose is a non-sulfated, non-water-soluble linear galactan that is composed of repeating disaccharide units of D-galactose (D-Gal) and 3,6-anhydro-L-galactose (L-AnGal). The repeating disaccharide unit is called agarobiose or neoagarobiose depending on (1) the position of each sugar in the disaccharide, (2) the bond that links the monomers within the disaccharide, and (3) the bond that links the disaccharides to form agarose. Agarobiose consists of a D-Gal residue followed by L-AnGal that is linked by a β-(1→4) glycosidic bond. In contrast, neoagarobiose consists of α-(1→3)-linked L-AnGal and D-Gal residues. Agarobiose and neoagarobiose units are linked by α-(1→3) and β-(1→4), respectively (**Table 2**) (Renn, 1997; Fu and Kim, 2010; Delattre et al., 2011).

Carrageenan is a sulfated water-soluble linear galactan of carrabiose (or neocarrabiose) subunits (**Table 2**). Carrabiose and neocarrabiose are similar in structure and linkage to agarobiose and neoagarobiose, respectively (De Ruiter and Rudolph, 1997; Renn, 1997; Delattre et al., 2011).

In brown algae, alginate is one of the major cell wall sugars, besides cellulose, accounting for approximately 30–60% of the total sugars. The main function of alginate is to provide the cell wall with elasticity and rigidity to survive aquatic habitats (Dornish and Rauh, 2006). Additionally, alginate is found in the matrix of some bacterial biofilms. Although its function in bacterial biofilms is not yet fully understood, alginate has been shown to play a role in bacterial pathogenesis and epiphytism (Halverson, 2009). Alginate is a linear polysaccharide composed of α-L-guluronate (G) and β-D-mannuronate (M) subunits. The G and M subunits are linked *via* 1,4-glycosidic bonds and are arranged in the form of polyM, polyG, and polyMG blocks (**Table 2**) (Renn, 1997; Draget et al., 2005; Hashimoto et al., 2009).

Ulvan is a water-soluble cell wall polysaccharide, which is found in green seaweeds, such as *Ulva* and *Enteromorpha* sp. (Jiao et al., 2011). The dry weight percentage of Ulvan varies between 8 and 29% (Lahaye and Robic, 2007; Robic et al., 2008). It is comprised of sulfated rhamnose, glucuronic acid, iduronic acid, xylose, and sulfated xylose. The ratio and linkage of ulvan constituent monosaccharides vary among species [refer to Lahaye and Robic (2007)]. Nevertheless, ulvan general structure is comprised of repeating disaccharide units that are linked by α-(1→) glycosidic bond. Ulvan disaccharides are formed from [glucuronic acid – sulfated rhamnose], [iduronic acid – sulfated rhamnose], [xylose – sulfated rhamnose], and

[sulfated xylose – sulfated rhamnose] monomers, which are linked by either α- or β-(1→4) glycosidic bonds (**Table 2**) (Jiao et al., 2011; Yanagisawa et al., 2013; Collén et al., 2014).

Fucoidan is a cell wall polysaccharide that is found in the members of family Laminariaceae of brown algae. The dry weight percentage of fucoidan ranges normally from 5 to 20%. Fucoidan structure is heterogeneous and varies among algal species. Nevertheless, it displays a general backbone consisting of predominantly α-(1→2)-linked sulfated L-fucose units and, in smaller amounts, α-(1→3)- and α-(1→4)-linked sulfated fucose units (Percival, 1979; Davis et al., 2003; Li et al., 2008; Ale and Meyer, 2013).

ENZYMATIC CONVERSION OF ALGAL SUGARS INTO BIOETHANOL

The process of bioethanol production from algal polysaccharides consists of three major steps: biomass pretreatment, enzymatic hydrolysis of algal polysaccharides, and fermentation of sugar monomers to ethanol. The pretreatment step disrupts algal cell and releases intracellular sugars. Additionally, the pretreatment step reduces algal cell wall crystallinity making its polysaccharides accessible for enzymatic hydrolysis. Algal biomass is pretreated by physical, chemical, or biological methods. This review will not cover the pretreatment step since it has been covered in detail by other reviews [refer to Harun et al. (2014)].

Subsequently, algal biomass is degraded by lytic enzymes into simple sugars and uronic acid monomers. Hydrolysis of α-glucans (i.e., starch and floridean starch), β-glucans (i.e., laminarin, chrysolaminarin, and cellulose), and galactans (i.e., agarose and carrageenan) releases several fermentable sugars, which enter glycolysis and are fermented into ethanol. In contrast, industrial production of ethanol from alginate requires engineering of alginate degradation, uptake, and metabolic pathways (See Fermentation of Algal Simple Sugars and Uronic Acids to Bioethanol). Simple sugars, which are released during the conversion steps, are required for the organism's growth, while ethanol is produced as a byproduct of the fermentation process.

Conversion of Starch-Type Polysaccharides to Glucose
Conversion of Starch to Glucose

Starch enzymatic hydrolysis is complex and requires the activity of several enzymes, such as α-amylase, isoamylase, pullulanase, β-amylase, and glucoamylase. Isoamylases (E.C.3.2.1.68) and pullulanases (EC 3.2.1.41) debranch amylopectin into amylose *via* α-(1→6) glycosidic bond cleavage. The endo-acting α-amylases (EC 3.2.1.1) hydrolyze amylose and amylopectin α-(1→4) glycosidic bonds randomly, producing glucose, maltose (two glucose residues), and maltodextrins of 10–20 glucose units. The exo-acting β-amylases (EC 3.2.1.2) hydrolyze amylose from the non-reducing ends to produce predominantly maltose, which is cleaved into two glucose residues by glucoamylases (EC 3.2.1.3) (**Figure 2A**) [reviewed by Van Zyl et al. (2012)].

At the industrial scale, the enzymatic hydrolysis of starch is carried out at elevated temperatures, and it is divided into

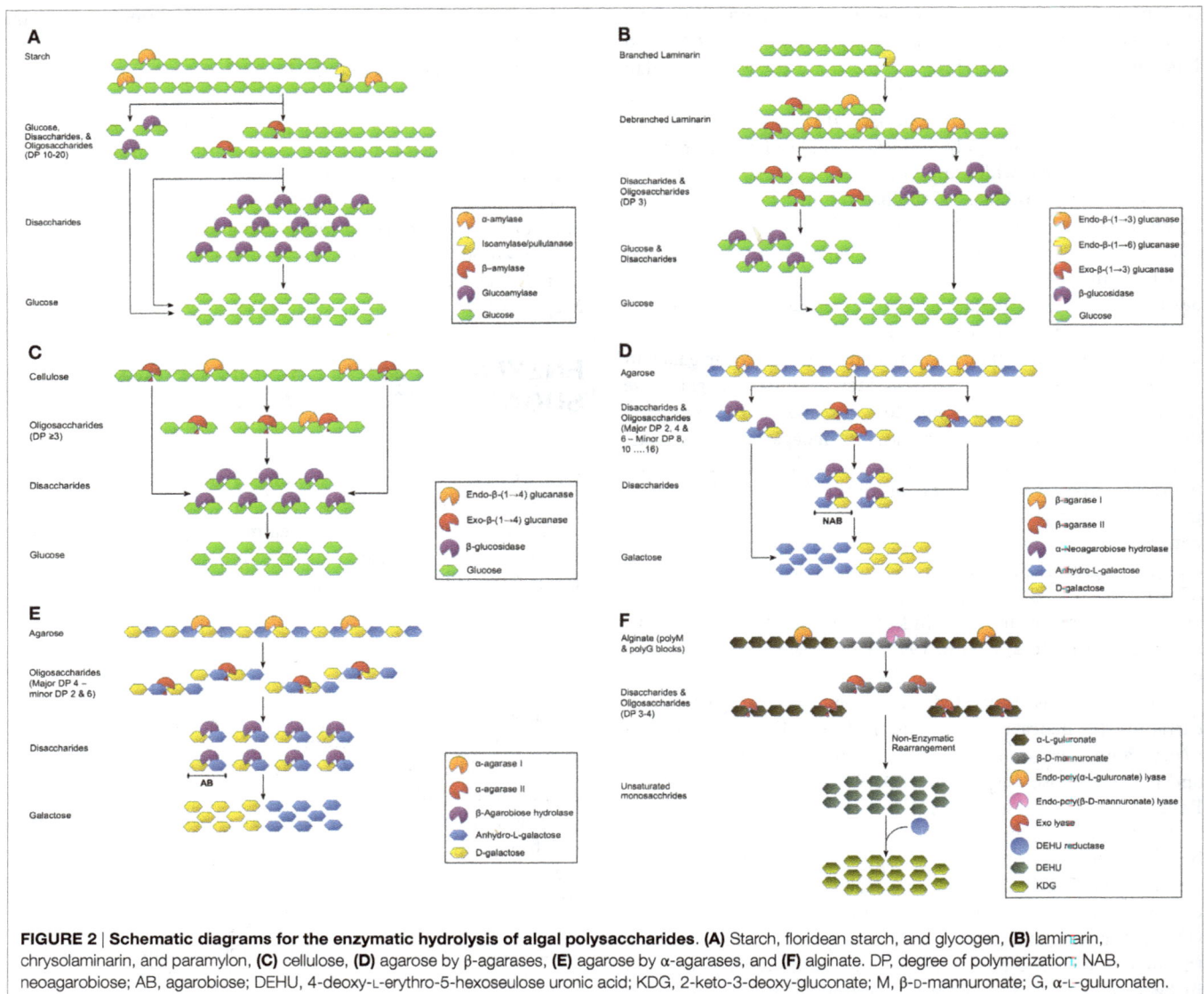

FIGURE 2 | Schematic diagrams for the enzymatic hydrolysis of algal polysaccharides. (A) Starch, floridean starch, and glycogen, **(B)** laminarin, chrysolaminarin, and paramylon, **(C)** cellulose, **(D)** agarose by β-agarases, **(E)** agarose by α-agarases, and **(F)** alginate. DP, degree of polymerization; NAB, neoagarobiose; AB, agarobiose; DEHU, 4-deoxy-L-erythro-5-hexoseulose uronic acid; KDG, 2-keto-3-deoxy-gluconate; M, β-D-mannuronate; G, α-L-guluronaten.

three steps: starch gelatinization, liquefaction, and saccharification. Starch gelatinization and liquefaction involves breaking down starch granules into a gelatinized suspension at 105°C followed by converting the gelatinized starch into oligosaccharides at 95°C, respectively, using thermostable α-amylase. Saccharification is the conversion process of oligosaccharides to primarily glucose along with other disaccharides (i.e., maltose and isomaltose) at very low concentrations. During this process, glucoamylase and isoamylase are added to hydrolyze α-(1→4) as well as α-(1→6) glycosidic bonds at 65°C (Kearsley and Dziedzic, 1995; Kulp and Ponte, 2000; Ratnayake and Jackson, 2009).

Hydrolysis of algal starch is similar to plant starch degradation steps and requires the same amylolytic enzymes [for reviews on microbial amylases, refer to Nielsen and Borchert (2000), De Souza and de Oliveira Magalhães (2010), Naidu (2013), Polizeli et al. (2013), and Saranraj and Stella (2013)].

Conversion of Floridean to Glucose

Similar to starch, gelatinized floridean starch is easier to degrade than granules and requires the same procedure of starch degradation described above (Yu et al., 2002).

Conversion of Glycogen to Glucose

Glycogen is similar in structure to amylopectin, and therefore the hydrolysis of glycogen requires the same amylolytic enzymes, which are used to breakdown starch. In fact, two reports found that only two enzymes, i.e., α-amylase and glucoamylase, are sufficient to partially hydrolyze glycogen and allow subsequent fermentation to ethanol. However, neither the synergistic effect of these isoamylases on glycogen hydrolysis nor the final products of hydrolysis were analyzed in these studies. Therefore, the amount of glycogen degradation achieved using a combination of only α-amylase and glucoamylase is unknown. The hydrolysis products are fermented to ethanol using budding yeast. Additionally, extraction of glycogen from cyanobacteria is simpler than algae as it only requires breaking down the cyanobacterial cell wall with lysozyme (Aikawa et al., 2013; Möllers et al., 2014).

Conversion of Laminarin-Type Polysaccharides to Glucose

Hydrolysis of laminarin and chrysolaminarin is catalyzed by the action of four enzymes: endo-β-(1→6) glucanases (EC 3.2.1.75), endo-β-(1→3) glucanases (EC 3.2.1.39), exo-β-(1→3) glucanases (EC 3.2.1.58), and β-glucosidases (EC 3.2.1.21). Endo-β-(1→6) glucanases debranch laminarin at the β-(1→6) linkages. Exo-β-(1→3) glucanases and endo-β-(1→3) glucanases degrade linear laminarin into laminaritriose and laminaribiose. Subsequently, laminarin oligosaccharides are lysed into glucose by β-glucosidases (**Figure 2B**). Degradation of the M-chain type of laminarin (i.e., with mannitol attached to the reducing end) generates small quantities of a mannitol-containing β-D-glucan (or 1-O-β-D-glucosyl-D-mannitol) (Chesters and Bull, 1963; Martin et al., 2007; Michel et al., 2010).

Paramylon is another laminarin-like starch; however, it is a linear polysaccharide and lacks β-(1→6)-linked branches. Therefore, degradation of paramylon to glucose is simpler than laminarin and requires the same enzymatic arsenal, with the exception of the debranching endo-β-(1→6) glucanases (Takeda et al., 2015).

Conversion of Mannitol to Fructose and/or Mannose

Mannitol is readily dissolved from algal biomass, and therefore conversion of mannitol to ethanol requires no pretreatment steps, which simplify bioethanol production process (Wang et al., 2013). In order to be fermented, mannitol must be converted to fructose-6-phosphate (fructose-6-P). Mannitol metabolic pathways vary among organisms. For example, non-lactic-acid bacteria and homofermentative lactic-acid bacteria assimilate mannitol *via* phosphoenolpyruvate-dependent mannitol phosphotransferase system to mannitol-1-phosphate (mannitol-1-P), which is dehydrogenated to fructose-6-P by mannitol-1-phosphate dehydrogenase (M1PDH, EC 1.1.1.17). In fungi, algae, and heterofermentative lactic-acid bacteria, mannitol is dehydrogenated by mannitol 2-dehydrogenase [M2DH, EC 1.1.1.67 (NAD), and 1.1.1.138 (NADP)] to fructose, which is phosphorylated to fructose-6-P by hexokinase (EC 2.7.1.1). In plants, mannitol is converted by mannitol 1-dehydrogenase (M1DH, EC 1.1.1.255) to mannose, which is phosphorylated to mannose-6-phosphate (mannose-6-P) by hexokinase (EC 2.7.1.1). Mannose-6-P is then converted to fructose-6-P by mannose-6-P isomerase (EC 5.3.1.8) [see Figure 1 in Iwamoto and Shiraiwa (2005)]. Finally, fructose-6-P is fermented to ethanol by ethanogenic microorganisms (See Fermentation of Algal Simple Sugars and Uronic Acids to Bioethanol).

Conversion of Cellulose to Glucose

Similar to plants, the commercial production of biofuel from algal cellulose remains a challenge since cellulose is embedded in a multilayered intricate rigid matrix of sugars and polymers, which protect cellulose from enzymatic degradation (Domozych et al.,

2012). However, the hydrolysis process of algal cell wall to ethanol remains simpler than lignocellulosic biomass because algal cell wall lacks (or contains low amounts of) lignin. Nevertheless, algal pretreatment is required to remove non-cellulosic cell wall matrix and reduce algal cellulose crystallinity making it accessible for enzymatic hydrolysis. Subsequently, cellulose is hydrolyzed by cellulolytic enzymes into glucose, which is fermented into ethanol.

Cellulose hydrolysis requires the action of several cellulolytic enzymes that cleave β-1,4-glycosidic linkages synergistically. These cellulases are endoglucanase, exoglucanase, and β-glucosidase. Endoglucanase (EC 3.2.1.4) cleaves randomly the internal β-1,4-glycosidic bonds. Exoglucanase (EC 3.2.1.91) hydrolyzes the β-1,4-glycosidic linkages of cellulose from the ends of the cellulose chain. Cellulose hydrolysis by endoglucanase and exoglucanase releases D-glucose dimer, β-cellobiose. Once β-cellobiose molecules are released, a third type of cellulases, i.e., β-glucosidase (EC 3.2.1.21) attacks β-cellobiose disaccharides and cleaves them into two glucose molecules (**Figure 2C**) (Pérez et al., 2002; Kuhad et al., 2011; Gupta et al., 2013).

Hydrolysis of Agarose to Galactose

The enzymatic breakdown of agarose is carried out by α-agarases (EC 3.2.1.158) and β-agarases (EC 3.2.1.81), which hydrolyze agarose α-1,3 and β-1,4 linkages and produce agarobiose and neoagarobiose as final hydrolysis products, respectively (**Table 1**). Agarobiose and neoagarobiose are further hydrolyzed by β-agarobiose hydrolase (EC is not available) and α-neoagarobiose hydrolase (EC is not available), respectively, to produce D-galactose and anhydro-L-galactose (**Figures 2D,E**) (Fu and Kim, 2010; Chi et al., 2012). With exception of α-agarases secreted by *Alteromonas agaralytica* and *Thalassomonas* sp. JAMB-A33, all characterized agarases exhibit β-agarolytic activities (Michel et al., 2006; Fu and Kim, 2010; Chi et al., 2012).

However, enzymatic hydrolysis of agarose yields low concentrations of galactose due to low solubility of agarose (Yun et al., 2015). To increase the yield, an agarose acid and heat pretreatment step (also known as chemical liquefaction) has been introduced preceding the enzymatic hydrolysis step. The chemical liquefaction process utilizes the fact that α-1,3 linkages are hydrolyzed by acid and heat pretreatment. Such chemical liquefaction process generates slightly large (degree of polymerization >12), water soluble agarooligosaccharides, which can be quickly degraded into neoagarobiose [see Figure 1 in Kim et al. (2012)].

Conversion of Carrageenan to Galactose

The degradation of carrageenan is one of the least studied among the major cell wall carbohydrates in algae. Carrageenolytic enzymes are not common among microbes. Only few microbes have been reported to excrete carrageenases. The majority of these microbes are marine bacteria (Michel et al., 2006).

Conversion of Alginate to Uronic Acid Monomers

The degradation of alginate into unsaturated alginate oligomers (degree of polymerization 2–4) is carried out by endo-alginate lyases. These enzymes are classified as poly(β-D-mannuronate) lyase (EC 4.2.2.3) and poly(α-L-guluronate) lyase (EC 4.2.2.11) based on their ability to hydrolyze alginate at the poly β-D-mannuronate (polyM) blocks or poly α-L-guluronate (polyG) blocks, respectively [reviewed by Hashimoto et al. (2009) and Wong et al. (2000)]. In the next step, alginate oligomers are cleaved into unsaturated monosaccharides [4-deoxy-α-L-erythro-hex-4-enopyranuronate (DEHEP)] by exo-acting oligoalginate lyases (EC 4.2.2.26). Subsequently, DEHEP is non-enzymatically rearranged into 4-deoxy-L-erythro-5-hexoseulose uronic acid (DEHU), which is reduced to 2-keto-3-deoxy-gluconate (KDG) by DEHU reductase (EC is not available) (**Figure 2F**) (Wargacki et al., 2012).

Degradation of Ulvans and Fucoidan

The enzymatic hydrolysis of ulvan and fucoidan has been studied less intensively than other polysaccharides, due to several reasons. First, both sugars can be easily degraded into monomeric sugars by acid treatment (Davis et al., 2003; Feng et al., 2011). Second, ulvan and fucoidan are species-specific and family-specific cell wall polysaccharides, respectively, and therefore only few microbes display activities against ulvan (Lahaye et al., 1997; Delattre et al., 2005; Nyvall Collén et al., 2011; Jung et al., 2013) and fucoidan [cited by Silchenko et al. (2013)].

FERMENTATION OF ALGAL SIMPLE SUGARS AND URONIC ACIDS TO BIOETHANOL

The hydrolysis of major algal polysaccharides releases several simple sugars, such as glucose, mannose, fructose, galactose, and uronic acids. These monomers are fermented to produce ethanol. Simple sugars are readily fermented to ethanol using microbial wild-type strains. In contrast, fermentation of uronic acid monomers requires genetically engineered microbes that can hydrolyze alginate to KDG and ferment it to ethanol.

Microbial Fermentation of Glucose, Galactose, Fructose, and Mannose to Ethanol

The classical budding yeast *Saccharomyces cerevisiae* is the most commonly used microbe for fermenting sugars to bioethanol. Additionally, the Gram-negative rod-shaped bacterium *Zymomonas mobilis* is used for fermentation, but to a lesser extent than the budding yeast.

Owing to their diversity, several metabolic pathways are required to convert algal sugars to ethanol. While glucose enters glycolysis directly, galactose must be converted to glucose 6-phosphate (glucose-6-P) *via* the Leloir pathway before entering glycolysis. Similarly, glucose isomers, such as mannose and fructose, are converted to fructose-6-phosphate (fructose-6-P), which is further metabolized through glycolysis. The conversion of fructose to fructose-6-P is simple and requires one enzyme (Hexokinase, EC 2.7.1.1). In contrast, mannose must

be first phosphorylated by hexokinase (EC 2.7.1.1) to mannose-6-phosphate (mannose-6-P), then isomerized to fructose-6-P by mannose-6-P isomerase (EC 5.3.1.8) [refer to Figure 2 in Van Maris et al. (2006) and Figure 1.2 in Zamora (2009)].

Once phosphorylated sugars enter glycolysis, they are metabolized to pyruvate. The major microbial glycolysis pathways are the Embden–Meyerhof–Parnas (EMP) and the Entner–Doudoroff (ED) pathway, depending on the microorganism [reviewed by Wolfe (2015)]. While *S. cerevisiae* utilizes the EMP pathway for metabolizing glucose, ED pathway is the common pathway for glucose metabolism in *Z. mobilis*. The EMP and ED pathways are divided into two multistep stages. In the first stage, glucose is converted to glyceraldehyde-3-phosphate (glyceraldehyde-3-P), which is further metabolized to pyruvate in the second stage (**Figures 3A,B**).

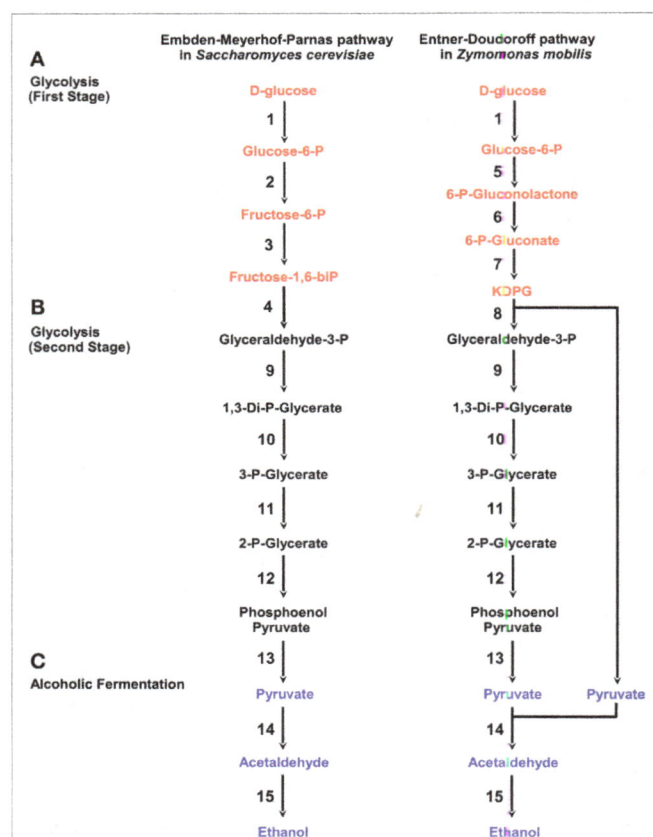

FIGURE 3 | The production of ethanol from glucose by Embden–Meyerhof–Parnas pathway in *Saccharomyces cerevisiae* and Entner–Doudoroff pathway in *Zymomonas mobilis*. The catalytic enzymes are denoted with numbers. The catalytic enzymes are (1) hexokinase, (2) Glucose-6-P isomerase, (3) 6-phosphofructokinase, (4) fructose-biP aldolase, (5) glucose-6-P dehydrogenase (NADP+), (6) 6-phosphogluconolactonase, (7) phosphogluconate dehydratase, (8) KDPG aldolase, (9) glyceraldehyde 3-P dehydrogenase (EC 1.2.1.12), (10) phosphoglycerate kinase (EC 2.7.2.3), (11) phosphoglycerate mutase (EC 5.4.2.11), (12) phosphopyruvate hydratase (enolase; EC 4.2.1.11), (13) pyruvate kinase (EC 2.7.1.40), (14) pyruvate decarboxylase, and (15) alcohol dehydrogenase. **(A)** Glycolysis (first stage). **(B)** Glycolysis (second stage). **(C)** Alcoholic fermentation.

In the first stage of the EMP pathway, glucose is phosphorylated by hexokinase (EC 2.7.1.1) to glucose-6-P, which is transformed into its keto isomer fructose-6-P by glucose-6-P isomerase (EC 5.3.1.9). In the next step, the enzyme 6-phosphofructokinase (EC 2.7.1.11) phosphorylates fructose-6-P to fructose-1,6-biphosphate, which is converted to glyceraldehyde-3-P by fructose-bisphosphate aldolase (EC 4.1.2.13) (**Figure 3**) (Flamholz et al., 2013). The first stage of the ED pathway begins with phosphorylation of glucose by hexokinase (EC 2.7.1.1) to glucose-6-P, which is oxidized to 6-phosphogluconolactone by glucose-6-P dehydrogenase (NADP⁺) (EC 1.1.1.49). Once oxidized, 6-phosphogluconolactone is hydrolyzed by the enzyme 6-phosphogluconolactonase (EC 3.1.1.31) to 6-phosphogluconate, which is dehydrated to 2-keto-3-deoxy-6-phosphogluconate (KDPG) by a phosphogluconate dehydratase (EC 4.2.1.12). Finally, KDPG is converted to pyruvate and glyceraldehyde-3-P by KDPG aldolase (EC 4.1.2.14) (**Figure 3A**) (Flamholz et al., 2013; Spaans et al., 2015).

In contrast to the first stage of EMP and ED pathways, the second stage of these two pathways is identical. During the second stage, glyceraldehyde-3-P is converted to pyruvate (summarized in **Figure 3B**). Following glycolysis, pyruvate is converted to ethanol primarily by a two-step alcoholic fermentation (**Figure 3C**). In the first step, pyruvate is converted by pyruvate decarboxylase (EC 4.1.1.1) to acetaldehyde, which is reduced to ethanol by alcohol dehydrogenase (EC 1.1.1.1).

Microbial Fermentation of 3,6-Anhydro-Galactoses to Ethanol

In addition to galactose, the enzymatic hydrolysis of agarose and carrageenan produces the non-fermentable sugars, 3,6-anhydro-L-galactose (L-AnG) and 3,6-anhydro-D-galactose (D-AnG), respectively. To increase bioethanol yield from algal biomass, L-AnG and D-AnG must be converted into fermentable sugar. Metabolic pathways for L-AnG and D-AnG are only common in agar- and carrageenan-degrading microorganisms. In these organisms, L-AnG and D-AnG are converted to 2-keto-3-deoxy-6-phospho-D-galactonate (D-KDPGal) through six and four enzyme-catalyzed reactions, respectively [see Figure 5 in Lee et al. (2014, 2016)]. D-KDPGal enters the DeLey–Doudoroff pathway, where it is converted to the fermentable sugars glyceraldehyde-3-phosphate and pyruvate (Lee et al., 2014, 2016). Glyceraldehyde-3-phosphate is then converted by glycolysis to pyruvate, which is metabolized to ethanol by alcoholic fermentation (**Figures 3B,C**).

Engineered Microbes to Metabolize Uronic Acid Monomers to Ethanol

Several microorganisms, which can metabolize alginate, have been identified [reviewed by Wong et al. (2000)]. In these microbes, the hydrolysis of alginate results in DEHU. To ferment it to ethanol, DEHU must be first reduced to KDG by DEHU reductase (EC is not available) (Wargacki et al., 2012). Then, KDG is phosphorylated to KDPG, i.e., KDPG by KDG kinase (EC 2.7.1.45). Finally, KDPG enters the ED pathway, where

it is cleaved by KDPG aldolase (EC 4.1.2.14) to pyruvate and glyceraldehyde-3-phosphate and fermented to ethanol [refer to Supplementary Figure 1 in Enquist-Newman et al. (2014)].

However, the commercial production of ethanol from alginate is challenged by the lack of robust microorganisms that can simultaneously digest, metabolize, and ferment alginate to ethanol at the industrial level. While alginolytic microbes lack the robustness for the production of ethanol at large scale, major ethanologenic microbes, such as *S. cerevisiae*, *Z. mobilis*, and *E. coli*, cannot degrade and metabolize alginate. To overcome this challenge, strains that can simultaneously degrade, metabolize, and ferment alginate to ethanol were engineered. The molecular engineering of alginate and mannitol metabolic pathways in *S. cerevisiae* results in a strain that can simultaneously ferment up to 83% of brown algae theoretical total sugars (i.e., alginate, mannitol, and glucose) to ethanol. The strain, which expresses the DEHU transporter gene from the marine fungus *Asteromyces cruciatus* and alginate metabolism genes from bacterial origin, is capable of degrading alginate to uronic acid monomers. These monomers are then converted to ethanol (Enquist-Newman et al., 2014).

Similarly, a plasmid-based *E. coli* strain, which is capable of simultaneous degradation and fermentation of alginate to ethanol, was engineered using alginate metabolic pathway genes from *Vibrio splendidus*. This strain, which can metabolize alginate, mannitol, and laminarin, fermented 80% of the maximum theoretical yield and produces 0.281 weight ethanol/weight dry macroalgae (Wargacki et al., 2012). Ethanol productivity of this strain was further enhanced using recombinase-assisted genome engineering (RAGE). The resulting engineered strain, which integrates alginate metabolism genes into its genome, produced a 330% higher titer than the canonical plasmid-based *E. coli* counterpart after culturing both strains on synthetic seaweed medium for 50 generations (Santos et al., 2013).

CONCLUSION AND FUTURE DIRECTIONS

Algae accumulate high amount of carbohydrates, which can be used to produce bioethanol. However the hydrolysis process of algal polysaccharides requires more enzymatic mixtures than plants. Significant amount of research has been done to decipher the hydrolysis processes of different algal sugars and identify their corresponding hydrolytic enzymes. However, one of the major challenges – within the context of the enzymatic hydrolysis of algal polysaccharides – is to identify hydrolytic enzymes that can breakdown algal-specific sugars, such as agarose, carrageenan, ulvans, and fucoidan. This requires the identification of new marine microorganisms, which thrive growing on algae. Additionally, identification of their enzymatic systems *via* a combination of genomics, transcriptomics, and proteomics approaches will enable high-throughput protein analysis in these microbes. Furthermore, traditional breeding and genetic engineering can be implemented to generate microbial strains that can hydrolyze algal carbohydrates. Protein engineering is another powerful tool to generate enzymes with novel properties. A wider range of pH-, temperature-, and salinity-tolerance levels, as well as broader substrate activities and higher conversion rates are desirable properties for bioethanol production from algae. Owing to their diversity, the fermentation of algal monosaccharides requires several metabolic pathways. Therefore, engineered strains, which can ferment several monosaccharides to ethanol, will simplify the conversion processes. Indeed, the best situation is to generate microbial strains, which can simultaneously hydrolyze and ferment total algal sugars to ethanol.

AUTHOR CONTRIBUTIONS

QA is the main author. JF and BN contributed substantially to the work reported.

FUNDING

The study was funded by U.S. Department of Energy; prime DOE award: DE-SC0006838.

REFERENCES

Aikawa, S., Joseph, A., Yamada, R., Izumi, Y., Yamagishi, T., Matsuda, F., et al. (2013). Direct conversion of Spirulina to ethanol without pretreatment or enzymatic hydrolysis processes. *Energy Environ. Sci.* 6, 1844–1849. doi:10.1039/C3EE40305J

Ale, M. T., and Meyer, A. S. (2013). Fucoidans from brown seaweeds: an update on structures, extraction techniques and use of enzymes as tools for structural elucidation. *RSC Adv.* 3, 8131–8141. doi:10.1039/C3RA23373A

Avigad, G., and Dey, P. M. (1997). "4 – carbohydrate metabolism: storage carbohydrates," in *Plant Biochemistry*, eds P. M. Dey and J. B. Harborne (London: Academic Press), 143–204.

Badger, P. C. (2002). "Ethanol from cellulose: a general review," in *Trends in New Crops and New Uses*. Proceedings of the Fifth National Symposium, eds J. Janick and A. Whipkey (Alexandria, VA: ASHS Press), 17–21.

Ball, S., Colleoni, C., Cenci, U., Raj, J. N., and Tirtiaux, C. (2011). The evolution of glycogen and starch metabolism in eukaryotes gives molecular clues to understand the establishment of plastid endosymbiosis. *J. Exp. Bot.* 62, 1775–1801. doi:10.1093/jxb/erq411

Barsanti, L., Passarelli, V., Evangelista, V., Frassanito, A. M., and Gualtieri, P. (2011). Chemistry, physico-chemistry and applications linked to biological activities of β-glucans. *Nat. Prod. Rep.* 28, 457–466. doi:10.1039/C0NP00018C

Beattie, A., Hirst, E. L., and Percival, E. (1961). Studies on the metabolism of the Chrysophyceae. Comparative structural investigations on leucosin (chrysolaminarin) separated from diatoms and laminarin from the brown algae. *Biochem. J.* 79, 531–537. doi:10.1042/bj0790531

Behera, S., Singh, R., Arora, R., Sharma, N. K., Shukla, M., and Kumar, S. (2015). Scope of algae as third generation biofuels. *Front. Bioeng. Biotechnol.* 2:90. doi:10.3389/fbioe.2014.00090

Benemann, J. (2013). Microalgae for biofuels and animal feeds. *Energies* 6, 5869–5886. doi:10.3390/en6115869

Borowitzka, M. A. (2013). "Energy from microalgae: a short history," in *Algae for Biofuels and Energy Developments in Applied Phycology*, eds M. A. Borowitzka and N. R. Moheimani (Netherlands: Springer), 1–15.

Brennan, L., and Owende, P. (2010). Biofuels from microalgae – a review of technologies for production, processing, and extractions of biofuels and co-products. *Renew. Sustain. Energy Rev.* 14, 557–577. doi:10.1016/j.rser.2009.10.009

Brodie, J., and Lewis, J. (2007). *Unravelling the Algae: The Past, Present, and Future of Algal Systematics*. Boca Raton, FL: CRC Press.

Brown, R. M., and Saxena, I. M. (2000). Cellulose biosynthesis: a model for understanding the assembly of biopolymers. *Plant Physiol. Biochem.* 38, 57–67. doi:10.1016/S0981-9428(00)00168-6

Bruton, T., Lyons, H., Lerat, Y., Stanley, M., and Rasmussen, M. B. (2009). *A Review of the Potential of Marine Algae as a Source of Biofuel in Ireland*. Dublin, Ireland: Sustainable Energy Ireland.

Busi, M. V., Barchiesi, J., Martín, M., and Gomez-Casati, D. F. (2014). Starch metabolism in green algae. *Starch - Stärke* 66, 28–40. doi:10.1002/star.201200211

Carlsson, A. S., van Beilen, J. B., Möller, R., and Clayton, D. (2007). *Micro- and Macro-Algae: Utility for Industrial Applications: Outputs from the EPOBIO Project*. Chippenham, England: CPL Press.

Chen, C.-Y., Zhao, X.-Q., Yen, H.-W., Ho, S.-H., Cheng, C.-L., Lee, D.-J., et al. (2013). Microalgae-based carbohydrates for biofuel production. *Biochem. Eng. J.* 78, 1–10. doi:10.1016/j.bej.2013.03.006

Chen, P., Min, M., Chen, Y., Wang, L., Li, Y., Chen, Q., et al. (2010). Review of biological and engineering aspects of algae to fuels approach. *Int. J. Agric. Biol. Eng.* 2, 1–30. doi:10.3965/ijabe.v2i4.200

Chesters, C. G. C., and Bull, A. T. (1963). The enzymic degradation of laminarin. 1. The distribution of laminarinase among micro-organisms. *Biochem. J.* 86, 28–31. doi:10.1042/bj0860028

Chi, W.-J., Chang, Y.-K., and Hong, S.-K. (2012). Agar degradation by microorganisms and agar-degrading enzymes. *Appl. Microbiol. Biotechnol.* 94, 917–930. doi:10.1007/s00253-012-4023-2

Chisti, Y. (2016). "Large-scale production of algal biomass: raceway ponds," in *Algae Biotechnology: Products and Processes*, eds F. Bux and Y. Chisti (Cham: Springer International Publishing), 21–40. doi:10.1007/978-3-319-12334-9_2

Ciferri, O. (1983). Spirulina, the edible microorganism. *Microbiol. Rev.* 47, 551–578.

Collén, P. N., Jeudy, A., Sassi, J.-F., Groisillier, A., Czjzek, M., Coutinho, P. M., et al. (2014). A novel unsaturated β-glucuronyl hydrolase involved in ulvan degradation unveils the versatility of stereochemistry requirements in family GH105. *J. Biol. Chem.* 289, 6199–6211. doi:10.1074/jbc.M113.537480

Dahiya, A. (2015). *Bioenergy: Biomass to Biofuels*, 1st Edn. Boston: Academic Press.

Davis, T. A., Volesky, B., and Mucci, A. (2003). A review of the biochemistry of heavy metal biosorption by brown algae. *Water Res.* 37, 4311–4330. doi:10.1016/S0043-1354(03)00293-8

De Ruiter, G. A., and Rudolph, B. (1997). Carrageenan biotechnology. *Trends Food Sci. Technol.* 8, 389–395. doi:10.1016/S0924-2244(97)01091-1

De Souza, P. M., and de Oliveira Magalhães, P. (2010). Application of microbial α-amylase in industry – a review. *Braz. J. Microbiol.* 41, 850–861. doi:10.1590/S1517-83822010000400004

Delattre, C., Fenoradosoa, T. A., and Michaud, P. (2011). Galactans: an overview of their most important sourcing and applications as natural polysaccharides. *Braz. Arch. Biol. Technol.* 54, 1075–1092. doi:10.1590/S1516-89132011000600002

Delattre, C., Michaud, P., Keller, C., Elboutachfaiti, R., Beven, L., Courtois, B., et al. (2005). Purification and characterization of a novel glucuronan lyase from *Trichoderma* sp. GL2. *Appl. Microbiol. Biotechnol.* 70, 437–443. doi:10.1007/s00253-005-0077-8

Deschamps, P., Haferkamp, I., d'Hulst, C., Neuhaus, H. E., and Ball, S. G. (2008). The relocation of starch metabolism to chloroplasts: when, why and how. *Trends Plant Sci.* 13, 574–582. doi:10.1016/j.tplants.2008.08.009

Dexter, J., Armshaw, P., Sheahan, C., and Pembroke, J. T. (2015). The state of autotrophic ethanol production in Cyanobacteria. *J. Appl. Microbiol.* 119, 11–24. doi:10.1111/jam.12821

Domozych, D. S., Ciancia, M., Fangel, J. U., Mikkelsen, M. D., Ulvskov, P., and Willats, W. G. T. (2012). The cell walls of green algae: a journey through evolution and diversity. *Front. Plant Sci.* 3:82. doi:10.3389/fpls.2012.00082

Dornish, M., and Rauh, F. (2006). "Alginate," in *An Introduction to Biomaterials*, eds S. A. Guelcher and J. O. Hollinger (Boca Raton, FL: CRC Press), 261–272.

Draget, K. I., Smidsrød, O., and Skjåk-Braek, G. (2005). Alginates from algae. *Biopolymers* 6. doi:10.1002/3527600035.bpol6008

Enquist-Newman, M., Faust, A. M. E., Bravo, D. D., Santos, C. N. S., Raisner, R. M., Hanel, A., et al. (2014). Efficient ethanol production from brown macroalgae sugars by a synthetic yeast platform. *Nature* 505, 239–243. doi:10.1038/nature12771

Feng, D., Liu, H., Li, F., Jiang, P., and Qin, S. (2011). Optimization of dilute acid hydrolysis of Enteromorpha. *Chin. J. Oceanol. Limnol.* 29, 1243–1248. doi:10.1007/s00343-011-0298-x

Fengel, D., and Gert, W. (1989). *Wood: Chemistry, Ultrastructure, Reactions*. Berlin, Germany: Walter de Gruyter & Co.

Flamholz, A., Noor, E., Bar-Even, A., Liebermeister, W., and Milo, R. (2013). Glycolytic strategy as a tradeoff between energy yield and protein cost. *Proc. Natl. Acad. Sci. U.S.A.* 110, 10039–10044. doi:10.1073/pnas.1215283110

Fu, X. T., and Kim, S. M. (2010). Agarase: review of major sources, categories, purification method, enzyme characteristics and applications. *Mar. Drugs* 8, 200–218. doi:10.3390/md8010200

Grant Reid, J. S. (1997). "5 – carbohydrate metabolism: structural carbohydrates," in *Plant Biochemistry*, eds P. M. Dey and J. B. Harborne (London: Academic Press), 205–236.

Groom, M. J., Gray, E. M., and Townsend, P. A. (2008). Biofuels and biodiversity: principles for creating better policies for biofuel production. *Conserv. Biol.* 22, 602–609. doi:10.1111/j.1523-1739.2007.00879.x

Gupta, R., Mehta, C., Deswal, D., Sharma, S., Jain, K. K., Kuhad, R. C., et al. (2013). "Cellulases and their biotechnological applications," in *Biotechnology for Environmental Management and Resource Recovery*, eds R. C. Kuhad and A. Singh (India: Springer), 89–106.

Habib, M. A. B., Parvin, M., Huntington, T. C., and Hasan, M. R. (2008). *A Review on Culture, Production and Use of Spirulina as Food for Humans and Feeds for Domestic Animals and Fish*. Rome: FAO.

Halverson, L. J. (2009). "Role of alginate in bacterial biofilms," in *Alginates: Biology and Applications*, ed. B. H. A. Rehm (Berlin, Heidelberg: Springer), 135–151.

Hannon, M., Gimpel, J., Tran, M., Rasala, B., and Mayfield, S. (2010). Biofuels from algae: challenges and potential. *Biofuels* 1, 763–784. doi:10.4155/bfs.10.44

Harun, R., Yip, J. W. S., Thiruvenkadam, S., Ghani, W. A. W. A. K., Cherrington, T., and Danquah, M. K. (2014). Algal biomass conversion to bioethanol – a step-by-step assessment. *Biotechnol. J.* 9, 73–86. doi:10.1002/biot.201200353

Hashimoto, W., Maruyama, Y., Itoh, T., Mikami, B., and Murata, K. (2009). "Bacterial system for alginate uptake and degradation," in *Alginates: Biology and Applications*, ed. B. H. A. Rehm (Berlin, Heidelberg: Springer), 73–94.

Hughes, A. D., Kelly, M. S., Black, K. D., and Stanley, M. S. (2012). Biogas from macroalgae: is it time to revisit the idea? *Biotechnol. Biofuels* 5, 86. doi:10.1186/1754-6834-5-86

Iwamoto, K., and Shiraiwa, Y. (2005). Salt-regulated mannitol metabolism in algae. *Mar. Biotechnol.* 7, 407–415. doi:10.1007/s10126-005-0029-4

Jiao, G., Yu, G., Zhang, J., and Ewart, H. S. (2011). Chemical structures and bioactivities of sulfated polysaccharides from marine algae. *Mar. Drugs* 9, 196–223. doi:10.3390/md9020196

John, R. P., and Anisha, G. S. (2012). "Macroalgae and their potential for biofuel," in *Plant Sciences Reviews 2011*, ed. D. Hemming (Wallingford, UK: CABI), 151–162.

John, R. P., Anisha, G. S., Nampoothiri, K. M., and Pandey, A. (2011). Micro and macroalgal biomass: a renewable source for bioethanol. *Bioresour. Technol.* 102, 186–193. doi:10.1016/j.biortech.2010.06.139

Jung, K. A., Lim, S.-R., Kim, Y., and Park, J. M. (2013). Potentials of macroalgae as feedstocks for biorefinery. *Bioresour. Technol.* 135, 182–190. doi:10.1016/j.biortech.2012.10.025

Kadam, S. U., Tiwari, B. K., and O'Donnell, C. P. (2015). Extraction, structure and biofunctional activities of laminarin from brown algae. *Int. J. Food Sci. Technol.* 50, 24–31. doi:10.1111/ijfs.12692

Kearsley, M. W., and Dziedzic, S. Z. (1995). *Handbook of Starch Hydrolysis Products and Their Derivatives*. New York: Springer.

Kim, H. T., Lee, S., Kim, K. H., and Choi, I.-G. (2012). The complete enzymatic saccharification of agarose and its application to simultaneous saccharification and fermentation of agarose for ethanol production. *Bioresour. Technol.* 107, 301–306. doi:10.1016/j.biortech.2011.11.120

Kim, S.-K. (2015). *Handbook of Marine Microalgae: Biotechnology Advances*. Boston: Academic Press.

Kuhad, R. C., Gupta, R., and Singh, A. (2011). Microbial cellulases and their industrial applications. *Enzyme Res.* 2011, 280696. doi:10.4061/2011/280696

Kulp, K., and Ponte, J. G. (2000). *Handbook of Cereal Science and Technology*. New York, NY: Marcel Dekker Inc.

Lahaye, M., Brunel, M., and Bonnin, E. (1997). Fine chemical structure analysis of oligosaccharides produced by an ulvan-lyase degradation of the water-soluble cell-wall polysaccharides from *Ulva* sp, (Ulvales, Chlorophyta). *Carbohydr. Res.* 304, 325–333. doi:10.1016/S0008-6215(97)00270-X

Lahaye, M., and Robic, A. (2007). Structure and functional properties of ulvan, a polysaccharide from green seaweeds. *Biomacromolecules* 8, 1765–1774. doi:10.1021/bm061185q

Lee, S. B., Cho, S. J., Kim, J. A., Lee, S. Y., Kim, S. M., and Lim, H. S. (2014). Metabolic pathway of 3,6-anhydro-L-galactose in agar-degrading microorganisms. *Biotechnol. Bioprocess Eng.* 19, 866–878. doi:10.1007/s12257-014-0622-3

Lee, S. B., Kim, J. A., and Lim, H. S. (2016). Metabolic pathway of 3,6-anhydro-D-galactose in carrageenan-degrading microorganisms. *Appl. Microbiol. Biotechnol.* 100, 4109–4121. doi:10.1007/s00253-016-7346-6

Lee, Y.-K. (1997). Commercial production of microalgae in the Asia-Pacific rim. *J. Appl. Phycol.* 9, 403–411. doi:10.1023/A:1007900423275

Li, B., Lu, F., Wei, X., and Zhao, R. (2008). Fucoidan: structure and bioactivity. *Molecules* 13, 1671–1695. doi:10.3390/molecules13081671

Lüning, K., and Pang, S. (2003). Mass cultivation of seaweeds: current aspects and approaches. *J. Appl. Phycol.* 15, 115–119. doi:10.1023/A:1023807503255

Martin, K., McDougall, B. M., McIlroy, S., Chen, J., and Seviour, R. J. (2007). Biochemistry and molecular biology of exocellular fungal β-(1,3)- and β-(1,6)-glucanases. *FEMS Microbiol. Rev.* 31, 168–192. doi:10.1111/j.1574-6976.2006.00055.x

Mata, T. M., Martins, A. A., and Caetano, N. S. (2010). Microalgae for biodiesel production and other applications: a review. *Renew. Sustain. Energy Rev.* 14, 217–232. doi:10.1016/j.rser.2009.07.020

Metting, F. B. (1996). Biodiversity and application of microalgae. *J. Ind. Microbiol.* 17, 477–489. doi:10.1007/BF01574779

Michel, G., Nyval-Collen, P., Barbeyron, T., Czjzek, M., and Helbert, W. (2006). Bioconversion of red seaweed galactans: a focus on bacterial agarases and carrageenases. *Appl. Microbiol. Biotechnol.* 71, 23–33. doi:10.1007/s00253-006-0377-7

Michel, G., Tonon, T., Scornet, D., Cock, J. M., and Kloareg, B. (2010). Central and storage carbon metabolism of the brown alga *Ectocarpus siliculosus*: insights into the origin and evolution of storage carbohydrates in Eukaryotes. *New Phytol.* 188, 67–81. doi:10.1111/j.1469-8137.2010.03345.x

Möllers, K. B., Cannella, D., Jørgensen, H., and Frigaard, N.-U. (2014). Cyanobacterial biomass as carbohydrate and nutrient feedstock for bioethanol production by yeast fermentation. *Biotechnol. Biofuels* 7, 64. doi:10.1186/1754-6834-7-64

Monfils, A. K., Triemer, R. E., and Bellairs, E. F. (2011). Characterization of paramylon morphological diversity in photosynthetic euglenoids (Euglenales, Euglenophyta). *Phycologia* 50, 156–169. doi:10.2216/09-112.1

Moroney, J. V., and Ynalvez, R. A. (2001). Algal photosynthesis. *eLS*. doi:10.1002/9780470015902.a0000322.pub2

Motone, K., Takagi, T., Sasaki, Y., Kuroda, K., and Ueda, M. (2016). Direct ethanol fermentation of the algal storage polysaccharide laminarin with an optimized combination of engineered yeasts. *J. Biotechnol.* 231, 129–135. doi:10.1016/j.jbiotec.2016.06.002

Murphy, F., Devlin, G., Deverell, R., and McDonnell, K. (2013). Biofuel production in Ireland – an APPROACH to 2020 targets with a focus on algal biomass. *Energies* 6, 6391–6412. doi:10.3390/en6126391

Naidu, M. A. (2013). Bacterial amylase a review. *Int. J. Pharm. Biol. Arch.* 4, 274–287.

Naik, S. N., Goud, V. V., Rout, P. K., and Dalai, A. K. (2010). Production of first and second generation biofuels: a comprehensive review. *Renew. Sustain. Energy Rev.* 14, 578–597. doi:10.1016/j.rser.2009.10.003

Nielsen, J. E., and Borchert, T. V. (2000). Protein engineering of bacterial α-amylases. *Biochim. Biophys. Acta* 1543, 253–274. doi:10.1016/S0167-4838(00)00240-5

Nozzi, N. E., Oliver, J. W. K., and Atsumi, S. (2013). Cyanobacteria as a platform for biofuel production. *Front. Bioeng. Biotechnol.* 1:7. doi:10.3389/fbioe.2013.00007

Nyvall Collén, P., Sassi, J.-F., Rogniaux, H., Marfaing, H., and Helbert, W. (2011). Ulvan lyases isolated from the flavobacteria *Persicivirga ulvanivorans* are the first members of a new polysaccharide lyase family. *J. Biol. Chem.* 286, 42063–42071. doi:10.1074/jbc.M111.271825

Ota, A., Kawai, S., Oda, H., Iohara, K., and Murata, K. (2013). Production of ethanol from mannitol by the yeast strain *Saccharomyces paradoxus* NBRC 0259. *J. Biosci. Bioeng.* 116, 327–332. doi:10.1016/j.jbiosc.2013.03.018

Packer, M. (2009). Algal capture of carbon dioxide; biomass generation as a tool for greenhouse gas mitigation with reference to New Zealand energy strategy and policy. *Energy Policy* 37, 3428–3437. doi:10.1016/j.enpol.2008.12.025

Peat, S., and Turvey, J. R. (1965). "Polysaccharides of marine algae," in *Fortschritte der Chemie Organischer Naturstoffe/Progress in the Chemistry of Organic Natural Products/Progrès dans la Chimie des Substances Organiques Naturelles*, ed. L. Zechmeister (Vienna: Springer), 1–45.

Percival, E. (1970). "Algal polysaccharides," in *Algal Polysaccharides The Carbohydrates: Chemistry and Biochemistry*, eds W. Pigman and D. Horton (New York, NY: Academic Press, Inc), 537–568.

Percival, E. (1979). The polysaccharides of green, red and brown seaweeds: their basic structure, biosynthesis and function. *Br. Phycol. J.* 14, 103–117. doi:10.1080/00071617900650121

Pérez, J., Muñoz-Dorado, J., de la Rubia, T., and Martínez, J. (2002). Biodegradation and biological treatments of cellulose, hemicellulose and lignin: an overview. *Int. Microbiol.* 5, 53–63. doi:10.1007/s10123-002-0062-3

Polizeli, M., de, L. T. M., and Rai, M. (2013). *Fungal Enzymes.* New York, NY: CRC Press.

Pruvost, J., Cornet, J.-F., and Pilon, L. (2016). "Large-scale production of algal biomass: photobioreactors," in *Algae Biotechnology: Products and Processes,* eds F. Bux and Y. Chisti (Cham: Springer International Publishing), 41–66. doi:10.1007/978-3-319-12334-9_3

Pu, Y., Hu, F., Huang, F., Davison, B. H., and Ragauskas, A. J. (2013). Assessing the molecular structure basis for biomass recalcitrance during dilute acid and hydrothermal pretreatments. *Biotechnol. Biofuels* 6, 15. doi:10.1186/1754-6834-6-15

Quintana, N., Van der Kooy, F., Van de Rhee, M. D., Voshol, G. P., and Verpoorte, R. (2011). Renewable energy from Cyanobacteria: energy production optimization by metabolic pathway engineering. *Appl. Microbiol. Biotechnol.* 91, 471–490. doi:10.1007/s00253-011-3394-0

Radakovits, R., Jinkerson, R. E., Darzins, A., and Posewitz, M. C. (2010). Genetic engineering of algae for enhanced biofuel production. *Eukaryot. Cell* 9, 486–501. doi:10.1128/EC.00364-09

Rajkumar, R., Yaakob, Z., and Takriff, M. S. (2014). Potential of micro and macro algae for biofuel production: a brief review. *BioResources* 9, 1606–1633. doi:10.15376/biores.9.1.1606-1633

Ramaiah, N. (2006). A review on fungal diseases of algae, marine fishes, shrimps and corals. *Indian J. Mar. Sci.* 35, 380–387.

Ratnayake, W. S., and Jackson, D. S. (2009). Starch gelatinization. *Adv. Food Nutr. Res.* 55, 221–268. doi:10.1016/S1043-4526(08)00405-1

Renn, D. (1997). Biotechnology and the red seaweed polysaccharide industry: status, needs and prospects. *Trends Biotechnol.* 15, 9–14. doi:10.1016/S0167-7799(96)10069-X

Robic, A., Bertrand, D., Sassi, J.-F., Lerat, Y., and Lahaye, M. (2008). Determination of the chemical composition of ulvan, a cell wall polysaccharide from *Ulva* spp. (Ulvales, Chlorophyta) by FT-IR and chemometrics. *J. Appl. Phycol.* 21, 451–456. doi:10.1007/s10811-008-9390-9

Santos, C. N. S., Regitsky, D. D., and Yoshikuni, Y. (2013). Implementation of stable and complex biological systems through recombinase-assisted genome engineering. *Nat. Commun.* 4, 2503. doi:10.1038/ncomms3503

Saranraj, P., and Stella, D. (2013). Fungal amylase – a review. *Int. J. Microbiol. Res.* 4, 203–211. doi:10.5829/idosi.ijmr.2013.4.2.75170

Sarsekeyeva, F., Zayadan, B. K., Usserbaeva, A., Bedbenov, V. S., Sinetova, M. A., and Los, D. A. (2015). Cyanofuels: biofuels from cyanobacteria. Reality and perspectives. *Photosynth. Res.* 125, 329–340. doi:10.1007/s11120-015-0103-3

Satyanarayana, K. G., Mariano, A. B., and Vargas, J. V. C. (2011). A review on microalgae, a versatile source for sustainable energy and materials. *Int. J. Energy Res* 35, 291–311. doi:10.1002/er.1695

Scaife, M. A., Merkx-Jacques, A., Woodhall, D. L., and Armenta, R. E. (2015). Algal biofuels in Canada: status and potential. *Renew. Sustain. Energy Rev.* 44, 620–642. doi:10.1016/j.rser.2014.12.024

Searchinger, T., Heimlich, R., Houghton, R. A., Dong, F., Elobeid, A., Fabiosa, J., et al. (2008). Use of U.S. Croplands for biofuels increases greenhouse gases through emissions from land-use change. *Science* 319, 1238–1240. doi:10.1126/science.1151861

Sheath, R. G., Hellebust, J. A., and Sawa, T. (1981). Ultrastructure of the floridean starch granule. *Phycologia* 20, 292–297. doi:10.2216/i0031-8884-20-3-292.1

Silchenko, A. S., Kusaykin, M. I., Kurilenko, V. V., Zakharenko, A. M., Isakov, V. V., Zaporozhets, T. S., et al. (2013). Hydrolysis of fucoidan by fucoidanase isolated from the marine bacterium, Formosa algae. *Mar. Drugs* 11, 2413–2430. doi:10.3390/md11072413

Singh, A., Nigam, P. S., and Murphy, J. D. (2011). Mechanism and challenges in commercialisation of algal biofuels. *Bioresour. Technol.* 102, 26–34. doi:10.1016/j.biortech.2010.06.057

Spaans, S. K., Weusthuis, R. A., van der Oost, J., and Kengen, S. W. M. (2015). NADPH-generating systems in bacteria and archaea. *Front. Microbiol.* 6:742. doi:10.3389/fmicb.2015.00742

Suutari, M., Leskinen, E., Fagerstedt, K., Kuparinen, J., Kuuppo, P., and Blomster, J. (2015). Macroalgae in biofuel production. *Phycol. Res.* 63, 1–18. doi:10.1111/pre.12078

Synytsya, A., Čopíková, J., Kim, W. J., and Park, Y. I. (2015). "Cell wall polysaccharides of marine algae," in *Springer Handbook of Marine Biotechnology,* ed. S.-K. Kim (Berlin, Heidelberg: Springer), 543–590.

Takeda, T., Nakano, Y., Takahashi, M., Konno, N., Sakamoto, Y., Arashida, R., et al. (2015). Identification and enzymatic characterization of an endo-1,3-β-glucanase from *Euglena gracilis.* *Phytochemistry* 116, 21–27. doi:10.1016/j.phytochem.2015.05.010

Van Maris, A. J. A., Abbott, D. A., Bellissimi, E., van den Brink, J., Kuyper, M., Luttik, M. A. H., et al. (2006). Alcoholic fermentation of carbon sources in biomass hydrolysates by *Saccharomyces cerevisiae*: current status. *Antonie Van Leeuwenhoek* 90, 391–418. doi:10.1007/s10482-006-9085-7

Van Zyl, W. H., Bloom, M., and Viktor, M. J. (2012). Engineering yeasts for raw starch conversion. *Appl. Microbiol. Biotechnol.* 95, 1377–1388. doi:10.1007/s00253-012-4248-0

Viola, R., Nyvall, P., and Pedersen, M. (2001). The unique features of starch metabolism in red algae. *Proc. Biol. Sci.* 268, 1417–1422. doi:10.1098/rspb.2001.1644

Vreeland, V., and Kloareg, B. (2000). Cell wall biology in red algae: divide and conquer. *J. Phycol.* 36, 793–797. doi:10.1046/j.1529-8817.2000.36512.x

Wang, J., Kim, Y. M., Rhee, H. S., Lee, M. W., and Park, J. M. (2013). Bioethanol production from mannitol by a newly isolated bacterium, *Enterobacter* sp. JMP3. *Bioresour. Technol.* 135, 199–206. doi:10.1016/j.biortech.2012.10.012

Wang, Y., Xie, N., and Wang, W. (2009). Effects of algal concentration and initial density on the population growth of *Diaphanosoma celebensis* Stingelin (Crustacea, Cladocera). *Chin. J. Oceanol. Limnol.* 27, 480. doi:10.1007/s00343-009-9267-z

Wargacki, A. J., Leonard, E., Win, M. N., Regitsky, D. D., Santos, C. N. S., Kim, P. B., et al. (2012). An engineered microbial platform for direct biofuel production from brown macroalgae. *Science* 335, 308–313. doi:10.1126/science.1214547

Wikfors, G. H., and Ohno, M. (2001). Impact of algal research in aquaculture. *J. Phycol.* 37, 968–974. doi:10.1046/j.1529-8817.2001.01136.x

Wolfe, A. J. (2015). Glycolysis for the microbiome generation. *Microbiol. Spectr.* 3(3). doi:10.1128/microbiolspec.MBP-0014-2014

Wong, T. Y., Preston, L. A., and Schiller, N. L. (2000). ALGINATE LYASE: review of major sources and enzyme characteristics, structure-function analysis, biological roles, and applications. *Annu. Rev. Microbiol.* 54, 289–340. doi:10.1146/annurev.micro.54.1.289

Yanagisawa, M., Kawai, S., and Murata, K. (2013). Strategies for the production of high concentrations of bioethanol from seaweeds. *Bioengineered* 4, 224–235. doi:10.4161/bioe.23396

Yu, S., Blennow, A., Bojko, M., Madsen, F., Olsen, C. E., and Engelsen, S. B. (2002). Physico-chemical characterization of floridean starch of red algae. *Starch - Stärke* 54, 66–74. doi:10.1002/1521-379X(200202)54:2<66::AID-STAR66>3.0.CO;2-B

Yun, E. J., Choi, I.-G., and Kim, K. H. (2015). Red macroalgae as a sustainable resource for bio-based products. *Trends Biotechnol.* 33, 247–249. doi:10.1016/j.tibtech.2015.02.006

Zamora, F. (2009). "Biochemistry of alcoholic fermentation," in *Wine Chemistry and Biochemistry,* eds M. V. Moreno-Arribas and M. C. Polo (New York: Springer), 3–26.

Zemke-White, W. L., and Ohno, M. (1999). World seaweed utilisation: an end-of-century summary. *J. Appl. Phycol.* 11, 369–376. doi:10.1023/A:1008197610793

Zheng, Y., Pan, Z., and Zhang, R. (2009). Overview of biomass pretreatment for cellulosic ethanol production. *Int. J. Agric. Biol. Eng.* 2, 51–68. doi:10.3965/ijabe.v2i3.168

Conflict of Interest Statement: The authors declare that the research was conducted in the absence of any commercial or financial relationships that could be construed as a potential conflict of interest.

8

Power Harvesting from Human Serum in Buckypaper-Based Enzymatic Biofuel Cell

Güray Güven[1]*, Samet Şahin[2], Arcan Güven[3] and Eileen H. Yu[2]*

[1] *Giner, Inc., Newton, MA, USA,* [2] *Chemical Engineering and Advanced Materials, Merz Court, Newcastle University, Newcastle upon Tyne, UK,* [3] *Pennsylvania State University College of Medicine, Hershey, PA, USA*

The requirement for a miniature, high density, long life, and rechargeable power source is common to a vast majority of microsystems, including the implantable devices for medical applications. A model biofuel cell system operating in human serum has been studied for future applications of biomedical and implantable medical devices. Anodic and cathodic electrodes were made of carbon nanotube-buckypaper modified with PQQ-dependent glucose dehydrogenase and laccase, respectively. Modified electrodes were characterized electrochemically and assembled in a biofuel cell setup. Power density of $16.12\ \mu W\ cm^{-2}$ was achieved in human serum for lower than physiological glucose concentrations. Increasing the glucose concentration and biofuel cell temperature caused an increase in power output leading up to $49.16\ \mu W\ cm^{-2}$.

Keywords: biofuel cell, power production, laccase, PQQ-dependent glucose dehydrogenase, human serum, implantable medical device

Edited by:
Mihri Ozkan,
University of California, USA

Reviewed by:
Lei Li,
Shanghai Jiao Tong University, China
Chunlei Wang,
Florida International University, USA

*Correspondence:
Güray Güven
gguven@ginerinc.com;
Eileen H. Yu
eileen.yu@newcastle.ac.uk

Specialty section:
This article was submitted to
Nanoenergy Technologies and
Materials,
a section of the journal
Frontiers in Energy Research

Citation:
Güven G, Şahin S, Güven A and
Yu EH (2016) Power Harvesting from
Human Serum in Buckypaper-Based
Enzymatic Biofuel Cell.
Front. Energy Res. 4:4.

INTRODUCTION

The power requirements are one of the key limiting factors for implantable electronic devices as well as all portable and autonomous electronic systems. Although secondary power cells (e.g., lithium ion) can still be developed to improve on the existing state of the art, biofuel cells introduce a breakthrough concept as recharging can be theoretically achieved with virtually any available biomaterial. Biofuel cells can enable a permanent implantation without the need for surgical replacement, which clearly strengthens the community's competence in this area.

Recently, the possibility of developing implantable enzymatic biofuel cells (EBFCs) has come into focus since these devices would serve as a low-power source to miniature systems within the human body. EBFCs employ glucose as the substrate and oxygen or hydrogen peroxide as oxidizers, in which the power is gained by an oxidation of the biofuel at the anode (**Figure 1**). The enzymes involved in the process are usually glucose oxidase (GOx), horseradish peroxide (HPR), bilirubin oxidase (BOD), and laccase among the others (Willner et al., 1998; Pizzariello et al., 2002; Mano et al., 2003a,b). The human body is a storehouse of energy. An average person of 68 kg with 15% body fat stores energy approximately equivalent to 384540 kJ (Starner and Paradiso, 2004). However, the body has an abundant supply of energy from the food that we eat. For example, the power consumption of the brain is approximately 0.29 kcal/min (about 20 W) (Angel, 1990). Thus, if even a small fraction of this stored energy could be scavenged, an electronic device would have a large and renewable resource to draw on (Starner and Paradiso, 2004). This may be accomplished through the use of the energy stored within chemicals present in biological fluids, such as glucose (Barrière et al., 2004).

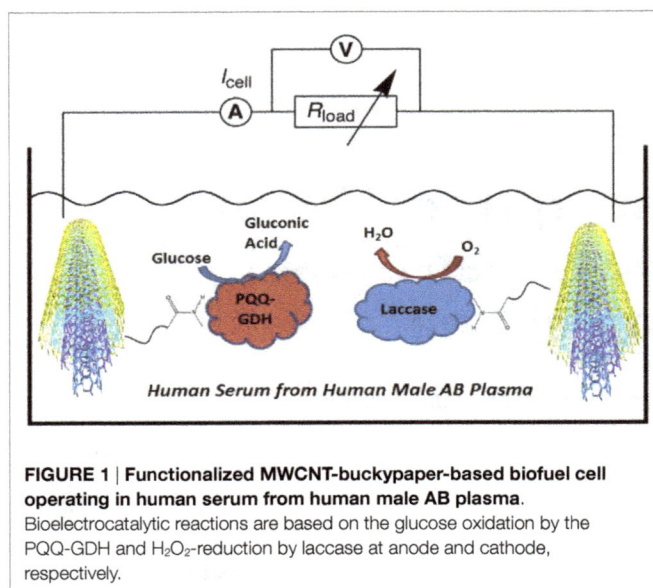

FIGURE 1 | Functionalized MWCNT-buckypaper-based biofuel cell operating in human serum from human male AB plasma. Bioelectrocatalytic reactions are based on the glucose oxidation by the PQQ-GDH and H_2O_2-reduction by laccase at anode and cathode, respectively.

A variety of electron mediators can be used for the electrical contacting of the biocatalysts and the electrodes. The functionalization of the electrode surfaces with assembled monolayers of redox enzymes, electrocatalysts, and bioelectrozymes (i.e., genetically engineered biocatalysts) (Güven et al., 2010; Yu et al., 2011) can lead to very effective electrical contact (Willner et al., 1998). However, most of these systems would involve the use of diffusional mediators which are dissolved in their respective compartments, and this poses an additional problem for the devices that are intended to be used as power sources for implantable medical devices. Bioelectrocatalytic electrodes, providing efficient non-mediated electric wiring for immobilized enzymes, must be simple in fabrication and stable in electrochemical reaction. A physiological biofuel cell-powered device is in need of constant, reliable, long-term power provided by a body-based biofuel. First *in vivo* trials with fuel cells, implanted subcutaneously in right flank of adult dog, yielded a power density performance of 2 µW cm⁻² (Drake et al., 1970) and in a rat (Wan and Tseung, 1974), which was generating 2.9 µW cm⁻² of power for a time period of at least 4 h. Although, these earlier results were encouraging with respect to the electricity produced for low-power medical implants, preliminary studies were operated with abiotically catalyzed glucose rather than bioelectrocatalyst reactions. With the aim of developing an EBFC operating *in vivo*, researchers demonstrated power production from the hemolymph of snails (Halámková et al., 2012), "cyborg" lobsters (MacVittie et al., 2013), and serially connected clams (Szczupak et al., 2012). In addition, a biofuel cell was placed in an insect (Rasmussen et al., 2012). Although first animal experiments were not performed with mammals, studies demonstrated that EBFCs can produce electricity out of living organisms.

Toward human implantable applications, the first demonstration of an implanted biofuel cell (fully biological), generating electricity inside a living organism and, in particular, implanted in a mammal, was in a rat (Cinquin et al., 2010). It was reported that the biofuel cell is able to generate electric power in a rat, using glucose and oxygen contained in its body fluids with an open circuit potential (OCP) of 0.275 V and maximum power output of 6.5 µW (Cinquin et al., 2010). Then, more sturdily constructed biofuel cell prototype was partially (anode compartment only) implanted in rabbit ear (Miyake et al., 2011), which was reflected in the maximum cell current (1.50 µA), whereas the power of this cell reached 0.42 µW at 0.56 V and in brain of a living rat with a maximum power of 2 µW cm⁻² at a cell voltage of 0.4 V (Andoralov et al., 2013). Later, with the recent improvements in terms of carbon nanotube (CNT) compression and direct electron transfer, researchers were successful to increase the power (38.7 µW) obtained from the implanted biofuel cell in a rat, using a specially designed electronic circuit to charge a capacitor, to run a LED, or a digital thermometer (Zebda et al., 2013). Recently, researchers reported a glucose/oxygen biofuel cell using FAD-dependent glucose dehydrogenase enzyme at the anode side operating in human serum, which produces maximum power densities of 39.5 ± 1.3 and 57.5 ± 5.4 µW cm⁻² for EFCs at 21 and 37°C, respectively (Milton et al., 2015).

In our study, the strategy of direct electrochemical communication has been successfully applied with nanostructured buckypaper composed of multi-wall CNTs (Narváez Villarrubia et al., 2011; Strack et al., 2011; Halámková et al., 2012; MacVittie and Katz, 2014). Enzymes were immobilized onto buckypaper by using a heterobifunctional cross-linker, 1-pyrenebutanoic acid succinimidyl ester (PBSE), which provides covalent binding with amino groups of protein lysine residues through the formation of amide bonds and interacts with CNTs *via* π–π stacking of the polyaromatic pyrenyl moieties (Strack et al., 2011). PBSE results in a random orientation of the enzyme molecules relative to the electrode surface because of the large number of amino acid groups randomly positioned in the protein structure (Katz, 1994). In the case of a flat, robust electrode surface, direct electrode transfer would not be possible because active centers are buried inside the enzymes and therefore would be far away from the transducer surface (Katz, 1994). However, in the case of a buykypaper, electron access channels of the enzymes can find closer electron transport pathways regardless of the enzyme orientation, allowing efficient communication between the CNTs and the enzyme. Selection of the enzymes to immobilize on the buckypaper is also an important issue. In this study, pyrroloquinoline quinone-dependent glucose dehydrogenase (PQQ-GDH) was selected for the anodic bioelectrode because of the successful bioelectrocatalytic generation of anodic currents in the presence of glucose, when enzyme is tethered to CNT materials (Strack et al., 2013). Eventually, a dehydrogenase-based anode operating in the presence of oxygen did not require complex molecular ensemble to achieve a kinetically preferred electron transfer to the bioelectrode instead of oxygen (Katz et al., 1999). In order to perform potentially favorable biofuel cell operation based on glucose oxidation and oxygen reduction at the anode and cathode electrodes, respectively, oxygen-reducing laccase was selected for the cathodic reaction. Laccase is a well-known biocatalyst that is commonly used in EBFCs (Calabrese Barton et al., 2004), and it is compatible with the buckypaper electrode (Hussein et al., 2011).

In this study, a model system is demonstrated for the biofuel cells that are operated in human serum. After the preliminary attempts to address the problems of integrated circuitries in a biofuel cell setup, with systems mimicking human physiology (Coman et al., 2010; Southcott et al., 2013), cyclic voltammetry and biofuel cell polarization that were applied for the characterization of the power generated from human serum is presented in this study.

MATERIALS AND METHODS

Enzyme Preparation

Pyrroloquinoline quinone-dependent glucose dehydrogenase (E.C. 1.1.5.2) from *Microorganism* (Toyobo Co., Japan) was used as supplied. Laccase (E.C. 1.10.3.2, from *Trametes versicolor*) was obtained from Sigma-Aldrich and used in experiments after the purification method as described elsewhere (Strack et al., 2011). Briefly, laccase was dissolved (5 mL, 20 mg mL^{-1}) in potassium phosphate buffer (20 mM, pH 7.3) and dialyzed using a Slide-A-Lyzer dialysis cassette (10 kDa molecular weight cutoff; Thermo Fisher Scientific, Inc.) against a series of sequential buffer exchanges (10 mM potassium phosphate, pH 7.0 for 12 h at 4°C) containing 1 mM CuSO$_4$, 1 mM EDTA, and buffer only. Following dialysis, the protein concentration of the preparation was determined using a bicinchoninic acid protein assay kit (Thermo Fisher Scientific, Inc.). Protein activity in respect to catalytic oxidation was determined for syringaldazine, using the supplier's standard method (Sigma-Aldrich). The laccase preparation was normalized to a defined protein concentration (1.5 mg mL^{-1}).

Electrode Preparation

Carbon nanotube-buckypaper prepared from 100% multi-walled CNTs (Buckeye Composites; NanoTechLabs, Yadkinville, NC, USA) was used as the electrode material (0.25 cm^2 geometric area). Electrodes were incubated with 10 mM PBSE (AnaSpec Inc.) in DMSO with moderate shaking for 1 h at room temperature (RT) and subsequently rinsed in DMSO (5 min) to remove excess PBSE and then in potassium phosphate buffer (10 mM, pH 7.0; 5 min). The PBSE-functionalized electrodes were immediately incubated with enzyme solutions, for 1 h at RT with moderate shaking, before washing extensively with potassium phosphate buffer (10 mM, pH 7.0). Enzyme solutions consist of 1 mg mL^{-1} PQQ-GDH dissolved in potassium phosphate buffer (10 mM, pH 7.0) supplemented with 1 mM CaCl$_2$ and 1.5 mg mL^{-1} of purified laccase for the modification of anode and cathode, respectively. The activity of PQQ-GDH and laccase immobilized on the anode and cathode were reported as 250 and 460 mU per electrode, respectively (Halámková et al., 2012), by performing optical assays as described elsewhere (Holwerda and Gray, 1975; Dewanti and Duine, 1998). The modified electrodes were used immediately or stored (4°C) in potassium phosphate electrolyte (0.1M, pH 7.0) until used in biofuel cell or electrochemical measurements.

Electrochemical Measurements

Enzyme (PQQ-GDH or laccase) and cross-linker (PBSE)-modified buckypaper electrodes were connected through a copper conductive tape (3M, SPI supplies) to a potentiostat (ECO Chemie Autolab PASTAT 10 electrochemical analyzer, using the GPES 4.9 software package) and used as the working electrode in electrochemical measurements where a Pt wire (Metrohm) and a Ag|AgCl|KCl 3M (Metrohm) were used as counter and reference electrodes, respectively. Cyclic voltammograms (CVs) were obtained in the potential range between 0 and 0.8 V (vs. Ag/AgCl) at a scan rate of 1 mV s^{-1}. CVs of bioanode were tested in sterile-filtered human serum from human male AB plasma (Sigma-Aldrich, Product# H4522) supplemented with glucose (19.89 mM, final concentration) and a no glucose buffer solution (22 mM NaHCO$_3$, 40 mg mL^{-1} BSA, 6.7 mM MgCl$_2$, 5 mM KCl; pH 7.4) mimicking *in vitro* physiological conditions (Roach, 1963), whereas biocathode measurements were achieved in the absence and presence of oxygen (in equilibrium with air). Anaerobic conditions were achieved by purging the solution with nitrogen.

Biofuel Cell Measurements

The electrodes were modified for the biofuel cell measurements as it is explained above. Biocatalytic electrodes were assembled in human serum solutions. Glucose concentration of the human serum used in this study is 4.89 mM (Ammam and Fransaer, 2010; Ammam and Easton, 2011). The effect of glucose concentration on power output was measured by adding D-glucose (from 4.89 to 24.89 mM, final concentration) into human serum solution. The voltage and current generated by the biofuel cell were measured by a digital multimeter (Meterman 37XR). For the determination of the power output, a variable resistance was used as an external load. The measurements were carried out at ambient temperature (23 ± 2°C). In order to evaluate the performance of the biofuel cell for *in vivo* use, it was tested in human serum at 37°C.

RESULTS AND DISCUSSION

The biofuel cell anode was characterized using CVs to validate whether the functionalized biocatalytic buckypaper electrodes (0.25 cm^2) are capable of electron transfer to the electrodes. Peaks in the voltage–current curves would be observed at potentials corresponding to the potentials of the enzyme active centers, such as PQQ and T1 in GDH (Katz et al., 1999) and laccase (Calabrese Barton et al., 2004), respectively, in order to confirm the direct electrochemistry between the enzyme active center and electrode surface.

Cyclic voltammograms of the PQQ-GDH-modified electrodes were obtained in human serum supplemented with glucose (24.89 mM, final concentration) (**Figure 2**, blue curve) and in mimicking solution free of glucose (**Figure 2**, red curve). The anodic currents produced by PQQ-GDH (**Figure 2**) were observed only in the presence of glucose, which shows the successful immobilization of PQQ-GDH onto PBSE-functionalized CNT-buckypaper electrode. The anodic current produced by the oxidation reaction of the PQQ-GDH electrode was developed at around 0.10–0.15 V vs. Ag/AgCl. Previous works have reported that the anodic current produced by the PQQ-GDH electrode appeared at potentials more positive than −0.1 V vs. Ag/AgCl in a solution mimicking the physiological conditions

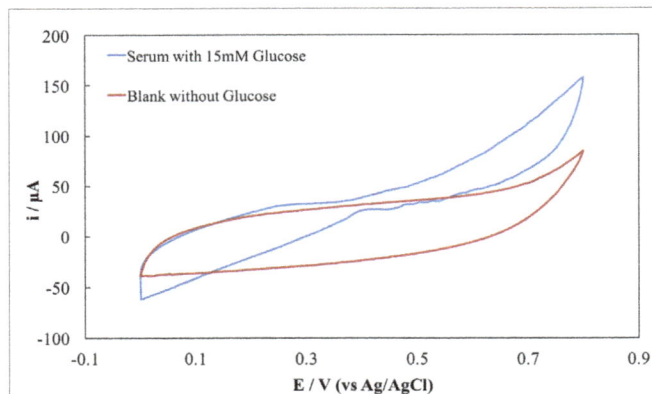

FIGURE 2 | CVs of the PQQ-GDH anode in human serum supplemented with 20.89 mM glucose (blue) and glucose-free mimicking solution (red).

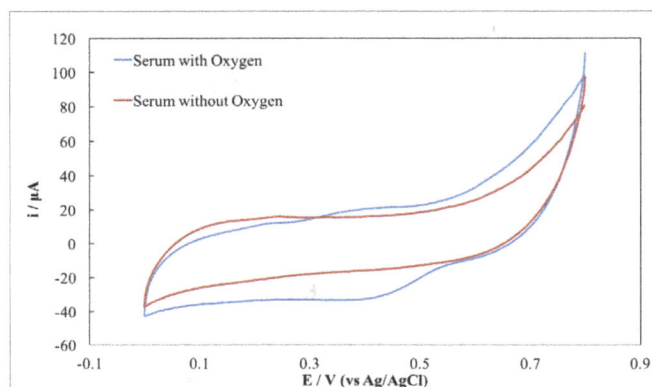

FIGURE 3 | CVs of laccase cathode in human serum in the presence (blue) and absence (red) of oxygen.

TABLE 1 | Summary of the biofuel cell performance results obtained from different human serum samples.

	OCP (V)	Maximum power (µW)
Serum	0.286	4.03
Serum + 5 mM glucose	0.336	6.28
Serum + 10 mM glucose	0.388	9.04
Serum + 15 mM glucose	0.413	11.50
Serum + 20 mM glucose	0.436	12.29
Serum at 37°C	0.396	8.27

potentials, i.e., EMF; however, in practice, the OCP is substantially lower than the theoretical value due to various potential losses, particularly activation/kinetic loss. The anodic and cathodic potential, obtained by the PQQ-GDH and laccase electrodes, were produced at 0.1 and 0.55 V vs. Ag/AgCl, respectively. Therefore, the OCP of ca. 0.45 V between the reactions at the anode and the cathode can be obtained.

After demonstrating the electrochemical activity of biocatalytic oxidation and reduction reactions, electrodes were used for harvesting electrical power from human serum *in vitro*. The overall performance of the biofuel cell is evaluated by its power output. Power is calculated as

$$P = i \times E_{cell} \tag{1}$$

To evaluate the power generation, a decreasing resistance from 300 kΩ to 100 Ω at a time interval of 10 s in each load was applied, and the current and cell potentials for each resistance were measured. Later, the applied resistor, cell potential, and current were used to calculate power. Therefore, it can be said that power generated by biofuel cell was computed as a product of current passing through an external load and voltage drop across the resistor.

The results of the different biofuel cell measurements operating in human serum and different additions of glucose concentrations are summarized in **Table 1**. The purpose of this experiment was to demonstrate the EBFC operation under different physiological conditions representing different glucose concentrations in human blood. With an attempt to demonstrate the biofuel cell in biomedical applications, human serum solution was spiked with various concentrations of glucose. Thus, final glucose concentrations between 4.89 and 24.89 mM were realized, corresponding to various physiological conditions (i.e., diabetes mellitus and health ranges) (Nelson et al., 2008). The biofuel cell operated in human serum showed an OCP value of 286 mV, and it was observed that upon an increase of the load resistance, the recorded power from human serum increased up to a maximum net value of 4.03 µW (power density of 16.12 µW cm⁻²) at an external load of 10 kΩ. The OCPs of the other human serum samples showed increasing values with increasing glucose concentrations as well as the power output, in which the maximum OCP and power density were 436 mV and 12.29 µW (power density of 49.16 µW cm⁻²), respectively.

Figures 4 and **5** show the cell voltage as a function of current density, also known as the polarization curve of the biofuel cell, and the power output of the biofuel cell operating in human serum and different glucose additions, respectively. The power

(Halámková et al., 2012). The anodic voltage obtained in human serum approximately correspond the values demonstrated in buffer solutions for similar electrodes. Therefore, this result can be interpreted as the electrochemical activity of bioelectrodes. A positive correlation between the glucose concentration and bioelectrocatalytic current (data not shown) reveals the role of the glucose oxidation reaction as a significant source of electrons. The cathodic reaction of laccase-modified electrode is shown in **Figure 3**. The presence of oxygen in human serum clearly resulted in the formation of bioelectrocatalytic currents (**Figure 3**, blue curve) where the oxygen reduction peak starts at around 0.50–0.55 V vs. Ag/AgCl. The appearance of the cathodic current obtained in human serum is similar to the previously reported currents appearing at ca. 0.58 V vs. Ag/AgCl in other related systems (Strack et al., 2011; Halámková et al., 2012).

The cell electromotive force (EMF) is the potential difference between the anode and the cathode or the voltage of the electrochemical cell. Theoretically, open circuit potential (OCP) should be the difference between anodic and cathodic generation from biofuel cell reached its maximum values where external resistance was equal to internal resistance (Menicucci et al., 2006), on the load of 10 kΩ followed by a gradual decay due to mass transport losses as the external load is increased, indicative of typical fuel cell behavior (Bockris and Srinivasan, 1969). At high current density values, the reaction is limited by mass transport of reactants. In most of the cases, it is a sharp decrease of the voltage to 0 at the certain current density value or around it as seen in our data for lower glucose concentrations. With higher glucose concentrations, after peak power is achieved, it was possible to continue polarizing the cell until the voltage dropped to a low value of 0.05 at 20 mM glucose. Further polarizing the cell did not increase current density as voltage dropping, a bend back on polarization curve, can be observed due to mass transport limitation. This is common in biological fuel cells known as "power overshoot."

The shapes of power curves for elevated sugar values in this model system were almost identical where the maximum power output was obtained with different glucose concentrations at the optimum external load of 10 kΩ. Since the variations at power generation represent the current/voltage dependence on the

glucose (Castorena-Gonzalez et al., 2013), obtained net power has increased from 4.03 to 12.29 µW when the glucose levels were increased from 4.89 to 24.89 mM. However, linearity between the power output and glucose concentration was achieved between 4.89 and 19.89 mM (final concentrations) as the saturation level of the glucose concentration was reached for the biofuel cell operation. **Figure 6** shows the maximum power generation by the biofuel cell as a function of the glucose concentration at the external load of 10 kΩ. The glucose-sensitive (0.50 µW/mM) biofuel cell is a redox pair of electrodes that generates power proportional to glucose concentration. Such a system can achieve simultaneous power transmission to implantable glucose sensors (Rai and Varadan, 2012) that measure real-time blood glucose as compared to conventional techniques involving drawing blood samples and *in vitro* processing.

The temperature dependence of the biofuel cell operating in human serum at RT and at 37°C was examined without further optimization of the system. The power output showed sensitivity against temperature, and the power production has increased up to 8.27 µW at 37°C, which is almost twofold higher than it is recorded at RT (**Figure 7**). Moreover, not only similar power curve behavior was observed at 37°C but also with applying an

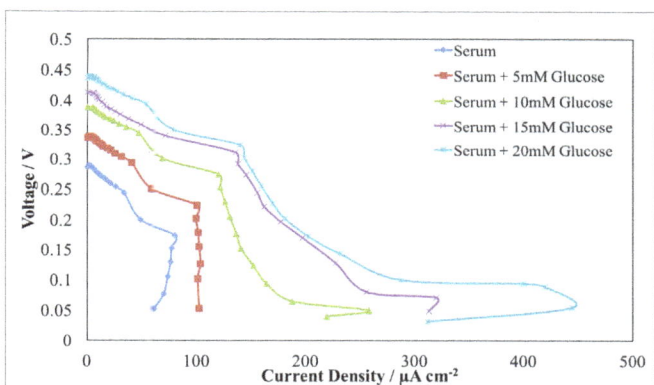

FIGURE 4 | Cell voltage vs. current density curve of the biofuel cell operating in human serum and different glucose additions, at room temperature.

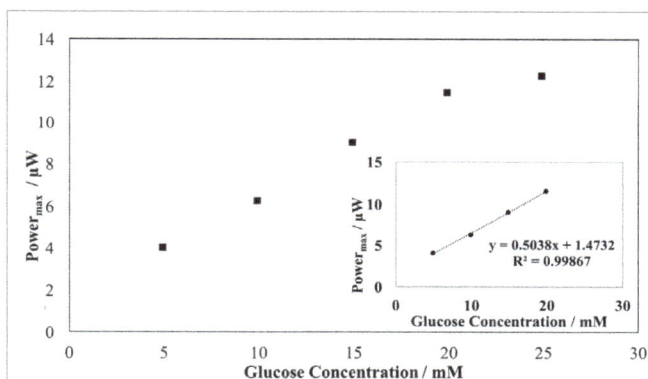

FIGURE 6 | Maximum produced power by the biofuel cell as a function of the glucose concentration at external load of 10 kΩ.

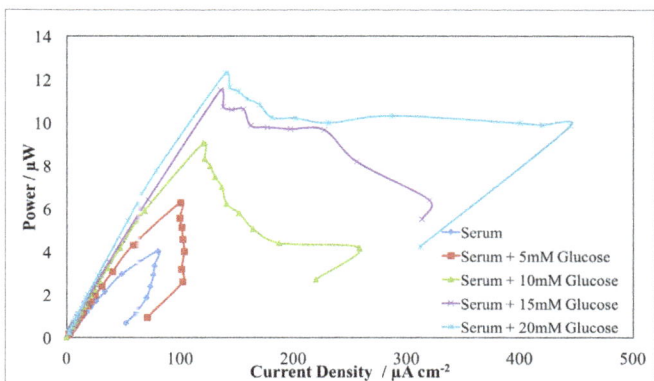

FIGURE 5 | The power output of the fuel cell operating in human serum and different glucose additions, at room temperature.

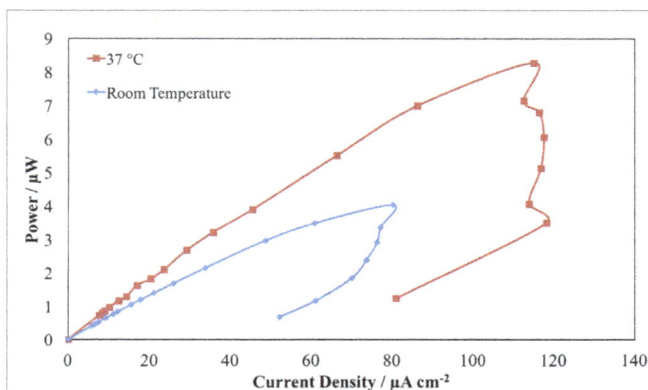

FIGURE 7 | Temperature dependence of the power output of the biofuel cell operating in human serum at RT (blue) and 37°C (red).

external load of 10 kΩ, the biofuel cell produced electrical power demonstrated a relatively long period of stability (at least 3 h) at 37°C, with no medium replacement during the entire experiment. However, the problem about whether the enzymes remain stable for a long period of time is a concern. The efficiency of electron transfer between enzymes and electrodes should be improved.

The enhancement on power output at human body temperature makes biofuel cell operating in human serum an attractive development prospect for use in future applications where power is needed for implantable medical devices.

CONCLUSION

In this study, a model enzymatic fuel cell was developed CNT-buckypaper with PQQ-GDH on the anode for glucose oxidation and laccase on the cathode for oxygen reduction. The maximum OCP and power density were obtained as 0.44 V and 12.29 μW (power density of 49.16 μW cm^{-2}), respectively. Additional studies

for electrochemical characterization of enzyme-immobilized buckypaper electrodes have been achieved. Current studies are underway for further identification of the operational stability of the energy production process but what is undeniable is the proof of principle that human serum can be used as an energy source for biofuel cells.

AUTHOR CONTRIBUTIONS

All authors listed have made substantial, direct, and intellectual contribution to the work and approved it for publication.

ACKNOWLEDGMENTS

EY would like to thank EPSRC for the research fellowship (Life Science Interface EP/C535456/1) to support this work. SŞ would like to thank the Turkish Government for the financial support to his Ph.D. study at Newcastle University.

REFERENCES

Ammam, M., and Easton, E. B. (2011). High-performance glucose sensor based on glucose oxidase encapsulated in new synthesized platinum nanoparticles supported on carbon Vulcan/Nafion composite deposited on glassy carbon. *Sens. Actuators B Chem.* 155, 340–346. doi:10.1016/j.snb.2011.04.016

Ammam, M., and Fransaer, J. (2010). Glucose microbiosensor based on glucose oxidase immobilized by AC-EPD: characteristics and performance in human serum and in blood of critically ill rabbits. *Sens. Actuators B Chem.* 145, 46–53. doi:10.1016/j.snb.2009.11.015

Andoralov, V., Falk, M., Suyatin, D. B., Granmo, M., Sotres, J., Ludwig, R., et al. (2013). Biofuel cell based on microscale nanostructured electrodes with inductive coupling to rat brain neurons. *Sci. Rep.* 3, 3270. doi:10.1038/srep03270

Angel, A. (1990). Textbook of physiology. 21st edition. two volumes. edited by H. D. Patton, A. F. Fuchs, B. Hille, A. M. Scher and R. Steiner. vol. 1, pp. 769 (1–769); vol. 2, pp. 826 (771–1596). (Harcourt Brace Jovanovich, 1989.). Each volume £64.00 hardback. vol. 1 ISBN 0 7216 2523 1; vol. 2 ISBN 0 7216 2524 X; both vols 1 and 2 together ISBN 0 7216 1990 8. *Exp. Physiol.* 75, 281–281.

Barrière, F., Ferry, Y., Rochefort, D., and Leech, D. (2004). Targetting redox polymers as mediators for laccase oxygen reduction in a membrane-less biofuel cell. *Electrochem. Commun.* 6, 237–241. doi:10.1016/j.elecom.2003.12.006

Bockris, J.O'M., Srinivasan, S. (1969). Fuel Cells: Fundamentals and Applications. New York, NY: McGraw-Hill Book Corporation.

Calabrese Barton, S., Gallaway, J., and Atanassov, P. (2004). Enzymatic biofuel cells for implantable and microscale devices. *Chem. Rev.* 104, 4867–4886. doi:10.1021/cr020719k

Castorena-Gonzalez, J. A., Foote, C., Macvittie, K., Halámek, J., Halámková, L., Martinez-Lemus, L. A., et al. (2013). Biofuel cell operating in vivo in rat. *Electroanalysis* 25, 1579–1584. doi:10.1002/elan.201300136

Cinquin, P., Gondran, C., Giroud, F., Mazabrard, S., Pellissier, A., Boucher, F., et al. (2010). A glucose biofuel cell implanted in rats. *PLoS ONE* 5:e10476. doi:10.1371/journal.pone.0010476

Coman, V., Ludwig, R., Harreither, W., Haltrich, D., Gorton, L., Ruzgas, T., et al. (2010). A direct electron transfer-based glucose/oxygen biofuel cell operating in human serum. *Fuel Cells* 10, 9–16. doi:10.1002/fuce.200900121

Dewanti, A. R., and Duine, J. A. (1998). Reconstitution of membrane-integrated quinoprotein glucose dehydrogenase apoenzyme with PQQ and the holoenzyme's mechanism of action. *Biochemistry* 37, 6810–6818. doi:10.1021/bi9722610

Drake, R. F., Kusserow, B. K., Messinger, S., and Matsuda, S. (1970). A tissue implantable fuel cell power supply. *ASAIO J.* 16, 199–205.

Güven, G., Prodanovic, R., and Schwaneberg, U. (2010). Protein engineering – an option for enzymatic biofuel cell design. *Electroanalysis* 22, 765–775. doi:10.1002/elan.200980017

Halámková, L., Halámek, J., Bocharova, V., Szczupak, A., Alfonta, L., and Katz, E. (2012). Implanted biofuel cell operating in a living snail. *J. Am. Chem. Soc.* 134, 5040–5043. doi:10.1021/ja211714w

Holwerda, R. A., and Gray, H. B. (1975). Kinetics of the reduction of *Rhus vernicifera* laccase by ferrocyanide ion. *J. Am. Chem. Soc.* 97, 6036–6041. doi:10.1021/ja00854a014

Hussein, L., Rubenwolf, S., Von Stetten, F., Urban, G., Zengerle, R., Krueger, M., et al. (2011). A highly efficient buckypaper-based electrode material for mediatorless laccase-catalyzed dioxygen reduction. *Biosens. Bioelectron.* 26, 4133–4138. doi:10.1016/j.bios.2011.04.008

Katz, E. (1994). An international journal devoted to all aspects of electrode kinetics, interfacial structure, properties of electrolytes, colloid and biological electrochemistry application of bifunctional reagents for immobilization of proteins on a carbon electrode surface: oriented immobilization of photosynthetic reaction centers. *J. Electroanal. Chem.* 365, 157–164. doi:10.1016/0022-0728(93)02975-N

Katz, E., Willner, I., and Kotlyar, A. B. (1999). A non-compartmentalized glucose| O 2 biofuel cell by bioengineered electrode surfaces. *J. Electroanal. Chem.* 479, 64–68. doi:10.1016/S0022-0728(99)00425-8

MacVittie, K., Halamek, J., Halamkova, L., Southcott, M., Jemison, W. D., Lobel, R., et al. (2013). From "cyborg" lobsters to a pacemaker powered by implantable biofuel cells. *Energy Environ. Sci.* 6, 81–86. doi:10.1039/C2EE23209J

MacVittie, K., and Katz, E. (2014). Self-powered electrochemical memristor based on a biofuel cell – towards memristors integrated with biocomputing systems. *Chem. Commun.* 2014, 4816–4819. doi:10.1039/C4CC01540A

Mano, N., Mao, F., and Heller, A. (2003a). Characteristics of a miniature compartment-less glucose – O2 biofuel cell and its operation in a living plant. *J. Am. Chem. Soc.* 125, 6588–6594. doi:10.1021/ja0346328

Mano, N., Mao, F., Shin, W., Chen, T., and Heller, A. (2003b). A miniature biofuel cell operating at 0.78 V. *Chem. Commun.* 4, 518–519. doi:10.1039/B211796G

Menicucci, J., Beyenal, H., Marsili, E., Veluchamy, R. A., Demir, G., and Lewandowski, Z. (2006). Procedure for determining maximum sustainable power generated by microbial fuel cells. *Environ. Sci. Technol.* 40, 1062–1068.

Milton, R. D., Lim, K., Hickey, D. P., and Minteer, S. D. (2015). Employing FAD-dependent glucose dehydrogenase within a glucose/oxygen enzymatic fuel cell operating in human serum. *Bioelectrochemistry* 106(Pt A), 56–63. doi:10.1016/j.bioelechem.2015.04.005

Miyake, T., Haneda, K., Nagai, N., Yatagawa, Y., Onami, H., Yoshino, S., et al. (2011). Enzymatic biofuel cells designed for direct power generation from biofluids in living organisms. *Energy Environ. Sci.* 4, 5008–5012. doi:10.1039/C1EE02200H

Narváez Villarrubia, C. W., Rincón, R. A., Radhakrishnan, V. K., Davis, V., and Atanassov, P. (2011). Methylene green electrodeposited on SWNTs-based "Bucky" papers for NADH and l-malate oxidation. *ACS Appl. Mater. Interfaces* 3, 2402–2409. doi:10.1021/am2003137

Nelson, D. L., Lehninger, A. L., and Cox, M. M. (2008). *Lehninger Principles of Biochemistry*. Macmillan.

Pizzariello, A., Stredansky, M., and Miertuš, S. (2002). A glucose/hydrogen peroxide biofuel cell that uses oxidase and peroxidase as catalysts by composite bulk-modified bioelectrodes based on a solid binding matrix. *Bioelectrochemistry* 56, 99–105. doi:10.1016/S1567-5394(02)00026-9

Rai, P., and Varadar, V. K. (2012). "Wireless glucose monitoring watch enabled by an implantable self-sustaining glucose sensor system," in *Proceedings SPIE 8548, Nanosystems in Engineering and Medicine*, 85481G-85481G-85488. doi:10.1117/12.946334

Rasmussen, M., Ritzmann, R. E., Lee, I., Pollack, A. J., and Scherson, D. (2012). An implantable biofuel cell for a live insect. *J. Am. Chem. Soc.* 134, 1458–1460. doi:10.1021/ja210794c

Roach, D. K. (1963). Analysis of the haemolymph of *Arion ater* L. (Gastropoda: Pulmonata). *J. Exp. Biol.* 40, 613–623.

Southcott, M., Macvittie, K., Halamek, J., Halamkova, L., Jemison, W. D., Lobel, R., et al. (2013). A pacemaker powered by an implantable biofuel cell operating under conditions mimicking the human blood circulatory system – battery not included. *Phys. Chem. Chem. Phys.* 15, 6278–6283. doi:10.1039/C3CP50929J

Starner, T., and Paradiso, J. (2004). "Human generated power for mobile electronics," in *Low Power Electronics Design* (CRC Press), 1–35.

Strack, G., Babanova, S., Farrington, K. E., Luckarift, H. R., Atanassov, P., and Johnson, G. R. (2013). Enzyme-modified buckypaper for bioelectrocatalysis. *J. Electrochem. Soc.* 160, G3178–G3182. doi:10.1149/2.028307jes

Strack, G., Luckarift, H. R., Nichols, R., Cozart, K., Katz, E., and Johnson, G. R. (2011). Bioelectrocatalytic generation of directly readable code: harnessing cathodic current for long-term information relay. *Chem. Commun.* 47, 7662–7664. doi:10.1039/C1CC11475A

Szczupak, A., Halamek, J., Halamkova, L., Bocharova, V., Alfonta, L., and Katz, E. (2012). Living battery – biofuel cells operating *in vivo* in clams. *Energy Environ. Sci.* 5, 8891–8895. doi:10.1039/C2EE21626D

Wan, B. Y. C., and Tseung, A. C. C. (1974). Some studies related to electricity generation from biological fuel cells and galvanic cells, *in vitro* and *in vivo*. *Med. Biol. Eng.* 12, 14–28. doi:10.1007/BF02629831

Willner, I., Katz, E., Patolsky, F., and Buckmann, F. A. (1998). Biofuel cell based on glucose oxidase and microperoxidase-11 monolayer-functionalized electrodes. *J. Chem. Soc. Perkin Trans.* 2, 1817–1822. doi:10.1039/A801487F

Yu, E., Prodanovic, R., Güven, G., Ostafe, R., and Schwaneberg, U. (2011). Electrochemical oxidation of glucose using mutant glucose oxidase from directed protein evolution for biosensor and biofuel cell applications. *Appl. Biochem. Biotechnol.* 165, 1448–1457. doi:10.1007/s12010-011-9366-0

Zebda, A., Cosnier, S., Alcaraz, J. P., Holzinger, M., Le Goff, A., Gondran, C., et al. (2013). Single glucose biofuel cells implanted in rats power electronic devices. *Sci. Rep.* 3. doi:10.1038/srep01516

Conflict of Interest Statement: The authors declare that the research was conducted in the absence of any commercial or financial relationships that could be construed as a potential conflict of interest.

9

Major Lipid Body Protein: A Conserved Structural Component of Lipid Body Accumulated during Abiotic Stress in *S. quadricauda* CASA-CC202

Anand Javee[1†], Sujitha Balakrishnan Sulochana[1,2†], Steffi James Pallissery[1] and Muthu Arumugam[1,2]*

[1] *Biotechnology Division, National Institute for Interdisciplinary Science and Technology (NIIST), Council of Scientific and Industrial Research (CSIR), Trivandrum, India,* [2] *Academy of Scientific and Innovative Research (AcSIR), New Delhi, India*

Edited by:
Peer Schenk,
University of Queensland, Australia

Reviewed by:
Yu-Shen Cheng,
National Yunlin University of
Science and Technology, Taiwan
Jaime Puna,
Instituto Superior de Engenharia
de Lisboa, Portugal
Ihsan Hamawand,
University of Southern
Queensland, Australia

*Correspondence:
Muthu Arumugam
arumugam@niist.res.in,
aasaimugam@gmail.com

[†]*Anand Javee and Sujitha*
Balakrishnan Sulochana
are joint first authors and
contributed equally.

Specialty section:
This article was submitted to
Bioenergy and Biofuels,
a section of the journal
Frontiers in Energy Research

Citation:
Javee A, Sulochana SB, Pallissery SJ
and Arumugam M (2016) Major Lipid
Body Protein: A Conserved Structural
Component of Lipid Body
Accumulated during Abiotic Stress in
S. quadricauda CASA-CC202.
Front. Energy Res. 4:37.

Abiotic stress in oleaginous microalgae enhances lipid accumulation, which is stored in a specialized organelle called lipid droplets (LDs). Both the LDs or lipid body are enriched with major lipid droplet protein (MLDP). It serves as a major structural component and also plays a key role in recruiting other proteins and enzymes involved in lipid body maturation. In the present study, the presence of MLDP was detected in two abiotic stress condition namely nitrogen starvation and salt stress condition. Previous research reveals that nitrogen starvation enhances lipid accumulation. Therefore, the effect of salt on growth, biomass yield, and fatty acid profile is studied in detail. The specific growth rate of *Scenedesmus quadricauda* under the salt stress of 10mM concentration is about 0.174 μ and in control, the SGR is 0.241 μ. An increase in the doubling time of the cells shows that the rate of cell division decreases during salt stress (2.87–5.17). The dry biomass content also decreased drastically at 50mM salt-treated cells (129 mg/L) compared to control (236 mg/L) on the day 20. The analysis of fatty acid composition also revealed that there is a 20% decrease in the saturated fatty acid level and 19.9% increment in monounsaturated fatty acid level, which makes salt-mediated lipid accumulation as a suitable biodiesel precursor.

Keywords: major lipid droplet protein, *Scenedesmus quadricauda*, abiotic stress, nitrogen deprivation, salt stress, FAME profile

INTRODUCTION

Microalgae are unicellular photosynthetic organisms, gaining importance as a feedstock for high-value nutraceuticals, bioactive compounds, and biodiesel. The microalgae have various advantages over other organisms due to its fast growth rate, ability to adapt to varying environmental condition, can be cultured in a season-independent manner, and do not compete with the inputs deployed for agriculture. Oleaginous microalgae accumulate the lipid in the form of triacylglycerol (TAG) in distinct organelles called lipid droplets (LDs) located in the cytoplasm.

Abbreviations: BBM, bold basal medium; FAME, fatty acid methyl ester; FITC, fluorescein iso thio cyanate; LD, lipid droplets; MLDP, major lipid droplet protein; PBS, phosphate buffer saline; SGR, specific growth rate; TAG, triacylglycerol; TDS, total dissolved solids; TLC, thin layer chromatography.

The increased level of LDs was observed during abiotic stresses, such as nutrient deprivation and high light exposure, in microalgae. As a result, they gained an attention in exploring as a source for high-value edible oil or renewable transportation fuel (Moellering and Benning, 2010). Like microalgae, plants also store oil in specialized compartments known asLDs, oil bodies/ globules or oleosomes. The neutral storage lipids (mainly TAGs) are stored in these spherical organelles enclosed by a membrane lipid monolayer coated with proteins (Davidi et al., 2012).

Major lipid droplet protein (MLDP) is found in lipid bodies. The MLDP forms a proteinaceous coat surrounding mature LDs. The characterization of LD proteins has been reported in terrestrial oil seed plants and in certain mammalian tissues (Zweytick et al., 2000; Murphy et al., 2001). Downregulation of MLDP through RNA interference affects LD size without any compromise in the TAG level or its metabolism in *Chlamydomonas reinhardtii* (Moellering and Benning, 2010). Similarly, LDs' size and number vary while silencing or knocking down of oleosin in oleaginous plants (Siloto et al., 2006; Schmidt and Herman, 2008). MLDP stabilizes mature LDs and acts as a recruiting platform for other proteins and enzymes required for the maturation of LDs. In *C. reinhardtii*, MLDP is identified as an abundant protein on the outer surface of the LDs proteome (Moellering and Benning, 2010; Nguyen et al., 2011). MLDP abundance positively correlates with the accumulation of TAGs (Tsai et al., 2014). During nitrogen starvation, both MLDP and TAG were over accumulated, and the inhibition of TAG biosynthesis impairs MLDP level, suggesting that MLDP induction is coregulated with TAG accumulation in *Dunaliella* (Davidi et al., 2012). Besides MLDP being a major structural protein of LDs proteome, it also contains many regulatory enzymes, such as acyl-activating enzymes, acyl transferase, or lipases, involved in lipid biosynthesis as well as the maturation of LDs (Huang, 1992; Brown, 2001; Athenstaedt and Daum, 2006).

Although the plant oleosins and animal perilipins are well characterized, the structure and function of the green algal MLDP are still not clear. MLDP has been reported in green algal lineages, such as *Chlorella vulgaris*, *Volvox carteri*, *Dunaliella salina*, and *Haematococcus pluvialis*, however not mapped in diatoms, red algae, or seed plants for which sequenced genomes are available (Moellering and Benning, 2010; Peled et al., 2011; Davidi et al., 2012). *D. salina* MLDP shares high homology with conserved proline-rich domain in C-terminal end of green algal linage. In addition, MLDP expression also correlated with the high-lipid accumulation (Davidi et al., 2012). Multiple sequence alignment of major oil globule proteins from *Dunaliella*, *C. reinhardtii*, and *H. pluvialis* revealed the consistence of 21 amino acids conserved motif, with four proline residues close to the C-terminal end. MLDP of green algae does not show any sequence homology to plant oleosins, to mammalian perilipins, and not even to diatoms or to any other alga whose genome is sequenced, suggesting that MLDPs are unique to the green algal lineage (Davidi et al., 2012).

In this present study, we made an attempt to detect the presence of MLDP in *Scenedesmus quadricauda* CASA-CC202 (KM250077) exposed to two practically feasible abiotic stress conditions namely, nitrogen starvation and salt stress. Since *Scenedesmus* genome sequence is not available, the MLDP cloning and its bio characterization or its antibody is not available. Therefore, we have used the other closely related to green microalgae *C. reinhardtii* MLDP antibody to validate the presence of MLDP during the abiotic stress condition. The results reveal that MLDP of *C. reinhardtii* cross-react with *Scenedesmus* MLDP as evident from the confocal images. In concurrence with the increment of MLDP, the current work envisioned to investigate the lipid yield and its associated attributes, such as cell number, specific growth rate, total biomass yield, photosynthetic pigments, and fatty acid composition during salt stress and nitrogen stress. The effect of nitrogen stress triggered lipid accumulation and other parameters were described by Anand and Arumugam (2015). Recently, we have demonstrated that during nitrogen starvation *Scenedesmus* leads to 2.27-fold more lipid accumulation with enlarged cell size. In addition to that, a decrease in protein synthesis, photosynthetic pigments, arrest in cell division, morphological change, and enlarged LDs were also observed during nitrogen-starved condition (Anand and Arumugam, 2015). Also, stress-responsive hormone abscisic acid (ABA) shoots up during nitrogen starvation within 24 h, and the level is falling down subsequently. ABA under nitrogen starvation has played an important role to cope-up the stress and adaptation to the stress environment (Sulochana and Arumugam, 2016). In the present study, the effect of salt on cell number, dry biomass, total lipid yield, and its fatty acid composition are discussed in detail. In general, MLDP a structural protein found in surface of lipid bodies of oleaginous microalgae, no information was reported for *S. quadricauda*, thus the present work aims to show the proof that MLDP indeed present in *S. quadricauda* during nitrogen and salt stress, the general stress condition applied to enhance the lipid yield.

MATERIALS AND METHODS

Strain, Growth Conditions, and Salt Stress Induction

The *S. quadricauda* CASA-CC202 (KM250077) cultures were grown in optimized conditions as described by Anand and Arumugam, 2015. The algal cells were grown at 25 ± 2°C with 14:10 h light–dark period. The flasks were gently shaken twice daily to avoid adherence of the cells to the walls, and there was no air and CO_2 was supplemented during the experimental period. The salt stress was induced when the cells grown to its log phase harvested by centrifugation (10000 rpm for 10 min), and the pellets were washed with double distilled water. Then the pellets were transferred to the stress media with different NaCl concentration (10–50mM), and subsequently, the nitrogen stress was induced (Sulochana and Arumugam, 2016).

Immunostaining of *Scenedesmus quadricauda* MLDP

The experimental algal cells were prefixed with 4% paraformaldehyde, subsequently, the fixed cells were suspended with sensitization solution composed of 0.1M Tris-7.5, 0.15M NaCl, and 0.1% Triton X-100 for 30 min. Sensitized cells were washed twice with 0.1 M Tris-7.5 with 0.1% Triton X-100 in the same tube. MLDP polyclonal antibody was added in 1: 1000 dilution ratio, and incubated for 3 h in rocker at 4°C. Cells were washed thrice with wash solution followed by incubation with secondary antibody (anti-rabbit antibody) conjugated with FITC for two more hours. The cells were washed three times and suspended in 1× PBS and viewed under a confocal microscope.

Growth Characteristics and Biomass Yield under Salt Stress

At every 5-day interval, the samples were drawn aseptically from the stressed and control flasks, and the growth was measured spectrophotometrically at 540 nm. Cell number, a direct microscopic enumeration, was performed at every 5-day intervals as described in Anand and Arumugam (2015). The specific growth rate was measured, number of generation (doubling time), that occurs per unit of time in an exponentially growing culture. The phase of growth was carefully determined, and the specific growth rate was obtained using the following equation (Guillard and Ryther, 1962).

$$\mu = Ln \ (N_t/N_0)/T_t - t_0$$

N_t = number of cells at the end of the log phase; N_0 = number of cells at the start of log phase; T_t = final day of log phase; and T_0 = starting day of log phase. If T expressed in days from the growth rate (μ) can be converted to division or doubling per day (K) by dividing (μ) by the natural log of 2(0.6931). In order to determine the dry biomass yield, 10 ml samples were filtered on to pre-dried and weighed GF/C fiber filters every third day of intervals. The filtered biomasses were oven dried overnight at 60°C along with filter paper, and reweighed using an analytical balance. The dry biomass of *Scenedesmus* under salt stress condition was expressed in grams per liter.

Estimation of Total Lipids

Total lipid was estimated by Folch et al. (1957) method for different NaCl-treated *S. quadricauda*. In brief, 5 ml of culture was taken from 400 ml sample in a 1000 ml Erlenmeyer flask. The biomass was collected by centrifugation at 10000 rpm for 10 min and suspended in 6 ml of chloroform:methanol (2:1) for 5 h, then 2 ml of 0.9% saline was added and mixed vigorously, and incubated for 12 h. From the lipid-containing lower phase, 0.5 ml was pipetted out and transferred to a fresh 2 ml Eppendorf tube and allowed for evaporation. After that, 0.5 ml of concentrated sulfuric acid was added and mixed well, and the tubes were kept in a water bath for 15 min and cooled to room temperature. From that, 0.2 ml was pipetted out and transferred to the fresh test tube containing 5 ml of vanillin reagent, was mixed well, and incubated for 20 min at room temperature. After incubation, the appearance of pink color indicates the presence of lipid, and then the sample was read at 520 nm.

Estimation of Total Chlorophyll Pigments

The total photosynthetic algal sample was taken and centrifuged at 8000 rpm for 10 min and the supernatant was discarded. Then, the pellet was resuspended in the 100% acetone, sonicated

for 2 min, and covered with aluminum foil and kept at 4°C for overnight. The next day, it was centrifuged for 10 min at 8000 rpm and the supernatant was collected in a fresh test tube and the reading was taken at subsequent wavelength. The supernatant was collected, and its optical density was measured and recorded at 644.8 λ, 661.6 λ, and 470 λ in UV-Visible spectrophotometer (Lichtenthaler, 1987).

Scanning Electron Microscopy

The salt-stressed algal cells were withdrawn at day 20 time point, and the algal cells were centrifuged at 8000 rpm for 5 min. Cell pellets were fixed using a fixative containing 2.5% glutaraldehyde and 2% paraformaldehyde in 0.1 M sodium phosphate (pH 7.0). Scanning electron micrographs were captured using a Scanning Electron microscope operating at 100 kV.

Lipid Extraction, Transesterification, and FAME Analysis

Cells were filtered after harvesting with GC filter paper; the pellet was dried in an oven overnight at 80°C for drying. From that, 1 g of dry algal biomass was processed further for total lipid extraction method. 1 g of dried algal biomass was suspended in 200 ml of chloroform:methanol (2:1), with 500 ml round bottom flask. This setup ran with soxhlet apparatus, for 12 h. Afterward, within the solvent mixture 40 ml of 0.9% saline solution was added, mixed well, and allowed for phase separation. The lipid-containing lower phase was collected in fresh beaker and washed well with methanol and water (1:1) ratio and allowed for phase separation to occur in separating funnel. Lipid-containing lower phase was collected in pre-weighed glass container and kept at room temperature to remove the residual solvent. The lipid yield was quantified gravimetrically, and the same was subjected to TLC or Gas chromatography (Folch et al., 1957; Cristie, 1982).

The total lipid sample was resolved by TLC along with appropriate standards. The transesterification of total lipid was performed as described by Marinkovic and Tomasevic (1998), with little modification. In brief, the cell pellet was re-suspended in 5% of sulfuric acid and methanol (V/V), followed by vigorous vortex for 30 s, and kept in a water bath for 2 h at 90°C, then allowed to cool. 1.5 ml of 0.9% saline solution (w/v) was added to the cooled mixture followed by 1 ml of hexane, vortexed well. The resulting mixture was centrifuged briefly and the upper hexane phase was transferred to a fresh tube. The resulting hexane containing FAME was sequentially washed with sodium bicarbonate followed by water and dissolved using anhydrous sodium sulfate. The dried sample was dissolved in hexane, and fatty acid profile was analyzed by gas chromatography (Su et al., 2011; Xin et al., 2011; Anand and Arumugam, 2015).

RESULTS AND DISCUSSION

Conservation of MLDP in Green Microalgae

The LD mainly comprises a globular neutral lipid core surrounded by a membrane lipid monolayer and it is highly conserved in different species (Murphy et al., 2001). A recent study of LDs in *C. reinhardtii* showed the abundance of MLDP (Moellering and Benning, 2010; James et al., 2011). The phylogenetic analysis shows that algal MLDP was highly conserved among the aligned sequence of *Dunaliella*, *H. pluvialis*, *C. reinhardtii*, and *V. carteri* (Davidi et al., 2012). The secondary structure prediction of MLDP shows that they are highly structured α-helices and a short unstructured domain toward the C-terminal end (Guermeur et al., 1998). MLDP comprises a highly conserved motif proline-rich domain of about 21 amino acids in their C-terminal end. All the known MLDPs have two hydrophobic stretch first at 30 and 50 and the second region between 150 and 250 amino acid residues. Even though it is well studied in other green algae, the presence of MLDP, or like proteins was not yet elucidated in *S. quadricauda* being a suitable strain of biofuel application. The current study confirms the presence of MLDP-like protein conserved in *S. quadricauda* genome.

Detection of MLDP during Abiotic Stress-Immunostaining

Major lipid droplet protein was found out to be the most abundant protein in *C. reinhardtii* (Moellering and Benning, 2010; Nguyen et al., 2011). Tsai et al. (2014) have reported that MLDP abundance positively correlates with the TAG accumulation, indicating the coregulation of LD and TAG. On the other hand, the lipid MLDP-RNAi silenced lines show about 40% increased LD size in *C. reinhardtii* under nitrogen-starved condition (Moellering and Benning, 2010). Collectively, in microalgae LD formation, maturation during abiotic stress is tightly regulated through MLDP and its associated partners. Not much information available for *S. quadricauda*, regarding the presence of MLDP and its abundance during the nitrogen starvation or salt stress condition.

Here, we report that *C. reinhardtii* MLDP polyclonal antibody (Huang et al., 2013) recognizes and cross-react with the *S. quadricauda* MLDP, indicating the presence and conservation of MLDP-like protein in green microalga linage (**Figure 1**). Additionally, we also tested the relative abundance of MLDP in two important stress conditions like nitrogen starvation and salt stress condition; confocal imaging reveals the accumulation of MLDP in both the stress condition (**Figure 1**). Evolutionarily, MLDP or MLDP-like proteins are functionally conserved among green microalgae. However, it is interesting to investigate more about the regulation of MLDP-like protein and its other partners in lipid body maturation during abiotic stress.

Influence of Salt Stress on Cell Growth in *S. quadricauda*

The increase in salt concentration exhibits negative effects on the growth of *S. quadricauda*. The maximum cell density occurs in the control (0.583) (BBM medium having 0.43mM salt concentration) and in 10mM (0.451) salt-treated cells. Increasing concentrations of salt (30, 40, and 50mM) lead to a significant amount of reduction in cell density. A similar effect was also observed in the case of cell number. The maximum cell numbers were recorded at 715×10^4 cells/ml in control, followed by 10 and 20 mM NaCl concentration with 497 and 391×10^4 cells/ml, respectively (**Figure 2**).

Similarly, Hyder and Greenway (1965) found out that salinity arrests the growth of microalgae. The salt-tolerant studies on *C. vulgaris* by Talebi et al. (2013) show decrement in growth and moderate resistance up to 0.5M NaCl concentration. Also, salinity more than 0.1M completely inhibited the growth of *C. emersoni*,

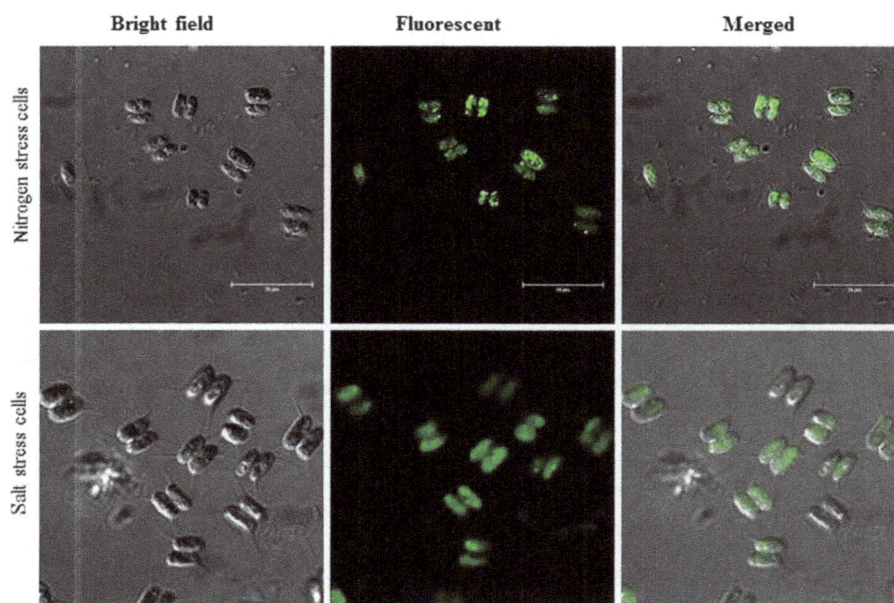

FIGURE 1 | Immunostaining lipid body structural protein fluorescent images under nitrogen-starved and salt stress cells. Scale bar represent 20 μm in each panel.

and cell density and biomass becomes completely vanished during prolonged stress.

The control *S. quadricauda* sample had significant SGR compared to higher NaCl-treated sample (0.241–0.134), respectively. An increase (30, 40, 50mM) in the NaCl concentration showed a decline in the SGR and doubling/d (*K*) occurred in the culture (0.134–0.193), respectively (**Table 1**). Although the doubling time under salt stress is negatively correlated with the SGR, there was a raise in doubling time (4.98, 5.02, 5.17) in the all higher (30, 40, 50mM) NaCl concentration, respectively. Whereas control had very short doubling time (2.87)(Tt) compared to other higher (30, 40, 50mM) NaCl concentration (**Table 1**).

Effect of Salt Stress on Biomass of *S. quadricauda*

The *S. quadricauda* biomass content was considerably higher at the salinities of control (0.43mM) and 10mM compared to 40 and 50 mM NaCl concentration. The maximum dry biomass content of *S. quadricauda* was obtained of about 236 mg/L dry biomass in 20th day in control, whereas in 50 mM NaCl concentration the biomass content of 20th day obtained at the 129 mg/L

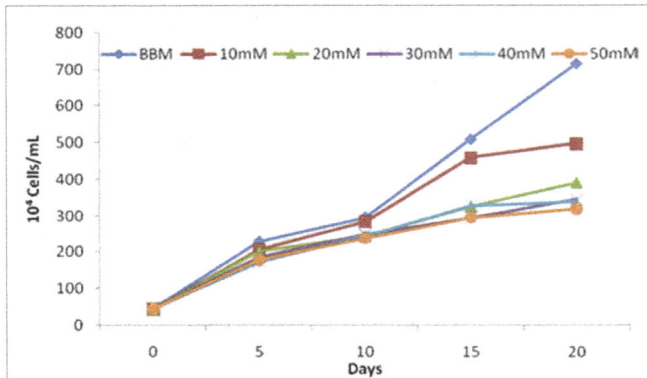

FIGURE 2 | *Scenedesmus quadricauda* cell numbers under salt stress condition. Each bar represent different concentration of salt levels.

TABLE 1 | Specific growth rate of *S. quadricauda* under different Salt stress.

	Control	10mM	20mM	30mM	40mM	50mM
Specific growth rate (μ)	0.241	0.174	0.161	0.139	0.138	0.134
Doubling/day (*K*)	0.348	0.253	0.233	0.201	0.200	0.193
Doubling time (Tt)	2.87	3.96	4.30	4.98	5.02	5.17

dry biomass (**Table 2**). The algal biomass production and NaCl concentration are negatively correlated each other in such a way that there was an increment in the biomass production observed at lower NaCl concentrations.

The *Chlorella* showed the maximum biomass content of about 1.021 g/L under 0.2 M NaCl concentrations. Similar effects were also observed in *Chlamydomonas mexicana* and *Scenedesmus obliquus* (Salama et al., 2013; Mohan and Devi, 2014). In contrast, the biomass yield was increased in the salinity stress, i.e., 17–34 mM in green microalgae, *Botryococcus braunii*, was also reported (Rao et al., 2007).

Influence of Salt Stress on Lipid in *S. quadricauda*

The maximum lipid yield was observed at 20th day in salt-treated cells compared to control. But the algal growth was slightly inhibited at higher NaCl concentration (40 and 50mM). The total lipid yield was increased in 30, 40, 50 mM NaCl-treated cultures (191 and 205 mg/L) compared to control (118 mg/L) on 15th day (**Figure 3**). As the trend, the lipid yield was also found to be increased in higher concentration on 20th day. It was about 208, 195, and 201 mg/L in 30, 40, and 50 mM, respectively (**Figure 3**). It was also further confirmed by TLC with appropriate standard for TAG (Supplementary Image 1).

Similarly, the total lipid content increased under higher salt concentration in *Dunaliella* were reported by Takagi et al. (2006), and it is possibly due to the cells adapted to the stress condition and accumulation of more lipids. The higher concentration may

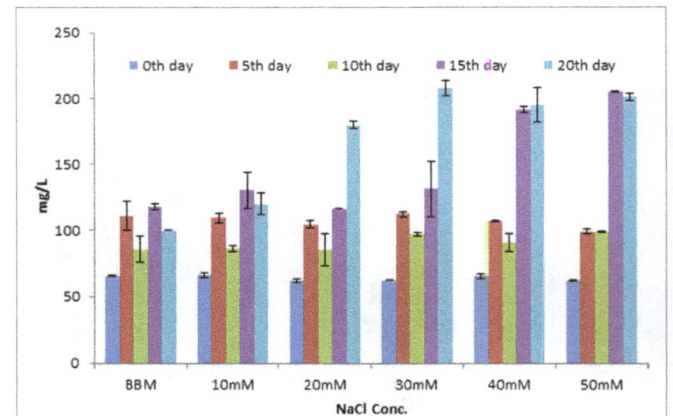

FIGURE 3 | Estimation of total lipid content in *Scenedesmus quadricauda* in different salt stress condition.

TABLE 2 | Biomass content of *S. quadricauda* under different salt stress condition.

	0th day	5th day	10th day	15th day	20th day
BBM	21.5 ± 1.5	48.0 ± 1.0	109.7 ± 3.1	146.29 ± 2.3	236.61 ± 3.2
10mM	19.5 ± 1.5	35.0 ± 3.0	78.95 ± 3.6	118.07 ± 3.7	178.61 ± 1.6
20mM	21.5 ± 1.5	32.5 ± 2.3	71.75 ± 3.2	99.8 ± 1.8	145.96 ± 2.6
30mM	22.5 ± 1.5	32.75 ± 0.7	67.75 ± 1.5	90.3 ± 2.3	148.10 ± 1.8
40mM	20.0 ± 1.0	30.8 ± 2.2	54.15 ± 1.8	82.81 ± 3.1	138.26 ± 2.2
50mM	20.0 ± 0.0	29.5 ± 0.5	48.38 ± 0.61	81.5 ± 1.5	129.5 ± 0.5

mediate lipid accumulation by nutrient starvation. And also, the algal cells incubated under saline condition for a longer time period revealed an enhancement in the lipid accumulation within the stressed algal cells. During salt stress condition, pronounced variation in pH, TDS, salinity, and conductivity were observed in *Scenedesmus* (Supplementary Table 1). According to Rao et al. (2007), the pH of the algal cells varies with respect to salt concentration in green microalgae *B. braunii*.

Influence of Salt Stress on Photosynthetic Pigment in *S. quadricauda*

The similar pattern of decrease in photosynthetic pigment was also seen in different NaCl concentration. The total photosynthetic pigment level varies gradually *viz* Chl-a, Chl-b, carotenoids (0.110 ± 0.01, 0.065 ± 0.02, 0.069 ± 0.00 mg/L) in 50 mM NaCl concentration on 20th day (**Table 3**). The photosynthetic pigments in salt-treated algal cells (50 mM) showed that there is a reduction from 0.377 (Control) to 0.110 g/L of Chl-a on the 20th day of stress. Also, the similar trend was observed with chlorophyll b and carotenoids during salt stress.

There was a report suggesting that the reduction in photosynthetic rate is directly proportional to that of reduction in the growth rate of microalgae (Ben-Amotz et al., 1985). The salt tolerance studies in *Chlorella* sp. showed that they can tolerate the lower NaCl concentration, and it can grow same as normal cells, but with an increased NaCl concentration revealed degradation of chlorophyll, and it leads to cell death (Salama et al., 2013; Mohan and Devi, 2014). Thus, an optimum NaCl concentration is required for the normal growth of microalgae, higher or lower concentration of salt concentration leads to a decrement in their growth rate (Takagi et al., 2006; Ruangsomboon, 2012).

Influence of Salt Stress in Fatty Acid Analysis of *S. quadricauda*

The fatty acid composition of *S. quadricauda* under salt stress revealed that there is a marked difference in the saturated: monounsaturated fatty acid composition. The gas chromatogram of control showed that about 79.2% of saturated and 20.8% monounsaturated fatty acids (**Figure 4**). But salt-treated (50 mM) has about 59.2% of saturated fatty acids, i.e., 20% reduction in saturated fatty acid level. Also, the monounsaturated level is 40.7% which is about 19.9% increment in monounsaturated fatty acids (**Figure 4**; Supplementary Image 2).

Here, the FAME composition reveals that the altered fatty acid profile is an appropriate proportion for biofuel production. Also, we can mix appropriate proportion of sea water to the media to cultivate the algae in defined concentration, which makes a reduction in biomass production cost and decreases the

TABLE 3 | Photosynthetic pigment content of *S. quadricauda* under salt stress.

Days	Photosynthetic pigment content of *S. quadricauda* under NaCl salt stress (mg/L)					
	Control	10mM	20mM	30mM	40mM	50mM
0th						
Chl-a	0.007 ± 0.00	0.008 ± 0.00	0.018 ± 0.01	0.008 ± 0.00	0.049 ± 0.04	0.049 ± 0.04
Chl-b	0.074 ± 0.01	0.026 ± 0.01	0.032 ± 0.02	0.026 ± 0.01	0.016 ± 0.00	0.016 ± 0.00
Carotenoid	0.006 ± 0.00	0.009 ± 0.00	0.018 ± 0.00	0.006 ± 0.00	0.021 ± 0.00	0.023 ± 0.00
Chl-a + b	0.081	0.034	0.05	0.034	0.065	0.065
Chl-a/b	0.094	0.307	0.562	0.307	3.062	3.062
5th						
Chl-a	0.164 ± 0.01	0.226 ± 0.03	0.196 ± 0.01	0.155 ± 0.01	0.164 ± 0.01	0.139 ± 0.00
Chl-b	0.064 ± 0.00	0.089 ± 0.01	0.071 ± 0.00	0.048 ± 0.00	0.064 ± 0.00	0.054 ± 0.00
Carotenoid	0.136 ± 0.010	0.149 ± 0.03	0.143 ± 0.00	0.125 ± 0.01	0.125 ± 0.00	0.104 ± 0.00
Chl-a + b	0.228	0.315	0.267	0.203	0.228	0.193
Chl-a/b	2.562	2.539	2.760	3.229	2.562	2.574
10th						
Chl-a	0.419 ± 0.05	0.343 ± 0.08	0.164 ± 0.00	0.105 ± 0.00	0.090 ± 0.00	0.068 ± 0.00
Chl-b	0.132 ± 0.03	0.093 ± 0.00	0.054 ± 0.00	0.018 ± 0.00	0.015 ± 0.01	0.013 ± 0.00
Carotenoid	0.312 ± 0.04	0.254 ± 0.04	0.151 ± 0.00	0.096 ± 0.00	0.081 ± 0.00	0.056 ± 0.00
Chl-a + b	0.551	0.436	0.218	0.123	0.105	0.081
Chl-a/b	3.17	3.688	3.037	5.833	6	5.230
15th						
Chl-a	0.287 ± 0.01	0.228 ± 0.00	0.180 ± 0.01	0.139 ± 0.01	0.123 ± 0.02	0.074 ± 0.01
Chl-b	0.066 ± 0.01	0.021 ± 0.00	0.019 ± 0.00	0.006 ± 0.00	0.002 ± 0.00	0.011 ± 0.00
Carotenoid	0.268 ± 0.01	0.244 ± 0.03	0.161 ± 0.00	0.088 ± 0.00	0.075 ± 0.00	0.059 ± 0.01
Chl-a + b	0.353	0.249	0.199	0.145	0.125	0.085
Chl-a/b	4.348	10.85	9.473	23.166	61.5	6.727
20th						
Chl-a	0.377 ± 0.04	0.227 ± 0.02	0.221 ± 0.01	0.167 ± 0.00	0.151 ± 0.01	0.110 ± 0.01
Chl-b	0.302 ± 0.04	0.165 ± 0.04	0.231 ± 0.03	0.178 ± 0.00	0.088 ± 0.00	0.065 ± 0.01
Carotenoid	0.300 ± 0.06	0.218 ± 0.00	0.175 ± 0.00	0.098 ± 0.01	0.096 ± 0.00	0.069 ± 0.00
Chl-a + b	0.679	0.392	0.452	0.345	0.239	0.175
Chl-a/b	1.248	1.375	0.956	0.938	1.715	1.692

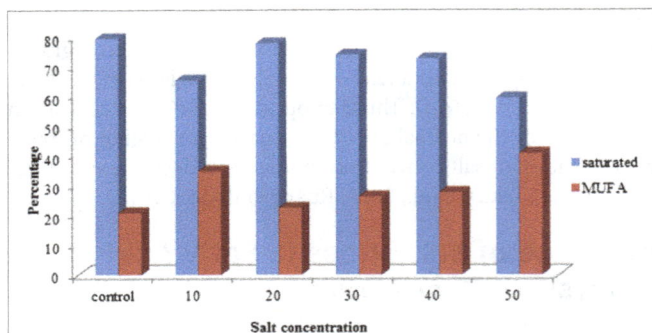

FIGURE 4 | Comparison of FAME concentration of saturated and monounsaturated fatty acid profile of *Scenedesmus quadricauda* under salt stress.

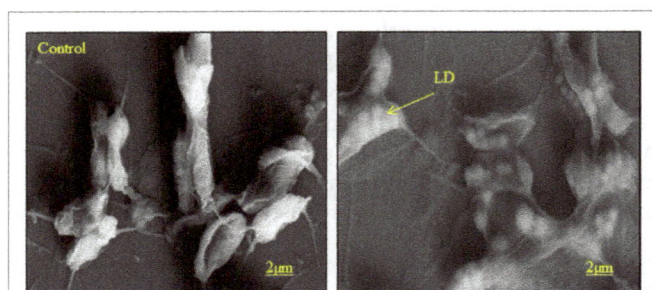

FIGURE 5 | Scanning electron microscope images of *S. quadricauda* under salt stress condition (LD) lipid droplet. Scale bars represent 2 μm size.

fresh water requirement for algal cultivation. Abundant availability of seawater can be used for algal cultivation with biofuel application. According to Rao et al. (2007), the presence of monounsaturated and polyunsaturated fatty acids was detected in *B. braunii* during salt stress with palmitoleic and oleic acids as major fatty acids.

Influence of Salt Stress in SEM Analysis of *S. quadricauda*

A time course LD formation during nitrogen-starved condition was performed in *C. reinhardtii* (Moellering and Benning, 2010). The results reveal that first LD was observed at 12 h of N deprivation, followed by the number as well as the large size

of LDs formed during 24 or 48 h of N deprivation (Moellering and Benning, 2010). In the present study, we also observed an LD-like structure in scanning electron microscopy shows (**Figure 5**) more lipid body formation in higher NaCl concentration (50 mM), such structure was not observed in cells grown in normal condition.

The previous reports describe that the *Chlamydomonas* wild and mutant has accumulated more storage lipid bodies inside the cell by an external environmental factor (Li et al., 2010). In the green microalgae, *C. minutissima* shows in the oil body formation in environmental stress factors (Wang et al., 2011).

CONCLUSION

Presence of major LD protein was detected in *S. quadricauda* in salt stress and nitrogen-starved condition. Thus, the MLDP is highly conserved in green algae, and the size and number are increased during the abiotic stress condition like nitrogen starvation and salt stress. Conclusively, MLDP-like protein can be used as a marker for stress-mediated lipid accumulation during abiotic stress.

AUTHOR CONTRIBUTIONS

Design and plan of experiments: MA. Performed the experiments: AJ, SS, and SP. Analyzed the data: MA, SS, and AJ. Contributed reagents: Dr. Anthony Huang. Wrote the paper: MA, SS, and JA.

FUNDING

The work was supported by the grant from DST-SERB, Government of India through the fast track project entitled "SB/YS/LS-13/2013": "Biochemical and Molecular investigation on stress-mediated lipid accumulation and biomass productivity in microalgae" to MA. We are thankful to Dr. Anthony Huang of Institute of Plant and Microbial Biology, University of California, Riverside for providing MLDP antibody. We are also grateful to Dr. R. Arumugam for confocal imaging at Rajiv Gandhi Centre for Biotechnology, Trivandrum, India.

REFERENCES

Anand, J., and Arumugam, M. (2015). Enhanced lipid accumulation and biomass yield of *Scenedesmus quadricauda* under nitrogen starved condition. *Bioresour. Technol.* 188, 190–194. doi:10.1016/j.biortech.2014.12.097

Athenstaedt, K., and Daum, G. (2006). The life cycle of neutral lipids: synthesis, storage and degradation. *Cell. Mol. Life Sci.* 63, 1355–1369. doi:10.1007/s00018-006-6016-8

Ben-Amotz, A., Tornabene, T. G., and Thomas, W. H. (1985). Chemical profile ofselected species of microalgae with special emphasis on lipids. *J. Phycol.* 21, 72–81. doi:10.1111/j.0022-3646.1985.00072

Brown, D. A. (2001). Lipid droplets: proteins floating on a pool of fat. *Curr. Biol.* 11, R446–R449. doi:10.1016/S0960-9822(01)00257-3

Cristie, W. W. (1982). *Lipid Analysis, Isolation, Separation, Identification and Structural Analysis of Lipids*. Oxford: Pregamon Press.

Davidi, L., Katz, A., and Pick, U. (2012). Characterization of major lipid droplet proteins from *Dunaliella*. *Planta* 236, 19–33. doi:10.1007/s00425-011-1585-7

Folch, J., Lees, M., and Sloane-Stanley, G. (1957). A simple method for the isolation and purification of total lipids from animal tissues. *J. Biol. Chem.* 226, 497–509.

Guermeur, Y., Paugam-Moisy, H., and Gallinari, P. (1998). "Multivariate linear regression on classifier outputs: a capacity study," in *ICANN98* (London: Springer), 693–698.

Guillard, R. R. L., and Ryther, J. H. (1962). Studies on marine planktonic diatoms I. *Cyclotella nana* Hustedt and *Detonulaconfervacea* (Cleve) Gran. *Can. J. Micro.* 8, 229–239. doi:10.1139/m62-029

Huang, A. H. (1992). Oil bodies and oleosins in seeds. *Ann. Rev. Plant Physiol.* 43, 177–200. doi:10.1146/annurev.pp.43.060192.001141

Huang, C. N., Huang, M. D., Chen, T. L., and Huang, A. H. (2013). Oleosin of subcellular lipid droplets evolved in green algae. *Plant Physiol.* 161, 1862–1874. doi:10.1104/pp.112.212514

Hyder, S. Z., and Greenway, H. (1965). Effects of Ca++ on plant sensitivity to high NaCl concentrations. *Plant Soil* 23, 258–260. doi:10.1007/BF01358351

James, G. O., Hocart, C. H., Hillier, W., Chen, H., Kordbacheh, F., Price, G. D., et al. (2011). Fatty acid profiling of *Chlamydomonas reinhardtii* under nitrogen deprivation. *Bioresour. Technol.* 102, 3343–3351. doi:10.1016/j.biortech.2010.11.051

Li, Y., Han, D., Hu, G., Dauvillee, D., Sommerfeld, M., Ball, S., et al. (2010). *Chlamydomonas* starchless mutant defective in ADP-glucose pyrophosphorylase hyper-accumulates triacylglycerol. *Metab. Eng.* 12, 387–391. doi:10.1016/j.ymben.2010.02.002

Lichtenthaler, H. K. (1987). "Chlorophylls and carotenoids, the pigments of photosynthetic biomembranes," in *Methods Enzymol*, Vol. 148, eds R.Douce and L.Packer (New York, NY: Academic Press Inc.), 350–382.

Marinkovic, S. S., and Tomasevic, A. (1998). Transesterification of sunflower oil in situ. *Fuel* 77, 1389–1391. doi:10.1016/S0016-2361(98)00028-3

Moellering, E. R., and Benning, C. (2010). RNA Interference silencing of a major lipid droplet protein affects lipid droplet size in *Chlamydomonas reinhardtii*. *Eukaryot. Cell* 9 97–106. doi:10.1128/EC.00203-09

Mohan, S. V., and Devi, M. P. (2014). Salinity stressinduced lipid synthesis to harness biodiesel duringdual mode cultivation of mixotrophicmicroalgae. *Bioresour. Technol.* 165, 288–294. doi:10.1016/j.biortech.2014.02.103

Murphy, W. J., Eizirik, E., O'Brien, S. J., Madsen, O., Scally, M., Douady, C. J., et al. (2001). Resolution of the early placental mammal radiation using Bayesian phylogenetics. *Science* 294, 2348–2351. doi:10.1126/science.1067179

Nguyen, H. M., Baudet, M., Cuine, S., Adriano, J. M., Barthe, D., Billon, E., et al. (2011). Proteomic profiling of oil bodies isolated from the unicellular green microalga *Chlamydomonas reinhardtii*: with focus on proteins involved in lipid metabolism. *Proteomics* 11, 4266–4273. doi:10.1002/pmic.201100114

Peled, E., Leu, S., Zarka, A., Weiss, M., Pick, U., Khozin-Goldberg, I., et al. (2011). Isolation of a novel oil globule protein from the green alga *Haematococcus pluvialis* (Chlorophyceae). *Lipids* 46, 851–861. doi:10.1007/s11745-011-3579-4

Rao, A. R., Dayananda, C., Sarada, R., Shamala, T. R., and Ravishankar, G. A. (2007). Effect of salinity on growth of green alga *Botryococcus braunii* and its constituents. *Bioresour. Technol.* 98, 560–564. doi:10.1016/j.biortech.2006.02.007

Ruangsomboon, S. (2012). Effect of light, nutrient, cultivation time and salinity on lipid production of newly isolated strain of the green microalga, *Botryococcus braunii* KMITL 2. *Bioresour. Technol.* 109, 261–265. doi:10.1016/j.biortech.2011.07.025

Salama, E. S., Kim, H. C., Abou-Shanab, R. A., Ji, M. K., Oh, Y. K., Kim, S. H., et al. (2013). Biomass, lipid content, and fatty acid composition of freshwater *Chlamydomonas mexicana* and *Scenedesmus obliquus* grown under salt stress. *Bioprocess Biosyst. Eng.* 36, 827–833. doi:10.1007/s00449-013-0919-1

Schmidt, M. A., and Herman, E. M. (2008). Suppression of soybean oleosin produces micro-oil bodies that aggregate into oil body/ER complexes. *Mol. Plant* 1, 910–924. doi:10.1093/mp/ssn049

Siloto, R. M., Findlay, K., Lopez-Villalobos, A., Yeung, E. C., Nykiforuk, C. L., and Moloney, M. M. (2006). The accumulation of oleosins determines the size of seed oil bodies in *Arabidopsis*. *Plant Cell* 18, 1961–1974. doi:10.1105/tpc.106.041269

Su, C.-H., Chien, L.-J., Gomes, J., Lin, Y. S., Yu, Y. K., Liouand, J. S., et al. (2011). Factors affecting lipid accumulation by *Nannochloropsis oculata* in a two-stage cultivation process. *J. Applphycol.* 23, 903–908. doi:10.1007/s10811-010-9609-4

Sulochana, S. B., and Arumugam, M. (2016). Influence of abscisic acid on growth, biomass and lipid yield of *Scenedesmus quadricauda* under nitrogen starved condition. *Bioresour. Technol.* 213, 198–203. doi:10.1016/j.biortech.2016.02.078

Takagi, M., Karseno, Y., and Yoshida, T. (2006). Effect of salt concentration on intracellular accumulation of lipids and triacylglyceride in marine microalgae *Dunaliella* cells. *J. Biosci. Bioeng.* 101, 223–226. doi:10.1263/jbb.101.223

Talebi, A. F., Tabatabaei, M., Mohtashami, S. K., Tohidfar, M., and Moradi, F. (2013). Comparative salt stress study on intracellular ion concentration in marine and salt-adapted freshwater strains of microalgae. *Not. Sci. Biol.*, 5, 309–315. doi:10.15835/nsb.5.3.9114

Tsai, C. H., Warakanont, J., Takeuchi, T., Sears, B. B., Moellering, E. R., and Benning, C. (2014). The protein compromised hydrolysis of triacylglycerols 7 (CHT7) acts as a repressor of cellular quiescence in *Chlamydomonas*. *Proc. Natl. Acad. Sci. U.S.A* 111, 15833–15838. doi:10.1073/pnas.1414567111

Wang, S. T., Pan, Y. Y., Liu, C. C., Chuang, L. T., and Chen, C. N. N. (2011). Characterization of a green microalga UTEX 2219-4: effects of photosynthesis and osmotic stress on oil body formation. *Bot. Stud.* 52, 305–312.

Xin, L., Hong-yingand, H., and Yu-ping, Z. (2011). Growth and lipid accumulation properties of a freshwater microalga *Scenedesmus* sp. under different cultivation temperature. *Bioresour. Technol.* 102, 3098–3102. doi:10.1016/j.biortech.2010.10.055

Zweytick, D., Leitner, E., Kohlwein, S. D., Yu, C., Rothblatt, J., and Daum, G. (2000). Contribution of Are1p and Are2p to steryl ester synthesis in the yeast *Saccharomyces cerevisiae*. *Eur. J. Biochem.* 267, 1075–1082. doi:10.1046/j.1432-1327.2000.01103

Conflict of Interest Statement: The authors declare that the research was conducted in the absence of any commercial or financial relationships that could be construed as a potential conflict of interest.

10

Design of a Stand-Alone Photovoltaic Model for Home Lightings and Clean Environment

Vincent Anayochukwu Ani*

Department of Electronic Engineering, University of Nigeria, Nsukka, Nigeria

This paper gives a well-documented health risk of fuel-based lighting (kerosene lamps and fuel-powered generators) and proposed a design of a stand-alone solar PV system for sustainable home lightings in rural Nigerian area. The design was done in three different patterns of electricity consumptions with energy efficient lightings (EELs) using two different battery types (Rolls Surrette 6CS25PS and Hoppecke 10 OpzS 1000) on; (i) judicious power consumption, (ii) normal power consumption, and (iii) excess power consumption; and compared them with the incandescent light bulb consumption. The stand-alone photovoltaic energy systems were designed to match the rural Nigerian sunlight and weather conditions to meet the required lightings of the household. The objective function and constraints for the design models were formulated and optimization procedures were used to demonstrate the best solution (reliability at the lowest lifecycle cost). Initial capital costs as well as annualized costs over 5, 10, 15, 20, and 25 years were quantified and documented. The design identified the most cost-effective and reliable solar and battery array among the patterns of electricity consumption with EEL options (judicious power consumption, normal power consumption, and excess power consumption).

Keywords: energy cost, power consumption, electricity, kerosene lamps, fuel-powered generators

Edited by:
Joni Jupesta,
Asean Centre For Energy, Indonesia

Reviewed by:
Behnam Mohammadi-Ivatloo,
University of Tabriz, Iran
Payman Dehghanian,
Texas A&M University, USA

*Correspondence:
Vincent Anayochukwu Ani
vincent_ani@yahoo.com

Specialty section:
This article was submitted to Energy
Systems and Policy,
a section of the journal
Frontiers in Energy Research

Citation:
Ani VA (2016) Design of a Stand-
Alone Photovoltaic Model for Home
Lightings and Clean Environment.
Front. Energy Res. 3:54.

INTRODUCTION

Access to adequate, reliable, sustainable, and affordable modern energy services facilitates basic household comforts. Electricity is an important resource to support economic and social development of any society; it is, in fact, one of the discoveries that have transformed mankind. Modern society's reliance on electrical power is so great that it is considered a basic need (Foster et al., 2010). As there are people who cannot imagine life without electricity, there are others who have never enjoyed the beauty of electricity. To them, they cannot imagine what electricity is? In Nigeria, especially in rural areas, many villagers are not connected to national grid and it will take time for them to be connected to the grid due to great expense in grid expansion and inefficiency of the total national installed capacity. It is very unfortunate that people in rural areas and among the poorest populations have to use kerosene lamps that have very poor light. Most of the people living in rural areas of Nigeria need electricity mostly for lighting, and lighting is the dominant use of kerosene in rural areas and among the poorest populations. Although, kerosene lamps are comparatively cheaper and easier to replace and viable alternative lighting sources already exist in rural areas and among the poorest populations, they are expensive to operate. Kerosene is costly both for low income households that buy it and for governments that subsidize it. Kerosene costs make up 10–25% of household monthly budgets (Lighting Africa, 2013). Moreover, the widespread use of kerosene for lighting in households poses health risks (Lam et al., 2012a; Mills, 2012a), and these chronic exposure to pollution from kerosene lamps is thus a concern for households.

Fuel-based lighting (kerosene lamps and fuel-powered generators) contributes to climate change, which is also a health and safety risk by releasing substantial amounts of greenhouse-gas emissions (Mills, 2005) and black carbon (Lam et al., 2012a). Research has shown that kerosene lamps are significant sources of atmospheric black carbon. Combustion of kerosene emits many health-damaging pollutants, including particulate matter (PM), carbon monoxide (CO), formaldehyde (CH_2O), polycyclic aromatic hydrocarbons (PAH), sulfur dioxide (SO_2), nitrogen oxides (NOx), and various volatile organic carbons (VOCs) (Mills, 2012b). Kerosene lamps emit both carbon dioxide (CO_2) and black carbon. Black carbon is the result of incomplete combustion of fossil fuel. Black carbon particles absorb sunlight and heat the atmosphere, increasing radiative forcing, and have a powerful but short-lived warming influence, known as a "short-lived climate pollutant" (SLCP). The particulate emissions of kerosene lanterns represent significant amounts of black carbon, strongly implicated in climate change (Zai et al., 2006; Lam et al., 2012a).

Well-documented health risks of kerosene lamps show that exposure to the lamps, which are used indoors and in close proximity to people pose significant health impacts – chronic illness resulting from inhalation of fumes; impairs lung function and increases the risk for respiratory disease, cancer, eye problems, and infectious disease, including tuberculosis, a major health issue in Nigeria (Lam et al., 2012a; Mills, 2014). Inhalation of particulates resulting from indoor combustion causes a range of adverse health effects ranging from tuberculosis to cancer (Dominici et al., 2003; Bai et al., 2007) and asthma (Lam et al., 2012b). Potentially harmful effects include impairment of ventilatory function (Behera et al., 1991) and acute lower respiratory infection (Sharma et al., 1998). Many households use kerosene for lighting (UNEP/GEF en.lighten initiative, 2013), while fuel-based lanterns are burned largely indoors and in close proximity to people, but no corresponding estimates of mortality have been made. Kerosene is highly flammable and there is a high risk of accidents, burns, and even fatalities associated with lamp use. There is no national estimate of burn-injuries attributable to fuel-based lighting. However, it is known that more than 95% of deaths nationwide from fires and all types of burns occur in the rural areas where kerosene lamps are commonly used for lighting. For instance, three multi-year reviews of admissions to Nigerian hospitals attributed approximately 30% of all burn cases to kerosene lamps (Olaitan et al., 2007; Asuquo et al., 2008; Oludiran and Umebese, 2009). Also, thousands of people are maimed each year by lamp explosions, with a 13% fatality rate (Lighting Africa, 2010). Moreover, poor light quality from kerosene lamps, which are often the sole source of lighting after daylight, limits productivity and opportunities for studying (insufficient illumination can lead to poor visual performance, fatigue, and eye strain) as shown in **Figure 1**.

Another type fuel-based power generation, known as stand-alone fuel-powered generators, a common off-grid source of

FIGURE 1 | A student studying with kerosene lamp at night.

light, emits concentrations of CO_2 and particulates per unit of power generation (Natural Resources Canada, 2008; Gilmore et al., 2010; Edenhofer et al., 2011; Ani and Emetu, 2013). Most households use fuel-based electricity generators to augment the shortfall of electricity supply, and these generators emit gaseous pollutants that negatively affect the environment. Information from Scientific Advisory Panel (2013) shows that fuel-based generators are an important source of black carbon emissions in Nigeria where public power supply lags behind electricity demand. Nationally, both stand-alone generators and kerosene lamps are significant sources of particles of black carbon, a SLCP whose contribution to total carbon emissions is particularly significant in rural areas of Nigeria. This is a growing concern to climate scientists in light of the widespread use of such generators for electrification in off-grid areas (Lam et al., 2012b; Scientific Advisory Panel, 2013). Controlling these sources (stand-alone generators and kerosene lamps) would not only reduce air pollution but also provide clear climate benefits.

Modern off-grid lighting alternative such as solar photovoltaic (PV) systems are generally safer and healthier than fuel-based lighting (kerosene lamps and fuel-powered generators), and have longer product lives, and lower lifecycle costs. Thus, alternatives to fuel-based lighting are an attractive area for achieving quick and cost-effective climate benefits. Moreover, in addition to mitigating climate change, there are major health and development co-benefits to be attained by upgrading from fuel-based to solar PV systems.

Nigeria being in equatorial region, very close to the Equator, has a very high potential of solar energy but, due to lack of awareness many villagers in rural areas do not even think of installing solar PV systems. Very few villagers have taken initiatives and installed PV systems from which they harvest electricity mostly for lighting. However, most of the projects implemented are inefficient and some unsuccessful, the main reason is the little knowledge and awareness on PV solar systems, such as load demand estimation, how to size the system, and which components are needed.

The aim of this study, therefore, is to provide a design models and guideline for a reliable, sustainable, clean, and cost-efficient stand-alone PV energy system that will replace wholly kerosene lamps and fuel-powered generators for lighting in rural areas. Three patterns of daily profile electricity consumption with energy efficient lightings (EELs) (judicious power consumption, normal power consumption, and excess power consumption), and another consumption with incandescent light bulbs, were designed and compared.

FEASIBILITY STUDY OF DIFFERENT PATTERNS OF ELECTRICITY CONSUMPTION ON LIGHTINGS IN RURAL NIGERIAN AREA

The Reference Household

From the acquired data, different lighting profiles of the household were created. These profiles consist of the household lighting (load) variations and usage patterns within the household. **Figures 2A–C** show the three patterns of daily profile electricity consumption with EELs (judicious power consumption, normal power consumption, and excess power consumption), whereas **Figure 2D** shows the consumption with incandescent light bulbs, in a household located in the rural area in Nsukka (Enugu State, Nigeria). Large quantities of electrical energy used for lightings are rarely found in solar home systems (SHSs) for a rural household application. Therefore, this household in the rural area in Nsukka is simple (three bedrooms and a living-room) and does not require large quantities of electrical energy used for

lightings [interior lighting (four lighting bulbs; one bulb for each bedroom including the living-room) and exterior lighting (two lighting bulbs; one bulb for the house frontage and another one for the wash-room)]. Tables S1A–S6A in Supplementary Material show an estimation of each energy efficient bulb's rated power, its quantity, and the hours of use by the three options of household in a single day, whereas (Tables S7A and S8A in Supplementary material) show an estimation of each incandescent bulb's rated power, its quantity, and hours of use.

DESCRIPTION OF DIFFERENT PATTERN OF ELECTRICITY CONSUMPTION (LIGHTINGS) IN A HOUSEHOLD

Consumption with Energy Efficient Lightings – Judicious Power Consumption

[This can be done by switching off lightings that were not in use at any given time. Additionally, exterior lighting can be installed with motion detectors to further reduce energy consumption]. The lights in the house will always be ON as from 6.00 a.m. (06:00 h) to 06:59 a.m. (06:59 h). By 7.00 a.m., the light will go off, since the rays of light come in through the windows during day time [7.00 a.m. to 5.59 p.m. (07:00–17:59 h)]. Once it is 6.00 p.m. (18:00 h), both the exterior and the interior lights will come ON and remain ON till 9.59 p.m. (21:59 h). Once it is 10.00 p.m. (22:00 h), there will be light out only in the house (interior lights) leaving the exterior lights ON till 7.00 a.m. (07:00 h) before it goes OFF. Meanwhile, the lights in the house comes ON again by 6.00 a.m. (06:00 h) till 7.00 a.m. (07:00 h) before it goes OFF

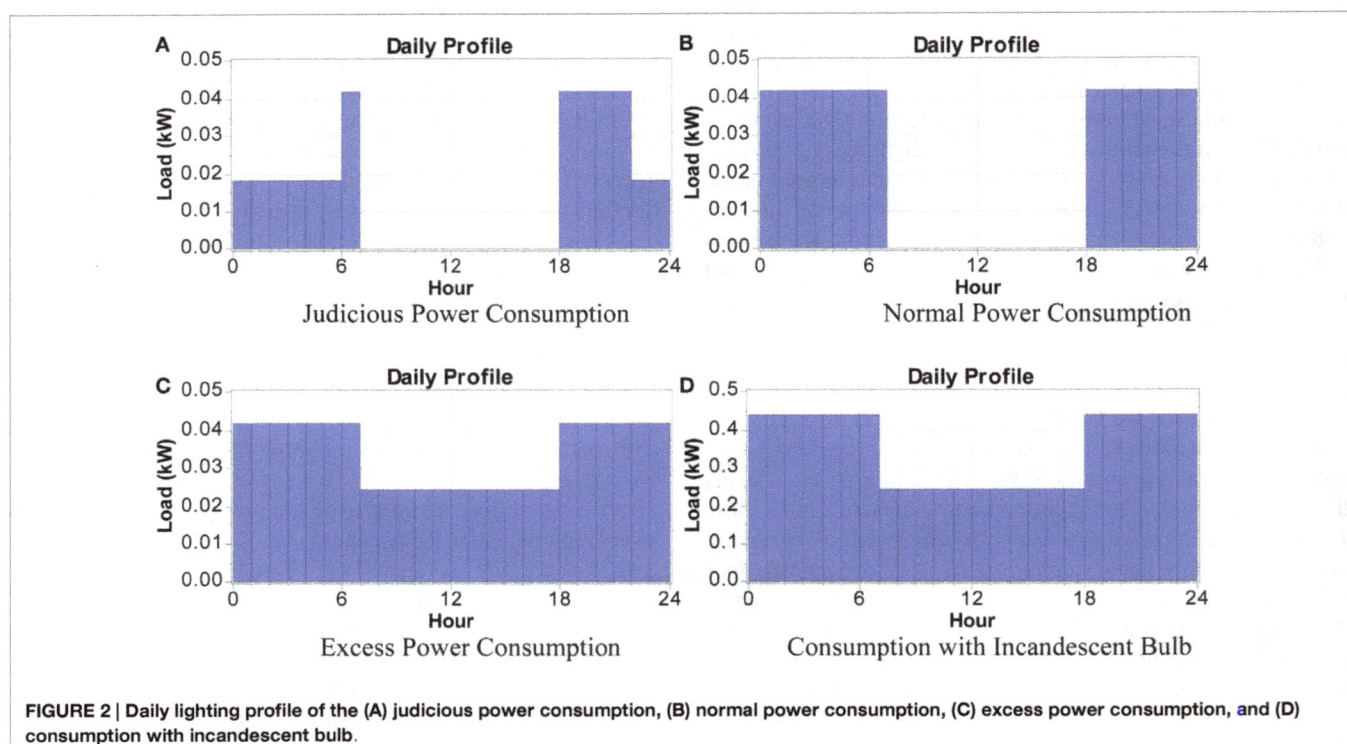

FIGURE 2 | Daily lighting profile of the (A) judicious power consumption, (B) normal power consumption, (C) excess power consumption, and (D) consumption with incandescent bulb.

as shown in **Figure 2A** as well as in Tables S1A and S2A in Supplementary Material.

Consumption with Energy Efficient Lightings – Normal Power Consumption

All the lightings (both the interior and the exterior) will be ON throughout the night [18:00–06:59 h (6.00 p.m. to 6.59 a.m.)] as shown in **Figure 2B** as well as in Tables S3A and S4A in Supplementary Material.

Consumption with Energy Efficient Lightings – Excess Power Consumption

The lights in the house always be ON for 24 h [00:00–23:59 h (12.00 a.m. to 12.00 p.m.)], while the exterior lights will always be ON throughout the night, i.e., 12 h [18:00–06:59 h (6.00 p.m. to 6.59 a.m.)] as shown in **Figure 2C** as well as in Tables S5A and S6A in Supplementary Material.

Consumption with Incandescent Light Bulbs

The lights in the house (interior lights) always be ON for 24 h [00:00–23:59 h (12.00 a.m. to 12.00 p.m.)], while the exterior lights will always be ON throughout the night, i.e., 12 h [18:00–06:59 h (6.00 p.m. to 6.59 a.m.)], as shown in **Figure 2D** as well as in Tables S7A and S8A in Supplementary Material, which is obtainable in the rural areas of Nigeria.

SOLAR POTENTIAL

The solar potential of a place can be evaluated using various sources, such as solar potential maps, data from the nearest meteorological station, and data from different research institutions. From the latitude (6°51′N) and longitude (7°35′E) of the location of the household sited in Nsukka (Longitude-Latitude-Maps, 2014; The GPS Coordinates, 2014), these data for solar resource were obtained from the National Aeronautics and Space Administration (NASA) Surface Meteorology and Solar Energy web site[1]. As a benchmark, the monthly mean daily insolation for Nsukka is given in **Figure 3**, whose average is 4.92 kWh/m²/day. This means that regions, such as North-Central (Benue), Northeast (Borno), Northwest (Kano), South–South (Delta), and Southwest (Ekiti), have a mean daily solar irradiation >4.92 kWh/m²/day.

METHODOLOGY

Mathematical Formulation of the Objective Function and Constraints

Hybrid optimization model for electric renewable (HOMER) software is a computer modeling tool that can evaluate different situations to determine the system configuration that will provide acceptable reliability at the lowest lifecycle cost, which is known as optimization. In the optimization procedure, the sizes of system components are decision variables, and their costs are the objective function. The objective function and constraints for the design model were defined and formulated below.

[1]http://eosweb.larc.nasa.gov/ (accessed October 10, 2014).

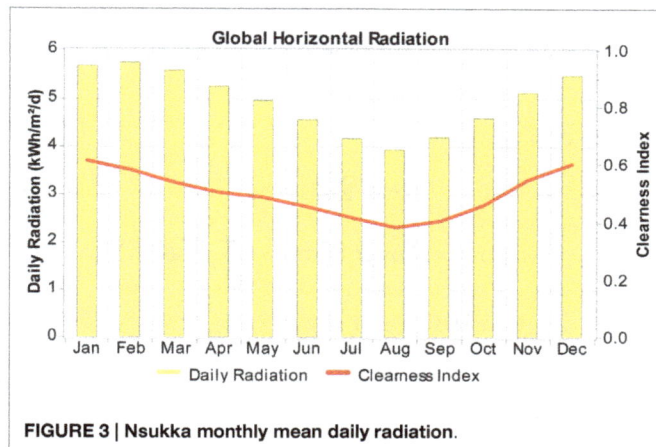

FIGURE 3 | Nsukka monthly mean daily radiation.

Objective Function

Objective function (costs objective function) to be minimized is the total net present cost (*NPC*) of the system, which includes the cost of the initial investment plus the discounted present values of all future costs throughout the total life of the installation; subjected to the *Constraints* which are the load to be met, maximum power wattage, and the state of charge (SOC) of battery. The life of the system is considered to be the life of the PV panels – which are the elements that have a longer lifespan. The costs taken into account are indicated as follows:

- Cost for purchasing the PV panels, the batteries, the inverter, and the charge regulator.
- Costs of replacing the components throughout the life of the system.
- Costs of operation and maintenance of components throughout the life of the system.

A more detailed description of its calculation can be found in Dufo-López and Bernal-Agustín (2005), Bernal-Agustín et al. (2006), Dufo-Lopez et al. (2007), Dufo-López (2007), and at http://www.homerenergy.com/.

Mathematical Formulation – Optimization

Mathematical calculations involve process of optimization; true meaning of optimization is to find the best answer for a particular problem. For example, problems dealing with the cost will require the best cost to be as less as possible. On the other hand, problems dealing with profit will see the maximum value as the best answer. So, "Optimum" is the word which is used to demonstrate the meaning of best, and the process of finding the best solution to a particular problem is known as the process of optimization (Antoniou and Lu, 2007; Waqas, 2011). To solve an optimization problem, an optimization algorithm is required. An optimization algorithm is the algorithm which is used to define an optimized solution for a particular function. Thus, for the stand-alone PV/battery system with constraints (load, maximum power wattage, and SOC of battery), the following optimization algorithm were derived.

Minimize

$$\text{Cost} = \sum_i C_{PV_i} N_{PV_i} + \sum_j C_{B_j} N_{B_j} + \sum_k C_{C_k} N_{C_k} \qquad (1)$$

Subject to the constraints

$$\text{Load} \leq \sum_i E_{PV_i} N_{PV_i} \qquad (2)$$

$$\text{Maximum Power Wattage} \leq \sum_k P_{C_k} N_{C_k} \qquad (3)$$

$$\text{SOC}_{min} \leq \text{SOC}(t) \leq \text{SOC}_{max} \qquad (4)$$

where

C_{PV_i} = Cost of a PV module

C_{B_j} = Cost of a battery

C_{C_k} = Cost of a converter

N_{PV_i} = Number of PV modules

N_{B_j} = Number of battery bank to be use

N_{C_k} = Number of converters

E_{PV_i} = kilowatt·hour generated by PV module i

P_{C_k} = Maximum output power of converter k

SOC_{min} = State of battery charge at minimum

SOC_{max} = State of battery charge at maximum

Optimal Design of PV/Battery System

Apart from correct costing and optimization, the quality and accuracy of the model and its implementation in the algorithm, greatly determines the usefulness of the simulation results.

Given that, the values of irradiation on tilted planes and the consumption patterns previously described, the system behavior can be simulated using an hourly time step. Based on a system energy balance and on the storage continuity equation, the simulation method used here is similar to that used by others (Sidrach de Cardona and Mora Lopez, 1992; Kaye, 1994). Considering the battery charger output power $P_{charger}(t)$, the PV output power $P_p(t)$, and the load power $P_l(t)$ on the simulation step Δt, the battery energy benefit during a charge time Δt_1 is given by ($\Delta t_1 < \Delta t$):

$$C_1(t) = \rho_{ch} \int_{\Delta t_1} \left[P_p(t) + P_{charger}(t) - P_l(t) \right] dt \qquad (5)$$

The battery energy loss during a discharge time Δt_2 is given by ($\Delta t_2 < \Delta t$):

$$C_2(t) = \left(\frac{1}{\rho_{dch}} \right) \int_{\Delta t_2} \left[P_p(t) + P_{charger}(t) - P_l(t) \right] dt \qquad (6)$$

The SOC of the battery is defined during the simulation time-step Δt by:

$$C(t) = C(t - \Delta t) + C_1(t) + C_2(t) \qquad (7)$$

As an input of a simulation time-step Δt (taken as 1 h), several variables must be determined: PV output power, load power, and battery SOC. A battery energy balance indicates the operating strategy of the PV system: charge (energy balance positive) or discharge (energy balance negative).

DESIGN OF A STAND-ALONE PHOTOVOLTAIC ENERGY SYSTEM FOR THE POWER GENERATION

As shown in **Figure 4**, a stand-alone solar PV system for a home lighting in rural Nigerian area is comprised of the following:

- Solar module(s);
- Charge controller;
- Storage system (batteries);
- Inverter; and
- AC load(s).

The set-up works this way.

The PV and battery are connected to the DC bus (V_{DC}), while the AC appliances are connected to the load bus (I_{AC}) as shown in **Figure 4**. However, as the system has AC loads an inverter has to be included. An inverter (a DC-to-AC converter) is used to convert DC current (I_{inv_DC}) to AC current (I_{inv_AC}) to serve the AC load. The inverter needs to meet two needs: peak (or surge) power and continuous power. During day time (08:00–17:59 h), the PV charge the battery system, and as from 18:00 h the PV either serve the load alone and if there is extra power produced is stored in a battery system, or serve the load with the help of battery till 18:59 h when the battery will fully take over as in the case of judicious and normal power consumption, whereas in the case of energy waste and consumption with incandescent bulbs, during the day time (08:00–18:59 h), the PV serve the load as well as charge the battery system, and as from 19:00 h the battery will take over to supply to the load. Moreover, during night time, the battery is the only

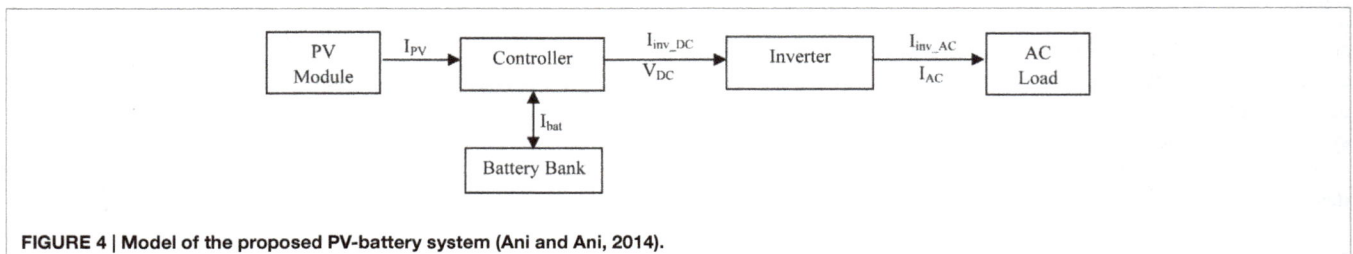

FIGURE 4 | Model of the proposed PV-battery system (Ani and Ani, 2014).

source of energy while the PV is off. The battery carries on all night till morning when the PV will start to charge the battery again.

COMPONENTS SYSTEM

Components of the Photovoltaic Energy System

Solar Module

Solar modules are made of several PV cells connected in series. Several PV cells make a module, and several modules make an array. The solar array is then a group of several modules as shown in **Figure 5**, connected in a series–parallel combination to generate the required current and voltage.

The underlying operating principle of a PV module is the photoelectric effect, by which radiation of photons of greater energy than the bandgap of the semiconductor material excite free electrons. Higher irradiation means more photons, hence more free electrons and thus higher currents. Therefore, the current generated by a PV module is directly dependent on the number of incoming photons and thus, the solar irradiation. The components added to the module constitute the balance of system (BOS). BOS components are as follows.

Charge Controller

The success of a stand-alone solar PV system depends to a large extent on the long-term performance of the batteries. When batteries are mishandled (overcharged or over-discharged) their performance and life span reduces dramatically. Normally, a charge controller is included in a stand-alone solar PV system to serve for the following purposes:

- Battery overcharge and over-discharge protection;
- Overload protection;
- Battery SOC monitoring; and
- Maximum power point tracking (MPPT).

A charge controller regulates power from a PV module to prevent batteries from overcharging; it also functions as a low-battery voltage disconnect to prevent the batteries from over-discharging.

Storage System

A storage battery is an electrochemical device. It stores electrical energy in the form of chemical energy that can later be released as electrical energy when a load is connected. Various types of batteries can be used to store electrical energy, i.e., lead-acid batteries, lithium-ion batteries, and so on. Although most other types of batteries have advantages such as high storage density or lower self-discharge, lead-acid batteries are used mostly in PV stand-alone systems due to its lower price as batteries make up the largest component cost over the lifetime of a stand-alone solar PV system.

Battery has been one of the problems in the deployment of a stand-alone PV power system in Nigeria. Substandard batteries (batteries not meant for solar systems such as car batteries) are been used in stand-alone PV power system. A normal car battery (starter battery), for instance, is operated in buffer mode. For most of the time, it is fully charged but occasionally must deliver short-term high currents to start the engine. Things are different with a battery to be used in a stand-alone solar PV system, which is charged during sun hours, and has to withstand deep discharge during the no-sun hours; such batteries are called deep cycle batteries. A car battery is not suitable as storage in a stand-alone solar plant as it would become defective in a short period due to the cycle operation. These batteries (car) will not work well with the system because it does not have the properties of solar battery and within 2 years it starts to fail. A good solar battery will last for 8–20 years. Table S9A in Supplementary Material shows the properties of the batteries meant for backup power and off-grid home systems[2]. The description of the battery variables can also be found in the Tables S1A–S13A in Supplementary Material.

Types of Backup Batteries Used for the Simulation

Two types of batteries were used for the simulation. They are:

Surrette/Rolls and Hoppecke Batteries

Rolls Surrette Batteries are the top of the line choice for backup power and off-grid home. These batteries (Surrette 6CS25P) are 1,156 amp/h@100 h rate[3], whereas Hoppecke Batteries are another choice for backup power and off-grid home. These batteries (Hoppecke 10 OpzS 1000) are 1,000 amp/h@100 h rate[4]. Their respective picture are shown below in **Figures 6A,B**.

---

[2]http://www.homerenergy.com/ (accessed April 23, 2014).
[3]http://www.dcbattery.com/rollssurrette_6cs25ps.html (accessed October 10, 2014).
[4]http://www.biotechx.com/

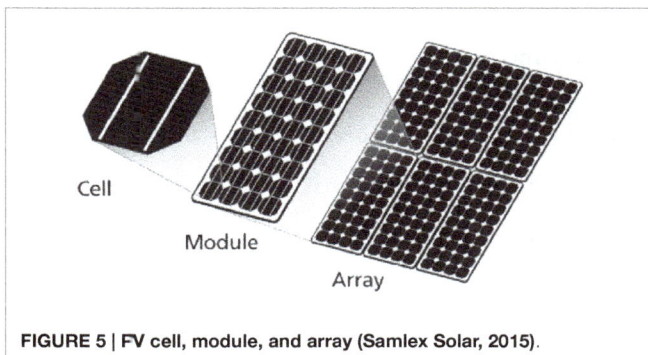

FIGURE 5 | FV cell, module, and array (Samlex Solar, 2015).

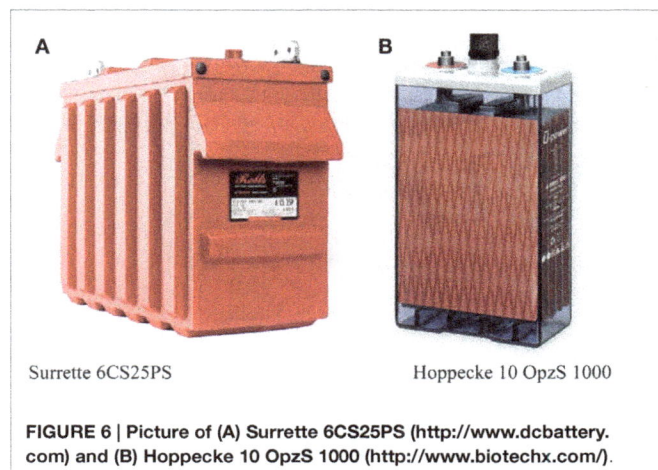

Surrette 6CS25PS

Hoppecke 10 OpzS 1000

FIGURE 6 | Picture of (A) Surrette 6CS25PS (http://www.dcbattery.com) and (B) Hoppecke 10 OpzS 1000 (http://www.biotechx.com/).

Inverter

An inverter is included in the stand-alone solar PV system to convert the DC into AC electricity.

SYSTEM ECONOMICS AND CONSTRAINTS

The capital costs for the PV module, and the battery are based on quotes from PV system suppliers in Nigeria[5]. These costs are estimates based on a limited number of internet enquiries and prices conducted as at 10th of October, 2014. They are likely to vary for the actual system quotes due to many market factors. The figures used in the analysis are therefore only indicative. The replacement costs of equipment are estimated to be 20% lower than the initial costs, but because decommissioning and installation costs need to be added, it was assumed that they are the same as the initial costs. The PV array, inverter, and battery maintenance costs are estimates based on approximate time required and estimated wages for this sort of work in a remote area of Nigeria. **Table 1** shows the capital cost of the energy system components used for the simulation. The project lifetime is estimated at 25 years, whereas the annual interest rate is fixed at 6%. There is no capacity shortage for the system and the operating reserve as a percentage of hourly load was 10%. Meanwhile, the operating reserve as a percentage of solar power output was 25%.

CONFIGURATION AND SIMULATION OF STAND-ALONE PHOTOVOLTAIC ENERGY SYSTEM

The sizing of the components of energy system is done using HOMERs design software developed by the National Renewable Energy Laboratory, accurate enough to reliably predict system performance. HOMER is an optimization model, which performs many hundreds or thousands of approximate simulations in order to design the optimal system. Using HOMER in designing a PV system reduces costs, time, risks, and errors associated with

[5]http://www.solarshopnigeria.com/(accessed October 10, 2014).

TABLE 1 | Capital cost of the energy system components.

Component	Initial capital cost in Nigerian Naira (₦)	Initial capital cost in United State Dollar ($)
130 W PV panel	62,000 (Nigeria Technology Guide, 2014)	376
Surrette 6CS25P	226,875	1,375[b]
Hoppecke 10 OpzS 1000	113,355	687[c]
1.0 kW Inverter	82,500[a]	500
0.6 kW Inverter	49,500[a]	300
0.4 kW Inverter	33,000	200 (Ani, 2014)

1 US dollar = 165 Nigerian Naira (www.themoneyconverter.com/USD/NGN.aspx, accessed October 10, 2014).
[a]http://www.solarshopnigeria.com/
[b]http://www.susitnaenergy.com/0437-surrette-6cs25ps.php
[c]http://www.biotechx.com/

preparing project prefeasibility studies; provide a low cost preliminary design method for project developers and industry, and thus increase the initiation of project studies that help to identify the best opportunities for successful PV project implementation. The system configuration is analyzed for various PV array sizes to operate in line with the storage (battery) system. The network architecture for the HOMER simulator of the completed stand-alone PV energy system for different pattern of load consumption can be seen in **Figures 7A–D**. The energy configuration of the household with different pattern of load consumption is shown in **Tables 2** and **3**.

RESULT ANALYSES AND DISCUSSION

An energy system is considered as an optimal solution for any pattern of electricity consumption for household lightings if it meets the required loads of the household at minimum total economic cost (NPC). Thus, the simulation results are collated and classified according to these two major factors.

1. Electric energy (kilowatt·hour) generated
2. Economic costs

Electric Energy Generated
Electricity Production
Energy from PV is greatly dependent on the availability of solar radiation and it differs from month to month with different pattern of load consumption as shown in **Figures 8A–D**. With these differences, excess electricity was observed in the months of January, February, March, November, and December. The PV array in this pattern of consumption (Judicious) generates 356 kWh of electricity per year which effectively powers the load demand of 129 kWh/year. The electrical production of PV energy system with different pattern of load consumption – with different type battery is shown in **Tables 2** and **3**.

The patterns of load consumption studied [Judicious (J), Normal (N), and Waste (W)] has different energy configuration. These different configurations show difference in electricity production from PV array due to difference in number of modules.

Judicious Pattern
Uses two modules (2) of 130 W solar panel which produces 356 kWh of electricity per year. This PV array charges the battery effectively which powers the load (129 kWh/year). There is excess electricity of 167 kWh/year, which occurs in the day time when there is no load, and the PV only charges the battery for evening/night use. This is due to the fact that in judicious load pattern, lighting is only use in the evening/night time.

Normal Pattern
Uses two-and-halve modules (2.5) of 130 W solar panel which produces 445 kWh of electricity per year. This PV array also charges the battery effectively which powers the load (199 kWh/year) the same way as in judicious pattern. Normal and judicious appears to be somehow the same only that normal allows its interior lighting to occur in line with that of exterior lighting, whereas judicious switches off its interior lighting when they were not in

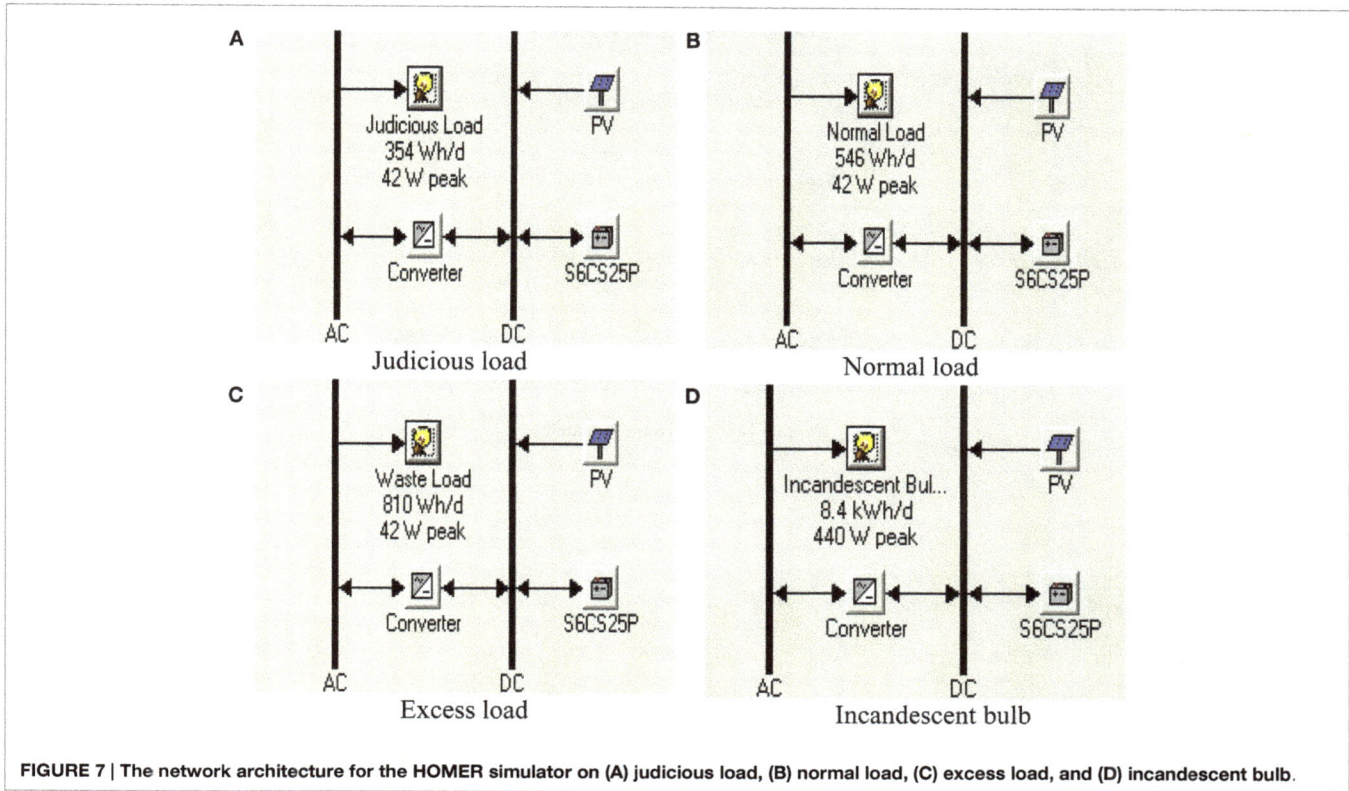

FIGURE 7 | The network architecture for the HOMER simulator on (A) judicious load, (B) normal load, (C) excess load, and (D) incandescent bulb.

TABLE 2 | Simulation results of electricity production, consumption, and losses (with Surrette battery).

Pattern of load consumption	Energy configuration	Electricity energy (kWh/year)			Surrette 6CS25P battery			Inverter (kWh/year)		
		Production from PV array	Consumption from the lightings	Excess electricity	Energy in	Energy out	Losses	Energy in	Energy out	Losses
Judicious	0.26 kW PV, 1 U battery, and 0.4 kW inverter	356	129	167	186	149	36	152	129	23
Normal	0.33 kW PV, 2 U battery, and 0.6 kW inverter	445	199	153	288	231	56	234	199	35
Excess (energy waste)	0.46 kW PV, 2 U battery, and 1.0 kW inverter	623	296	217	291	233	57	348	296	52

TABLE 3 | Simulation results of electricity production, consumption, and losses (with Hoppecke battery).

Pattern of load consumption	Energy configuration	Electricity energy (kWh/year)			Hoppecke 10 OpzS 1000			Inverter (kWh/year)		
		Production from PV array	Consumption from the lightings	Excess electricity	Energy in	Energy out	Losses	Energy in	Energy out	losses
Judicious	0.26 kW PV, 1 U battery, and 0.4 kW inverter	356	129	180	173	149	24	152	129	23
Normal	0.33 kW PV, 2 U battery, and 0.6 kW inverter	445	199	173	268	231	37	234	199	35
Excess (energy waste)	0.46 kW PV, 2 U battery, and 1.0 kW inverter	623	296	237	271	233	37	348	296	52

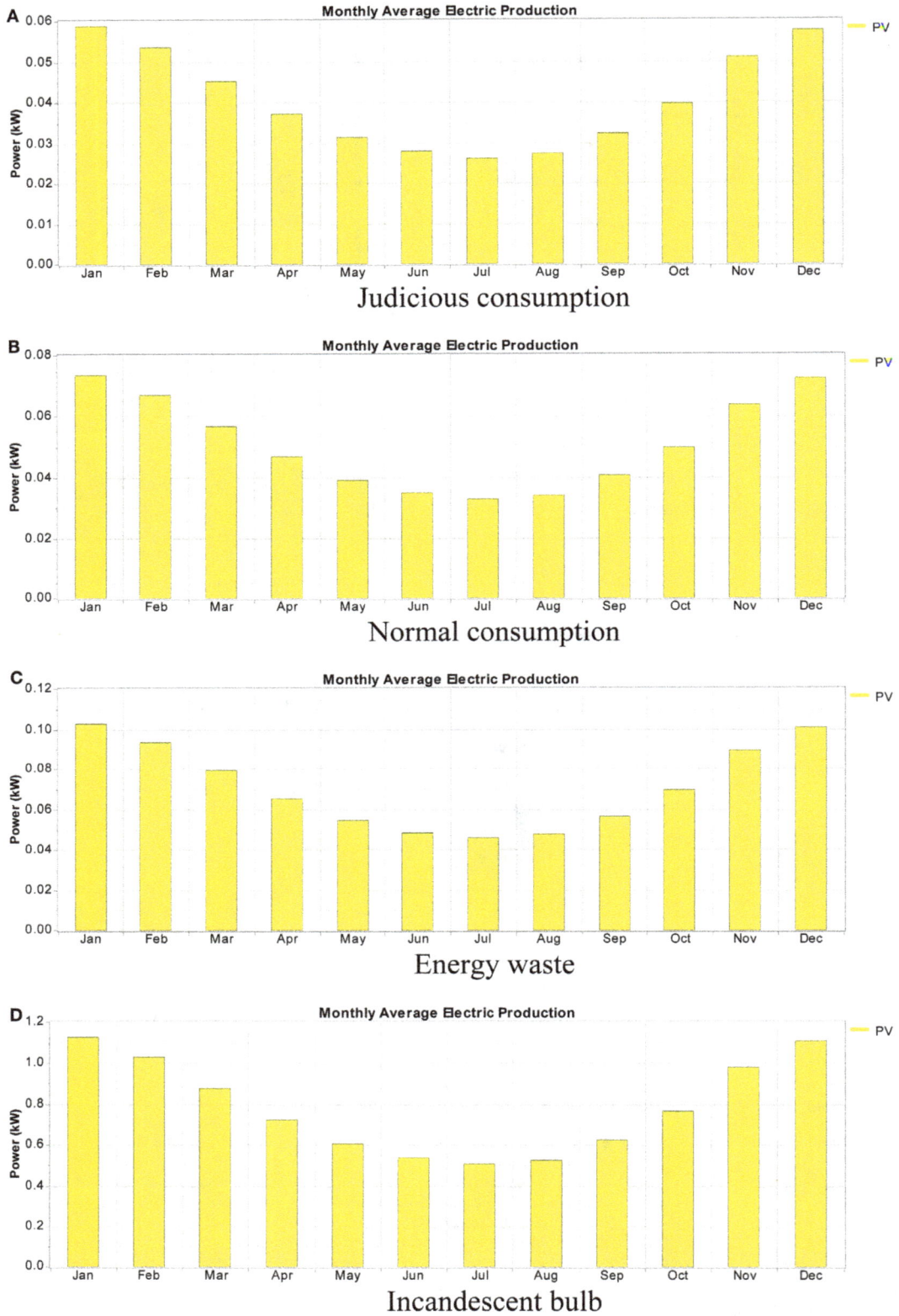

FIGURE 8 | Electrical production of PV energy system for **(A)** judicious consumption, **(B)** normal consumption, **(C)** energy waste, and **(D)** incandescent bulb.

use. There is excess electricity of 153 kWh/year which occurs in the day time when there is no load, and the PV only charges the battery for evening/night use.

Excess Pattern

Popularly known as "energy waste" uses three-and-halve modules (3.5) of 130 W solar panel, produces 623 kWh/year of electricity with 217 kWh/year of excess electricity. This pattern of load allows the interior lighting to be ON all day (24 h) and it is viewed as energy waste.

Energy Conservation

From the three load options studied, if one decides to go for the judicious pattern option, a conservation of 167 kWh/year will be earned when compared with the excess pattern option, while a 70-kWh/year can be conserved when compared the option (judicious pattern) with the normal pattern option. In the same comparison, if one choses to go for the normal pattern option, then a conservation of 97 kWh/year will be achieved when compared with the excess pattern option.

Battery Capacity

Battery contribute to the system design; different battery manufacturer, type, and properties (especially the battery capacity; see Table S9A in the Supplementary Material) contribute to the differences in battery charging and discharging which affects the electricity production (excess electricity) as were shown in **Tables 2** and **3**. Also, battery type changes (increases or decreases) the price of system configuration showing that battery makes the largest component cost in system sizing and design, and over the lifetime of a stand-alone solar PV system, as shown in **Tables 4** and **5**.

Economic Cost Analysis

One of the most important criteria for assessing an optimal solution for any particular household lightings is the economic cost of the system. The cost ratings here are discussed in terms of two major cost components: (1) the initial capital cost (ICC) and (2) the total NPC, the former being completely exclusive (i.e., ICC excludes other costs), while the latter is inclusive (i.e., includes the present value of all the costs that it incurs over its lifetime).

Initial Capital Cost

The ICC of a component is the total installed cost of that component at the beginning of the project; the initial cost results of PV

configuration with Surrette battery and PV configuration with Hoppecke battery are illustrated in **Tables 4** and **5**, respectively. Results show that energy design cost depends mainly on the load demand and pattern of energy consumption, configuration, and the type of components chosen (such as battery) just as studied here. Unnecessary consumption of energy increases the load demand, thereby increase the energy design. The capital cost of judicious configuration is nearly two times lower than that of excess configuration, and almost less than one-half of the normal configuration. This is due to the pattern of energy consumption which increases the load demand, thereby affect the sizing and system configuration.

Total Net Present Cost

The total NPC of a system has been described as the present value of all the costs that it incurs over its lifetime, minus the present value of all the revenue that it earns over its lifetime. Costs include capital costs, replacement costs, operation and maintenance costs, and the costs of buying power from the grid. Revenues include salvage value and grid sales revenue. However, the analysis presented here considers neither the costs of buying power from the grid nor grid sales revenue, since the focus of this study is on lightings in rural areas without grid connections. **Tables 4** and **5** show the NPC results for the configuration of PV system with Surrette, and Hoppecke, respectively, of different pattern of consumption. In this study, the NPC of excess option is nearly two times higher than that of judicious option as shown in **Tables 4** and **5**. Therefore, judicious option has the most cost-effective over time.

QUANTIFICATION OF INITIAL CAPITAL COSTS AS WELL AS ANNUALIZED COSTS OVER 5, 10, 15, 20, AND 25 YEARS

To appreciate the significance of the life cycle cost (NPC) of the energy system in the choice of energy components and pattern of energy consumption (as illustrated in **Tables 4** and **5**), further simulation runs were conducted at 5 years intervals (for a period of 25 years). The results of these simulations are illustrated in **Figures 9A,B** as well as in Tables S10A–S13A in Supplementary Material.

The simulation results for incandescent light bulbs with PV/Surrette battery system configuration is provided in Tables S1B–S4B in Supplementary Material.

TABLE 4 | Simulation results of economic costs (PV configuration with Surrette battery).

| Pattern of consumption | Economic costs | | | |
| | Economic cost USD ($) | | Economic cost NGN (₦) | |
	Initial capital cost (ICC)	Net present cost (NPC)	Initial capital cost (ICC)	Net present cost (NPC)
Judicious	2,327	2,935	383,955	484,275
Normal	3,990	5,206	658,350	858,990
Excess (energy waste)	4,566	5,782	753,390	954,030

TABLE 5 | Simulation results of economic costs (PV configuration with Hoppecke battery).

| Pattern of consumption | Economic costs | | | |
| | Economic cost USD ($) | | Economic cost NGN (₦) | |
	Initial capital cost (ICC)	Net present cost (NPC)	Initial capital cost (ICC)	Net present cost (NPC)
Judicious	1,639	1,733	270,435	285,945
Normal	2,614	2,802	431,310	462,330
Excess (energy waste)	3,190	3,378	526,350	557,370

A

B

FIGURE 9 | Economic cost difference in 5 years intervals for different pattern of load consumptions for energy efficient lightings with (A) PV/Surrette battery system configuration, and (B) PV/Hoppecke battery system configuration.

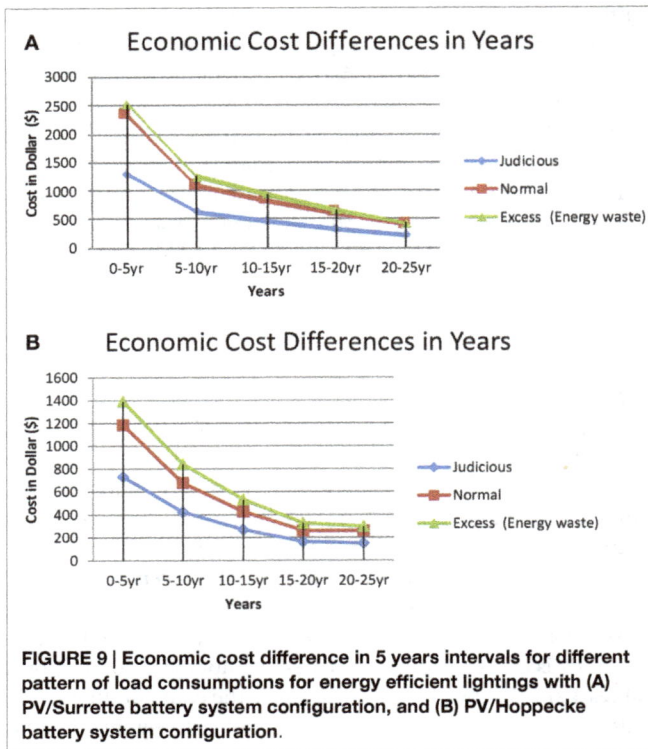

COMPARISON OF ENERGY EFFICIENT LIGHTINGS WITH INCANDESCENT LIGHT BULBS

The simulation results show that this house will conserve 2,753 W/year if EELs are used (excess option), and when configured with PV + Surrette 6CS25P battery + inverter will save $ 24 746 (₦ 4,083,090) of initial cost and $ 29 609 (₦ 4,885,485) for the NPCs, as shown in **Table 6**. These options [excess-energy waste (EEL) and incandescent light bulbs] were used for the comparison because they have the same energy consumption pattern; the same hour of actual utilization, but different rated power (wattage). Results also show that EEL reduces energy consumption due to lighting by as much as 90.3% (2,753 W). If judicious pattern option were used for the comparison, the conservation and the cost savings will be more.

CONCLUSION

The system configuration options were designed to run exclusively on solar power and require no diesel generator backup.

TABLE 6 | Comparison of energy efficient lightings (excess option) with incandescent light bulbs for PV + Surrette 6CS25P battery + Inverter configuration.

Load consumption	Lighting load (W)	Energy configuration	Economic costs	
			Initial cost ($)	Net present cost ($)
Energy efficient lightings	296	0.46 kW PV, 2 U battery, and 1.0 kW inverter	4,566	5,782
Incandescent light bulbs	3,049	5.0 kW PV, 10 U battery, and 1.0 kW inverter	29,312	35,391
Conservation/ cost savings	2,753	4.54 kW PV and 8 U battery	24,746	29,609

It is also designed to be simple – all resulting in zero OPEX, dramatically lower CAPEX, and near zero maintenance. Results show that the patterns of load consumption studied [Judicious (J), Normal (N), and Waste (W)] has different energy configuration. From the three load options studied, results also show that a conservation of 167 kWh/year will be earned with the use of judicious pattern option when compared with the excess pattern option, while a 70-kWh/year can be conserved when compared the option (judicious pattern) with the normal pattern option. In the same comparison, a conservation of 97 kWh/year will be achieved with the use of normal pattern option when compared with the excess pattern option. Comparing the three energy efficient load options (judicious power consumption, normal power consumption, and excess power consumption) and the incandescent light bulb consumption, results show that the initial and NPC of the incandescent light bulb option (8,360 W/h) is extremely expensive ($ 29,312; $ 35,391), excess power consumption option (810 W/h) is moderately expensive ($ 4,566; $ 5,782), normal power consumption option (546 W/h) is less expensive ($ 3,990; $ 5,206) than excess power consumption option, and the judicious power consumption option (354 W/h) is the least expensive ($ 2,327; $ 2,935). In summary, the results show that energy design cost depends mainly on the load demand and pattern of energy consumption; and the best way to cut costs is through the use of energy efficient bulbs and by switching off lightings that were not in use at any given time (judicious power consumption option).

REFERENCES

Ani, V. A. (2014). Feasibility and optimal design of a stand-alone photo-voltaic (PV) energy system for the orphanage. *J. Renew. Energy* 2014, 8. doi:10.1155/2014/379729

Ani, V. A., and Ani, E. O. (2014). Feasibility study and simulation of optimal power system for off-grid voter registration centres. *Int. J. Renew. Energy Res.* 4, 3. Available at: http://www.ijrer.org/ijrer/index.php/ijrer/article/view/1456

Ani, V. A., and Emetu, A. N. (2013). Simulation and optimization of photovoltaic diesel hybrid power generation systems for health service facilities in rural environments. *Electron. J. Energy Environ* 1, 57–70. doi:10.7770/ejee-V1N1-art521

Antoniou, A., and Lu, W. S. (2007). *Practical Optimization: Algorithms and Engineering Applications.* Berlin: Springer.

Asuquo, M. E., Ngim, O., and Agbor, C. (2008). A prospective study of burn trauma in adults at the University of Calabar Teaching Hospital, Calabar (South-Eastern Nigeria). *Open Access J. Plastic Surg.* 8, 370–376.

Bai, N., Khazaei, M., Eden, S., and Laher, I. (2007). The pharmacology of particulate matter air pollution-induced cardiovascular dysfunction. *Pharmacol. Ther.* 113, 16–29. doi:10.1016/j.pharmthera.2006.06.005

Behera, D., Dash, S., and Yadav, S. P. (1991). Carboxyhemoglobin in women exposed to different cooking fuels. *Thorax* 46, 344–346. doi:10.1136/thx.46.5.344

Bernal-Agustín, J. L., Dufo-López, R., and Rivas-Ascaso, D. M. (2006). Design of isolated hybrid systems minimizing costs and pollutant emissions. *Renew. Energy* 31, 2227–2244. doi:10.1016/j.renene.2005.11.002

Dominici, F., McDermott, A., Zeger, S., and Samet, J. (2003). Airborne particulate matter and mortality: timescale effects in four US cities. *Am. J. Epidemiol.* 157, 1055–1065. doi:10.1093/aje/kwg087

Dufo-López, R. (2007). *Design and Control of Hybrid Systems Using Evolutionary Algorithms*. PhD thesis, Universidad de Zaragoza, Zaragoza.

Dufo-López, R., and Bernal-Agustín, J. L. (2005). Design and control strategies of PV-diesel systems using genetic algorithms. *Solar Energy* 79, 33–46. doi:10.1016/j.solener.2004.10.004

Dufo-Lopez, R., Bernal-Agustín, J. L., and Contreras, J. (2007). Optimization of control strategies for stand-alone renewable energy systems with hydrogen storage. *Renew. Energy* 32, 1102–1126. doi:10.1016/j.renene.2006.04.013

Edenhofer, O., Pichs-Madruga, R., Sokona, Y., Seyboth, K., Matschoss, P., Kadner, S., et al. (eds) (2011). *Renewable Energy Sources and Climate Change Mitigation: Special Report of the Intergovernmental Panel on Climate Change (IPCC)*. Cambridge: Cambridge University Press.

Foster, R., Ghassemi, M., and Cota, A. (2010). *Solar Energy: Renewable Energy and the Environment*. Boca Raton, FL: CRC Press.

Gilmore, E. A., Adams, P. J., and Lave, L. B. (2010). Using backup generators for meeting peak electricity demand: a sensitivity analysis on emission controls, location, and health endpoints. *J. Air Waste Manag. Assoc.* 60, 523–531. doi:10.3155/1047-3289.60.5.523

Kaye, J. (1994). "Optimising the value of photovoltaic energy in electricity supply systems with storage," in *12th EC Photovoltaic Solar Energy Conference: Proceedings of the International Conference* (Amsterdam: H. S. Stephen & Associates), 431–434.

Lam, N. L., Smith, K. R., Gauthier, A., and Bates, M. N. (2012a). Kerosene: a review of household uses and their hazards in low- and middle income countries. *J. Toxicol. Environ. Health Sci.* 15, 396–432. doi:10.1080/10937404.2012.710134

Lam, N. L., Chen, Y., Weyant, C., Venkataraman, C., Sadavarte, P., Johnson, M. A., et al. (2012b). Household light makes global heat: high black carbon emissions from kerosene wick lamps. *Environ. Sci. Technol.* 46, 13531–13538. doi:10.1021/es302697h

Lighting Africa. (2010). "Solar lighting for the base of the pyramid – overview of an emerging market", in *2nd International Business Conference & Trade Fair*. (Nairobi: Lighting Africa).

Lighting Africa. (2013). *Lighting Africa Market Trends Report 2012*, 98. Available at: http://www.dalberg.com/documents/Lighting_Africa_Market_Trends_Report_2012.pdf (accessed September 6, 2015)

Longitude-Latitude-Maps. (2014). *Longitude and Latitude in Nsukka, Enugu, Nigeria – GPS Coordinates*. Available at: http://www.longitude-latitude-maps.com/city/156_701,Nsukka,Enugu,Nigeria (accessed October 10, 2014).

Mills, E. (2005). The specter of fuel-based lighting. *Science* 308, 1263–1264. doi:10.1126/science.1113090

Mills, E. (2012a). *Health Impacts of Fuel-Based Lighting*. California, CA: Lawrence Berkeley National Laboratory, University of California.

Mills, E. (2012b). Health impacts of fuel-based lighting. *Working Paper Presented at the 3rd International Off-Grid Lighting Conference*, 13-15 November, 2012, Dakar, Senegal.

Mills, E. (2014). *Light for Life: Identifying and Reducing the Health and Safety Impacts of Fuel-Based Lighting*. Nairobi: United Nations Environment Programme. Available at: http://www.enlighten-initiative.org/portals/0/documents/Resources/publications/Light%20for%20Life%20-%20Health%20and%20Safety%20Impacts%20of%20Fuel-Based%20Lighting.pdf

Natural Resources Canada. (2008). *Emissions Factors for Diesel Generator Systems for Three Different Levels of Load Factor [Online]*. Available at: http://www.retscreen.net/ang/emission_factors_for_diesel_generator_image.php (accessed September 6, 2015).

Nigeria Technology Guide. (2014). *Solar Power Systems Components – Solar Panels Prices in Nigeria*. Available at: http://www.naijatechguide.com/2008/11/solar-energy-system-components.html (accessed October 10, 2014).

Olaitan, P. B., Faidora, S. O., and Agodirin, O. S. (2007). Burn injuries in a young Nigerian teaching hospital. *Ann. Burns Fire Disasters* 20, 59–61.

Oludiran, O. O., and Umebese, P. F. A. (2009). Pattern and outcome of children admitted for burns in Benin City, mid-western Nigeria. *Indian J. Plast. Surg.* 42, 189–193. doi:10.4103/0970-0358.59279

Samlex Solar. (2015). *Solar (PV) Cell Module, Array*. Available at: www.samlexsolar.com/learning-center/solar-cell-module-array.aspx (accessed September 7, 2015).

Scientific Advisory Panel. (2013). *Scientific Advisory Panel 2013 Annual Science Update*. Paris: Climate and Clean Air Coalition to Reduce Short-Lived Climate Pollutants, United Nations Environment Programme.

Sharma, S., Sethi, G. R., Rohtagi, A., Chaudhary, A., Shankar, R., Bapna, J. S., et al. (1998). Indoor air quality and acute lower respiratory infection in Indian urban slums. *Environ. Health Perspect.* 106, 291–297. doi:10.1289/ehp.98106291

Sidrach de Cardona, M., and Mora Lopez, L. I. (1992). "Optimizing of hybrid photovoltaic generator systems for installations of rural electrification," in *11th E C. Photovoltaic Solar Energy Conference: Proceedings of the International Conference* (Montreux: Harwood Academic Publishers, Chur, Switzerland), 1287–1290.

The GPS Coordinates. (2014). *GPS Coordinates of Nsukka, Nigeria – latitude and Longitude*. Available at: www.thegpscoordinates.com/./nsukka/

UNEP/GEF en.lighten initiative. (2013). *Country Lighting Assessments*. Available at: http://luminanet.org/page/country-data-2# (accessed September 6, 2015).

Waqas, S. (2011). *Development of an Optimisation Algorithm for Auto Sizing Capacity of Renewable and Low Carbon Energy Systems*. Master of Science, Department of Mechanical Engineering, University of Strathclyde Engineering, Lanarkshire.

Zai, S., Zhen, H., and Jia-song, W. (2006). Studies on the size distribution, number and mass emission factors of candle particles characterized by modes of burning. *J. Aerosol Sci.* 37, 1484–1496. doi:10.1016/j.jaerosci.2006.05.001

Conflict of Interest Statement: The author declares that the research was conducted in the absence of any commercial or financial relationships that could be construed as a potential conflict of interest.

11

Thermal hydraulics of accelerator-driven system windowless targets

Bruno Panella*, Luigi Consalvo De Giorgi, Mario De Salve, Cristina Bertani and Mario Malandrone

Department of Energy, Politecnico di Torino, Torino, Italy

The study of the fluid dynamics of the windowless spallation target of an accelerator-driven system (ADS) is presented. Several target mockup configurations have been investigated: the first one was a symmetrical target, which was made by two concentric cylinders and the other configurations are not symmetrical. In the experiments, water has been used as hydraulic equivalent to lead–bismuth eutectic fluid. The experiments have been carried out at room temperature and flow rate up to 24 kg/s. The fluid velocity components have been measured by an ultrasound technique. The velocity field of the liquid within the target region either for the approximately axial–symmetrical configuration or for the not symmetrical ones as a function of the flow rate and the initial liquid level is presented. A comparison of experimental data with the prediction of the finite volume FLUENT code is also presented. Moreover, the results of a 2D–3D numerical analysis that investigates the effect on the steady state thermal and flow fields due to the insertion of guide vanes in the windowless target unit (TU) of the EFIT project ADS nuclear reactor are presented, by analyzing both the cold flow case (absence of power generation) and the hot flow case (nominal power generation inside the TU).

Keywords: accelerator-driven system, proton beam target, windowless spallation target, fluid dynamics, thermal hydraulics

Edited by:
Muhammad Zubair,
University of Engineering and
Technology, Pakistan

Reviewed by:
Yacine Addad,
Khalifa University, UAE
Maurizio Luigi Cumo,
Sapienza University of Rome, Italy

*Correspondence:
Bruno Panella,
Department of Energy,
Politecnico di Torino,
Corso Duca degli Abruzzi 24,
Torino 10129, Italy
bruno.panella@polito.it

Specialty section:
This article was submitted to Nuclear
Energy, a section of the journal
Frontiers in Energy Research

Citation:
Panella B, De Giorgi LC, De Salve M,
Bertani C and Malandrone M (2015)
Thermal hydraulics of accelerator-
driven system windowless targets.
Front. Energy Res. 3:32.

Introduction

The accelerator-driven system (ADS) is an interesting solution for incineration of long-lived nuclear waste coming from conventional nuclear reactors (Rubbia et al., 1995, 2001; Salvatores et al., 1997) and consists mainly of three parts: the subcritical core, the spallation target unit (TU), and the light particle accelerator (in most cases protons). The TU and the accelerator constitute the external neutron source that is necessary to sustain the fission chain in the subcritical core, as the accelerator provides an intense and energetic proton beam that impinges into the spallation fluid, a molten heavy metal, contained into the TU, generating the neutrons (Salvatores, 2006). The European R&D for nuclear waste management EUROTRANS Program has selected the XT-ADS, the EFIT (European Facility for Industrial Transmutation), and MYRRHA projects (Cinotti and Gherardi, 2002; Cinotti et al., 2003, 2004; Bianchi, et al., 2005; Knebel et al., 2006; Salvatores, 2006; Petrazzini, 2008; Aït Abderrahim et al., 2012; Mansani et al., 2012). The EFIT spallation TU is located in the center of the reactor where 800 MeV protons from the accelerator impinge on a free surface of lead exposed to vacuum (Mansani et al., 2012). Two types of connection between the accelerator tube and the target case are possible: window and windowless. In the windowless spallation target (WST) configuration, introduced by the Ansaldo Company nuclear division team for a demonstrative design (Cinotti and Gherardi, 2002; Bianchi, et al., 2005, 2008; Aït Abderrahim et al., 2012; Mansani et al., 2012), the proton beam impinges directly the target (lead or lead–bismuth eutectic) and the spallation region velocity field must be able to remove efficiently the rather high volumetric thermal power due to the proton interactions with the liquid metal nuclei (Abanades et al., 2008; Kumawat et al., 2008; Sordo et al., 2008). As regards the WST, a major issue is the study of the velocity and temperature fields because boiling phenomena are critical for a correct neutrons production: the spallation region velocity field must be able to remove the volumetric thermal power due to the proton interactions with lead–bismuth nuclei without the local boiling of the liquid. The flow rate should be able to remove a specific power of about 12.5 kW/cm³ at a temperature of about 670 K in the case of XT-ADS and 18.0 kW/cm³ in the case of EFIT at the same temperature. Material properties issues of Pb and T91 steel limit the maximum velocity near walls to 2.0–2.5 m/s and temperature to 950 K with oxygen control. Moreover, the WST limits the free surface temperature to around 900 K to avoid boiling or an excessive evaporation rate from the free surface at operating pressure, which is estimated to be in the range of 0.1–1 Pa. The shape of the interface, which should be stable with no vortex and/or other perturbations that could facilitate the evaporation, has to be carefully investigated. The flow is partially unconfined for the presence of a free surface (interface) at the top of the TU. The effective removal of the thermal power represents a major issue in the design of the windowless TU, as the metal vapors, which are produced either by the evaporation from the free surface or by local boiling into the bulk of the spallation fluid, can enter the proton beam tube with negative effects on the neutrons production and device integrity. In order to preserve high vacuum conditions into the accelerator tube, a thermal fluid-dynamic analysis of the flow in the TU must be performed, and, owing to the difficulty to perform thermal fluid-dynamic calculations at the free surface, experimental work is needed in order to simulate the flow of the molten lead or eutectic in the target region, for conditions similar to the real plant. In the present experimental research, three configurations of the windowless target, one symmetrical and two asymmetrical have been experimented, using water as hydraulic equivalent to lead or lead–bismuth eutectic fluid, by maintaining the same velocity values (the same Froude number) and about the same Weber number, whereas the different Reynolds number (the ratio is of the order of 7) seems less important owing to the prevailing inertial and gravity forces. A comparison of experimental data with the prediction of the commercial code Fluent 6.3.26 (Fluent Inc, 2005) is also presented. As a result of the investigation of the flow characteristics inside several theoretical and experimental mock-ups that simulate the EFIT reactor WST, but operating with water, it has been shown that important geometrical modifications, such as insertion of guide vanes inside the target and proper modifications of the container geometry, have to be introduced in order to avoid flow recirculation, vortices and fluid detachment from the walls, and to get suitable flow conditions within the spallation target heat deposition region. Moreover, the results of a computational fluid dynamics (CFD) numerical analysis with reference to the EFIT WST geometry are presented with the aim of evaluating the steady state effects on the flow and thermal fields induced by the insertion of the vanes as a function of the vanes number, dimensions, shape, and position in the TU and of the profile of smoothed corners. The CFD method and a statistical approach to the problem allows the development of accurate analytical models capable to predict with adequate precision the turbulent flow and the thermal field inside the heat deposition region of the EFIT WST equipped with guide vanes and, to identify possible geometries consistent with imposed design constraints that depend on the adopted optimization design criteria. The presence of the free surface is neglected (closed tube approximation) by assuming that it produces a small perturbation on the flow structure. An algorithm has been developed to generate the modified geometry of the TU and the geometrical features of the vanes depend only on the value of a discrete number of input parameters, which constitutes the design space that is explored by applying the design of experiments (DOE) technique (Sacks et al., 1989; Morris and Mitchell, 1995; Santner et al., 2003) and the response surface methodology (RSM) (Myers and Montgomery, 2002). The study takes into account both the cold flow case (absence of power generation) and the hot flow case (nominal power generation inside TU). Two turbulence models available with the commercial CFD Fluent 6.3.26 code have been tested: standard k-ϵ model and reynolds stress model.

Symmetrical Test Section

The first target geometry that has been investigated is characterized by a flow with axial–symmetrical properties. The test facility is an open loop with a centrifugal pump, a Venturi flow meter, a heat exchanger, to control the fluid temperature, the axial–symmetrical test section, and an air separator tank. The test section (**Figure 1**) is made by two concentric cylinders with inner cylinder diameter equal to 200 mm and outer cylinder diameter equal to 290 mm. The overall height is equal to 1750 mm. Water flows up in the annular region and flows down in the central region. Further details of the facility can be found in De Salve et al. (2002).

The most important experimental parameters are the imposed mass flow rate W (the Venturi flow rate measurement accuracy is 2%) and the initial fluid level h_0 measured from the top edge of the inner cylinder. The experiments have been carried out at room temperature equal to 298 K in the following range: flow rate from 2.5 to 20 kg/s with the average velocity ranging from 0.07 to 0.64 m/s; inner pipe Reynolds number from 15,000 to 127,000; initial fluid level from −50 to 190 mm. The velocity measurement method is based on the pulsed echo ultrasound technique by detecting the Doppler shift of the ultrasound wave reflected from moving particles that are suspended in the fluid (ADV Acoustic Doppler Velocimeter) and the ADV SonTek probe (SonTek Co, 1997) has been used, with a velocity measurement uncertainty of 5%. At zero or lower liquid levels, the free surface assumes a typical funnel shape and an air entrainment into the liquid flow at the free surface can be seen; the air remains inside the test section only if the flow rate is sufficiently low; at higher flow rate, the air is entrained. Runs with high levels show a low air entrainment if the flow rate is less than about 17.5 kg/s; in this case, the free surface is about flat with some surface waves and few vortices. **Figure 2A** shows a typical free surface profile on a vertical radial section obtained by measuring the free surface level by means a metric rule. **Figure 2B** shows the effect of inlet mass flow rate on the free surface profile. It can be seen that at higher mass flow rates, the level reaches a maximum near the inner pipe radius, $r = 100$ mm.

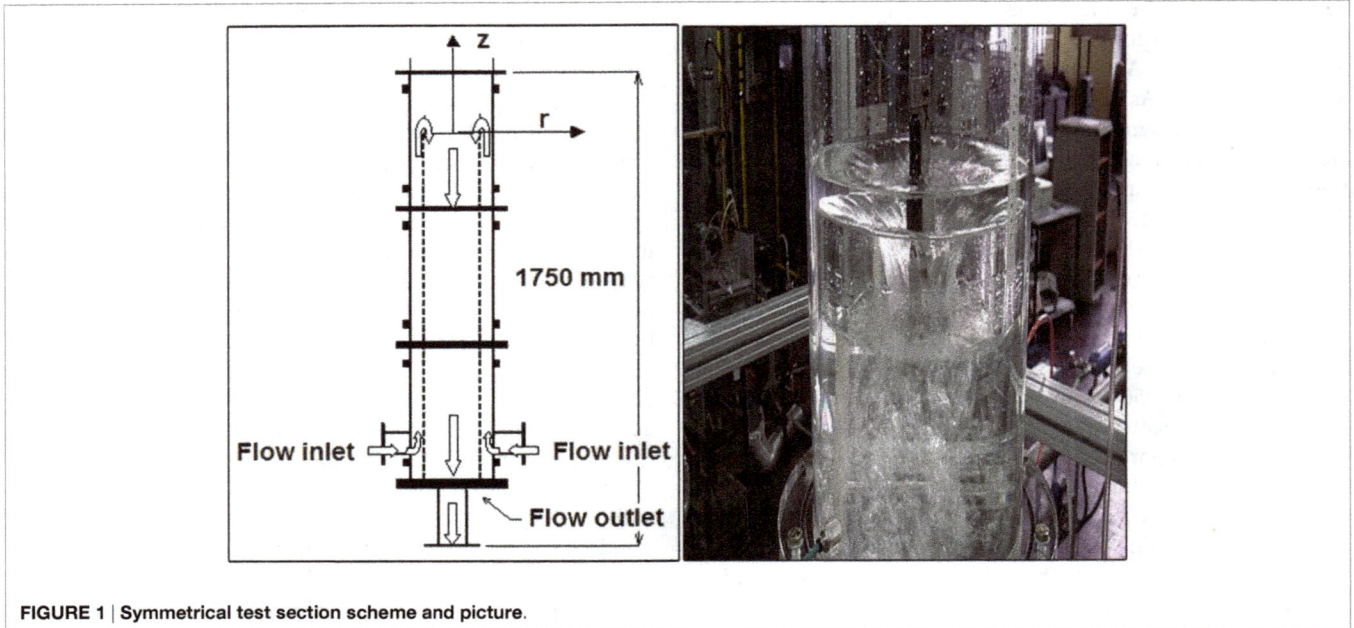

FIGURE 1 | Symmetrical test section scheme and picture.

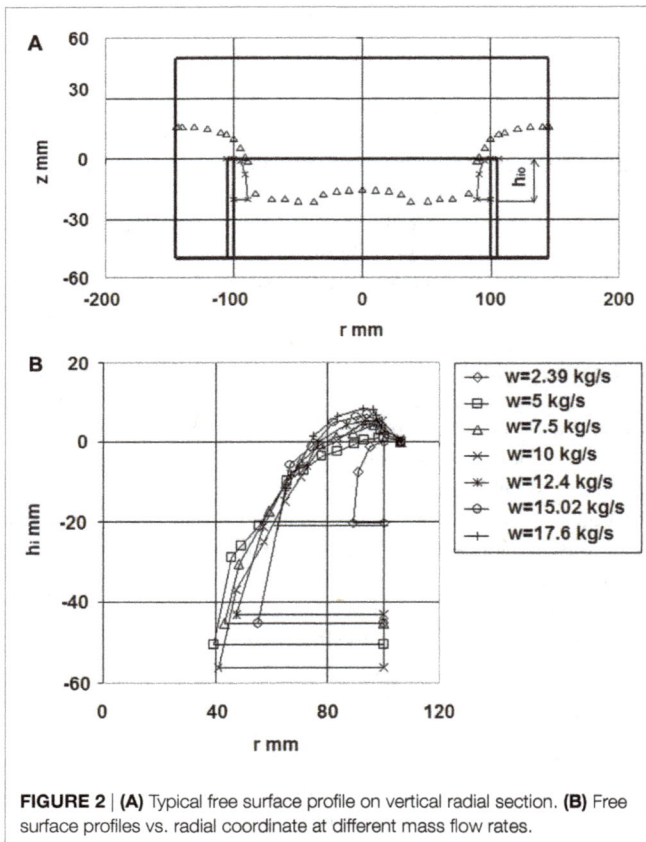

FIGURE 2 | **(A)** Typical free surface profile on vertical radial section. **(B)** Free surface profiles vs. radial coordinate at different mass flow rates.

Figures 3A,B show the effect of the mass flow rate on velocity field: the average velocity axial component V_z vs. the radius r for different mass flow rates, from 5.5 to 15.3 kg/s, is presented in **Figure 3A** at initial water level $h_0 = 176$ mm. The measuring plane elevation is $z = 45$ mm below the inner cylinder edge. These radial profiles show the V_z increase near the inner cylindrical wall in the region that is about 40 mm far from the wall, whereas the profile is rather flat within most of the channel; it seems that the maximum V_z value is reached about 60 mm far from the axis. It can be deduced that the flow is still developing in agreement with the measured velocities values, which are higher than the cross-averaged values that are derived from the measured flow rate. **Figure 3B** shows the velocity radial component V_r vs. the radius at different mass flow rates and for the same run conditions: the data show an increase in the radial component of the velocity as the flow rate increases; it can be deduced that the slight symmetry lack and the uncertainty of the probe position can cause rather high V_r values also on the axis, which are not very different from the values in correspondence of the other radial positions. As a conclusion, the main findings from experiments on the symmetrical target are that the axial velocity is very low (and even upward oriented) just above the inner pipe edge. The radial component velocity is dominant between $z = -50$ mm and $z = +30$ mm. In this region, some recirculation phenomena are also observed. The investigated geometry does not meet the design objectives of the WST due to the presence of a stagnant region of the fluid in upper part of the target where the thermal power released by protons is higher. So, the symmetrical WST design was dropped by Ansaldo Company nuclear team.

Asymmetrical Test Sections

Two asymmetrical WST geometries have been experimentally investigated, by changing the test section in the same facility. The first one consists of a rectangular tank whose dimensions are: $L = 760$ mm, $W = 360$ mm, $H = 500$ mm. **Figure 4** shows a scheme of the test section.

Two vertical pipes with the inner diameter of 203 mm connect the bottom of the tank with the heat exchanger and the pump by

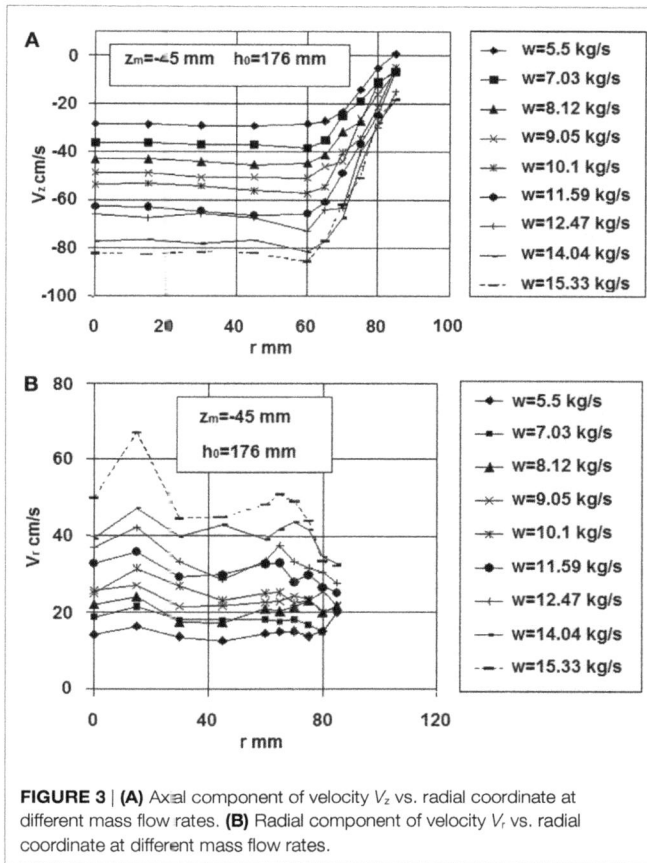

FIGURE 3 | (A) Axial component of velocity V_z vs. radial coordinate at different mass flow rates. (B) Radial component of velocity V_r vs. radial coordinate at different mass flow rates.

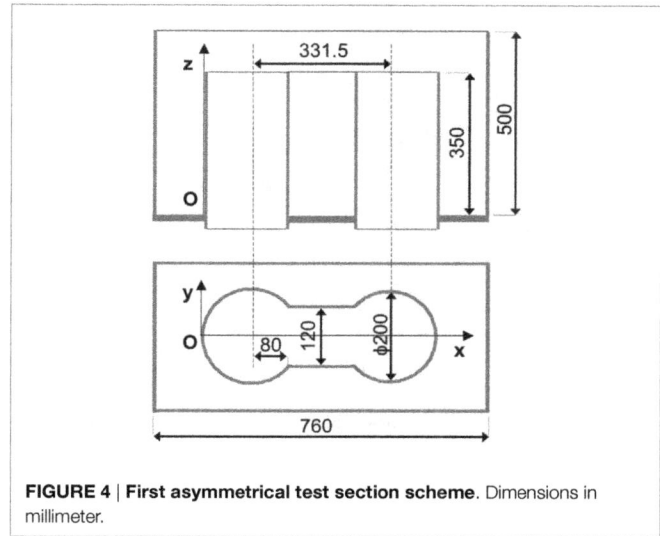

FIGURE 4 | First asymmetrical test section scheme. Dimensions in millimeter.

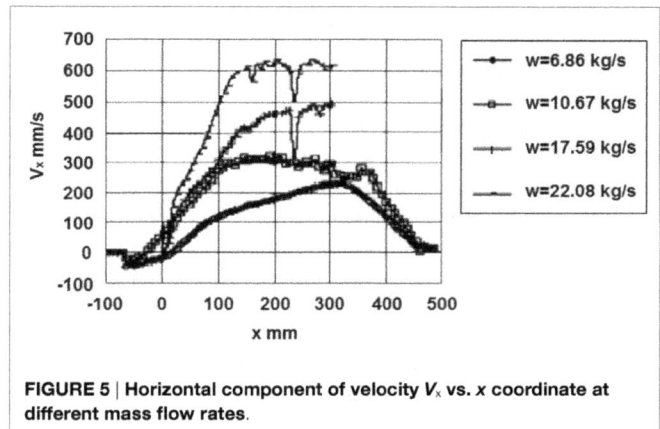

FIGURE 5 | Horizontal component of velocity V_x vs. x coordinate at different mass flow rates.

means of several conic reductions. The test section dimensions are those of the Ansaldo reference XT-ADS: the distance between the axes of the cylindrical tubes of the spallation region is 330 mm, the width of the channel is 120 mm, the inner diameter of the feed and discharge pipe is 200 mm. The pump can provide a mass flow rate of 22 kg/s at 14 m head. Experimental runs have been carried out, at room temperature, in the range of mass flow rates from 5 to 22 kg/s at the initial level of 350 mm. Measurements were performed using either the ultrasounds ADV SonTek, or the instrument DOP2000 (Signal Processing S. A., 2004) developed by Signal Processing, based on the Doppler effect too. The probes used by DOP 2000 are non-intrusive if compared with other ultrasound probes. The results show that by raising the flow rate from about 5 kg/s to about 22 kg/s, the flatness of the interface tends to overfill at the feed exit, accompanied by a phenomenon of level growing and a vortex at the entrance of the discharge pipe. In the suction region, there is a depression of the liquid level. Fluid dynamics is complex due to the surface irregularities, as well as to the presence of oscillations of the interface and of the velocity.

Figure 5 presents the velocity components in x direction along the symmetry plan of the test section with the probe located at $x_0 = 110$ mm, $y_0 = 0$ mm, $z_0 = 290$ mm at low mass flow rates (6.86 and 10.67 kg/s) and at high mass flow rates (17.59 and 22.08 kg/s). One can see that in the rectangular region of the test section for x between 119 and 350 mm, the velocity in x direction is high and constant enough. Around $x = 350$ mm at the inlet of the downcomer region, the component V_x is reduced and the flow changes direction.

Figure 6 shows the component V_x for a mass flow rate of 17.6 kg/s with the probe located at $x_0 = 0$ mm, $y_0 = 0$ mm, and for variable z_0 with a 25 mm step: for the same x, V_x rises with z both in the feed pipe (x <105 mm) and in the rectangular zone of the test section. The reduction of V_x where the flow changes direction, as well as the existing maximum values for V_x in the region situated between z = 200 and 400 mm is significant. The position of the maximum depends on the flow rate as well as on the surface level. The analysis of these profiles shows that V_x increases with x and presents a high x derivative in the first 180 mm, whereas the variation between 180 and 220 mm is rather low. An abrupt variation occurs at approximately 240 mm, which probably characterizes the region of transition between two big vortices. V_x increases with z until about 269 mm and then decreases. At z = 294 mm, V_x is about 600 mm/s. The tests show that at flow rates higher than 15 kg/s, there is a little height difference of the interface between the region of flow feeding and the region of the target. The experimental results and the visual observation show: (a) characteristic behavior of the horizontal component of the velocity V_x in the exit region,

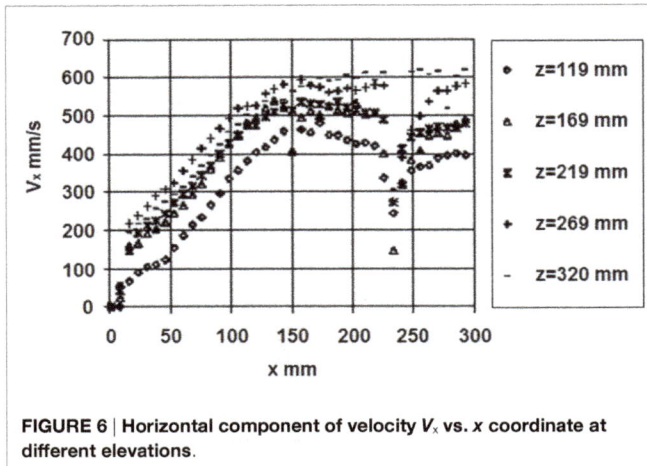

FIGURE 6 | Horizontal component of velocity V_x vs. x coordinate at different elevations.

reaching a maximum at about $z = 200$ mm, (b) recirculation near the tank walls and near the riser and downcomer pipe walls, (c) oscillations, which cannot be neglected at the interface at high mass flow rates.

The second experimental section that has been investigated introduces two major modifications with respect to the first asymmetrical target: the first one concerns the dimensions that refers to EFIT target geometry design (scaled 1–1), the second concerns the geometry of the feed and discharge pipes, which are prismatic channels. The test section, which is sketched in **Figure 7**, consists of two Plexiglas prismatic channels. At the bottom, two conic reductions connect them to the rest of the loop. The target is located at the top between the two channels. During all the tests, a constant free surface liquid level 470 mm high (equal to the level of the interface of the reference reactor) has been fixed. The hydraulic circuit and instrumentation are the same as seen previously.

Experimental tests have been carried out, at room temperature, in the range of flow rates from 10 to 20 kg/s and initial level h_0 of 470 mm. The ultrasound DOP2000 probe has been inserted on the outer surface of the Plexiglas walls at elevations of 50, 145, 240, and 420 mm. **Figure 8A** shows the mass flow rate ($W = 10, 15, 20$ kg/s) effect on x velocity component at the height $z = 240$ mm. It can be seen that, with good agreement, horizontal velocity is proportional to the flow rate. **Figure 8B** shows the horizontal component of flow field at the heights 50, 240, 420 mm: the flow is characterized by the presence of a vortex in the lower part of the spallation region that is determined by the separation of fluid vein in the feed pipe entering the spallation region. Moreover, the flow is characterized by low velocities near the interface.

CFD Simulations

Numerical experiments have been carried out by means of the Fluent 6.3.26 code with the following main aims:

- validation of the numerical code with respect to the flow in the second asymmetrical test section (**Figure 7**) and identification of related issues in terms of flow characteristics;

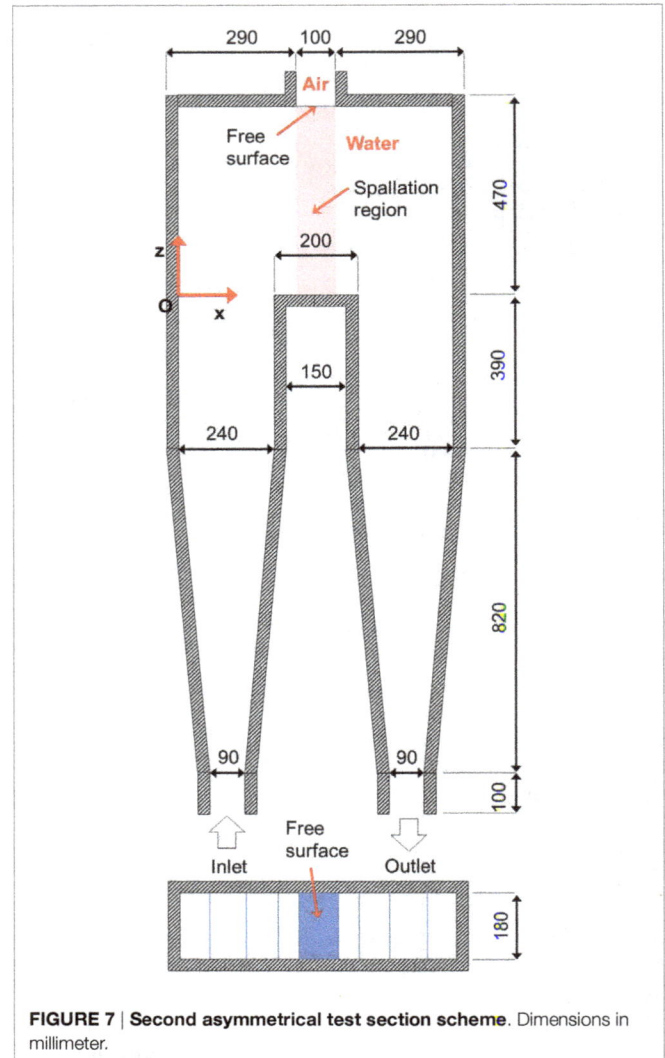

FIGURE 7 | Second asymmetrical test section scheme. Dimensions in millimeter.

- development of a statistical tool to be used to predict the flow characteristics in more complex geometries;
- identification of possible geometries, to get the desired flow conditions, once appropriate criteria have been adopted (numerical solution of an inverse Navier–Stokes problem).

Concerning the first goal, the flow into the second asymmetrical facility at the mass flow rate of 20 kg/s has been simulated. A 2D approximation of the test section presented in **Figure 7**, discretized with quadrilateral cells having a mean mesh size of 5 mm (which assures the grid independence of the solution), has been adopted. The simulated flow is isothermal, multiphase, for the presence of a free surface separating air and water, and turbulent. An initial level of the water equal to 470 mm has been chosen. A transient calculation has been performed and the steady solution depends upon the initial height of water in the target region. The interface between water and upper air has been tracked thanks to the geo-reconstruct scheme, which is an explicit method within the volume of fluid (VOF) implicit algorithm, a well known scheme for modeling multiphase flows. The average time step is of the order of 10^{-3} s. The

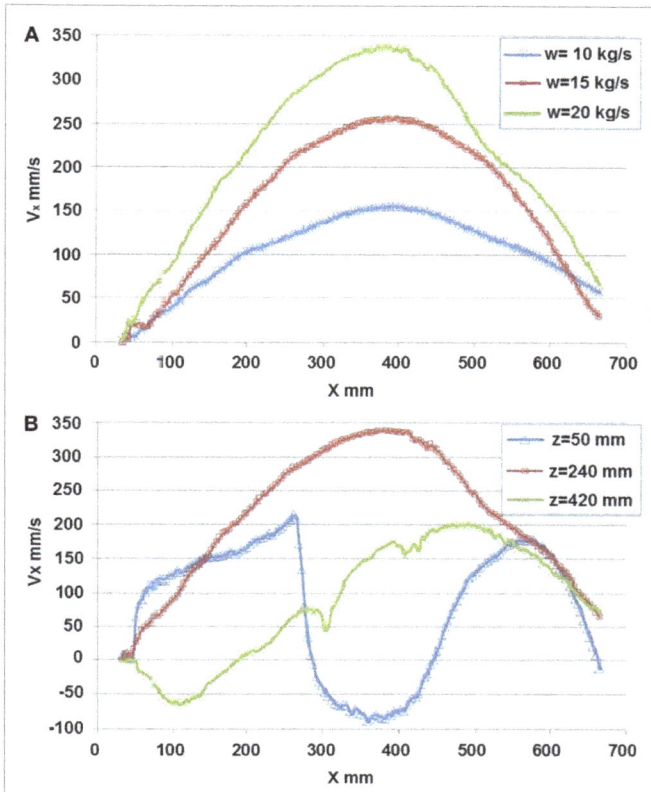

FIGURE 8 | (A) Horizontal component of velocity vs. x coordinate at different mass flow rates and height z = 240 mm. **(B)** Horizontal component of velocity vs. x coordinate at different heights and mass flow rate W = 20 kg/s.

FIGURE 9 | Interface profile at different times at flow rate = 20 kg/s and initial level = 470 mm.

$k-\epsilon$ "Realizable" model, widely used in industrial applications, has been used for modeling the turbulence. The second order accurate discretization scheme QUICK has been used for momentum and turbulence equations and PRESTO! for pressure evaluation. PISO algorithm has been adopted for velocity–pressure coupling. At inlet and outlet sections of uniform and equal velocity profiles have been imposed by assuming that water is the only phase that enters and exits the domain. For the operating feed condition of 20 kg/s, the value of velocity at inlet–outlet sections is 1.24 m/s. At the interface section, the static pressure equal to the atmospheric pressure has been imposed. The boundary conditions for turbulent quantities k and ϵ concern the specifications of intensity of turbulence. Standard wall functions have been used to model the flow near the wall. Residuals, which define the convergence within each time step, have been set equal to 10^{-5} for continuity, momentum, and turbulence equations. **Figure 9** shows the interface profile at different times for an initial water level of 470 mm. A stable condition is reached after about 10 s.

Figure 10 presents the velocity field simulated at operating conditions and for $z > -390$ mm, which is the more interesting target subdomain. It can be seen that the flow field can be subdivided in three main regions with peculiar characteristics. The first region is a tube flux in which the entire mass flow rate is driven and at $x = 340$ mm (in correspondence of the plane of symmetry of the target), the horizontal velocity is higher in the lower part

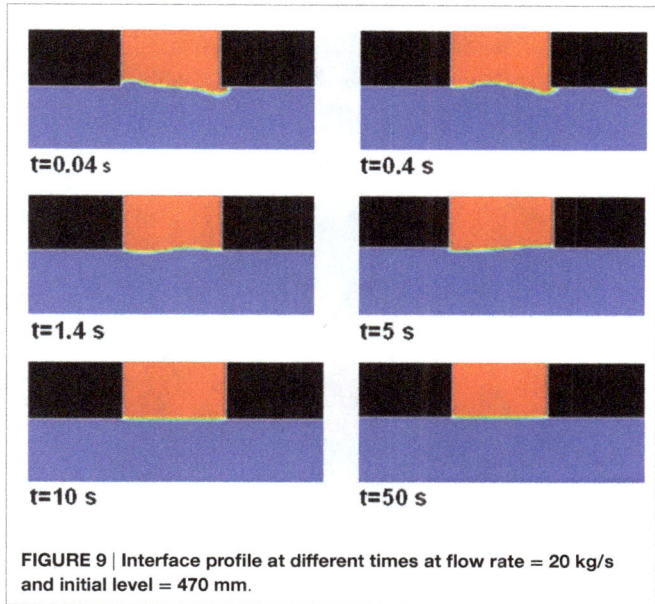

of this tube flux. The tube flux in the spallation region extends from 100 to 350 mm with respect to the z coordinate, whereas from 0 to 100 mm there is an asymmetric recirculating zone that is located at the bottom of the spallation region. It originates from the detachment of the fluid vein when the flow direction suddenly changes from vertical to horizontal one. Finally, the third region (at $350 \leq z \leq 470$ mm in the spallation region) is characterized by very low flow velocities and two weak recirculating fluid structures, placed at the upper corners of the target region, can be identified. **Figures 11** and **12** show the comparison between the Fluent simulated profiles of the velocity horizontal component at the elevation $z = 145$ and $z = 50$ mm and the test data, which present a velocity range in each location due to the fluid turbulence (the thick blue line represent the average of the data).

The low velocities at the interface and the recirculating region at the bottom of the spallation region constitute the main drawback that characterizes the investigated geometry. In order to obtain a more suitable flow field, it is necessary to introduce some modifications on the target's geometry, and the choice has been the introduction of internal vanes that drive the flow together with the rounding of target's corners with a suitable profile. A numerical study has been performed to evaluate the effect of this type of modifications. The introduction of guide vanes produces higher velocities at the interface while smooth corners prevent the occurrence of recirculating regions. A first attempt has been done considering a vane's profile geometrically designed starting from a basic elliptic arc (**Figure 13**) with the corners that are smoothed with elliptic arcs. Lower elliptic arcs have minor radius equal to 50 mm and major radius equal to 100 mm. Upper elliptic arcs have minor radius equal to 145 mm and major radius equal to 290 mm. The distribution of vanes is uniform in vertical direction and the distance between two contiguous vanes has been adopted equal to 78 mm. In order to redistribute more suitably in horizontal direction, the mass flow rate entering the spallation target, starting from the left lower corner as a reference origin and considering,

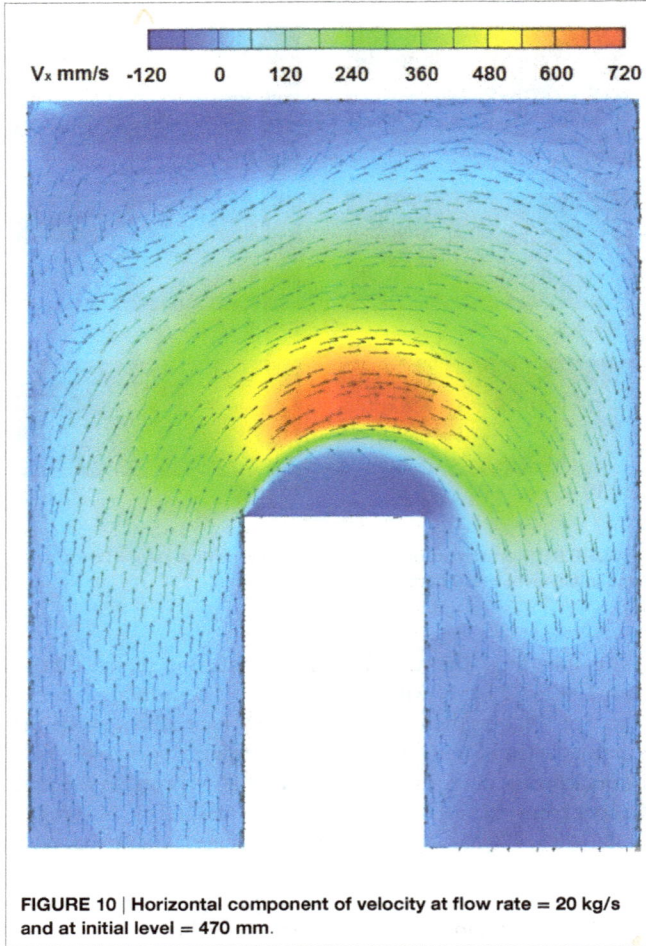

FIGURE 10 | Horizontal component of velocity at flow rate = 20 kg/s and at initial level = 470 mm.

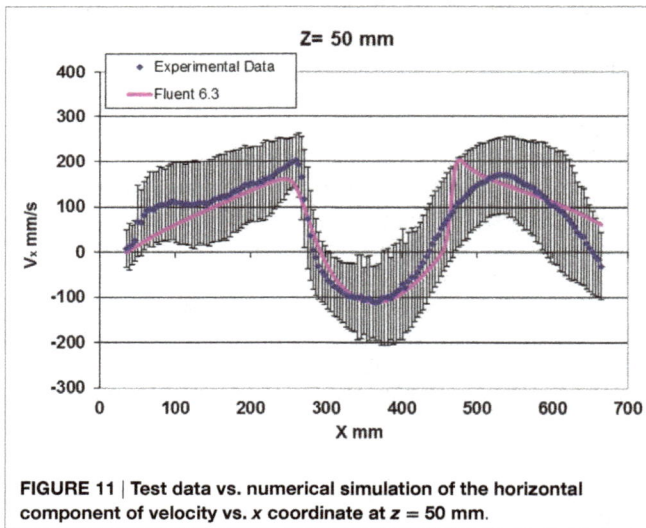

FIGURE 11 | Test data vs. numerical simulation of the horizontal component of velocity vs. x coordinate at z = 50 mm.

the vanes are distributed according to a geometrical progression rule with ratio equal to 1.3. The modified geometry is discretized with a triangular mesh with mean size of 2 mm. All other settings in the code, which does not include multiphasic specifications, are taken as for the previous case.

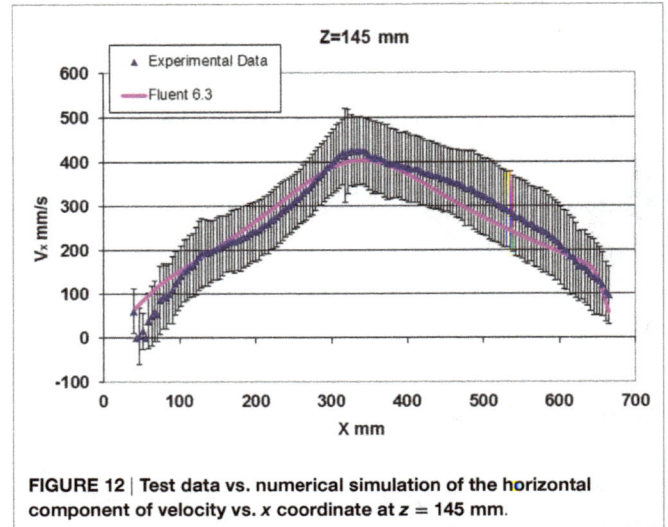

FIGURE 12 | Test data vs. numerical simulation of the horizontal component of velocity vs. x coordinate at z = 145 mm.

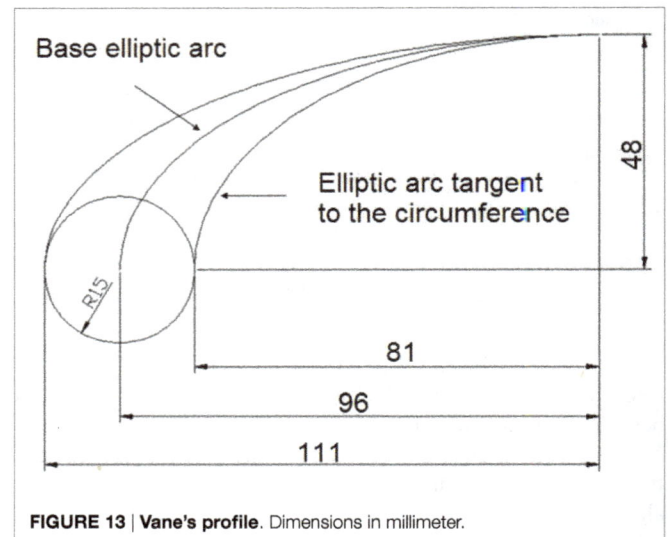

FIGURE 13 | **Vane's profile**. Dimensions in millimeter.

Figure 14 shows the flow field for the modified geometry that has been simulated by the Fluent code: it can be seen that the geometrical modifications produce significant effects on the velocity. In particular (**Figure 15**), the horizontal velocity in the upper part of the spallation region is much higher than the reference case and it is comparable with the mean velocity at the inlet section of the target that is equal to 470 mm/s at W = 20 kg/s.

Optimization of the WST-Modified Geometry Applied to EFIT Project

In order to identify better solutions, a WST-modified geometry (for 2D–3D approximation) has been designed by means of an *ad hoc* algorithm implemented on a commercial CAD software. The geometrical features of the internal vanes and those of the target container are parameterized in terms of a limited number of input parameters whose range of variation is determined by preliminary simulations. The design space of the input parameters

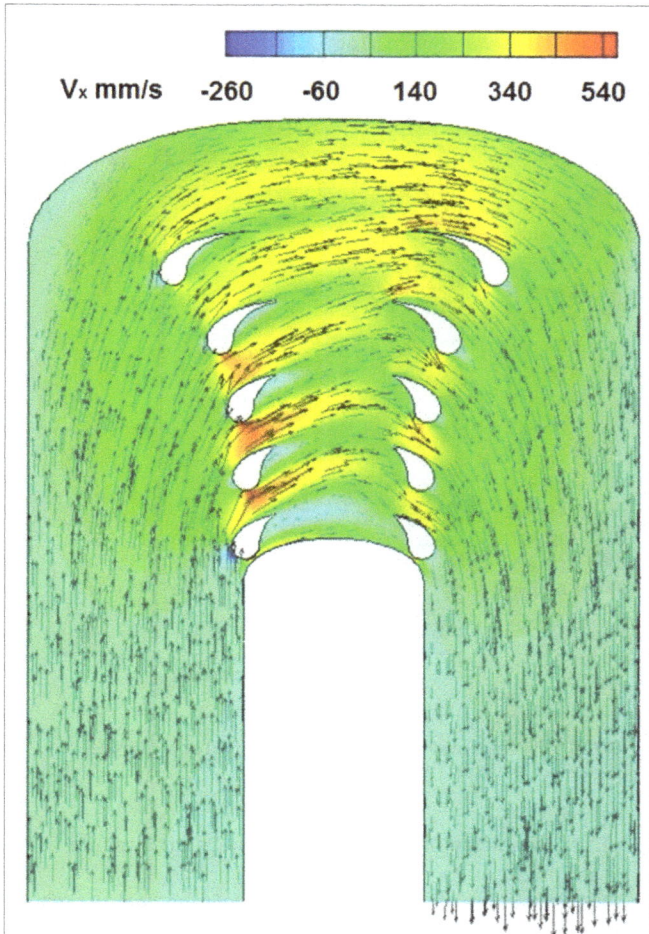

FIGURE 14 | Horizontal component of velocity simulated in a geometry modified by insertion of internal vanes and rounding of the corners. Mass flow rate = 20 kg/s.

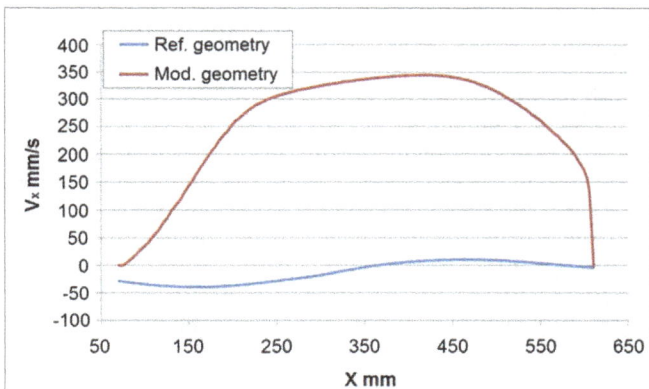

FIGURE 15 | Comparison of simulated horizontal velocities near the interface at $z = 430$ mm between the reference facility geometry (lower blue line) and the modified geometry (upper red line) at $W = 20$ kg/s.

has been explored by generation of a sample of the population according to DOE methodology and successive refinement; over 1500 sample points have been generated for each number of guide vanes; three configurations of the WST geometry have been studied considering, respectively, 8, 6, and 4 internal guide vanes. Each point of the modified geometry has been simulated with the Fluent 6.3.26 code, considering nominal operating conditions of EFIT WST unit (**Figure 16**; **Tables 1** and **2**).

The thermal source has a characteristic trapezoidal profile with respect to the elevation, with higher values at higher height, as the Bragg peak at the operating 800 MeV proton beam energy is expected to be weak (Kumawat et al., 2008). In **Table 1**, the reference conditions and the main characteristics of the thermal source for the EFIT TU are reported.

Table 2 presents the boundary conditions and the main assumptions, which have been adopted in the CFD solver. The thermal source in the numerical study has been conservatively considered to have a triangular shape with an integral thermal power released into the heat deposition region (**Figure 16**) equal to 11.6 MW.

Output parameters (OP) have been defined to synthesize the physical information arising from the single CFD simulation. OP is user defined and can be local and/or integral quantities of

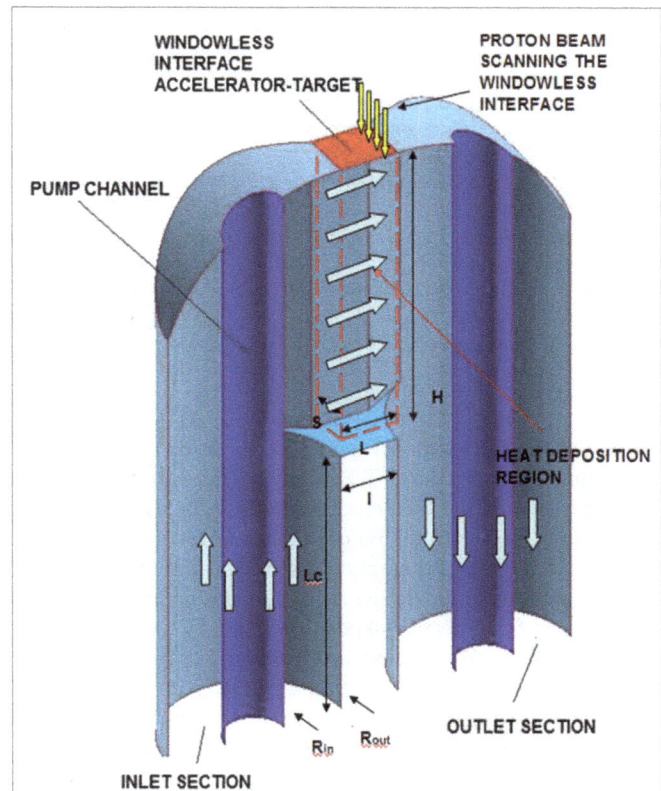

FIGURE 16 | EFIT target unit (all dimensions in millimeter): $H = 470$, $L = 100$, $S = 45$, $I = 140$, $R_{in} = 42$, $R_{out} = 152$, $Lc = 420$.

TABLE 1 | Operating conditions and thermal source characteristics for EFIT TU (Mansani et al., 2012).

Target unit type	Windowless with two mechanical pumps in series
Proton beam	20 mA, 800 MeV Max power: 16 MW Proton travel depth: ~ 440 mm
Thermal power deposited	70% of Max power: ~11 MW in a parallelepiped 440 mm × 140 mm × 100 mm
Reference conditions	Inlet temperature: ~700 K Mass flow rate: 800 kg/s

TABLE 2 | Assumptions and boundary conditions in the numerical parametric study.

Geometry	2D: quadrilateral mesh of average size equal to 5 mm in the bulk of the domain with a refinement near the walls with quadrilateral mesh with an average size of 1 mm 3D: tetrahedral mesh of 10 mm average size in the bulk with refinement with size of 5 mm
Hypothesis	The effects induced by the presence of the free surface have been neglected
Operating conditions	Spallation material: leads with constant thermal–physical properties, evaluated at T = 700 K
Thermal source	Triangular shaped source with respect to the elevation: maximum value at top, zero at bottom Heat deposition domain: for 2D 470 mm × 100 mm, for 3D 470 mm × 100 mm × 140 mm The integral thermal power is equal to 11.6 MW (in 3D) Boussinesq approximation for momentum and energy equations coupling
Boundary conditions	Inlet: inlet velocity = 1.15 m/s Outlet: outlet pressure = 0 Pa (gage pressure) Wall: no slip flow condition
Turbulence	k-ε standard model Inlet conditions: turbulence intensity = 3%, hydraulic diameter = 304 mm
Convergence criterion	Maximum of the residuals <10^{-6} (2D), <10^{-4} (3D)
Numerical schemes	Second Order Upwind for continuity, momentum and energy equations SIMPLE algorithm for pressure-velocity coupling
Solution	Steady state

interest of the turbulent flow and thermal fields in the thermal deposition region of the WST. Second order polynomial models, capable to predict the OP determinations in the design space of input parameters with a 15% of maximum relative error, have been adopted. One of the output parameter that can be chosen is the maximum temperature $T_{MAX,H}$ in the upper part of the exit section of the thermal deposition region, which is the most critical part of the WST (it extends from the free surface to about 5 cm below it). Another output parameter that can be considered is the pressure drop in the system. The performance objectives of the WST geometry design required the minimization of $T_{MAX,H}$ while keeping the pressure drop below the maximum allowable value of 20 kPa. The analytical models were used in the identification of a potential geometry that allows suitable flow conditions, expressed in terms of the given performance objectives within the thermal deposition region. **Figure 17** shows the simulated temperature field, in 2D approximation of the flow for such an

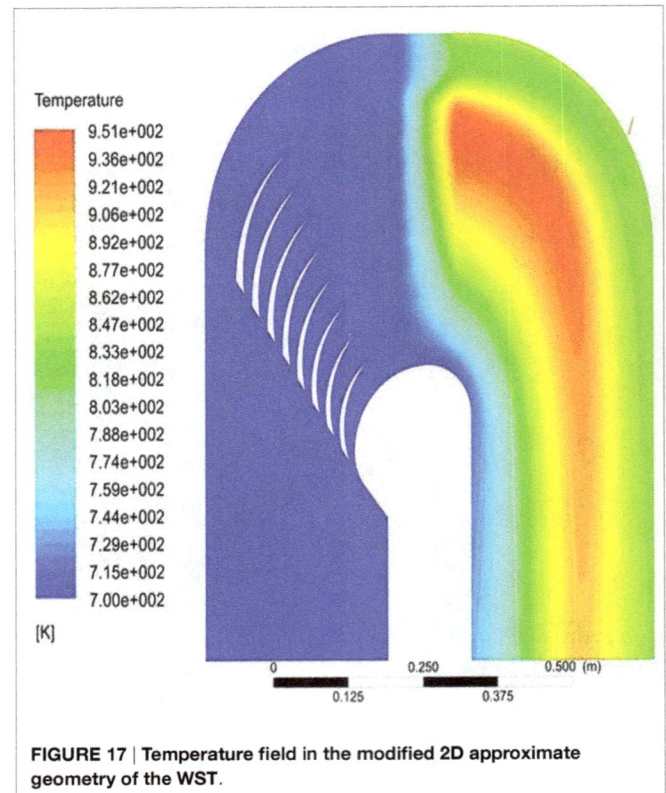

FIGURE 17 | Temperature field in the modified 2D approximate geometry of the WST.

identified geometry in the case that eight internal guide vanes were adopted.

Figure 18 shows the effect of the 3D calculation on the temperature distribution in the mid-plane of the TU. The analysis of 2D vs. 3D simulations shows that 3D effects (pump channel, cylindrical geometry of the feed and return channels, restriction of the cross section area in the heat deposition region) are important but a 2D geometry approximation, which is computationally less expensive, is satisfactory.

The 2D simulations results are conservative in terms of calculated $T_{MAX,H}$, although with a little increase of pressure drop in the TU; this is true for all 3D simulated sample points. In a 2D approximation, the velocity field has been predicted, with the **Table 2** assumptions, by investigating mainly the effect of the number of vanes on the velocity. In **Figure 19**, the horizontal component of the velocity evaluated at the symmetry section vs. the height is presented. The number of vanes, higher than 4, does not affect qualitatively the flow field and the calculated OP depend weakly on it. A benefit of using a lower number of vanes is that the pressure drop is lower for the same set of GIP parameters, but as a main drawback the control of the velocity profile by the insertion of a lower number of vanes is more difficult to be performed. In the case of 8 vanes and with heat generation, the turbulence Reynolds Stress Model has been adopted to investigate the effect of the turbulence models on the average velocity by comparison with k–ε model: there is a significant effect at lower heights, near the bottom wall ($z < 100$ mm), while at higher elevation, the curves shape and the numerical results are comparable. In **Table 3**, the calculated OP for the optimal configuration of geometrical

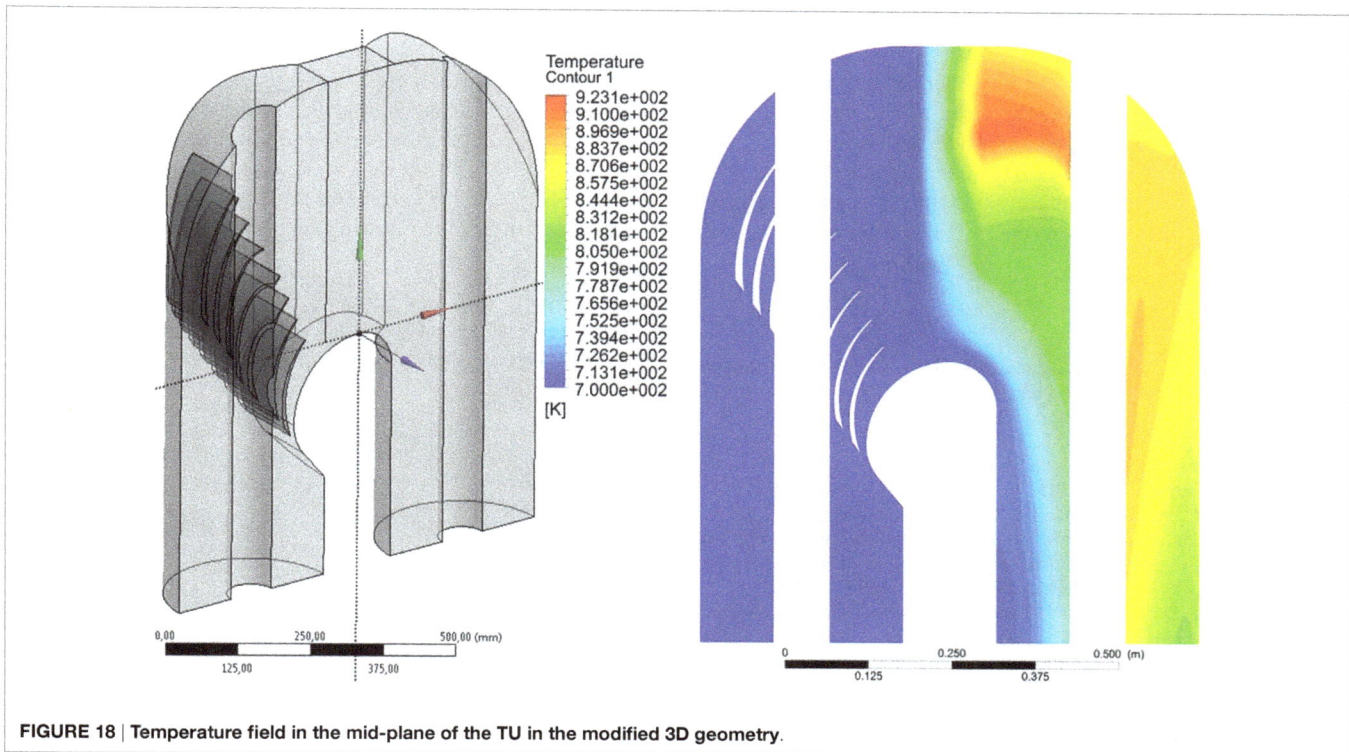

FIGURE 18 | Temperature field in the mid-plane of the TU in the modified 3D geometry.

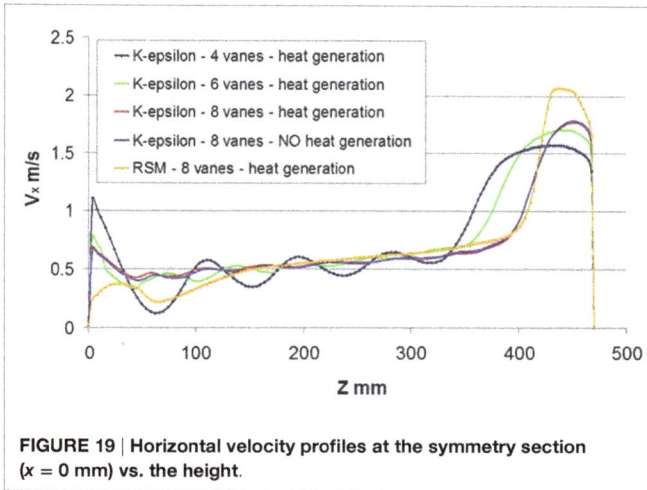

FIGURE 19 | Horizontal velocity profiles at the symmetry section (x = 0 mm) vs. the height.

TABLE 3 | Output parameters determinations for the optimal configuration of geometrical input parameters.

	Pressure drop (kPa)	T_{Max} (K)
2D – k–ε – heat generation – 8 vanes	18	963
3D – k–ε – heat generation – 8 vanes	25	923
2D – k–ε – no heat generation – 8 vanes	17.5	–
2D – k–ε – heat generation – 6 vanes	16.5	961
2D – k–ε – heat generation – 4 vanes	15	960
2D –RSM – heat generation – 8 vanes	24	934

input parameters is reported. The details of the algorithm that implements the DOE techniques and the RSM can be found in Ref. De Giorgi et al. (2011).

Conclusion

In the present paper, the results of experiments and CFD prediction on different test sections, which have been studied as possible solutions for the WST of an ADS, are presented. Both symmetrical and asymmetrical configurations have been investigated. The fluid dynamics of the symmetrical target shows the presence of a fluid stagnant region at the proton beam entrance, which suggests that a symmetrical configuration cannot be considered a suitable geometry for the windowless target design. The asymmetrical configuration fixes the problem of stagnant fluid in the spallation region, but the analysis of the flow field highlights that the required flow conditions inside the target, to efficiently remove the heat generated by spallation reactions, are not achieved. The presence of a recirculating region, where an important heat generation is expected, may potentially lead to boiling. Low velocities have been measured and simulated in proximity of the interface where velocities must be much higher since heat generation is higher at top of the target. So, experimental and numerical results suggest that the asymmetrical configuration is to be considered a promising geometry for a windowless target, but it should be carefully redesigned in order to face the investigated issues. The adopted solution has been to insert some guide vanes in the riser, upstream of the protons target, and the fluid dynamics into the EFIT TU equipped with guide vanes has been investigated. It has been shown that an effective heat removal of the power deposited by the proton beam is possible by means of proper shaping and positioning of the vanes into the TU. By means of an algorithm for the geometry generation, the issue of the fluid dynamics for a complex geometry has been addressed and the number and the position of the vanes have been changed in order to minimize the maximum temperature in the upper part of the exit section of thermal deposition region. The number of vanes can

be lowered to a minimum value equal to four with a comparable performance, but, with less fine control of the velocity profile in the heat deposition region. The turbulence k–ϵ and Reynolds Stress models in such a complex geometry give results different but comparable. Preliminary experiments have been carried out with the same facility but with a test section (always scaled 1.1 to the real plant) similar to the one shown in **Figure 18** equipped with guide vanes: the first results are encouraging.

References

Abanades, A., Sordo, F., Lafuente, A., and Marinez-Val, J. M. (2008). Thermal features of spallation window targets. *Energy Convers. Manage.* 49, 1934–1945. doi:10.1016/j.enconman.2007.08.016

Aït Abderrahim, H., Baeten, P., De Bruyn, D., and Fernandez, R. (2012). MYRRHA – A multi-purpose fast spectrum research reactor. *Energy Convers. Manag.* 63, 4–10. doi:10.1016/j.enconman.2012.02.025

Bianchi, F., Artioli, C., Burn, K. W., Gherardi, G., Monti, S., Mansani, L., et al. (2005). "Status and trend of core design activities for heavy metal cooled accelerator driven system," in *Proceedings of the 12th International Conference on Emerging Nuclear Energy Systems (ICENES'2005)*; 2005 August 21-26. Brussels: CD-ROM.

Bianchi, F., Ferri, R., and Moreau, V. (2008). Thermo-hydraulic analysis of the windowless target system. *Nucl. Eng. Des.* 238, 2135–2145. doi:10.1016/j.nucengdes.2007.10.026

Cinotti, L., and Gherardi, G. (2002). The Pb-Bi cooled XADS status of development. *J. Nucl. Mater.* 301, 8–14. doi:10.1016/S0022-3115(01)00721-8

Cinotti, L., Giraud, B., and Ait Abderrahim, H. (2004). The experimental accelerator driven system (XADS) in the EURATOM 5th framework program. *J. Nucl. Mater.* 335, 148–155. doi:10.1016/j.jnucmat.2004.07.006

Cinotti, L., Saccardi, G., Traverso, M., Garbarini, M., Gherardi, G., Orden, A., et al. (2003). "XADS cooled by Pb-Bi system description," in *Proceedings of the International Workshop on P&T and ADS Development*; 2003 October 6-8. Mol: SCK-CEN.

De Giorgi, L. C., De Salve, M., and Panella, B. (2011). "Thermal hydraulics in the ADS windowless target equipped with internal guide vanes," in *Proc. XXIX National Heat Transfer Conference* (Torino: ETS Pisa), 359–365.

De Salve, M., Malandrone, M., and Panella, B. (2002). "Fluiddynamics in a windowless symmetrical target for accelerator driven hybrid nuclear reactors," in *Proceedings of the Twelfth International Heat Transfer Conference*, Vol. 4; 2002 August 18-23. (Grenoble: Elsevier), 597–602.

Fluent Inc. (2005). *Fluent 6.2 User's Guide.* Lebanon, NH: Centerra Resource Park.

Knebel, J., Aït Abderrahim, H., Cinotti, L., Mansani, L., Delage, F., Fazio, C., et al. (2006). *EUROTRANS: European Research Programme for the Transmutation of High-level Nuclear Waste in Accelerator-driven System, 9th Information Exchange.* Nimes.

Kumawat, H., Dutta, D., Mantha, V., Mohanty, A. K., Satyamurthy, P., Choudhury, R. K., et al. (2008). Heat deposition in thick targets due to interaction of high energy protons and thermal hydraulics analysis. *Nucl. Instrum. Methods Phys. Res.* 266, 604–612. doi:10.1016/j.nimb.2007.11.064

Mansani, L., Artioli, C., Schikorr, M., Rimpault, G., Angulo, C., and De Bruyn, D. (2012). The European Lead-cooled EFIT Plant: an industrial-scale accelerator-driven system for minor actinide transmutation. *Nucl. Technol.* 180, 241–263.

Morris, M. D., and Mitchell, T. J. (1995). Exploratory designs for computational experiments. *J. Stat. Plan Inference* 43, 381–402. doi:10.1016/0378-3758(94)00035-T

Myers, R., and Montgomery, D. (2002). *Response Surface Methodology.* New York, NY: Wiley.

Petrazzini, M. (2008). Main components functional sizing of EFIT, Ansaldo EUROTRANS, 121 SMFX 007 Rev. 1, Work package 1.2, Id. No: D1.24.

Rubbia, C., Abderrahim, H. A., Björnberg, M., Carluec, B., Gherardi, G., and Romero, E. G. (2001). *A European Roadmap for Developing Accelerator Driven Systems (ADS) for Nuclear Waste Incineration.* Rome: ENEA.

Rubbia, C., Rubio, J. A., Buono, S., Carminati, F., Fieter N., Galvez, J., et al. (1995). *A Conceptual Design of a Fast Neutron Operated High Power Energy Amplifier.* CERN/AT/95-44.

Sacks, J., Welch, W. J., Mitchell, T. J., and Wynn, H. P. (1989). Design and analysis of computer experiments. *Stat. Sci.* 4, 409–435. doi:10.1214/ss/1177012413

Salvatores, M. (2006). Fuel cycle strategies for the sustainable development of nuclear energy: the role of accelerator driven systems. *Nucl. Instrum. Methods Phys. Res.* 562, 578–584. doi:10.1016/j.nima.2006.02.013

Salvatores, M., Slessarev, I., Tchistiakov, A., and Ritter, G. (1997). The potential of accelerator driven systems for transmutation or power production using thorium or uranium fuel cycles. *Nucl. Sci. Eng.* 126, 333–340.

Santner, T., Williams, B., and Notz, W. (2003). *The Design and Analysis of Computer Experiments.* New York, NY: Springer.

Signal Processing, S. A. (2004). *DOP 2000 Model 2125 User Manual.* Switzerland: Signal Processing S.A.

SonTek Co. (1997). *Acoustic Doppler Velocimeter Technical Documentation.* San Diego, CA: SonTek Co.

Sordo, F., Leon, P. T., and Martinez-Val, J. M. (2008). Nuclear and thermal-hydraulics analysis of spallation window targets. *Nucl. Instrum. Methods Phys. Res.* 574, 232–243. doi:10.1016/j.nima.2007.02.018

Conflict of Interest Statement: The authors declare that the research was conducted in the absence of any commercial or financial relationships that could be construed as a potential conflict of interest.

12

Potential of Sugarcane in Modern Energy Development in Southern Africa

Simone P. Souza[1]*, Luiz A. Horta Nogueira[1], Helen K. Watson[2], Lee Rybeck Lynd[3], Mosad Elmissiry[4] and Luís A. B. Cortez[5]

[1] Interdisciplinary Center for Energy Planning, University of Campinas (UNICAMP), Campinas, SP, Brazil, [2] School of Agricultural, Earth and Environmental Sciences, University of KwaZulu-Natal, Durban, KZN, South Africa, [3] Dartmouth College, Thayer School of Engineering, Dartmouth, NH, USA, [4] New Partnership for Africa's Development (NEPAD), Johannesburg, GT, South Africa, [5] Faculty of Agricultural Engineering, University of Campinas (UNICAMP), Campinas, SP, Brazil

Edited by:
Fu Zhao,
Purdue University, USA

Reviewed by:
Subbu Kumarappan,
Ohio State University, India
Karl Haapala,
Oregon State University, USA

*Correspondence:
Simone P. Souza
sp.souza@yahoo.com.br

Specialty section:
This article was submitted to
Energy Systems and Policy,
a section of the journal
Frontiers in Energy Research

Citation:
Souza SP, Horta Nogueira LA,
Watson HK, Lynd LR, Elmissiry M
and Cortez LAB (2016) Potential of
Sugarcane in Modern Energy
Development in Southern Africa.
Front. Energy Res. 4:39.

For more than half of the Southern African population, human development is limited by a lack of access to electricity and modern energy for cooking. Modern bioenergy merits consideration as one means to address this situation in areas where sufficient arable land is available. While numerous studies have concluded that Africa has significant biomass potential, they do not indicate by how much it can effectively reduce the use of traditional biomass and provide more accessible energy, especially at a country level. Here, we evaluate the potential of sugarcane to replace traditional biomass and fossil fuel and enlarge the access to electricity in Southern Africa. By using its current molasses for ethanol production, Swaziland could increase electricity generation by 40% using bagasse and replace 60% of cooking fuel or 30% of liquid fossil fuel. Sugarcane expansion over 1% of the pasture land in Angola, Mozambique, and Zambia could replace greater than 70% of cooking fuel. Bioelectricity generation from modest sugarcane expansion could be increased by 10% in Malawi, Mozambique, and Zambia and by 20% in Angola. Our results support the potential of sugarcane as a modern energy alternative for Southern Africa.

Keywords: sustainability, bioenergy, bioelectricity, sugarcane ethanol, traditional biomass

INTRODUCTION

Most of the population of Southern African lacks access to electricity and modern energy for cooking (FAO, 2012c; IEA, 2014a). Their heavy dependence on the traditional biomass for cooking direct affect the living conditions in this region. For example, charcoal and firewood supply more than 95% of the cooking energy consumption in Mozambique and Malawi. By contrast, in South Africa, only 13% of the population relies on the traditional use of biomass (IEA, 2014b). The use of charcoal and firewood has been related to approximately 600,000 premature death per year in Africa (WHO, 2013). The stoves are typically inefficient and placed in poorly ventilated spaces, causing indoor air pollution (IEA, 2014a). Also, the use of these traditional biomass leads to household energy accidents, such as burns, scalds, fires, and poisonings (Kimemia et al., 2014).

Electricity access is lacking for 60% of the Southern African population. In Malawi, less than 10% of the population is supplied with electricity (IEA, 2014b). In some cases, countries are highly dependent on imported electricity. For example, 70% of the electricity in Swaziland is imported. Also, all of the Southern African countries are net importers of gasoline and distillate fuel oil (EIA, 2012) (**Table 1**), with all except South Africa and Zambia wholly dependent on imports. This scenario reduces their energy security and harms the national trade balance.

Although the Southern Africa economy is changing rapidly, with annual gross domestic product (GDP) growth of 5.7% from 2000 to 2012, attracting investments and opening up new opportunities (IEA, 2014a; Taliotis et al., 2016), the region must expand its population's access to modern, reliable, and affordable energy and improve the social indicators to maintain and consolidate the economic expansion observed during the last decade.

In regions with sufficient land resources, bioenergy can play this role, promoting energy access and rural development integrated with an improved food security and greater national energy sovereignty (Lynd and Woods, 2011). In terms of physical geography, much of Africa has the capacity to produce bioenergy crops without compromising biodiversity and water use (Lynd and Woods, 2011).

Sugarcane is one of the best feedstocks for bioenergy because of its semiperennial productive cycle, which involves replanting at intervals of 5 years or more (De Cerqueira Leite et al., 2009), and its efficient conversion of solar radiation into chemical energy (Zhu et al., 2010). Sugarcane bioenergy can be cost competitive, promote human development, and comply with strict sustainability indicators, reducing greenhouse gas (GHG) emissions by approximately 80% compared to gasoline (Seabra et al., 2011). Moreover, sugarcane can address the triple challenge of energy insecurity, climate change, and rural poverty in sub-Saharan Africa (Johnson and Seebaluck, 2012).

Approximately 40 million t of sugarcane are produced in Southern Africa, mainly in South Africa, Swaziland, Mauritius, Zimbabwe, Zambia, and Mozambique (FAO, 2012d). Despite the existence of suitable areas for sugarcane cultivation in African countries, the overall potential of the countries is different and depend on particular agricultural and economic issues. The Southeast region has the largest potential for rain-fed sugarcane production, with additional potential to grow this crop using irrigation (Hermann et al., 2014).

However, most of the studies about the potential of the biomass as renewable energy in Africa address only the geographic suitability of renewable energy sources (Watson, 2011; FAO, 2012b) or a general overview about the potential energy supply (IRENA-DBFZ, 2013). There is a lack of studies that quantify how much is the ability to replace the current and future uses of traditional biomass and fossil fuels or even enlarge the electricity access according to the current and projected demand. This intriguing question motivated this study.

By considering two scenarios to produce modern energy from sugarcane in Southern Africa, this study explores the potential of this crop to promote a cleaner and more accessible energy, including the required investment and GHG emissions savings. We assume that sugarcane ethanol will be used as a vehicle fuel and partially replace the traditional use of solid biomass for cooking, thereby contributing to reducing the deforestation associated with burning firewood and charcoal. We also evaluate the potential of cogenerating bioelectricity from bagasse. A probabilistic methodology is applied to nine Southern African countries: Angola, Malawi, Mauritius, Mozambique, South Africa, Swaziland, Tanzania, Zambia, and Zimbabwe. Other countries in Southern Africa (Botswana, Democratic Republic of the Congo, Lesotho, Madagascar, Namibia, and Seychelles) are excluded because of a lack of data or inadequate conditions for sugarcane production.

As the real potential of biomass in Africa is still not accurate—studies report an enormous range of suitable area for biomass production (IRENA-DBFZ, 2013), we assumed the use of only 1% of the pasture area, which the equivalent area is realistic in terms of suitable land (Watson, 2011; FAO, 2012b; Johnson and Seebaluck, 2012) (**Table 2**). Pasture lands are usually underutilized, and by using appropriated pasture management integrated with sustainable intensification practices, such as rotational grazing, incorporation of legumes and integrated crop–livestock–forestry systems, is possible to increase agricultural output (Latawiec et al., 2014) without compromising the grazing activity.

ETHANOL INITIATIVES: A LOOK TO SOUTHERN AFRICA

Ethanol is produced by the fermentation of a mash prepared from molasses (residual sugars from sugar production) or sugarcane juice and distillation, resulting in hydrous ethanol (containing approximately 6% water) or anhydrous ethanol. Pure hydrous ethanol can be used in dedicated or flex-fuel engines, which allow the use of any ethanol-gasoline blend, and as a diesel replacement in modified diesel engines (Nylund et al., 2013). Gasoline containing up to 10% anhydrous ethanol can be used in conventional gasoline vehicles without any modification, and higher blending levels (up to 30%) can be used after relatively simple changes (BNDES/CGEE, 2008). Processing sugarcane to produce either ethanol or sugar results in lignocellulosic residue, corresponding to approximately 27% (dry basis) of cane stalks (Rodrigues Filho, 2005). Bagasse is typically burned to produce power and heat for industrial needs and, increasingly, to generate

TABLE 1 | Net imports of electricity, gasoline and distillate fuel oil.

Country	Net imports of electricity (2009/12 average)	Net imports of gasoline (2009/12 average)	Net imports of distillate fuel oil (2009/12 average)
	TWh/year[a]	1,000 bl/d[a]	1,000 bl/d[a]
Angola	0	23.6	35.35
Malawi	0	1.8	2.90
Mauritius	0	2.7	6.15
Mozambique	−3.37	3.5	11.01
South Africa	−3.08	23.1	22.46
Swaziland	0.91	2.1	2.20
Tanzania	0.06	7.4	17.13
Zambia	−0.58	0.9	1.84
Zimbabwe	0.79	3.7	8.49

[a]U.S. Energy Information Administration (EIA, 2012).

TABLE 2 | Potentially suitable areas for sugarcane and current production.

Country	1% of the current pasture land (ha)[a]	Potentially suitable area for sugarcane (ha)	Current production 10³ t/year (2012)[b]	Additional area relative to the current cane area (%)[b,c]
Angola	571,170	1,127,000[d]	520	42
Malawi	19,568	206,000[d]	2,800	1
Mauritius	74	n.a.[e]	3,947	<1
Mozambique	465,397	2,338,000[d]	3,394	10
South Africa	887,725	5,080,000[f]	17,278	3
Swaziland	10,916	870,000[f]	5,400	<1
Tanzania	254	5,184,000[d]	2,717	<1
Zambia	211,544	1,178,000[d]	3,900	5
Zimbabwe	127,984	620,000[d]	3,929	2

[a]Related to the permanent meadows and pasture from FAOStat database (FAO, 2012a).
[b]From FAOStat database (FAO, 2012d).
[c]The additional area (1% of the pasture land) proposed in this study in relation to the current sugarcane area; (additional sugarcane area)/(current sugarcane area).
[d]Suitable and available areas. Excludes protected areas, crops and wetlands, existing sugarcane areas, slopes >16% and areas <500 ha; from Watson (2011).
[e]n.a. = not available.
[f]Does not exclude unavailable areas used for other activities; from Schulze et al. (1997).

surplus bioelectricity for the grid (Seabra and Macedo, 2011). In Brazil, sugarcane represents a relevant primary electricity source: for example, the power-generation capacity of sugarcane bagasse is 10,500 MW, corresponding to 7.5% of the total installed capacity (ANEEL Fontes de Energia: Biomassa, 2015).

More than 30 countries worldwide have ethanol-blending mandates motivated by various factors, including energy security, rural economic development, and GHG emission reduction (Munyinda et al., 2012; REN21, 2015a). Global ethanol production reached 94 Mm³ in 2014 and was mainly based on sugarcane, corn, and cassava. Biofuels, including biodiesel, represent less than 5% of the global road transport fuel demand on an energy basis (REN21 2014).

Several initiatives aimed at introducing sugarcane bioenergy in Southern Africa, including various concepts and scales of operation, have been attempted. In 1982, Malawi adopted E10 blending using biofuel locally produced from sugarcane molasses. By 2004, the total production capacity reached 36 million liters per year (Chakaniza, 2013), allowing for E20 blending and the use of pure hydrous ethanol. In 2012, the CleanStar project was launched in Mozambique to promote a transition away from nontraditional biomass and inefficient stoves by disseminating up to 30,000 clean-burning and efficient cooking stoves fueled with locally produced ethanol (UNFCCC, 2013). The project target was to produce 2 million liters per year of ethanol from cassava as cooking fuel supplied by local small farmers (Novozymes, 2012). However, the progress was impaired by feedstock supply and failed to achieve the required sales to sustain the manufacturing flow, leading to the end of the project in 2014 (REN21, 2015b). Challenges also included overcoming economic and cultural barriers (Dasappa, 2011).

In Angola, the BIOCOM Enterprise was created to diversify the economy by activating the sugarcane agroindustry and generating jobs and income. This project is a joint venture involving Angolan and Brazilian investors (worth US$ 750 million) and relies strongly on technology transferred from the Brazilian sugarcane agroindustry model. The commercial operation of the facility started in June 2015 (Macauhub, 2015). When fully implemented (planned for 2019/2020), 42,000 ha of sugarcane

will annually produce 30 million liters of anhydrous ethanol, 235 GWh of surplus bioelectricity, and 260,000 t of sugar (Biocom, 2015), which is sufficient to supply at least 50% of the domestic demand for sugar (FAO, 2013a). Currently, most sugar consumed in Angola is imported (FAO, 2013a).

MATERIALS AND METHODS

By using energy demand data (**Table 1**) and applying some assumptions to project a future scenario for 2030 (see Energy Demand Projection), we evaluated the potential of sugarcane to provide cleaner and more accessible energy in Southern Africa by considering two scenarios: Current Molasses (CM) and New Policies (NP). CM represents a short-term framework, in which ethanol is produced exclusively from molasses, considering the existing sugarcane production and the current technology access. NP refers to an enhanced approach likely to be deployed over the medium to long term based on the 2030 scenario proposed by the International Energy Agency (IEA, 2014a); in this case, sugarcane is cultivated over 1% of the pasture land, and ethanol is produced from molasses (existing sugarcane mills) and direct juice (additional sugarcane mills). We assessed the use of ethanol as cooking fuel and a displacer of fossil fuels, including gasoline and diesel. The use of ethanol in diesel engines is supported by the Scania technology, which allows the use of pure ethanol with 5% ignition improver in diesel engines [BioEthanol for Sustainable Transport project (EHA, 2011)]. Ethanol for cooking is likely to replace fuelwood because the latter is the main energy resource used as cooking fuel in Southern Africa (greater than 90%), except in South Africa, where electricity is the primary cooking fuel for approximately 60% of households (Adkins et al., 2012; IEA, 2014a).

Both scenarios correspond to a mill-crushing capacity of one million t of sugarcane and distillery consumption of 30 kWh/t cane (mechanical and electrical energy) (Dias et al., 2011). Further assumptions include the following:

- *CM Scenario:* Ethanol is produced exclusively from existing molasses, and no additional sugarcane production occurs. We assumed a low efficiency for the cogeneration system, with the ability to generate 60 kWh/t cane (42 bar, 450°C) (BNDES/CGEE, 2008). We consider that all the existing sugarcane industries will be able to deploy such as system, if it does not exist already.

- *NP Scenario:* This scenario is based on the 2030 scenario proposed by the International Energy Agency (IEA, 2014a). Ethanol is produced from molasses and additional sugarcane (direct juice), which is cultivated over 1% of the current pasture land. For each country, we assessed the availability of suitable areas for sugarcane cultivation. The efficiency for the cogeneration system in this scenario is higher, with a capacity of 110 kWh/t cane (65 bar, 480°C) (BNDES/CGEE, 2008).

Uncertainty Analysis

The overall uncertainty of ethanol and electricity production, beyond the potential to address cooking and transportation demand, was evaluated by applying a stochastic approach based on the Monte Carlo method. In this method, an appropriate probability distribution is associated with each of the input parameters subjected to uncertainties (**Table 3**). Values for these parameters are generated randomly and combined with other randomly generated values; we use 10,000 trials. The results are presented as an average value associated with a probability distribution for all possible outputs (ADB, 2002).

Greenhouse Gas Emission

We assessed the GHG emission savings given the replacement of fossil fuels by ethanol and the use of bagasse as bioelectricity source instead of the current electricity mix employed in each country. We considered the emissions throughout the life cycle. The carbon emission was evaluated for the NP scenario to identify the potential carbon savings if the countries invested in improving their energy generation profile for 2030, i.e., using ethanol as vehicle fuel instead of gasoline and diesel and bagasse for bioelectricity generation rather than maintaining the current scenario. We did not estimate the GHG emissions savings for the CM scenario; indeed, suggesting energy replacement in this scenario is not sensible because most of the analyzed countries still lack access to electricity or have a latent demand for fuels. In the NP scenario, fuels are not replaced; instead, the energy sector expansion in the coming years is reevaluated.

The GHG emission factors for electricity production were estimated based on the life cycle emissions of energy systems (**Table 3**) and the current electricity generation (**Table 4**). As for ethanol, gasoline, diesel, and electricity from bagasse cogeneration, the GHG emission factors correspond to the life cycle carbon emissions (**Table 5**).

Investment

We estimate the investment required to implement the additional sugarcane industrial plants for the NP scenario considering an industry crushing capacity of 1 million t per year and an investment of 212 US$/t cane annual crushing capacity (BNDES, 2014), including the agricultural sector, sugarcane mill, and cogeneration system. We assume the investment would occur over 10 years,

TABLE 3 | Parameters for uncertainty analysis.

	Distribution	Mean/SD or likeliest	Location/ scale/limits[k]
Both scenarios			
Cooking fuel use (L/household/year)[a,b]	Normal	360/36	–
Ethanol yield—direct juice (L/t cane)[c]	Triangular	81	68–85
Ethanol yield—molasses (L/t cane)[c]	Triangular	10	6–12
Electricity consumption[b,d]	Normal	See note	–
Gasoline consumption[b,d]	Normal	See note	–
Cane yield (t/ha/year)[e]			
Angola	Lognormal	38/0.7	35
Malawi	Logistic	108	1
Mauritius	Logistic	73	2
Mozambique	Logistic	60	11
South Africa	Lognormal	62/5	41
Swaziland	Lognormal	98/3	95
Tanzania	Lognormal	81/28	0
Zambia	Logistic	104	1
Zimbabwe	Lognormal	84/13	0
Current molasses scenario			
Surplus electricity (kWh/t cane)[f,b]	Normal	30/3	–
Household electricity demand[g]	Triangular	1,117	480–2,072
New policies scenario			
Pasture land (ha)[b,h]	Normal	See note	–
Surplus electricity (kWh/t cane)[f,b]	Normal	80/8	–
Household electricity demand[g]	Triangular	1,430	538–2,322
Electricity increasing rate (%)[i]			
Southern countries (excluding South Africa)	Triangular	73	64–82
South Africa	Triangular	47	40–54
Gasoline increasing rate[i]	Triangular	67	59–75
Diesel increasing rate[c,i]	Triangular	50	42–58
Life cycle GHG emissions from electricity			
Coal (g CO_2e/kWh)[j]	Triangular	1,001	675–1,689
Oil (g CO_2e/kWh)[j]	Triangular	840	510–1,170
Natural gas (g CO_2e/kWh)[j]	Triangular	469	290–930
Biopower (g CO_2e/kWh)[j]	Triangular	18	−633–75
Nuclear (g CO_2e/kWh)[j]	Triangular	16	1–220
Hydro (g CO_2e/kWh)[j]	Triangular	4	0–43
Solar PV (g CO_2e/kWh)[j]	Triangular	46	5–217
Wind (g CO_2e/kWh)[j]	Triangular	12	2–81

[a]From United Nations (2007).
[b]SD corresponds to 10% of the mean (adopted).
[c]From United Nations (2006).
[d]Mean values according to 2012 data (**Table 6**).
[e]Probability distribution fitting according to 2000–2014 period; from FAOStat database (FAO, 2012d).
[f]Estimated from BNDES/CGEE (2008) and Dias et al. (2011).
[g]From Castellano et al. (2015).
[h]Uncertainty assumed based on experts' judgments justified by errors involved in estimating pasture land. For details, see **Table 2**.
[i]According to IEA (2014a); minimum and maximum values corresponding to 5% of the mean (adopted).
[j]From Moomaw et al. (2012).
[k]Location for lognormal distributions, scale for logistic distributions, and limits for triangular distributions.

and thus divided the total investment by 10 and compared the annual investment with the GDP at market price and to the gross fixed capital formation. We do not consider investments in

TABLE 4 | Current generation and emission factors for electricity.

Country	Total electricity generation (GWh/year)[a]	Source[a]						g CO_2e/kWh[b]
		Coal (%)	Oil (%)	Natural gas (%)	Biofuels (%)	Nuclear (%)	Hydro (%)	
Angola	5,613	0	29	0	0	0	71	247
Malawi	2,179	0	6	6	0	0	87	87
Mauritius[c]	2,797	41	38	0	18	0	3	736
Mozambique	15,166	0	0	0	0	0	100	5
South Africa	257,919	94	0	0	0	5	2	938
Swaziland	425	71	0	0	0	0	29	708
Tanzania[d]	5,589	0	15	53	0	0	32	376
Zambia	12,387	0	0	0	0	0	100	6
Zimbabwe	9,124	40	1	0	1	0	59	406

[a]Estimated from IEA (2012).
[b]Based on life cycle GHG emissions for electricity from **Table 3**.
[c]Also 0.1% of wind.
[d]Also 0.2% of solar PV.

TABLE 5 | Emission factors for fuels and electricity from bagasse.

Energy source	Value
Electricity from bagasse (g CO_2e/kWh)[a]	66.5
Sugarcane ethanol (g CO_2e/MJ)[a,c]	18.5
Gasoline (g CO_2e/MJ)[b,c]	88.4
Diesel (g CO_2e/MJ)[b,c]	92.8

[a]GHG emissions (WTW) for electricity and ethanol were adapted from Souza et al. (2012) considering 30% mechanized harvesting and 70% burning harvesting. GHG emissions were allocated on an energy basis. Although Dunkelberg et al. (2014) estimated the GHG emission for sugarcane ethanol in Malawi, the assumptions adopted in the life cycle scope are not adequate for our requirement (e.g., it doesn't include bioelectricity as coproduct). We thus used data from ethanol sugarcane in Brazil due to lack of data for Southern African reality.
[b]The GHG emissions (WTW) refer to pure gasoline and conventional diesel and were modeled using the Argonne GREET Model 2014 (Wang et al., 2014).
[c]The avoided emissions attributable to gasoline replacement were 70 g CO_2e/MJ (88.4–18.5 g CO_2e/MJ). The lower heating values assumed for pure gasoline, ethanol, and diesel were 32.36 MJ/L, 21.27 MJ/L, and 35.8 MJ/L, respectively (Wang et al., 2014).

the distribution and transmission systems, assuming that these investments will happen regardless.

Energy Demand Projection

To identify the potential electricity consumption in 2030 for the NP scenario, we applied a 47% rate of increase for South Africa and a 73% rate of increase for the remaining countries relative to 2012 (annual growth rates of 2 and 3%, respectively) (IEA, 2014a) (**Table 6**). Gasoline and diesel consumption increase by 67 and 50% from 2012 to 2030, corresponding to 2 and 3% p.a., respectively (IEA, 2014b).

RESULTS AND DISCUSSION

We evaluated the potential of sugarcane to provide cleaner and more accessible energy in Southern Africa by considering a short-term framework, named CM scenario, and an enhanced approach likely to be deployed over the medium to long term, entitled NP scenario. Results show the potential of energy supply,

the GHG emissions savings, and the total investment required to enlarge the sugarcane production, with further discussion on challenges for implementing bioenergy systems in Africa.

Potential Modern Energy Supply and Fossil Fuel Displacement

We found good prospects for implementing modern sugarcane bioenergy in Southern Africa, with large differences among countries (**Table 7**). By using the existing molasses for ethanol production, Swaziland could meet 57 ± 6% of the household cooking energy demand or displace 30 ± 3% of fossil fuel use. Ethanol is already produced from molasses in this country but is mostly used in the beverage and pharmaceutical industries (IRENA, 2014).

By using 1% of pasture land for sugarcane cultivation, ethanol could meet up to 50% of the cooking fuel demand for most of Southern African countries. A low availability of pasture land reflects, in principle, a lower potential for sugarcane ethanol production and, thus, for displacement of fossil fuels, as found in Mauritius and Tanzania. In countries with low consumption of liquid fuels and large availability of land, such as Mozambique and Zambia, sugarcane expansion plays a high potential to replace fossil fuel. Even though the assumption of 1% of the pasture land may disadvantage countries with low availability of grazing area, it is more realistic than proposing a percentage of sugarcane expansion based on the energy demand, and also 1% is in accordance with the suitable area for sugarcane (**Table 2**), especially in a scenario of pasture intensification (Latawiec et al., 2014).

In some cases, the need for ethanol for a specific purpose is lower than its production; therefore, the excess could be used as a fuel for both cooking and vehicles (**Table 7**). In our analysis, we considered a progressive increase in the electricity and transportation fuel consumptions, according to IEA (2014a), which would be related to an economic growth and thus a higher energy consumption. Even assuming that, the ethanol supply would be able to attend a large share of the fuel demand in the future NP scenario (**Table 7**).

TABLE 6 | Current and projected consumption of electricity, gasoline, and diesel.

Country	Household size[a]	Population (1,000 people)[b]		Final electricity (GWh/year)		Gasoline (1,000 bl/d)		Diesel (1,000 bl/d)	
		Current (2012)	Projection (2030)	Current (2012)[c]	Projection (2030)	Current (2012)[c]	Projection (2030)	Current (2012)[c]	Projection (2030)
Angola	5.0	20,821	34,783	4,842	8,377	25.4	42.5	55.0	83
Malawi	4.2	15,906	25,960	2,027	3,507	1.8	3.0	2.9	4
Mauritius	4.2	1,240	1,288	2,472	4,277	2.9	4.8	6.5	10
Mozambique	4.4	25,203	38,876	11,284	19,521	4.1	6.9	11.0	17
South Africa	4.2	52,386	58,096	211,573	311,012	197.6	330.0	223.0	335
Swaziland	4.6	1,231	1,516	1,295	2,241	2.1	3.5	2.2	3
Tanzania	5.1	47,783	79,354	4,545	7,863	7,4	12,4	18	27
Zambia	5.1	14,075	24,957	8,327	14,406	5.4	9.0	9.3	14
Zimbabwe	4.1	13,724	20,292	6,831	11,818	4.1	6.8	9.2	14

[a]From DHS Program (ICF International, 2012).
[b]FAOStat database (FAO, 2012c).
[c]EIA (2012). Refers to distillate fuel oil, which includes diesel fuels and fuel oils.

TABLE 7 | Potential ethanol supply and fuel displacement.

Country	Current Molasses scenario				New Policies scenario			
	Ethanol production (10^6 L/year; mean $\pm$ SD)	Cooking Fuel	Ethanol as gasoline displacement	Ethanol as diesel displacement	Ethanol production (10^6 L/year; mean $\pm$ SD)	Cooking Fuel	Ethanol as gasoline displacement	Ethanol as diesel displacement
Angola	5 $\pm$ 1	0%	0.24 $\pm$ 0.02%	0%	1,727 $\pm$ 195	69 $\pm$ 11%	47 $\pm$ 7%	26 $\pm$ 3%
Malawi	27 $\pm$ 4	2 $\pm$ 0.2%	18 $\pm$ 2%	13 $\pm$ 0.3%	189 $\pm$ 19	9 $\pm$ 1%	72 $\pm$ 10%	55 $\pm$ 6%
Mauritius	37 $\pm$ 5	–	16 $\pm$ 2%	8 $\pm$ 0.4%	38 $\pm$ 5	34 $\pm$ 6%	9 $\pm$ 2%	5 $\pm$ 1%
Mozambique	26 $\pm$ 9	1 $\pm$ 0.5%	8 $\pm$ 3%	3 $\pm$ 1%	2,197 $\pm$ 754	69 $\pm$ 25%	368 $\pm$ 131%	168 $\pm$ 57%
South Africa	184 $\pm$ 29	4 $\pm$ 1%	1 $\pm$ 0.1%	1 $\pm$ 0.1%	4,433 $\pm$ 586	89 $\pm$ 15%	16 $\pm$ 3%	17 $\pm$ 2%
Swaziland	51 $\pm$ 7	57 $\pm$ 6%	30 $\pm$ 3%	31 $\pm$ 1%	134 $\pm$ 12	113 $\pm$ 16%	45 $\pm$ 6%	51 $\pm$ 5%
Tanzania	43 $\pm$ 16	1 $\pm$ 0.5%	7 $\pm$ 3%	3 $\pm$ 1%	44 $\pm$ 17	1 $\pm$ 0.3%	4 $\pm$ 2%	2 $\pm$ 1%
Zambia	38 $\pm$ 5	4 $\pm$ 0.4%	9 $\pm$ 1%	6 $\pm$ 0.1%	1,748 $\pm$ 194	99 $\pm$ 15%	224 $\pm$ 33%	159 $\pm$ 17%
Zimbabwe	42 $\pm$ 9	4 $\pm$ 1%	13 $\pm$ 2%	6 $\pm$ 1%	874 $\pm$ 164	49 $\pm$ 11%	149 $\pm$ 33%	80 $\pm$ 15%

The results correspond to the mean values from Monte Carlo simulations. Additional information about the uncertainty analysis is presented in Figures S1 and S2 in Supplementary Material. The New Policies scenario accounts for the increasing fuel consumption rate and population (**Table 6**). Ethanol is not utilized as a cooking fuel in Mauritius because there is no use of traditional biomass for cooking.

Using ethanol as a cooking fuel could help to substantially reduce or even eliminate the traditional use of biomass for cooking in most Southern African countries (**Figure 1**). Approximately 85 million t of firewood per year could be saved by implementing the NP scenario in all of the Southern African countries; this would reduce forest exploitation by 145 million ha [estimated assuming the use of 1.5 kg of firewood per capita per day (IEA, 2014b)], 0.9 ha per capita of forest for firewood gathering (IEA, 2014a), and the population projected for 2030 (**Table 6**). The forest area required for firewood extraction depends on management and biomass stocks, which are not uniform across countries.

Enlarging Electricity Access

We identified attractive opportunities to enhance electricity access by burning sugarcane bagasse in a cogeneration system, especially for Swaziland, where bioelectricity production under the CM scenario could increase current generation by 38 $\pm$ 4% (**Figure 2**) and using bagasse under the NP scenario could increase electricity production by 70 $\pm$ 8%. The existing bioelectricity produced from sugarcane bagasse in Swaziland is completely consumed by the industry itself. However, the sugarcane mills could provide

surplus power to the grid by implementing higher efficient boilers (IRENA, 2014), as in the NP scenario. In Angola, bagasse could contribute highly to power generation under the NP scenario (**Figure 2**). The lack of electricity for Southern African households would be significantly reduced, or even eliminated, by a slight expansion of sugarcane and the use of bagasse as an energy source (**Figure 3**).

GHG Implications

Because more than 90% of the electricity generation in South Africa is derived from coal (**Table 4**) the use of bagasse as an electricity source could eliminate 940-4,800 kt CO_2e/year (25th-75th percentiles), in addition to an annual reduction potential of 900-9800 or 550-6,000 kt CO_2e by displacing diesel or gasoline, respectively (**Figure 4A**). Electricity generation in Mozambique and Zambia is based on hydropower. Increasing bagasse use over hydropower would impair the current power profile because of higher life cycle GHG emissions from the biomass (**Tables 3** and **5**). However, the potential reduction in GHG emissions resulting from displacing fossil fuel compared with 2012 emissions could be as high as 50–70% in these countries. In Swaziland, where 70%

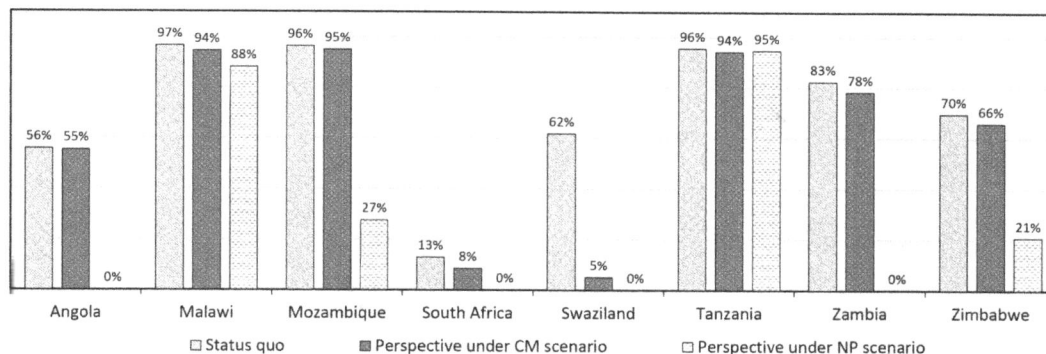

FIGURE 1 | Population relying on traditional use of biomass for cooking. The potential replacement of traditional biomass when ethanol is used exclusively as cooking fuel. *Status quo* as reported by the International Energy Agency (IEA, 2014a). 0% indicates total replacement of traditional biomass.

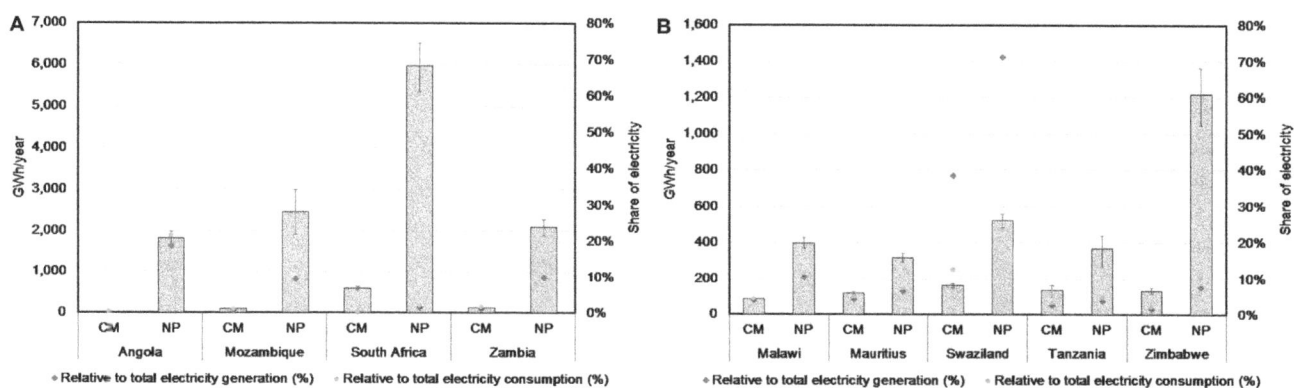

FIGURE 2 | Potential bioelectricity supply. The graphs are split to facilitate visualization: **(A)** larger producers and **(B)** smaller producers. The columns show the mean values, and bars represent the 25th and 75th percentiles from the Monte Carlo simulation. The difference between the red and green markers indicates the external dependency on electricity.

of the electricity production is based on coal (**Table 4**), sugarcane could also contribute to a cleaner energy sector by reducing the annual GHG emissions by more than 60% (**Figure 4B**).

Investments and Land Use

The investment needed to expand sugarcane production and processing over 1% of the pasture land varies according to the available area of each country. The required annual investment for all of the countries is insignificant compared to the GDP at the market price. However, in Mozambique, it represents more than 30% of the fixed capital formation and, thus, may be prohibitively high (**Table 8**). In South Africa, it would be infeasible to invest in 55 new sugarcane industries.

With regard to the sugarcane expansion in the NP scenario, the additional area at a national level is quite small compared to the current sugarcane production in the Southern Africa countries, except for Angola in which the required area would correspond to over 40% of the existing crop area and thus would place a barrier for the NP scenario (**Table 2**). However, a deeper analysis at local level is essential to identify the need for irrigation and the soil and climate conditions.

CHALLENGES IN IMPLEMENTING SUGARCANE ENERGY IN SOUTHERN AFRICA

Despite the potential of energy production using sugarcane, Southern Africa countries face a number of economic, social, and environmental challenges that can hamper the development of bioenergy systems. Among them are the need of adequate logistics infrastructure and the limited number of trained professionals, which reduces productivity. The situation is further aggravated by land acquisition schemes, in which the legal procedures make the investment processes quite difficult, by corruption, abuse of human rights and lack of governance transparency, which compromises the assurance of energy programs (Mwakasonda and Farioli, 2012), and by the access to affordable loans and sound business models (Rutz and Janssen, 2012).

Hence, policies must take the people's welfare and the sustainable supply as first priorities through implementing and integrating effective plans for land use, energy, agriculture, and rural development focused on employment opportunities, education, and energy security. However, policies must be aligned with strategies

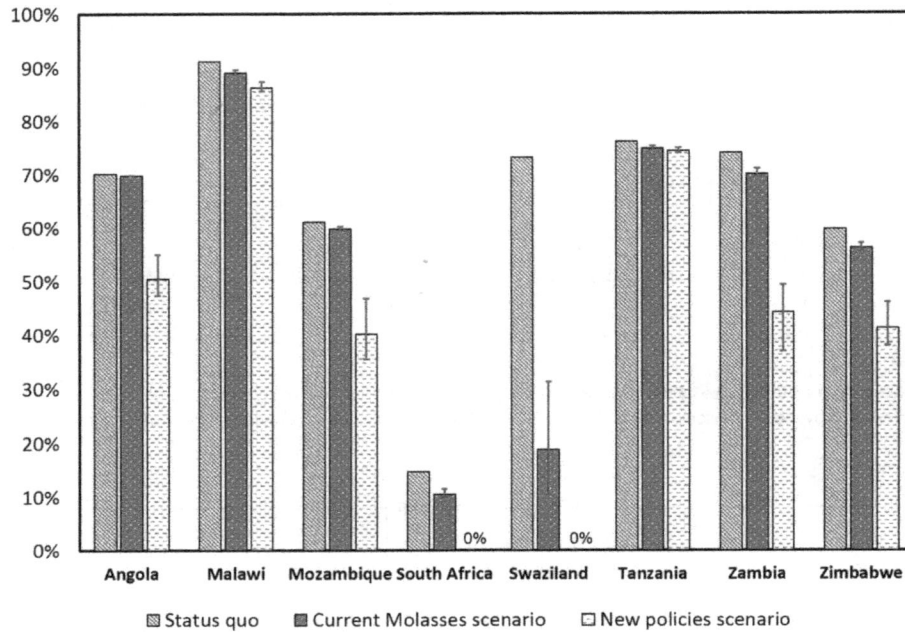

FIGURE 3 | Population without access to electricity. The new perspectives for electricity shortage when implementing the Current Molasses (CM) and New Policies (NP) scenarios. *Status quo* corresponds to the current situation as reported by the International Energy Agency (IEA, 2014a). Household electricity demands were estimated considering average annual consumptions of 1,120 and 1,430 kWh (Castellano et al., 2015) per household for the CM and NP scenarios, respectively. The columns show the mean values, and the bars represent the 25th and 75th percentiles from the Monte Carlo simulation. Swaziland and South Africa can fully meet the residential demand in the NP scenario.

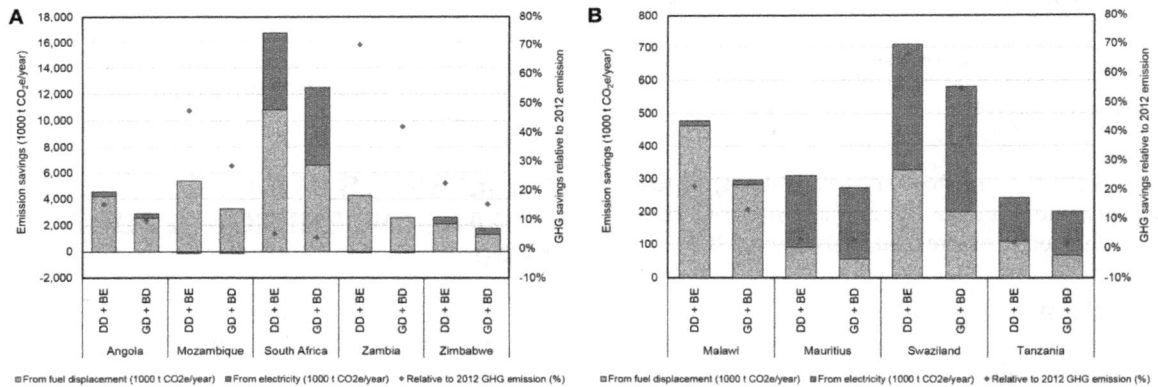

FIGURE 4 | Greenhouse gas (GHG) emission savings by introducing sugarcane as energy source (2030 scenario). The primary axis represents the emissions savings from promoting bagasse as an electricity source (BE), rather than keeping the current electrical system, and displacing diesel (DD) or gasoline (GD) by implementing ethanol use. The secondary axis corresponds to the relative GHG emissions savings compared with 2012 fossil fuel emissions attributed to goods and services discounted from cement manufacturing emissions; data (UNFCCC, 2014; Boden et al., 2015; British Petroleum, 2015) were retrieved from the Global Carbon Atlas (2012). The graphs are split to facilitate visualization: (A) larger scale emission and (B) smaller scale emission. The columns show the mean values from the Monte Carlo simulation.

to limit negative impacts of biofuels by promoting and applying sustainability standards and criteria based on aspects such as biodiversity, GHG emissions, land use, and water use. Furthermore, Africa is predicted to experience designed to undergo the adverse effects of the climate change on the agriculture systems, such as land degradation, water stress, increased occurrence of pests, droughts and floods, and loss of yield, which require adequate strategies to address such impacts, besides developing more accurate researches in this region to ensure the real effects of climate change (Malaviya and Ravindranath, 2012). Cooperation among African countries could boost the development of regional strategies and technologies suitable for their reality.

TABLE 8 | Number of new 1 million t sugarcane mills and total investment for the new policies scenario.

Countries	Number of mills (1 Mt)	Annual investment required (1,000 US$)[a,b]	Investment related to GDP at market price (%)[b]	Investment related to gross fixed capital formation (%)[c]
Angola	22	470,000	0.6%	1.9%
Malawi	2	44,400	0.8%	3.4%
Mauritius	<1	115	<0.01%	<1%
Mozambique	28	592,000	5.8%	32.6%
South Africa	55	1,162,500	0.3%	1.6%
Swaziland	1	22,600	0.6%	4.5%
Tanzania	<1	440	<0.01%	<1%
Zambia	22	467,600	2.3%	8.9%
Zimbabwe	11	227,600	2.4%	11.1%

[a]Corresponds to the total agricultural and industrial investments normalized by 10 years; USD 212/t cane (BNDES, 2014).
[b]2010 current price.
[c]Related to investments in fixed capital.

Under planning and monitoring actions, bioenergy can enhance agricultural and technological progress, boost social growth, and contribute to the development of the food sector and well-being of Africans at the local level. These results require efficient enterprises, financial mechanisms, and government support for research, education, and agriculture, such as extension services to guide farmers. International agreements, such as carbon emission reduction policies, can play an important role in promoting modern energy in developing countries.

Sugarcane can serve as a core bioenergy source for the growing energy demands of African countries and help to reduce GHG emissions and fossil fuel imports by changing their current energy generation profiles. However, displacing the use of fossil fuels and nonsustainable bioenergy requires new perspectives and solutions throughout the energy life cycle to ensure modern energy access. Additionally, implementing the scenarios proposed in this study are associated with improvements in biomass production, conversion technology, rural infrastructure, and societal integration (Lynd and Woods, 2011), especially for the NP scenario, which relies on more efficient technologies.

CONCLUSION

This study confirms the great potential for sugarcane ethanol production, the good prospects for using this biofuel in cooking stoves and transport sector in Southern Africa, and the opportunity for this sector to contribute to enhancing electricity access in the long term. As consequence of promoting the sugarcane sector for energy proposal, there are benefits regarding the reduction in fossil fuel consumption and the external dependency in fossil fuels imports, as well as reducing the GHG emission, in line with the voluntary national pledges assumed in COP21.

The development of sugarcane bioenergy is aligned with the goals proposed by the United Nations at the Sustainability Energy for All Program (SE4ALL): universal energy access, renewable energy, and energy efficiency. We believe that this study can help decision-makers and stakeholders on planning energy strategies in Southern Africa based on sugarcane as alternative to promote the modern energy supply. However, challenges such as financing the agroindustry, transport infrastructure, education and personnel training, and regulatory adjustment must be properly evaluated and solved, and a clear and effective government commitment is essential to make the sustainable sugarcane bioenergy a reality in Southern Africa.

AUTHOR CONTRIBUTIONS

SS was the primary author of the manuscript, performed the research, collected the data, analyzed the results, applied the uncertainty analysis, and elaborated the tables and charts. LN was the leader of the paper and an active coauthor, contributed in elaborating the scope, charts, and tables, and also played an important role in analyzing and discussing the results. HW contributed to the land use data and provided a deeper analysis and perspective of the sugarcane development in Southern Africa. LL played an important role in editing the manuscript and defining the scope of the study and also performed a deep analysis of the results. ME provided perspectives on bioenergy deployment in Southern Africa, especially the challenges for implementing sugarcane sector in this region. LC is the leader of the research project and was involved in the discussion of scope and results.

ACKNOWLEDGMENTS

Special thanks to professor Marcelo Cunha for his contribution to the discussion of the economic aspects and the uncertainty analysis.

FUNDING

This research was funded by São Paulo Research Foundation (FAPESP) under Grants 2012/00282-3 and 2015/02270-0.

REFERENCES

ADB. (2002). *Handbook for Integrating Risk Analysis in the Economic Analysis of Projects.* Manila: Asian Development Bank.

Adkins, E., Oppelstrup, K., and Modi, V. (2012). Rural household energy consumption in the millennium villages in Sub-Saharan Africa. *Energy Sustain. Dev.* 16, 249–259. doi:10.1016/j.esd.2012.04.003

ANEEL Fontes de Energia: Biomassa. (2015). *Banco de Informações de Geração.* Available at: http://www2.aneel.gov.br/aplicacoes/capacidadebrasil/CombustivelPorClasse_fn1.cfm

Biocom. (2015). *The Company: Biocom in Numbers.* Available at: http://www.biocom-angola.com/en/company/biocom-numbers

BNDES. (2014). *Estudo de Viabilidade de Produção de Biocombustíveis na UEMOA (União Econômica e Monetária do Oeste Africano).* Rio de Janeiro, RJ: Banco Nacional de Desenvolvimento Econômico e Social.

BNDES/CGEE. (2008). "Ethanol as vehicle fuel," in *Sugarcane-Based Bioethanol: Energy for Sustainable Development* (Rio de Janeiro, RJ: Banco Nacional de Desenvolvimento Econômico e Social, Centro de Gestão e Estudos Estratégicos), 304. Available at: http://sugarcane.org/resource-library/studies/BNDES%20-%20Sugarcane%20Based%20Bioethanol.pdf

Boden, T., Marland, G., and Andres, R. (2015). *Global, Regional, and National Fossil-Fuel CO2 Emissions.* Oak Ridge, TN, USA: Carbon Dioxide Information Analysis Center, Oak Ridge National Laboratory, U.S. Department of Energy.

British Petroleum. (2015). *BP Statistical Review of World Energy.* 48. Available at: http://scholar.google.com/scholar?hl=en&btnG=Search&q=intitle:BP+Statistical+Review+of+World+Energy#0

Castellano, A., Kendall, A., Nikomarov, M., and Swemmer, T. (2015). *Brighter Africa: The Growth Potential of the Sub-Saharan Electricity Sector.* New York, NY. Available at: http://www.mckinsey.com/~/media/mckinsey/dotcom/insights/energyresourcesmaterials/poweringafrica/brighter_africa_the_growth_potential_of_the_sub-saharan_electricity_sector.ashx

Chakaniza, L. (2013). *Production of Ethanol for National Consumption in Malawi.* Maputo, Mozambique: Workshop on Sustainable Biomass Production in Southern Africa. Available at: https://www.b2match.eu/system/biomass-workshop-2013-maputo/files/CHAKANIZA.pdf?1364980612

Dasappa, S. (2011). Potential of biomass energy for electricity generation in sub-Saharan Africa. *Energy Sustain. Dev.* 15, 203–213. doi:10.1016/j.esd.2011.07.006

De Cerqueira Leite, R. C., Verde Leal, M. R. L., Barbosa Cortez, L. A., Griffin, W. M., and Gaya Scandiffio, M. I. (2009). Can Brazil replace 5% of the 2025 gasoline world demand with ethanol? *Energy* 34, 655–661. doi:10.1016/j.energy.2008.11.001

Dias, M. O. S., Cunha, M. P., Jesus, C. D. F., Rocha, G. J. M., Pradella, J. G. C., Rossell, C. E. V., et al. (2011). Second generation ethanol in Brazil: can it compete with electricity production? *Bioresour. Technol.* 102, 8964–8971. doi:10.1016/j.biortech.2011.06.098

Dunkelberg, E., Finkbeiner, M., and Hirschl, B. (2014). Sugarcane ethanol production in Malawi: measures to optimize the carbon footprint and to avoid indirect emissions. *Biomass Bioenergy* 71, 37–45. doi:10.1016/j.biombioe.2013.10.006

EHA. (2011). *BioEthanol for Sustainable Transport – Results and Recommendations from the European Best Project,* eds P. Fenton, and H. Carlsson (Stockholm, Sweden: Environment and Health Administration (EHA)).

EIA. (2012). *International Energy Statistics.* Available at: http://www.eia.gov/cfapps/ipdbproject/IEDIndex3.cfm?tid=5&pid=5&aid=2

FAO. (2012a). *Inputs, Land. FAOSTAT.* Available at: http://faostat3.fao.org/download/R/RL/E

FAO. (2012b). *Natural Resource Assessment for Crop and Land Suitability: An Application for selected bioenergy crops in Southern Africa region.* Rome, Italy: Integrated.

FAO. (2012c). *Population, Annual Population. FAOSTAT.* Available at: http://faostat3.fao.org/download/O/OA/E

FAO. (2012d). *Production, Crops. FAOStat.* Available at: http://faostat3.fao.org/download/Q/QC/E

FAO. (2013a). *Food Balance Sheets. FAOStat.* Available at: http://faostat3.fao.org/download/FB/FBS/E

FAO. (2013b). *Trade. FAOStat.* Available at: http://faostat3.fao.org/download/T/*/E

Global Carbon Atlas. (2012). *Territorial Emission.* Available at: http://www.globalcarbonatlas.org/?q=en/emissions

Hermann, S., Miketa, A., and Fichaux, N. (2014). *Estimating the Renewable Energy Potential in Africa.* Abu Dhabi: IRENA-KTH working paper.

ICF International. (2012). *United States Agency for International Development. Mean Number of Household Members.* STAT Compil. – Demogr. Heal. Surv. Progr. Available at: http://www.statcompiler.com/en/index.html

IEA. (2012). *International Energy Agency: Statistics. Electr. Heat.* Available at: http://www.iea.org/statistics/statisticssearch/

IEA. (2014a). *Africa Energy Outlook: A Focus on Energy Prospects in Sub-Saharan Africa,* ed. R. Priddle (Paris, France: Organisation for Economic Co-operation and Development (OECD), International Energy Agency (IEA)).

IEA. (2014b). *World Energy Outlook 2014.* Paris, France: IEA Publications.

IRENA. (2014). *Swaziland: Renewables Readiness Assessment.* Abu Dhabi, United Arab Emirates: International Renewable Energy Agency (IRENA).

IRENA-DBFZ. (2013). *Biomass Potential in Africa,* eds K. Stecher, A. Brosowski, and D. Thrän (Abu Dhabi, United Arab Emirates: International Renewable Energy Agency (IRENA)). Available at: http://www.irena.org/DocumentDownloads/Publications/IRENA-DBFZ_Biomass%20Potential%20in%20Africa.pdf

Johnson, F. X., and Seebaluck, V. (2012). *Bioenergy for Sustainable Development and International Competitiveness: The Role of Sugar Cane in Africa.* London and. New York: Routledge Taylor & Francis Group.

Kimemia, D., Vermaak, C., Pachauri, S., and Rhodes, B. (2014). Burns, scalds and poisonings from household energy use in South Africa: are the energy poor at greater risk? *Energy Sustain. Dev.* 18, 1–8. doi:10.1016/j.esd.2013.11.011

Latawiec, A. E., Strassburg, B. B. N., Valentim, J. F., Ramos, F., and Alves-Pinto, H. N. (2014). Intensification of cattle ranching production systems : socioeconomic and environmental synergies and risks in Brazil. *Animal* 8, 1255–1263. doi:10.1017/S1751731114001566

Lynd, L. R., and Woods, J. (2011). Perspective: a new hope for Africa. *Nature* 474, S20–S21. doi:10.1038/474S020a

Macauhub. (2015). *Start of Ethanol and Sugar Production Increases Diversification of Angola's Economy.* Features. Available at: http://www.macauhub.com.mo/en/2015/06/01/start-of-ethanol-and-sugar-production-increases-diversification-of-angolas-economy/

Malaviya, S., and Ravindranath, N. H. (2012). "Implications of climate change on sustainable biofuel production in Africa," in *Bioenergy for Sustainable Development in Africa,* eds R. Janssen, and D. Rutz (Munich: Springer), 281–298.

Moomaw, W., Burgherr, P., Heath, G., Lenzen, M., Nyboer, J., and Verbruggen, A. (2012). "Annex II: methodology," in *Renewable Energy Sources and Climate Change Mitigation: Special Report of the Intergovernmental Panel on Climate Change,* eds O. Edenhofer, R. Pichs-Madruga, Y. Sokona, K. Seyboth, P. Matschoss, S. Kadner, et al. (New York, NY: Intergovernmental Panel on Climate Change), 1076. Available at: http://www.ipcc.ch/pdf/special-reports/srren/Annex%20II%20Methodology.pdf

Munyinda, K., Yamba, F. D., and Walimwipi, H. (2012). "Bioethanol potential and production in Africa: sweet Sorghum as a complementary feedstock," in *Bioenergy for Sustainable Development in Africa,* eds R. Janssen, and D. Rutz (Munich: Springer), 81–91. doi:10.1007/978-94-007-2181-4_8

Mwakasonda, S., and Farioli, F. (2012). "Social impacts of biofuel production in Africa," in *Bioenergy for Sustainable Development in Africa,* eds R. Janssen, and D. Rutz (Munich: Springer), 323–334.

Novozymes. (2012). *CleanStar Mozambique Launches World's First Sustainable Cooking Fuel Facility.* Bioenergy, Corp. News. Available at: http://novozymes.com/en/news/news-archive/Pages/CleanStar-Mozambique-launches-world's-first-sustainable-cooking-fuel-facility.aspx

Nylund, N.-O., Laurikko, J., Laine, P., Suominen, J., and Anttonen, M. P. A. (2013). Benchmarking heavy-duty ethanol vehicles against diesel and CNG vehicles. *Biomass Convers. Biorefinery* 3, 45–54. doi:10.1007/s13399-012-0049-z

REN21. (2014). *Renewables 2014 Global Status Report.* Paris: Renewable Energy Policy Network for the 21st Century. Available at: http://www.ren21.net/Portals/0/documents/Resources/GSR/2014/GSR2014_full report_low res.pdf

REN21. (2015a). *Renewables 2015 – Global Status Report.* France: Renewable Energy Policy Network for the 21st Century.

REN21. (2015b). *Status Report: SADC Renewable Energy and Energy Efficieny.* France: Renewable Energy Policy Network for the 21th Century.

Rodrigues Filho, J. E. (2005). "Trash recovery cost," in *Biomass power generation: Sugar cane bagasse and trash*, eds S. J. Hassuani, M. R. L. V. Leal, and I. de Carvalho Macedo (Piracicaba: PNUD – Programa das Nações Unidas para o Desenvolvimento, CTC – Centro de Tecnologia Canavieira), 74–85. Available at: http://www.ctcanavieira.com.br/site/images/downloads/biomassa.pdf

Rutz, D., and Janssen, R. (2012). "Opportunities and risks of bioenergy in Africa," in *Bioenergy for Sustainable Development in Africa*, eds R. Janssen, and D. Rutz (Munich: Springer), 391–400.

Schulze, R. E., Maharaj, M., Lynch, S. D., Howe, B. J., and Melvil-Thompson, B. (1997). *South African Atlas of Agrohydrology and Climatology*. Pretoria, South Africa: WRC Report No. TT 82/96.

Seabra, J. E. A., and Macedo, I. C. (2011). Comparative analysis for power generation and ethanol production from sugarcane residual biomass in Brazil. *Energy Policy* 39, 421–428. doi:10.1016/j.enpol.2010.10.019

Seabra, J. E. A., Macedo, I. C., Chum, H. L., Faroni, C. E., and Sarto, C. A. (2011). Life cycle assessment of Brazilian sugarcane products: GHG emissions and energy use. *Biofuels Bioprod. Biorefining* 5, 519–532. doi:10.1002/bbb.289

Souza, S. P., De Ávila, M. T., and Pacca, S. (2012). Life cycle assessment of sugarcane ethanol and palm oil biodiesel joint production. *Biomass Bioenergy* 44, 70–79. doi:10.1016/j.biombioe.2012.04.018

Taliotis, C., Shivakumar, A., Ramos, E., Howells, M., Mentis, D., Sridharan, V., et al. (2016). An indicative analysis of investment opportunities in the African electricity supply sector – using TEMBA (the electricity model base for Africa). *Energy Sustain. Dev.* 31, 50–66. doi:10.1016/j.esd.2015.12.001

UNFCCC. (2013). *CDM Project: Cleanstar Mozambique – Maputo Ethanol Cookstove and Cooking Fuel Project 1*. Mozambique: CleanStar Mozambique Community Bioinnovation Ltd; Merrill Lynch Commodities (Europe) Ltd.

UNFCCC. (2014). *National Inventory Submissions*. Available at: http://unfccc.int/national_reports/annex_i_ghg_inventories/national_inventories_submissions/items/8108.php

United Nations. (2006). *Costos y Precios para Etanol Combustible en América Central*, ed. L. A. H. Nogueira (México: Comisión Económica para América Latina y El Caribe (CEPAL)).

United Nations. (2007). *Feasibility Study for the Use of Ethanol as a Household Cooking Fuel in Malawi – Draft Final Report*, ed. Ethio Resource Group Pvt. Ltd., *Growing Sustainable Business for Poverty Reduction Program in Malawi* (Lilongwe, Malawi: The United Nations Development Program (UNDP))

Wang, M., Sabbisetti, R., Elgowainy, A., Dieffenthaler, D., Anjum, A., Sokolov, V., et al. (2014). *GREET Model: The Greenhouse Gases, Regulated Emissions, and Energy Use in Transportation Model*. Chicago, USA: Argonne National Laboratory.

Watson, H. K. (2011). Potential to expand sustainable bioenergy from sugarcane in southern Africa. *Energy Policy* 39, 5746–5750. doi:10.1016/j.enpol.2010.07.035

WHO. (2013). *Database on use of solid fuels*. World Heal. Organ. Available at: http://apps.who.int/gho/data/view.main.37000

Zhu, X.-G., Long, S. P., and Ort, D. R. (2010). Improving photosynthetic efficiency for greater yield. *Annu. Rev. Plant Biol.* 61, 235–261. doi:10.1146/annurev-arplant-042809-112206

Conflict of Interest Statement: The authors declare that the research was conducted in the absence of any commercial or financial relationships that could be construed as a potential conflict of interest.

13

Accelerated Carbonation of Steel Slags Using CO$_2$ Diluted Sources: CO$_2$ Uptakes and Energy Requirements

Renato Baciocchi[1]*, Giulia Costa[1], Alessandra Polettini[2], Raffaella Pomi[2], Alessio Stramazzo[2] and Daniela Zingaretti[1]

[1] *Department of Civil Engineering and Computer Science Engineering, University of Rome "Tor Vergata", Rome, Italy,*
[2] *Department of Civil and Environmental Engineering, University of Rome "La Sapienza", Rome, Italy*

This work presents the results of carbonation experiments performed on Basic Oxygen Furnace (BOF) steel slag samples employing gas mixtures containing 40 and 10% CO$_2$ vol. simulating the gaseous effluents of gasification and combustion processes respectively, as well as 100% CO$_2$ for comparison purposes. Two routes were tested, the slurry-phase (L/S = 5 l/kg, T = 100°C and Ptot = 10 bar) and the thin-film (L/S = 0.3–0.4 l kg, T = 50°C and Ptot = 7–10 bar) routes. For each one, the CO$_2$ uptake achieved as a function of the reaction time was analyzed and on this basis, the energy requirements associated with each carbonation route and gas mixture composition were estimated considering to store the CO$_2$ emissions of a medium size natural gas fired power plant (20 MW). For the slurry-phase route, maximum CO$_2$ uptakes ranged from around 8% at 10% CO$_2$, to 21.1% (BOF-a) and 29.2% (BOF-b) at 40% CO$_2$ and 32.5% (BOF-a) and 40.3% (BOF-b) at 100% CO$_2$. For the thin-film route, maximum uptakes of 13% (BOF-c) and 19.5% (BOF-d) at 40% CO$_2$, and 17.8% (BOF-c) and 20.2% (BOF-d) at 100% were attained. The energy requirements of the two analyzed process routes appeared to depend chiefly on the CO$_2$ uptake of the slag. For both process route, the minimum overall energy requirements were found for the tests with 40% CO$_2$ flows (i.e., 1400 – 1600 MJ / t$_{CO_2}$ for the slurry-phase and 2220 – 2550 MJ / t$_{CO_2}$ for the thin-film route).

Keywords: CO$_2$ capture and storage, mineral carbonation, steel slags, energy requirements, fluegas

Edited by:
Jennifer Wilcox,
Stanford University, USA

Reviewed by:
Anozie Ebigbo,
Imperial College London, UK
Hannu-Petteri Mattila,
Åbo Akademi University, Finland

*Correspondence:
Renato Baciocchi
baciocchi@ing.uniroma2.it

Specialty section:
This article was submitted to Carbon Capture, Storage, and Utilization, a section of the journal Frontiers in Energy Research

Citation:
Baciocchi R, Costa G, Polettini A, Pomi R, Stramazzo A and Zingaretti D (2016) Accelerated Carbonation of Steel Slags Using CO$_2$ Diluted Sources: CO$_2$ Uptakes and Energy Requirements. Front. Energy Res. 3:56.

INTRODUCTION

Accelerated carbonation of alkaline earth metal silicate ores has been proposed as a method for *ex situ* carbon dioxide storage in the early 90s (Seifritz, 1990; Lackner et al., 1995). Since then, many studies have extensively investigated at laboratory scale the applicability of this treatment to Ca and Mg bearing silicates, such as serpentine, olivine and wollastonite, in view of their high carbon sequestration potential and widespread availability (especially for the first two types of minerals). Different types of process routes, i.e., gas-solid or aqueous, each performed in direct or indirect mode, have been tested (see e.g., Sipilä et al., 2008; Baciocchi et al., 2014). The most applied route is the direct aqueous one, by which the alkaline feedstock is mixed with water to obtain a Liquid to

Solid (L/S) ratio above 2 l/kg, and the resulting slurry is contacted with pressurized CO_2 so that Ca or Mg dissolution and carbonate precipitation take place in the same reactor. It should be noted that the operating conditions that have proven effective to promote the carbonation kinetics of the tested minerals, and hence obtain high reaction yields in technically feasible operating times (<2 h), are energy intensive; for example, direct carbonation of minerals is found to be effective at a temperature of 100–200°C and pressures of 10–100 bar (O'Connor et al., 2005; Sipilä et al., 2008). In addition, regardless of the reaction route employed, to achieve a significant CO_2 uptake in short timeframes, the surface of the minerals needs to be activated by physical pretreatments, including size reduction, magnetic separation, and even thermal or steam treatment to eliminate chemically bound water for minerals like serpentine (Baciocchi et al., 2014). Hence, the energy intensive operating conditions typically required for accelerated mineral carbonation still hinder the industrial application of this technology (Zingaretti et al., 2014).

In parallel to the investigations carried out on minerals, the reactivity with CO_2 of alkaline industrial residues characterized by high calcium or magnesium (hydr)oxide or silicate contents has also been extensively examined, as reported in several reviews (Costa et al., 2007; Bobicki et al., 2012; Pan et al., 2012; Baciocchi et al., 2014). Although significantly less abundant in comparison to minerals, these materials have proven to be more reactive at mild operating conditions (Huijgen et al., 2006a). Furthermore, carbonation has shown to affect some of the properties of the tested materials (i.e., main mineralogy, porosity, and mobility of specific elements) improving their long-term technical performance and/or environmental behavior (Bobicki et al., 2012; Pan et al., 2012). Thus, the application of this treatment to alkaline industrial residues may be indeed considered as a CO_2 storage and utilization option, since accelerated carbonation may be also employed as a valorization technique for widening the reuse options of these materials and also to achieve other environmental benefits (Baciocchi et al., 2014). Steel manufacturing plants are the typical examples of industrial sites in which relevant flows of both CO_2 and alkaline industrial residues are generated and therefore represent one of the potentially most interesting contexts for the application of accelerated carbonation. Several studies have shown that a number of different types of steel slag present a significant reactivity with CO_2, allowing to achieve, for specific process routes and operating conditions, relevant CO_2 uptakes (Huijgen et al., 2005; Baciocchi et al., 2010, 2015; Uibu et al., 2011; Chang et al., 2012; Santos et al., 2013a,b). Furthermore, several types of residues generated in steel manufacturing plants, such as Basic Oxygen Furnace (BOF), Electric Arc Furnace (EAF), and argon oxygen decarburization (AOD) slag, are typically not valorized and generally landfilled, or employed only for low-end applications, owing for their significant content of free calcium and magnesium (hydr)oxides that may result in poor volumetric stability and hence in a low technical performance in construction applications (Morone et al., 2014). Therefore, accelerated carbonation of these types of materials may also represent a treatment strategy to improve their properties in view of valorization.

Despite all the potential assets mentioned above, industrial-scale applications of accelerated carbonation of residues such as

steel slag are also currently still missing owing to uncertainties on the feasibility of this process, especially as far as material and energy requirements are concerned. Specifically, with regard to the latter, significant CO_2 uptakes in reasonable timeframes were achieved via the slurry-phase route using residues with a fine particle size (typically below 100 µm), obtained either by milling or simple sieving of the material, and operating the carbonation reactor at enhanced temperature and CO_2 pressure (Huijgen et al., 2005; Chang et al., 2012; Santos et al., 2013a,b; Baciocchi et al., 2015). It should be noted, however, that in a recent study, milder operating conditions were shown to be effective employing a proper lixiviant (ammonium chloride) to enhance calcium dissolution (Mattila et al., 2014). Besides, promising results in terms of CO_2 uptake (even if lower than those attained through the former route) have been achieved also via the thin-film (or wet) route (Baciocchi et al., 2011, 2015; Santos et al., 2013b; Morone et al., 2014), which is operated at lower temperature and CO_2 pressure and also presents the advantage of not generating a liquid by-product that necessitates suitable management and treatment.

In the last years, several studies have assessed the energy requirements associated to direct carbonation of minerals and industrial residues adopting the slurry-phase route (e.g., Huijgen et al., 2006b; Kelly et al., 2011; Kirchofer et al., 2012), while to our knowledge this evaluation for the thin-film route was performed only by Zingaretti et al., 2014. In the latter study, based on the results of lab-scale tests carried out on different types of industrial residues using pure CO_2, a comparison of the energy requirements associated with both reaction routes was performed. The authors estimated that the energy requirement for CO_2 storage only was around 1300–2750 MJ/t_{CO_2} for the slurry-phase process and between 550 and 2600 MJ/t_{CO_2} for the thin-film one depending on the type of residues and operating conditions applied (Zingaretti et al., 2014). Besides, the feasibility of the thin-film carbonation route was somehow linked to the possibility of using diluted CO_2 sources, thus lumping capture and storage in a single step.

This work is aimed at further assessing the feasibility of performing accelerated carbonation using diluted CO_2 sources, extending the evaluation also to the slurry-phase route. To this aim, batch carbonation tests via both the slurry-phase and thin-film routes were performed on different types of BOF steel slags, using diluted CO_2 sources (i.e., containing 40 and 10% vol. CO_2) and a 100% CO_2 source as a reference case. Then, based on the CO_2 uptakes achieved in these experiments, the energy requirements for storing the CO_2 emissions of a medium size natural gas fired power plant (20 MW) were assessed and discussed in terms of the type of BOF slag treated, the carbonation reaction time (from 0.5 to 24 h), and the gas composition considered (10, 40, or 100% vol.).

MATERIALS AND METHODS

The slag samples tested in this work were collected from a steel plant employing the integrated steelmaking process both directly at the outlet of the BOF unit (BOF-a and BOF-c) after grinding (to $d < 1$ mm) and at the storage site after grinding (to $d < 1$ mm) for metals recovery (BOF-b and BOF-d). The slag samples were then oven-dried at 60°C and ball-milled to a final particle size of <150 μm in the case of samples a and b used for the slurry-phase route tests, or only sieved to a particle size below 125 μm, in the case of samples c and d employed in the thin-film tests. The slight differences in the preparation and final particle size of the samples tested in the two types of carbonations tests was due to the fact that the two routes were investigated in two distinct laboratories, equipped with different milling and sieving devices. Anyhow, this difference is not expected to significantly affect the obtained results, since the composition and mineralogy as well as CO_2 reactivity of the same type of slag milled <150 μm or sieved <125 μm showed to be quite similar (as indicated by the results of thin-film route experiments performed on BOF samples milled at 150 μm, Baciocchi et al., 2015).

The elemental composition of the four BOF slag samples is reported in **Table 1**. With regard to the main elements of interest for the carbonation process, the following concentration ranges were detected for the investigated materials: Ca 21.3 − 28.9% dry wt., Mg 3.3 − 6.9%, Fe 13.9 − 19.6%, Mn 2.4 − 3.8%. Such elements were present at comparable levels in the four slag samples investigated, although BOF-c and BOF-d exhibited slightly higher Ca and Mg concentrations and somewhat lower Fe contents. Based on the evidence provided by XRD analysis and other chemical characterization methods, we estimated the amounts of the main mineralogical phases of each sample (see **Table 2**). Making reference to Ca and Mg-bearing phases, the samples collected directly at the outlet of the BOF unit (a and c) were

characterized by a higher content of silicates, as larnite (Ca_2SiO_4) and hatrurite (Ca_3SiO_5) and oxides (MgO), while those collected from the storage site (b and d) presented basically only larnite as a Ca-silicate phase and a higher content of hydroxide phases, i.e., portlandite [$Ca(OH)_2$] and brucite [$Mg(OH)_2$], besides more calcite ($CaCO_3$), an indication of natural weathering.

Slurry-phase accelerated carbonation tests were performed on BOF-a and BOF-b slag samples in a pressurized stainless steel reactor equipped with an external magnetic stirring device and a heating jacket. Based on the results of previous investigations (Baciocchi et al., 2011, 2015), a liquid to solid ratio (L/S) of 5 l/kg, operating temperature of 100°C and total pressure of 10 bar were selected. The thin-film route carbonation tests were carried out on BOF slag samples c and d in a pressurized stainless steel reactor placed in a water bath for temperature control. The operating conditions selected for this route based on previous investigations (Baciocchi et al., 2011, 2015; Morone et al., 2014) were a L/S of 0.3 l/kg, a temperature of 50°C, and a total pressure of 7–10 bar. For both types of routes, different reaction times (ranging from 0.5 to 24 h) and CO_2 concentrations in the gas flow were tested. In particular, the slurry-phase tests were performed with 100, 40, or 10% vol. CO_2 flows to simulate not only a pure CO_2 flow deriving from a preliminary capture step but also diluted CO_2 flows from gasification or combustion processes. In the thin-film route only 100 and 40% vol. CO_2 flows were employed, since previous results obtained for steel slag adopting this configuration indicated that the CO_2 uptakes achievable using a gas composition similar to fluegas were significantly lower than those attained with higher CO_2 concentrations of the treated gas flow (Zingaretti et al., 2014). At the end of the slurry-phase carbonation tests, the reactor was immediately degassed and then rapidly cooled down to ambient temperature through immersion in a water bath. Then liquid/solid separation was subsequently accomplished through centrifuging at 4000 rpm followed by 0.45 μm filtration. The solid samples obtained from both types of routes were oven-dried at 105°C and analyzed to determine their carbonate content through inorganic carbon (IC) measurement employing a Shimadzu TOC analyzer equipped with a solids module. The CO_2 uptake, defined as the ratio between the mass (in grams) of CO_2 sequestered and the mass (in grams) of as-received dry material and expressed as a percentage, was calculated through Eq. 1 (see Supplementary Material) from the IC contents of the carbonated (CO_{2final}) and untreated ($CO_{2initial}$) samples reported as CO_2 weight percent contents. The IC contents of the untreated slag samples were: 0.11% (BOF-a), 1.01% (BOF-b), 0.48% (BOF-c), and 0.71% (BOF-d).

$$CO_2 \text{ uptake } (\%) = \frac{CO_{2final} \, (\%) - CO_{2initial} \, (\%)}{100 - CO_{2final} \, (\%)} \times 100 \quad (1)$$

The evaluation of the energy requirements for the carbonation process was carried out adopting the approach developed in our previous work (Zingaretti et al., 2014). Based on the mass flow rate of CO_2 to store, i.e., 1 kg/s, which corresponds to the typical emission value of a 20 MW natural gas fired power plant, the amount of residues required in the process for each carbonation condition was determined. Hence, the energy requirements associated to each route were calculated summing those necessary for

TABLE 1 | Major element composition of the BOF slags.

| Element | Concentration (g/kg) | | | |
	BOF_a	BOF_b	BOF_c	BOF_d
Al	10.79	6.91	3.38	7.19
Ca	225.71	213.38	289.05	257.45
Fe	249.12	195.61	210.35	138.69
Mg	33.30	37.82	41.16	69.07
Mn	37.88	24.85	30.59	23.95
Na	7.21	7.09	0.08	2.02
K	0.27	0.31	0.05	0.41
Si	56.43	40.61	22.38	24.62

TABLE 2 | Mineralogical composition assumed for the BOF slags on the basis of the results of XRD and other chemical analysis.

| Phase | Concentration t (%) | | | |
	BOF_a	BOF_b	BOF_c	BOF_d
Ca_2SiO_4	19.8	17.2	12.1	30.9
MgO	6.4	–	7.9	–
$CaCO_3$	0.9	8.4	4.0	5.9
Fe_3O_4	–	34.9	22.1	19.6
$Ca(OH)_2$	6.5	28.0	17.1	26.6
$Mg(OH)_2$	–	11.5	–	17.0
FeO	37.6	–	–	–
Ca_3SiO_5	26.3	–	32.1	–
MnO	–	–	4.6	–
Al_2O_3	2.4	–	–	–

all the unit operations foreseen for each configuration (expressed as MJ/t_{CO_2} stored), on the basis of the experimental data obtained for the type of investigated BOF slag (i.e., chemical composition, mineralogy, and CO_2 uptake). **Figure 1** reports the process layout adopted for the slurry-phase and thin-film configurations.

The process layout considered for the slurry-phase route is similar to the ones proposed by O'Connor et al. (2005) and Huijgen et al. (2006b). Namely, as can be seen in **Figure 1A**, in this route, the BOF slags are firstly ground to the desired particle size (unit A) and then mixed with water at a specific liquid to solid ratio (unit B). Then, the slurry is pumped (unit C) to a heat exchanger (unit D) where it is heated to 30°C below the reaction temperature. Before entering the carbonation reactor (unit F), the slurry is heated by a further heater (unit E) where it reaches the desired temperature. In the carbonation reactor, the slurry is contacted with carbon dioxide, already pressurized in a multi-stage compressor (unit G) to the desired pressure, and the carbonation reaction takes place. The slurry leaving the reactor passes through unit D to heat recovery and finally is separated (unit H) producing a solid product and a liquid stream. In the thin-film route (see **Figure 1B**), the BOF slags are fed to the carbonation reactor (unit K) after being ground to the specific particle size in unit A. In unit K, the material is humidified to the desired L/S ratio, heated, and contacted with CO_2 already pressurized in a multi-stage compressor (unit G) to the desired pressure. The carbonation reactor in this route is envisioned as a rotary drum and the associated energy requirements are due to the material heating and the rotation of the drum. **Table 3** reports the equations used to estimate the energy requirements associated to each unit operation considered in the two carbonation routes. It should be noted that, for unit D in the slurry-phase route, the overall energy requirement associated with the heating of the slurry was estimated considering the energy required to achieve the final carbonation temperature, the energy recovered in the heat exchanger by cooling the outlet stream from the carbonation reactor, the heat of reaction of the carbonation process, and the heat loss from the carbonation reactor. In the thin-film

route, instead, reaction heat recovery was not taken into account since it was considered to be unpractical due to the fact that in this case there is no liquid solution. For both process routes, when a concentrated CO_2 flow was assumed to be used (i.e. 100% CO_2), the energy demand associated to the preliminary capture step was also accounted for, and overall quantified as 4000 kJ/kg CO_2 captured [Intergovernmental Panel on Climate Change (IPCC), 2005].

RESULTS AND DISCUSSION

Carbonation Performance

The CO_2 sequestration yield attained from the experiments as a function of the reaction time for the two process routes and the four slag samples investigated is reported in **Figure 2**. Since the aim of the present paper is to provide an assessment of the energetic profile of CO_2 sequestration through mineral carbonation of steel manufacturing residues applying both the slurry-phase and the thin-film route, the energy demand of the various process units as affected by the operating conditions and the intrinsic reactivity of the material toward CO_2 was estimated. Thus, CO_2 uptake yields are provided in the following paragraphs and discussed in light of their expected influence on the whole energetic profile of the mineral carbonation process. For an in-depth discussion of the physical phenomena, chemical reactions and mineralogical changes occurring during the process, the reader is referred to previous publications (Baciocchi et al., 2009, 2010, 2011, 2015). A review of the carbonation performance of steel slag materials from different literature studies and of its dependence on the slag characteristics and process conditions can also be found in Baciocchi et al. (2015).

The CO_2 uptakes measured after 24 h carbonation applying the slurry-phase route were: ≈8% at 10% CO_2 for both types of slag, 21.1% (BOF-a) and 29.2% (BOF-b) at 40% CO_2, and 32.5% (BOF-a) and 40.3% (BOF-b) at 100% CO_2; the corresponding values resulting for the thin-film route were: 12.9% (BOF-c) and 19.5% (BOF-d) at 40% CO_2, and 17.8% (BOF-c)

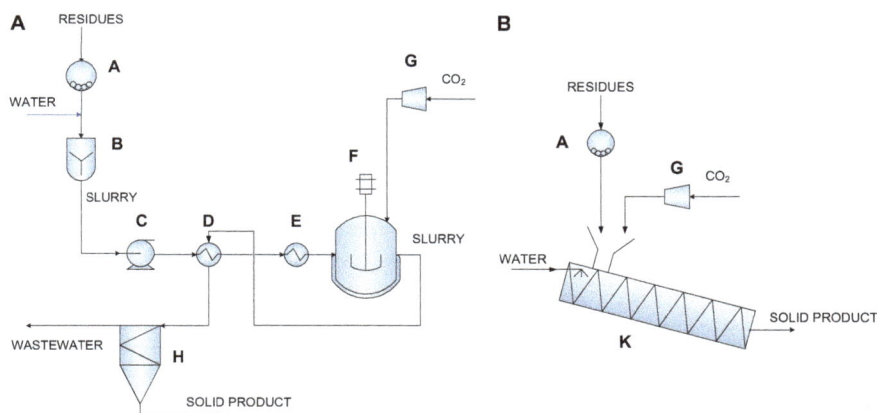

FIGURE 1 | Carbonation process layout for the slurry-phase (A) and the thin-film (B) routes [reprinted with permission from Zingaretti et al. (2014)**].** Copyright (2014) American Chemical Society. A = mill; B = rapid mixing unit; C = pump; D, E = heat exchangers; F, K = carbonation reactors; G = compressor; H = settling tank.

TABLE 3 | Equations used to assess the energy requirement associated to each unit considered for the two carbonation routes.

Unit	Operation	Associated energy requirement (MJ/t$_{CO2}$)
A	Grinding	$E = \dfrac{m_{RES}}{m_{CO_2}} \cdot 0.01 \cdot W_i \left(\dfrac{1}{\sqrt{d_1}} - \dfrac{1}{\sqrt{d_0}} \right)$
B	Mixing	$E = \dfrac{V \cdot \mu_{SLURRY} \cdot G^2}{m_{CO_2}}$
C	Pumping	$E = \dfrac{(Q_{RES} + Q_{WATER}) \cdot \Delta H \cdot \rho_{SLURRY} \cdot g}{m_{CO_2}}$
D	Heating	$E = E_{HEATING} - E_{COOLING}$ $E_{HEATING} = \dfrac{[m_{RES} \cdot C_{P,RES}(T) + m_{WATER} \cdot C_{PW}] \cdot (T_1 - T_0)}{m_{CO_2}}$ $E_{COOLING} = \dfrac{[m_{RES} \cdot C_{P,RES}(T) + m_{WATER} \cdot C_{PW}] \cdot (T_{CARB} - T_{OUT})}{m_{CO_2}}$
E	Heating	$E = \dfrac{[m_{RES} \cdot C_{P,RES}(T) + m_{WATER} \cdot C_{PW}] \cdot (T_{CARB} - T_1)}{m_{CO_2}}$
F	Mixing	$E = \dfrac{V \cdot \mu \cdot G^2}{m_{CO_2}}$
G	Multi-stage compression	$E_i = \left(\dfrac{Z_s \cdot R \cdot T_{in}}{M_{CO_2} \cdot \eta_{is}} \right) \cdot \left(\dfrac{k_s}{k_s - 1} \right) \cdot \left[(CR)^{\frac{k_s-1}{k_s}} - 1 \right]$
H	Mixing	$E = \dfrac{V \cdot \mu \cdot G^2}{m_{CO2}}$
K	Heating	$E = \dfrac{[m_{RES} \cdot C_{P,RES}(T) + m_{WATER} \cdot C_{PW}] \cdot (T_{CARB} - T_0)}{m_{CO_2}}$
	Drum rotation	$E = \dfrac{\tau \cdot \omega}{m_{CO_2}}$

E, energy requirement $\left(MJ / t_{CO2} \right)$; E_s energy required for each compression stage (MJ/t_{CO2}); $E_{HEATING}$, energy required to heat the slurry $\left(MJ / t_{CO_2} \right)$; $E_{COOLING}$, energy recovered from slurry cooling $\left(MJ / t_{CO_2} \right)$; m_{RES}, flowrate of the residues (t/s); m_{WATER}, flowrate of water (t/s); (m_{CO_2}), flowrate of carbon dioxide to be stored (t/s); Q_{RES}, volumetric flowrate of the residues (m³/s); Q_{WATER}, volumetric flowrate of water (m³/s); W_i, standard Bond's work index (kJ/t$_{RES}$); d_1, theoretical sieve size through which 80 wt.% of the residues pass (m); d_0, original particle size of the residues (m); V, total volume of water and residues to be mixed (m³); μ_{SLURRY}, dynamic viscosity of the slurry (Pa s); G, average velocity gradient (s⁻¹); ΔH, total dynamic head (m); ρ_{SLURRY}, slurry density (t/m³); g, gravitational acceleration (m/s²); $C_{P,RES}(T)$, specific heat capacity of the residues (kJ/t K); $C_{P,W}$, specific heat capacity of the water (kJ/t K); T_{CARB}, carbonation reaction temperature (K); T_1, slurry temperature reached in the heat exchanger (K); T_0, temperature of inlet residues and water flows (K); T_{OUT}, temperature of outlet residues and water flows (K); Zs, average CO_2 compressibility for each stage (–); R, gas constant (kJ/kmol K); T_{in}, CO_2 temperature at the compressor inlet (K); (M_{CO_2}), molecular weight of CO_2 (t/kmol); η_{is}, isentropic efficiency of the compressor(–); k_s, average ratio of the specific heats of CO_2 for each individual stage (–); CR, compression ratio of each stage (–); τ, resistive torque due to kinetic friction (kJ); ω, angular velocity of the drum (rad/s).

and 20.2% (BOF-d) at 100% CO_2. These results indicate the generally better performance of the BOF-b and BOF-d slag compared to BOF-a and BOF-c samples that can be related to the differences in the mineralogy of the slag; samples b and d being richer in readily reactive hydroxide phases such as portlandite and brucite compared to a and c that presented a higher content of silicates (see **Table 2**). It is interesting to note that, particularly for the thin-film carbonation route and for short reaction times for the slurry-phase process, the use of a 40% CO_2 flow instead of a pure one did not appear to significantly affect the CO_2 uptake of the slag (especially BOF-b and BOF-d). Instead, for a 10% CO_2 flow the uptake resulted below 10% and appeared not to be appreciably affected by the reaction time.

Although the direct comparison of the two process routes should be made with care owing to the slight differences in composition of the slag samples investigated (see **Tables 1** and **2**), the observed trends certainly highlight the higher carbonation yields of the slurry-phase process as opposed to the thin-film route, confirming the findings of our previous study (Baciocchi et al., 2015). Compared to the thin-film process, the slurry-phase conditions are believed to promote the dissolution of reactive metals from the solid phase (assuming that the dissolution/precipitation reactions approach thermodynamic equilibrium conditions during the process), as well as to prevent the coating effect exerted by the precipitated carbonate product onto the unreacted core of the slag particles (Santos et al., 2013a).

Energy Requirements Assessment

The specific energy requirements of each unit operation for the slurry-phase route are reported in **Figure 3** making reference to the experimental results achieved for the BOF-a and BOF-b slags. It may be noted that the various process units contributed in quite different extents to the overall energy demand. For improved clarity, the relative contributions of the process units to the total energy consumption are reported in **Figure 4**.

Several interesting features may be derived from the inspection of **Figures 3** and **4**. Specifically, the carbonation yield attained obviously turned out to play a relevant role on the energy requirements per unit mass of CO_2 sequestered. For a CO_2 concentration in the gaseous phase of 10% (simulating the typical content of combustion flue gases) the notably lower CO_2 sequestration performances achieved implied the use of larger amounts of slag and water for storing the target amount of CO_2 and therefore higher associated energy demands for almost all process units, in particular slurry heating and mixing. Comparing the energy requirements of the slurry-phase process using a combustion effluent or a concentrated CO_2 flow (CO_2 concentration = 100%), it is found that in the latter case the higher energy consumption associated to the preliminary capture stage is counterbalanced by the notably lower consumption of the other process operations. As a consequence, for very diluted gaseous flows a carbonation process including a pre-treatment stage aimed at concentrating CO_2 in the gaseous phase is deemed economically more viable compared to the option without CO_2 capture. On the other hand, for intermediate CO_2 concentrations in the gas phase (CO_2 = 40% vol., representing the typical CO_2 level in the gaseous effluent from gasification processes), the significant CO_2 uptake potential displayed by the slag, with sequestration yields comparable to those achieved using the concentrated stream, made the carbonation process energetically more competitive since in this case the duties related to the capture step were not considered.

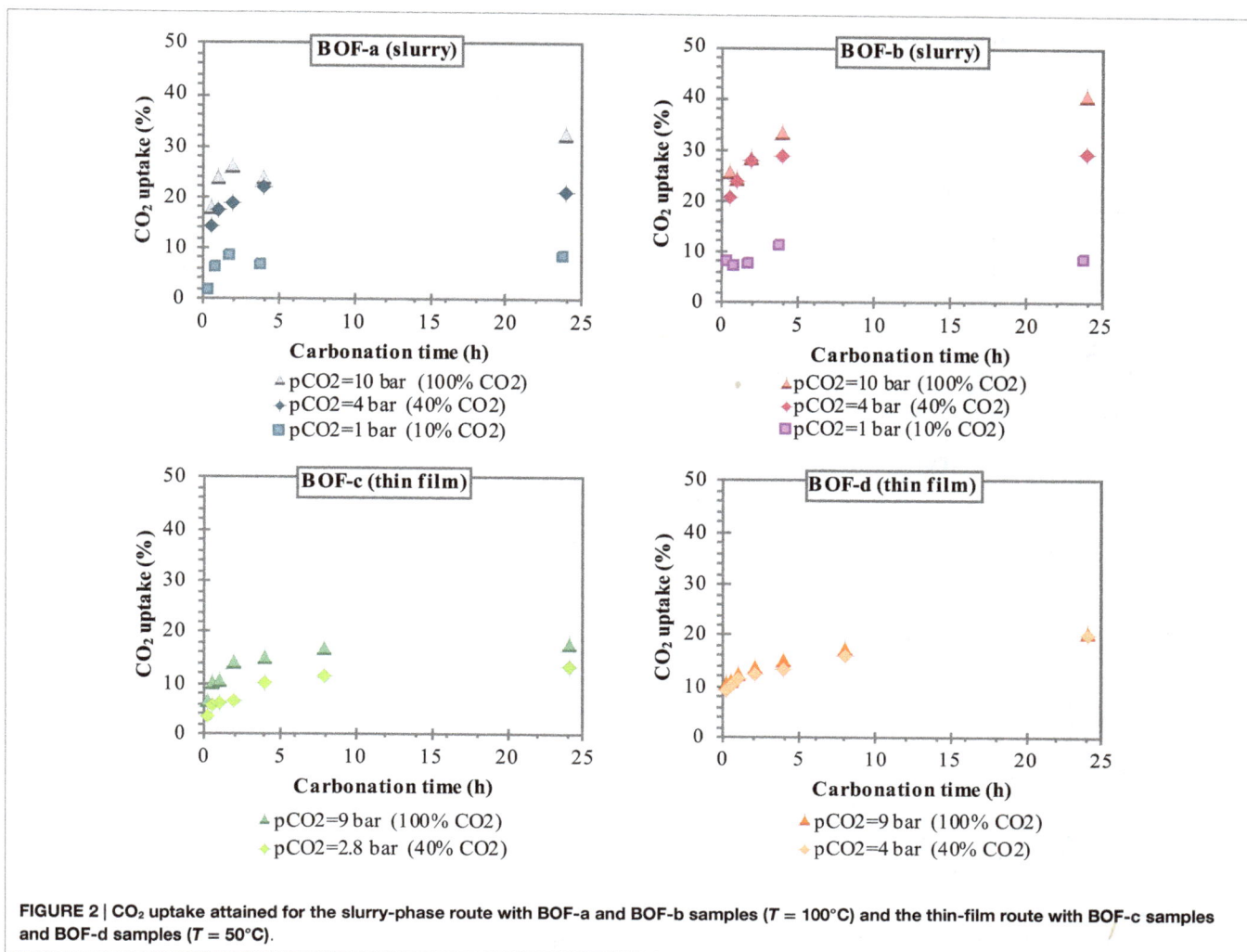

FIGURE 2 | CO_2 uptake attained for the slurry-phase route with BOF-a and BOF-b samples ($T = 100°C$) and the thin-film route with BOF-c samples and BOF-d samples ($T = 50°C$).

The relative contribution of the individual unit operations to the total energy consumption was also found to vary with the specific process condition adopted. In particular, at a CO_2 concentration of 10% the most energy-demanding stage was the heating phase, with energy consumptions for BOF-a ranging from 5900 to 58100 MJ / t_{CO_2} for 2-h and 0.5-h carbonation (corresponding to 69 and 86% of the total energy demand) and for BOF-b ranging from 3900 to 9700 MJ / t_{CO_2} for 4-h and 24-h carbonation (61 and 57% of the total energy demand). For long carbonation durations (24 h), the relative incidence of the energy consumption associated to slurry mixing inside the reactor became appreciable, with a contribution of ~ 5200 MJ / t_{CO_2} for both slag samples (~30% of total consumption). On the other hand, for BOF-a and a short carbonation time (0.5 h), for which the CO_2 uptake was of only 1.3%, the high amount of solid material to be used in the process implied a significant energy consumption (~ 6300 MJ / t_{CO_2}; 12% of total consumption) for slag milling.

Compared to the results described above, the addition of a preliminary capture stage to the carbonation process layout (i.e., the use of a 100% CO_2 gas flow) resulted in a significant decrease in the energy requirement of all process stages (with the exception of the compression unit), and particularly of the heating stage, which was of several orders of magnitude lower than that observed when working with a gas flow containing 10% CO_2. The overall energy requirement was reduced by 1.2–2.8 times for BOF-b and 1.6–12.2 times for BOF-a compared to the use of a 10%-CO_2 effluent. In absolute terms, the calculated ranges for the total energy consumption were 5400 – 6500 and 5300 – 6200 MJ / t_{CO_2} for BOF-a and BOF-b, respectively, with the highest values being in both cases associated to a carbonation time of 24 h. Using the concentrated gaseous effluent, the most relevant contribution to the overall energy demand was provided by the CO_2 capture stage, which accounted for 62–74% and 65–75% of the total requirements for BOF-a and BOF-b; the gas compression stage had also an appreciable energy duty (~ 830 MJ / t_{CO_2}), contributing by 12–15% and 13–16% of the overall demand for the two slag samples. The other terms were notably less relevant, except for the contribution of slurry mixing in the carbonation reactor for the longest residence time (24 h), which was of ~ 1300 and ~ 1100 MJ / t_{CO_2} for BOF-a and BOF-b, corresponding to 20 and 17% of the total duty.

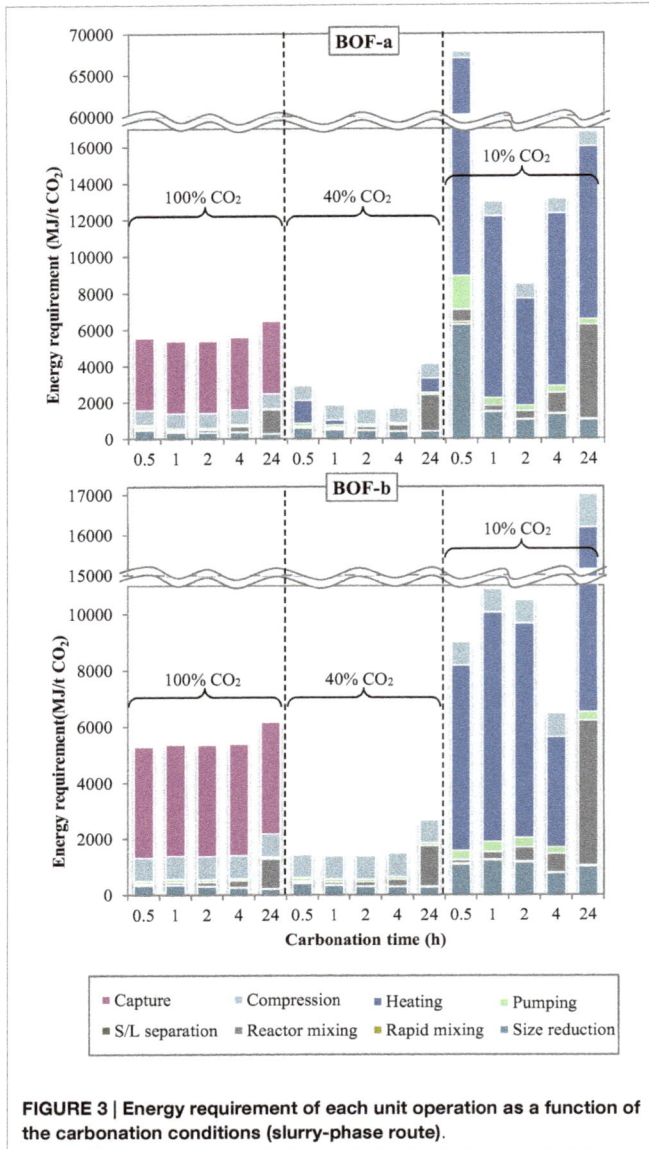

FIGURE 3 | Energy requirement of each unit operation as a function of the carbonation conditions (slurry-phase route).

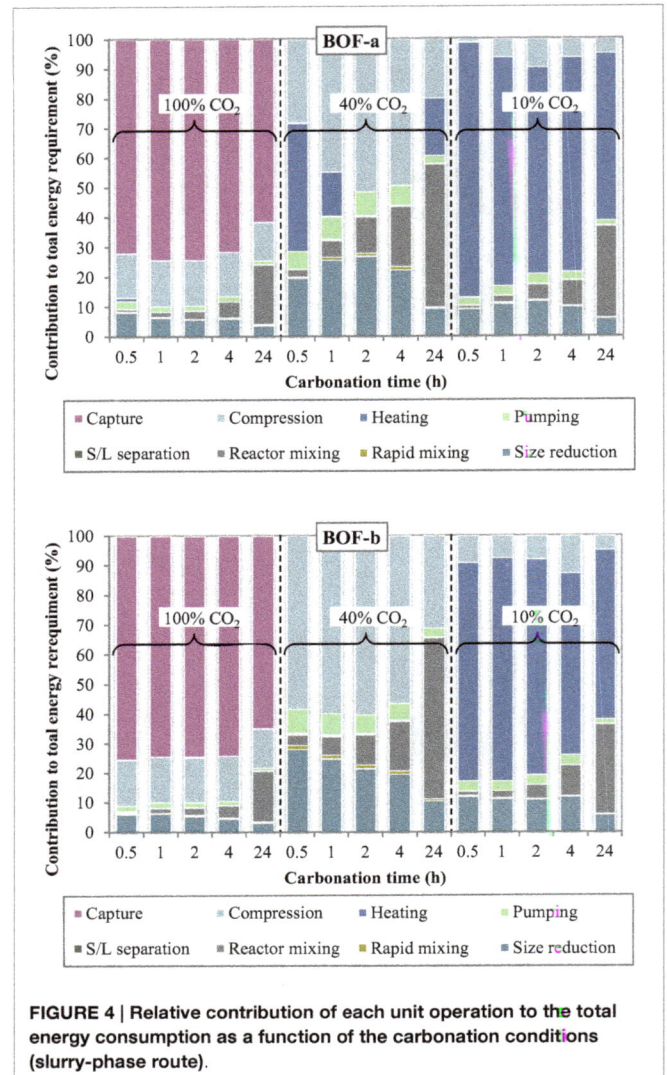

FIGURE 4 | Relative contribution of each unit operation to the total energy consumption as a function of the carbonation conditions (slurry-phase route).

When employing a 40% CO_2 concentration in the gas phase, the total energy requirement of the process was found to be considerably lower than that estimated for the two cases discussed above. As mentioned earlier, the improved energetic performance of the process using a 40% CO_2 effluent was due to the fact that, while the reactivity of the slag upon carbonation was comparable to that achieved with pure CO_2, the absence of the capture stage in the process layout allowed to save a significant share of energy. In this case, the overall energy requirements of the CO_2 sequestration process for the BOF-a and BOF-b slags lay within the ranges $1600 - 4200$ and $1400 - 2700$ MJ / t_{CO_2}. Relevant contributions to the energy penalties of the process were provided by the gas compression and size reduction units, while the duty of slurry mixing was only relevant for prolonged contact times; for the BOF-a slag the energy requirement of

the heating operation was also appreciable for 0.5-h and 24-h carbonation. In absolute terms, the BOF-a slag had the following associated main energy requirements: ~ 830 MJ / t_{CO_2} for gas compression, $380 - 580$ MJ / t_{CO_2} for size reduction, $\sim 70 - 2000$ MJ / t_{CO_2} for slurry mixing and $0 - 1300$ MJ / t_{CO_2} for heating. The corresponding values calculated for the BOF-b sample were as follows: ~ 830 MJ / t_{CO_2} for gas compression, $280 - 400$ MJ / t_{CO_2} for size reduction, $50 - 1400$ MJ / t_{CO_2} for slurry mixing, while heating theoretically required no net energy input thanks to the heat recovered from the exhaust slurry recycled to the first heater (see **Figure 1**).

The specific energy requirements of each unit operation for the thin-film route are reported in **Figure 5** making reference to the experimental results achieved for the BOF-c and BOF-d slags. The relative contributions of the process units to the total energy consumption are reported in **Figure 6**. As can be observed, for the case in which carbonation was assumed to be applied on a CO_2 pure stream obtained after capture, the total energy requirement

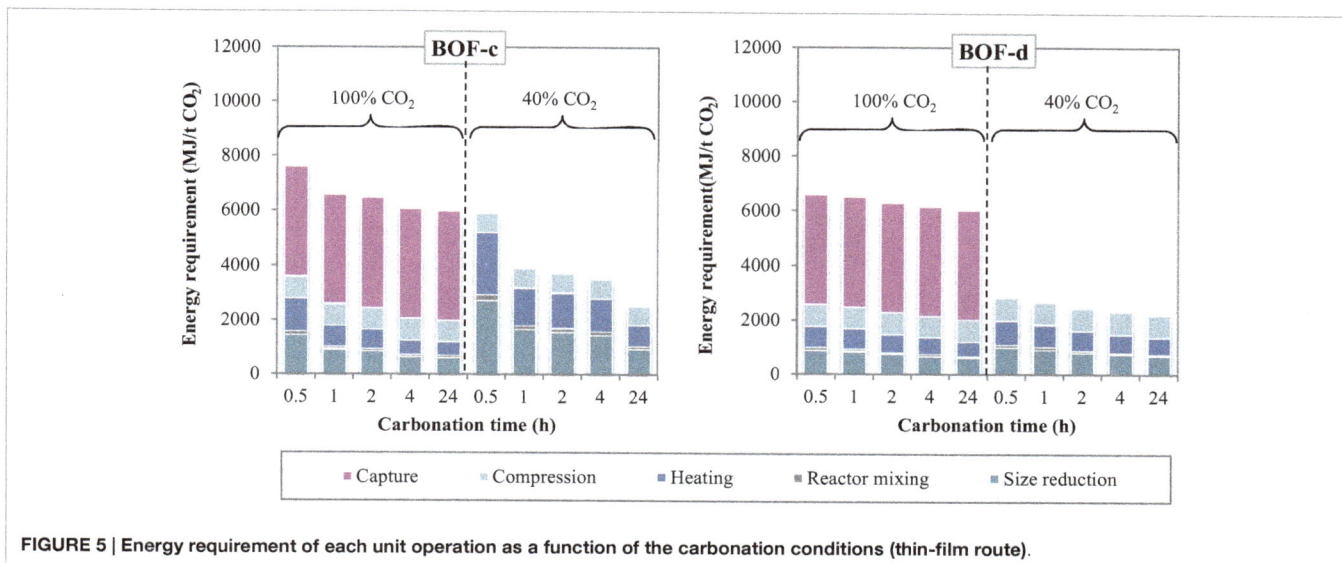

FIGURE 5 | Energy requirement of each unit operation as a function of the carbonation conditions (thin-film route).

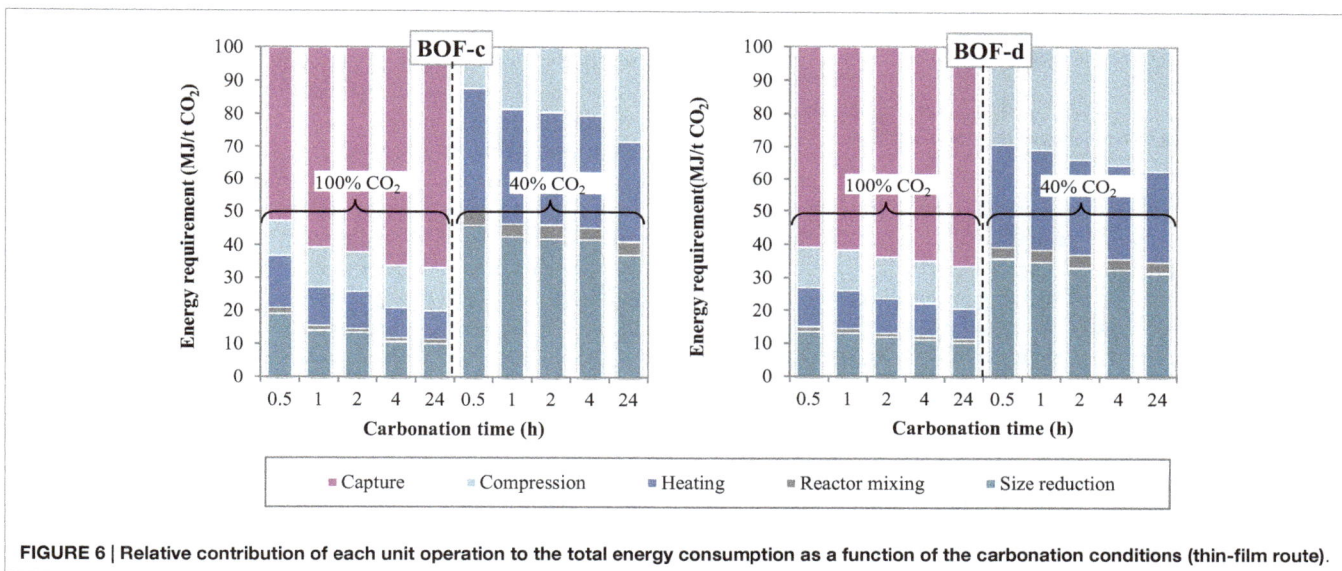

FIGURE 6 | Relative contribution of each unit operation to the total energy consumption as a function of the carbonation conditions (thin-film route).

as well as the contributions of the different unit operations for the two BOF slag samples analyzed appeared to be very similar. In particular, for BOF-c slag the overall energy requirements ranged from ~ 6000 (at 48h) to 7610 MJ / t_{CO_2} (at 0.5 h), while for sample BOF-d they varied from 6036 (at 24h) to 6590 MJ / t_{CO_2} (at 0.5 h). It should be noted that in this case, differently from the slurry-phase route, due to the increase in the CO_2 uptake of the slag for longer reaction times, the energy requirements showed to decrease progressively. This may be related to the fact that in this case the heating requirements resulted significantly lower compared to the slurry-phase route even though heat recovery was not taken into account for the thin-film route (Zingaretti et al., 2014). However, it should be acknowledged with reference to both the slurry-phase and thin-film route that for an industrial-scale application of carbonation as a CCS technique,

the adoption of long reaction times (i.e., above 1–2 h) would be unfeasible. As shown in **Figure 6**, the main contributions to the energy requirements were: CO_2 capture (~50–70%), milling and heating of the residues (~10–20%) and CO_2 compression (~13%).

As for the tests carried out with 40% CO_2 gas flows, similarly to what reported for the slurry-phase route, owing to the fact that capture was not assumed to be applied in this case, a notable reduction in the total energy requirements was achieved. In particular, the overall requirement ranged from 2550 to 5920 MJ / t_{CO_2} for sample BOF-c and from 2220 to 2850 MJ / t_{CO_2} for BOF-d, increasing also in this case progressively with reaction time. The higher requirements resulting for the first type of slag especially for short durations may be related to the lower CO_2 uptakes obtained which in turn may be ascribed to the differences in the mineralogy of the two slags (type c being mainly made up by silicate phases

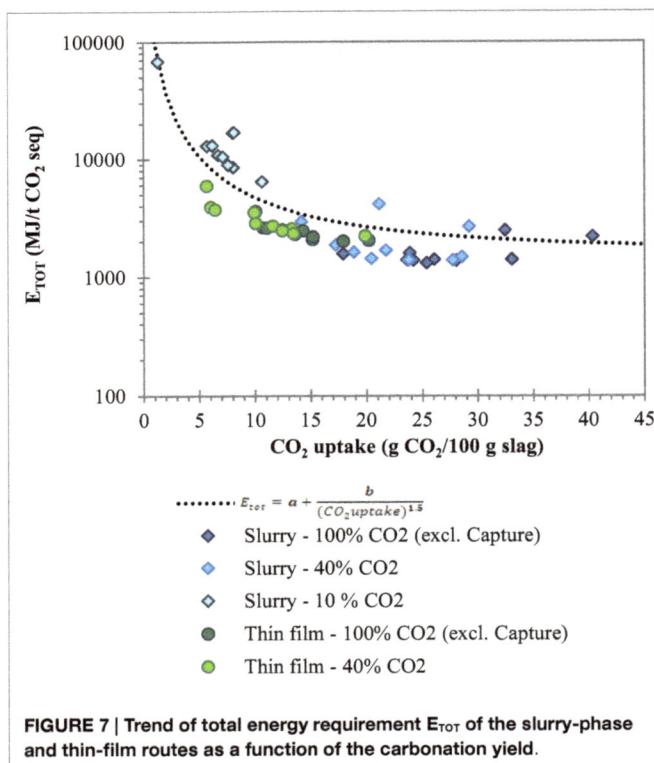

FIGURE 7 | Trend of total energy requirement E_TOT of the slurry-phase and thin-film routes as a function of the carbonation yield.

presenting a slower reaction kinetics compared to hydroxides). It should, however, be noted that for this type of slag the tests with 40% CO_2 were carried out applying a total pressure of 7 bar. As for the contributions of the single unit processes, the most relevant for BOF-c appeared to be milling (40–50%), followed by heating (25–35%) and CO_2 compression (15–30%), whereas for BOF-d the three above mentioned processes appeared to exert a similar weight (30–40%) on the total energy demand.

In order to derive some considerations of wider applicability concerning the dependence of the total energy requirement of slurry-phase and thin-film carbonation from the CO_2 uptake of the treated residues, the total energy requirements (excluding CO_2 capture) obtained for the four types of tested slags, the two routes and different gas compositions and reaction times tested were all reported together in the same graph as a function of the CO_2 uptake (see **Figure 7**). The trend of the data points showed to be adequately described by the following theoretical correlation function (see Eq. 2) that for a null CO_2 uptake tends to infinity and remains quite constant for CO_2 uptakes close to the maximum achievable ones that depend on the chemical and mineralogical composition of the material:

$$E_{tot} = a + \frac{b}{\left(CO_2\ uptake\right)^{1.5}} \quad (2)$$

with $a = 1524.2\ MJ/t_{CO_2}$ and $b = 101.5\ MJ/t_{CO_2}^{0.5}/tslag^{1.5}$. This trend highlights the key role of the CO_2 uptake in determining the overall energy requirements related to the carbonation process in particular for low reaction yields (<10% wt.). It is interesting to note that for the BOF slag samples analyzed in this study, regardless of

the type of process route applied and operating conditions selected, an estimation of the total energy requirement of the process can be directly made on the basis of the CO_2 uptake of the residues.

CONCLUSION

In this study, the possibility of directly storing the CO_2 contained in diluted gas flows so to skip the energy intensive capture step was assessed by accelerated carbonation of BOF steelmaking slags both via slurry-phase and thin-film route batch scale experiments. For the slurry-phase route, maximum CO_2 uptakes ranged from around 8% at 10% CO_2 to 21.1% (BOF-a) and 29.2% (BOF-b) at 40% CO_2 and 32.5% (BOF-a) and 40.3% (BOF-b) at 100% CO_2. For the thin-film route, maximum uptakes of 13% (BOF-c) and 19.5% (BOF-d) at 40% CO_2, and 17.8% (BOF-c) and 20.2% (BOF-d) at 100% were attained. The slurry-phase route hence showed to yield significantly higher uptakes than the thin-film route. However, it should be noted that this second process operates at lower temperature and makes use of significantly less water than the former one. Hence, the overall performance of these two routes was assessed also in terms of their total energy requirements on the basis of the energy demand of each their unit operations. The results of this evaluation indicated that working with a 40%-CO_2 effluent turned out to be the most energy efficient carbonation strategy for BOF steelmaking slag, at least under the experimental conditions tested, applying either the slurry-phase or the thin-film route. For the slurry-phase process the most suitable conditions in terms of specific energy requirements were found for both slag samples tested at a carbonation time of 2 h, and implied a total energy duty of 1610 and 1380 MJ/t_{CO_2} for BOF-a and BOF-b slag samples, respectively. For the thin-film route, the lowest energy requirements were obtained for the highest reaction times tested (2550 and 2200 MJ/t_{CO_2} after 24 h for BOF-c and BOF-d slag samples, respectively), although especially for BOF-d slags these values were not far from the ones estimated even for lower reaction times. In this respect it should be considered that for an industrial-scale application of carbonation, regardless of the type of process route selected, the reaction time should be limited to 1–2 h. Another interesting finding of this work was that the total energy requirement for both types of routes regardless of the specific properties of the residues and the operating conditions tested showed to be directly related to the CO_2 uptake of the slag. Hence, in conclusion the results of this study suggest that accelerated carbonation of BOF steel slag may be applied as a process to directly store the CO_2 content of diluted gas flows such as syngas or biogas. The most promising reaction route (from an energy requirement perspective) appeared to be the slurry-phase one. However, an overall assessment of the carbonation process should also take into account that the higher CO_2 sequestration performance attained by the slurry-phase route may be counterbalanced by an increased process complexity and the production of a final liquid effluent that needs further treatment prior to discharge. For this reason, prior to the selection of a specific reaction route, additional evaluations that take into consideration the potential environmental impacts of the process through a life cycle approach, as well as technical and economic aspects, are required.

AUTHOR CONTRIBUTIONS

RB contributed to the paper starting from the conception and design of the study to the analysis and interpretation of the results. In addition he carried out the critical revision of the final manuscript. GC contributed to the paper starting from the conception and design of the study to the carrying out of the wet route carbonation experiments and interpretation and analysis of the results. AP and RP contributed to the paper starting from the conception and design of the study to the analysis and interpretation of the results. In addition they prepared the draft manuscript.

AS carried out the slurry-phase experiments and contributed in the analysis and interpretation of the data. DZ contributed to the paper starting from the conception and design of the study to the analysis of the energy requirements and interpretation of the results.

REFERENCES

Baciocchi, R., Costa, G., Di Bartolomeo, E., Polettini, A., and Pomi, R. (2010). Carbonation of stainless steel slag as a process for CO_2 storage and slag valorization. *Waste Biomass Valor.* 1, 467–477. doi:10.1007/s12649-010-9047-1

Baciocchi, R., Costa, G., Di Bartolomeo, E., Polettini, A., and Pomi, R. (2011). Wet versus slurry carbonation of EAF steel slag. *Greenhouse Gases Sci. Technol.* 1, 312–319. doi:10.1002/ghg.38

Baciocchi, R., Costa, G., Di Gianfilippo, M., Polettini, A., Pomi, R., and Stramazzo, A. (2015). Thin-film versus slurry-phase carbonation of steel slag: CO_2 uptake and effects on mineralogy. *J. Hazard. Mater.* 283, 302–313. doi:10.1016/j.jhazmat.2014.09.016

Baciocchi, R., Costa, G., Polettini, A., and Pomi, R. (2009). Influence of particle size on the carbonation of stainless steel slag for CO_2 storage. *Energy Procedia* 1, 4859–4866. doi:10.1016/j.egypro.2009.02.314

Baciocchi, R., Costa, G., and Zingaretti, D. (2014). "Accelerated carbonation processes for carbon dioxide capture, storage and utilization," in *Transformation and Utilization of Carbon Dioxide, Green Chemistry and Sustainable Technology*, eds B. M. Bhanage and M. Arai (Berlin: Springer-Verlag). p. 263–299.

Bobicki, E. R., Liu, Q., Xu, Z., and Zeng, H. (2012). Carbon capture and storage using alkaline industrial wastes. *Prog. Energy Combust. Sci.* 38, 302–320. doi:10.1016/j.pecs.2011.11.002

Chang, E. E., Pan, S. Y., Chen, Y. H., Tan, C. S., and Chiang, P. C. (2012). Accelerated carbonation of steelmaking slags in a high-gravity rotating packed bed. *J. Hazard. Mater.* 22, 97–106. doi:10.1016/j.jhazmat.2012.05.021

Costa, G., Baciocchi, R., Polettini, A., Pomi, R., Hills, C. D., and Carey, P. J. (2007). Current status and perspectives of accelerated carbonation processes on municipal waste combustion residues. *Environ. Monit. Assess.* 135, 55–75. doi:10.1007/s10661-007-9704-4

Huijgen, W. J. J., Witkamp, G. J., and Comans, R. N. J. (2005). Mineral CO_2 sequestration by steel slag carbonation. *Environ. Sci. Technol.* 39, 9676–9682. doi:10.1021/es050795f

Huijgen, W. W. J., Witkamp, G. J., and Comans, R. N. J. (2006a). Mechanisms of aqueous wollastonite carbonation as a possible CO_2 sequestration process. *Chem. Eng. Sci.* 61, 4242–4251. doi:10.1016/j.ces.2006.01.048

Huijgen, W. J. J., Ruijg, G. J., Comans, R. N. J., and Witkamp, G. J. (2006b). Energy consumption and net CO_2 sequestration of aqueous mineral carbonation. *Ind. Eng. Chem. Res.* 45, 9184–9194. doi:10.1021/ie060636k

Intergovernmental Panel on Climate Change (IPCC). (2005). *IPCC Special Report on Carbon Dioxide Capture and Storage*. Cambridge, UK: Cambridge University Press.

Kelly, K. E., Silcox, G. D., Sarofim, A. F., and Pershing, D. W. (2011). An evaluation of ex situ, industrial-scale, aqueous CO2 mineralization. *Int. J. Greenhouse Gas Control* 5, 1587–1595. doi:10.1016/j.ijggc.2011.09.005

Kirchofer, A., Brandt, A., Krevor, S., Prigiobbe, V., and Wilcox, J. (2012). Impact of alkalinity sources on the life-cycle energy efficiency of mineral carbonation technologies. *Energy Environ. Sci.* 5, 8631–8641. doi:10.1039/C2EE22180B

Lackner, K. S., Wendt, C. H., Butt, D., Joyce, E. L. Jr., and Sharp, D. H. (1995). Carbon dioxide disposal in carbonate minerals. *Energy* 20, 1153–1170. doi:10.1016/0360-5442(95)00071-N

Mattila, H. P., Wyrsta, M. D., and Zevenhoven, R. (2014). Reduced limestone consumption in steel manufacturing using a pseudo-catalytic calcium lixiviant. *Energy Fuels* 28, 4068–4074. doi:10.1021/ef5007758

Morone, M., Costa, G., Polettini, A., Pomi, R., and Baciocchi, R. (2014). Valorization of steel slag by a combined carbonation and granulation treatment. *Miner. Eng.* 59, 82–90. doi:10.1016/j.mineng.2013.08.009

O'Connor, W. K., Dahlin, D. C., Rush, G. E., Gerdemann, S. J., Penner, L. R., Nilsen, D. N. (2005). *Aqueous Mineral Carbonation: Mineral Availability, Pretreatment, Reaction Parametrics, and Process Studies*. Report DOE/ARC-TR-04–002. Albany, OR: Albany Research Center.

Pan, S. Y., Chang, E. E., and Chiang, P. C. (2012). CO_2 capture by accelerated carbonation of alkaline wastes: a review on its principles and applications. *Aerosol Air Qual. Res.* 12, 770–791. doi:10.4209/aaqr.2012.06.0149

Santos, R. M., François, D., Mertens, G., Elsen, J., and Van Gerven, T. (2013a). Ultrasound-intensified mineral carbonation. *Appl. Therm. Eng.* 57, 154–163. doi:10.1016/j.applthermaleng.2012.03.035

Santos, R. M., Van Bouwel, J., Vandevelde, E., Mertens, G., Elsen, J., and Van Gerven, T. (2013b). Accelerated mineral carbonation of stainless steel slags for CO_2 storage and waste valorization: effect of process parameters on geochemical properties. *Int. J. Greenhouse Gas Control* 17, 32–45. doi:10.1016/j.ijggc.2013.04.004

Seifritz, W. (1990). CO_2 disposal by means of silicates. *Nature* 345, 486. doi:10.1038/345486b0

Sipilä, S., Teir, S., Zevenhoven, R. (2008). *Sequestration by Mineral Carbonation, Literature Review Update 2005-2007*. Report VT 2008-1. Turku: Åbo Akademi University Faculty of Technology Heat Engineering Laboratory.

Uibu, M., Kuusik, R., Andreas, L., and Kirsimäe, K. (2011). The CO_2-binding by Ca-Mg silicates by direct aqueous carbonation of oil shale ash and steel slag. *Energy Procedia* 4, 925–932. doi:10.1016/j.egypro.2011.01.138

Zingaretti, D., Costa, G., and Baciocchi, R. (2014). Assessment of accelerated carbonation processes for CO_2 storage using alkaline industrial residues. *Ind. Eng. Chem. Res.* 53, 9311–9324. doi:10.1021/ie403692h

Conflict of Interest Statement: The authors declare that the research was conducted in the absence of any commercial or financial relationships that could be construed as a potential conflict of interest.

14

Comparison of Chemical Composition and Energy Property of Torrefied Switchgrass and Corn Stover

Jaya Shankar Tumuluru*

Idaho National Laboratory, Idaho Falls, ID, USA

In the present study, 6-mm ground corn stover and switchgrass were torrefied in temperatures ranging from 180 to 270°C for 15- to 120-min residence time. Thermogravimetric analyzer was used to do the torrefaction studies. At a torrefaction temperature of 270°C and a 30-min residence time, the weight loss increased to >45%. At 180°C and 120 min, there was about 56 and 73% of moisture loss in the corn stover and switchgrass; further increasing the temperature to 270°C and 120 min resulted in about 78.8–88.18% moisture loss in both the feedstock. Additionally, at these temperatures, there was a significant decrease in the volatile content and increase in the fixed carbon content, and the ash content for both the biomasses tested. The ultimate composition like carbon content increased and hydrogen content decreased with increase in the torrefaction temperature and time. At 270°C and 120-min residence time, the carbon content observed was 56.63 and 58.04% and hydrogen content observed was 2.74 and 3.14%. Nitrogen and sulfur content measured at 270°C and 120 min were 0.98, 0.8, 0.076, and 0.07% for both the corn stover and switchgrass. The hydrogen/carbon and oxygen/carbon ratios calculated decreased to the lowest values of 0.59 and 0.64, and 0.71 and 0.76 for both biomasses. The van Krevelen diagram drawn for corn stover and switchgrass torrefied at 270°C indicated that H/C and O/C values are closer to coals like Illinois Basis and Powder River Basin. In the present study, the maximum higher heating value that was observed by corn stover and switchgrass was 21.51 and 21.53 MJ/kg at 270°C and a 120-min residence time. From these results, it can be concluded that corn stover and switchgrass, after torrefaction, shows consistent proximate, ultimate, and energy properties.

Keywords: corn stover, switchgrass, torrefaction temperature and time, chemical composition, energy property, mathematical model

Edited by:
Cherng-Yuan Lin,
National Taiwan Ocean University,
Taiwan

Reviewed by:
Shanmugaprakash Muthusamy,
Kumaraguru College of Technology,
India
Jaime Puna,
Instituto Superior Engenharia Lisboa,
Portugal

*Correspondence:
Jaya Shankar Tumuluru
jayashankar.tumuluru@inl.gov

Specialty section:
This article was submitted to
Bioenergy and Biofuels,
a section of the
journal Frontiers in Energy Research

Citation:
Tumuluru JS (2015) Comparison of
Chemical Composition and Energy
Property of Torrefied Switchgrass and
Corn Stover.
Front. Energy Res. 3:46.

INTRODUCTION

There is growing concern to reduce the use of fossil fuels due to the greenhouse gas emissions, which have a direct impact on the global warming temperatures. This has led researchers to explore alternative renewable energy sources. Biomass is considered a potential resource as it is considered carbon neutral because it is still part of the carbon cycle. According to the Bioenergy Technology office at the U.S. Department of Energy (DOE), biomass feedstock is defined as any renewable biological material that can be used directly as a fuel or converted to another form of fuel, such as ethanol, butanol, biodiesel, and other hydrocarbon fuels. Some examples of the biomass feedstocks that are widely used for bioenergy applications are corn starch, sugarcane juice, and crop residues (e.g., corn stover, sugarcane bagasse, purpose-grown grass crops, and woody plants). Corn stover is the largest quantity of biomass residue in the United States, and 120 million tons are available for biofuels production (U.S. Department of Energy, 2011). Corn stover is the leaf, husk, and cob remaining in the field after the harvest of cereal grain. The residue is the stalk of the leaves of maize (*Zea mays ssp. mays L.*) plants left in a field after harvest. Stover makes up about half of the yield of a crop and is similar to straw. Switchgrass (*Panicum virgatum*) is a warm-season perennial grass native to North America. It is a dedicated perennial crop and is considered a viable energy crop that could significantly increase the amount of biomass available for conversion to biofuel (Simmons et al., 2008). U.S. DOE in the 1990s selected switchgrass as a potential bioenergy feedstock due to adaptation to different growing conditions (Newman et al., 2014), and is successfully used for biopower application.

According to Nhuchhen et al. (2014), the lignocellulosic biomass, in spite of all its positive attributes, has different short-comings like structural heterogeneity, non-uniform physical properties, low-energy density, hygroscopic nature, and low bulk density. These challenges in terms of physical (lower mass density, high-moisture content, irregular size and shape, and hydrophilic in nature), chemical (low carbon and high hydrogen, oxygen, and high volatiles), and energy properties [high hydrogen/carbon (H/C) and oxygen/carbon (O/C), and lower heating values] limit the use of woody, herbaceous, and agricultural straws and other biomasses for energy application. These limitations create difficulties in transportation, handling, storage, and conversion processes (Arias et al., 2008; Medic et al., 2011; Phanphanich and Mani, 2011; Uemura et al., 2011; Wannapeera et al., 2011). Tumuluru et al. (2012) in their review indicated that raw biomass physical properties and chemical composition does not make them suitable for co-firing higher percentages with coal. These authors stated that boiler inefficiency, due to higher moisture and volatiles and lower energy content of the biomass fuels as compared to coal, is a major limitation to cofire higher percentages with coal.

To overcome the biomass challenges in terms of physical, chemical, and energy properties, a torrefaction process was developed. Torrefaction is a thermal pretreatment method where the biomass is thermally pretreated in a temperature range of 200–300°C for about 30 min in absence of air at atmospheric pressure (Tumuluru et al., 2011). Torrefaction makes biomass (a) brittle product making it easier to grind (better particle size and shape), (b) changes the chemical composition (removing the moisture and low-energy content of volatiles), and (c) increases the net energy content of the biomass. According to Lu and Chen (2014), biomass torrefaction helps to retain most of its energy and simultaneously loses its hygroscopic properties. Sarkar et al. (2014) and Yang et al. (2014) have successfully used torrefied switchgrass for gasification and pyrolysis application. Nhuchhen et al. (2014), in their review on torrefaction of biomass, suggested that the process makes biomass suitable for co-firing applications and can be promoted as an alternative to charcoal. They also stated that the torrefaction increases its hydrophobicity, grindability, and energy density making it more suitable for thermochemical applications like gasification and pyrolysis.

The common biomass reactions during torrefaction are dehydration, devolatilization, depolymerization, and carbonization. During initial heating at a temperature about 180°C, the dehydration reactions remove most of the moisture in the biomass. Furthermore, heating biomass in the temperature range of 180–270°C promotes the reactions like depolymerization, devolatilization, and carbonization of hemicellulose, cellulose, and lignin. For torrefaction, process temperatures over 270°C can lead to extensive devolatilization of the biomass due to the initiation of the pyrolysis process (Tumuluru et al., 2011). Also during these reactions, biomass loses some of the lipophilic compounds that make it hydrophobic.

The torrefaction process is influenced by process parameters like temperature, heating rate, absence of oxygen, residence time, ambient pressure, and the properties of feedstock like moisture content and particle size. Torrefaction temperature is typically between 200 and 300°C and the residence time is adjusted to produce a brittle, hydrophobic, and high-calorific value product. Typically, biomass is pre-dried to <10% moisture content prior to torrefaction. The particle size of the biomass will influence the reaction mechanisms, kinetics, duration, and its specific heating rate of the process. Several researchers have worked on the torrefaction of agricultural and woody biomass using a thermogravimetric analyzer (TGA) (Bridgeman et al., 2008; Chen and Kuo, 2010; Repellin et al., 2010). These researchers have successfully used TGA to estimate the effects of particle size, temperature, and moisture to better understand the torrefaction kinetics, chemical, proximate, and energy properties of both woody and herbaceous biomass.

The focus of this work is to understand the effect of the torrefaction time and temperature on proximate and ultimate composition and energy property of corn stover and switchgrass using a TGA. The specific objectives for this research is to (a) understand the effect of torrefaction temperature (180–270°C) and its residence time (15–120 min) on the weight loss, proximate composition (moisture, ash, volatiles, and fixed carbon), ultimate composition (hydrogen, carbon, sulfur, nitrogen, and oxygen), and energy property (higher heating value), (b) understand the significance of torrefaction temperature and time on weight loss, chemical composition and higher heating value, and (c) develop mathematical models for weight loss, proximate and ultimate composition, and higher heating value experimental data.

MATERIALS AND METHODS

Corn stover and switchgrass samples were used in the present torrefaction experiments. These biomass samples were harvested from farms in Iowa and Nebraska in the form of bales. It was initially ground to bigger particle sizes using a 50.8-mm screen with a Vermeer HG200 grinder (Vermeer Corporation–Agriculture, Pella, IA, USA). The ground material was evaluated for moisture content and stored in sealed plastic containers that were maintained at about 4°C until it was further size-reduced to 6 mm.

Thermogravimetric Analyzer

Torrefaction was performed on the LECO TGA701 (see **Figure 1**) in a batch procedure. Experiments were conducted in a temperature range of 180–270°C and a residence time of 15–120 min.

Biomass sample preparation includes grinding corn stover and switchgrass to 6 mm using a Retsch splitter. These samples were double-bagged and stored in air-tight containers and used for torrefaction studies. A method file was also developed to carry out the torrefaction experiments using TGA (Bridgeman et al., 2008). This method file includes the steps for drying and torrefaction, and its associated cooling steps.

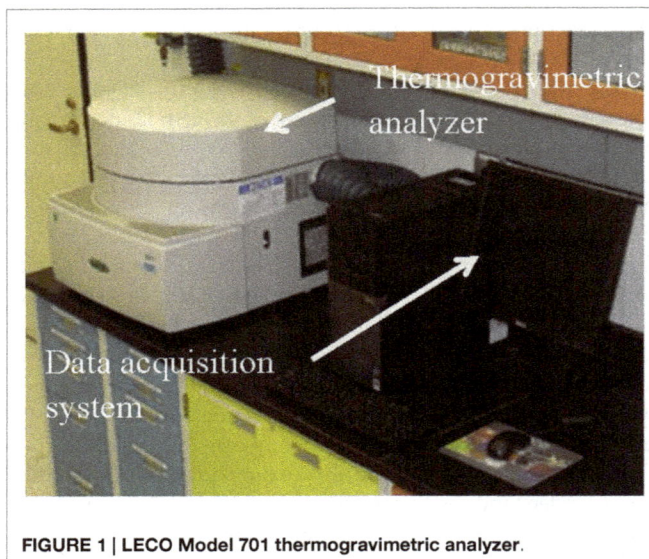

FIGURE 1 | LECO Model 701 thermogravimetric analyzer.

TABLE 1 | Methods followed for measurement of chemical and energy properties of corn stover and switchgrass biomass samples.

	Chemical composition	
	Proximate	
1	Moisture	ASTM International D3173 (2008)
2	Ash	ASTM International D3174 (2002)
3	Volatiles	ASTM International D3175 (2007)
4	Fixed carbon	Fixed carbon by difference method
	Ultimate composition	
1	Carbon	ASTM International D3178 (2002)
2	Hydrogen	ASTM International D3178 (2002)
3	Nitrogen	ASTM International D3179 (2002)
4	Sulfur	ASTM International D3177 (2002)
5	Oxygen	Oxygen by difference method
6	H/C ratio	H/C: Number of hydrogen atoms/number of Carbon atoms = (%H/1)/(%C/12)
7	O/C ratio	O/C: Number of oxygen atoms/number of Carbon atoms = (%O/8)/(%C/12)
8	HHV (higher heating value)	ASTM International D5865 (2010)

For the present study, the heating rate selected was 10°C/min. LECO instrumentation contains an easy-to-follow menu, driven by Microsoft™ Windows®-based software, which allows its analysis method. Temperature, temperature ramp rate, and atmosphere are selectable at each step. Analysis methods can also be entered to evaluate the moisture, volatiles, and ash content. American Society for Testing and Materials (ASTM) standard methods for estimating the chemical composition was used to measure the proximate and ultimate composition of the torrefied biomass (**Table 1**). **Table 2** shows the experimental design used for conducting the torrefaction tests.

Data Analysis

The experimental data on weight loss, proximate and ultimate composition, and higher heating value obtained at different torrefaction temperatures (180°, 230°, and 270°C) and different residence times (15, 30, 60, and 120 min) for both corn stover and switchgrass was used to draw the bar plots, develop multiple regression equations (Eq. 1), and understand the significance of the torrefaction process variables with respect to torrefied material properties studied. Statistica (Version 9) statistical software was used to develop the multiple regression models and analysis of variance (ANOVA) for the experimental data. Coefficient of determination was used to understand the model fit and the ANOVA was used to understand the significance of the process variables with respect to the weight loss, proximate, ultimate composition, and higher heating value.

$$y = b_0 + b_1 x_1 + b_2 x_2 \qquad (1)$$

where b_0, b_1, and b_2 are equation constants; x_1 and x_2 are torrefaction temperature (°C) and torrefaction residence time (min), respectively; and y is dependent variable (weight loss, proximate, ultimate composition, and higher heating value).

RESULTS AND DISCUSSION

The proximate and ultimate composition of raw corn stover and switchgrass are indicated in **Table 3**. Ash content of corn stover is slightly higher compared to switchgrass, and volatiles are slightly higher in corn stover when compared to switchgrass. The carbon content of the corn stover and switchgrass were in the range of 42–43%. The heating values of both feedstocks were close. The oxygen content of the biomass feedstocks are in the range of 40–41%. The H/C and O/C ratios calculated are slightly higher for corn stover (1.64 and 1.38) compared to switchgrass (1.52 and 1.45). **Table 4** indicates the ultimate composition and H/C and O/C ratios of Appalachian, Powder River Basin, Illinois Basin, and North Dakota lignite coals. The Appalachian coal has the highest carbon and lowest oxygen (66.93 and 7.55), and North Dakota lignite has the lowest carbon and highest oxygen values (31.8 and 26.35). The H/C and O/C ratios calculated for

TABLE 2 | Experimental design for torrefaction experiments using TGA.

Process	Temperatures (°C)	Residence time (min)	Particle size (mm)	Heating rate (°C/min)	Feedstocks
Torrefaction	180, 230, and 270	15, 30, 60, and 120	6	10	Corn stover and Switchgrass

TABLE 3 | Proximate, ultimate, and energy property of raw corn stover and switchgrass.

Chemical composition	Feedstock	
Proximate	Corn stover	Switchgrass
1 Moisture	4.01	6.01
2 Ash	5.13	4.01
3 Volatiles	75.63	73.32
4 Fixed carbon	15.23	16.66
Ultimate composition and higher heating value		
1 Carbon	43.92	42.08
2 Hydrogen	6.01	5.44
3 Nitrogen	0.42	0.36
4 Sulfur	0.07	0.05
5 Oxygen	40.44	41.38
6 H/C	1.64	1.52
7 O/C	1.38	1.45
8 HHV (higher heating value)	17.31	17.36

TABLE 4 | Ultimate composition and H/C and O/C ratio of different coals (source: Tillman et al., 2009; Tumuluru et al., 2012).

Coals	Hydrogen	Carbon	Oxygen	H/C	O/C
Coal Appalachian	4.43	66.93	7.55	0.79	0.17
Illinois Basin	4.77	60.68	13.61	0.94	0.33
Powder River Basin	3.55	51.89	12.77	0.82	0.37
North Dakota Lignite	4.51	31.8	26.35	1.70	1.24

the coals indicated that Appalachian coals has the lowest values of 0.79 and 0.17, and North Dakota lignite has the highest values of 1.7 and 1.34.

Analysis of Variance and Mathematical Models

The experimental data was analyzed further to understand the significance of the process variables, and to develop multiple regression models for weight loss, proximate, ultimate composition, and energy data. **Table 5** indicates the significance of the process variables, and **Table 6** shows the multiple regression equations fitted for the experimental data. For the weight loss for both corn stover and switchgrass, the torrefaction temperature was found to be significant at $P < 0.001$ and the torrefaction residence time at $P < 0.05$. In the case of moisture content, volatiles, and fixed carbon, torrefaction temperature was found to be significant at $P < 0.001$ for both switchgrass and corn stover. Torrefaction residence time was found to be significant for switchgrass at $P < 0.05$ for moisture content, volatiles, and fixed carbon at $P < 0.01$; whereas for corn stover, it was found to be non-significant. In case of ash content, torrefaction temperature was found to be significant at $P < 0.001$, and residence time was found to be non-significant for both corn stover and switchgrass. Ultimate composition, hydrogen, oxygen, and carbon content were influenced by the torrefaction

TABLE 5 | Significance of the torrefaction temperature and time with respect to proximate, ultimate, and energy property.

Proximate composition	Corn stover		Switchgrass	
	Process variables		Process variables	
	Torrefaction temperature (x_1)	Torrefaction time (x_2)	Torrefaction temperature (x_1)	Torrefaction time (x_2)
Moisture content (%, w.b.)	(−)***	(−)*	(−)***	(−)*
Ash (%)	(+)***	ns	(+)*	ns
Volatiles (%)	(−)***	ns	(−)***	(−)*
Fixed carbon (%)	(+)***	ns	(+)***	(+)**
Ultimate composition and higher heating value				
Hydrogen (%)	(−)***	(−)*	(−)***	(−)*
Carbon (%)	(+)***	ns	(+)***	ns
Oxygen (%)	(−)***	ns	(−)***	ns
H/C ratio	(−)***	(−)*	(−)***	ns
O/C ratio	(−)***	ns	(−)***	ns
Higher heating value (MJ/kg)	(+)***	ns	(+)***	ns
Weight loss (%)	(+)***	(+)*	(+)***	(+)*

*$p < 0.05$; **$p < 0.01$; ***$p < 0.001$.

TABLE 6 | Multiple regression equations for weight loss, proximate and ultimate composition, and energy properties of corn stover and switchgrass.

	Multiple regression equation	(R^2)
Corn stover		
Proximate composition		
Moisture content (%, w.b.)	$y = 4.45 - 0.011x_1 - 0.0057x_2$	0.81
Ash (%)	$y = -6.12 + 0.060x_1 + 0.013x_2$	0.75
Volatiles (%)	$y = 138.22 - 0.32x_1 - 0.057x_2$	0.88
Fixed carbon (%)	$y = 36.55 + 0.28x_1 + 0.050x_2$	0.91
Ultimate composition, weight loss and higher heating value		
Hydrogen (%)	$y = 10.89 - 0.025x_1 - 0.0073x_2$	0.92
Carbon (%)	$y = 24.76 + 0.108x_1 + 0.0098x_2$	0.90
Oxygen (%)	$y = 66.41 - 0.133x_1 - 0.014x_2$	0.83
H/C ratio	$y = 3.19 - 0.0084x_1 - 0.0018x_2$	0.95
O/C ratio	$y = 2.516 - 0.0062x_1 - 0.00052x_2$	0.89
Higher heating value (MJ/kg)	$y = 12.55 + 0.030x_1 + 0.0051x_2$	0.89
Weight loss (%)	$y = -82.10 + 0.474x_1 + 0.076x_2$	0.94
Switchgrass		
Proximate composition		
Moisture content (%, w.b.)	$y = -4.18 - 0.0107x_1 - 0.0047x_2$	0.76
Ash (%)	$y = -1.01 + 0.024x_1 + 0.0151x_2$	0.62
Volatiles (%)	$y = 111.30 - 0.20x_1 - 0.070x_2$	0.90
Fixed carbon (%)	$y = -14.47 + 0.187x_1 + 0.060x_2$	0.91
Ultimate composition, weight loss and higher heating value		
Hydrogen (%)	$y = 8.33 - 0.0138x_1 - 0.00636x_2$	0.75
Carbon (%)	$y = 27.12 + 0.099x_1 + 0.017x_2$	0.85
Oxygen (%)	$y = 61.15 - 0.100x_1 - 0.019x_2$	0.85
H/C ratio	$y = 2.48 - 0.0053x_1 - 0.0017x_2$	0.83
O/C ratio	$y = 2.31 - 0.0050x_1 - 0.00086x_2$	0.88
Higher heating value (MJ/kg)	$y = 12.41 + 0.031x_1 + 0.0046x_2$	0.87
Weight loss (%)	$y = -73.26 + 0.41x_1 + 0.090x_2$	0.88

x_1 = torrefaction temperature (degree Celsius) and x_2 = torrefaction time (min).

temperature at $P < 0.001$ for both corn stover and switchgrass. Torrefaction residence time was found to be significant only for hydrogen content at $P < 0.05$; whereas for carbon, it was found to be non-significant. For the H/C ratio and O/C ratio, the torrefaction temperature was found to be significant at $P < 0.001$; however, the residence time was significant only for corn stover at $P < 0.05$. In the case of the higher heating value, only the torrefaction temperature was found to be statistically significant at $P < 0.05$. The regression equation developed for the experimental data has adequately fitted based on the coefficient of determination values. Also, all the equations were found to be statistically significant at $P < 0.001$. These equations can help to predict the weight loss, proximate and ultimate composition, higher heating values, and H/C and O/C ratios of corn stover and switchgrass at different torrefaction temperatures and residence times studied in this paper.

Weight Loss

The weight loss for both corn stover and switchgrass at different torrefaction temperatures and residence times are indicated in **Figure 2**. At 180°C, and 15–120 min residence time, the weight loss was in the range of 8.1–9.3% for switchgrass and 9.1–10.04% for corn stover. Increasing the torrefaction temperature and residence time increased the weight loss in both corn stover and switchgrass. The maximum weight loss observed for corn stover at this temperature was about 33% and about 26% for switchgrass. At 270°C and 30 min, the weight loss observed for corn stover and switchgrass was 42.63 and 51.26%, and at 120 min the weight loss values increased to final values of 54.93–58.29%.

Proximate Composition

The loss of moisture was significant at all the torrefaction temperature and residence times. At lower temperature of 180°C and 30 min residence times, the loss of moisture was about 26% for corn stover and 54% for switchgrass compared to its original value. Furthermore, increasing the residence time to 120 min reduced the final moisture content in both corn stover and switchgrass to about 1.67 and 1.62% – a reduction of 56 and 73%

compared with their original value. The decrease in moisture content of the switchgrass and corn stover had increased with the increase in torrefaction temperature and residence time. At 270°C and 120-min residence time, a maximum moisture reduction of 78.8% for corn stover and 88.18% for switchgrass was observed (**Figure 3**). Tumuluru et al. (2011) in their review indicated that lower temperatures of <200°C contributes to the loss of moisture resulting from the dehydration reactions; whereas, at temperatures of >200°C, the loss of moisture is attributed to the complicated interactions of biomass components to temperature and residence time. Phanphanich and Mani (2011) in their studies indicated that torrefying the biomass at temperature >200°C results in improved grinding characteristics. The main reason for this behavior is due to loss of moisture and other low energy volatiles, which makes biomass brittle and easy to grind. **Figure 4** shows the photos of corn stover and switchgrass torrefied at 180°, 230°, and 270°C for 120-min residence time. It is very clear from the figure that the color changes significantly with the torrefaction temperature, which can be a good indicator for the degree of torrefaction. According to Tumuluru et al. (2011), a temperature regime of 150–200°C, also called the reactive drying range, initiates the breakage of hydrogen and carbon bonds and results in a structural deformity that does not result in a significant change in color of the biomass. The same authors indicated that at a temperature range of 200–300°C, also called as destructive drying, results in the disruption of most of the inter- and intramolecular hydrogen bonds and C–C and C–O bonds and emits condensable and non-condensable gases, which result in darkening of the biomass. At these temperatures, cell structure is completely destroyed as the biomass loses its fibrous nature, becomes brittle, and is easier to grind (Bergman and Kiel, 2005).

During torrefaction, the ash components in the biomass do not change. All ash components of biomass are still present in torrefied biomass. The change in the ash content is more relative to the change in the original biomass components. As the biomass loses some of the moisture and volatiles during the process, the ash content is more of a relative increase with respect to the original components. In the present study, the initial ash content

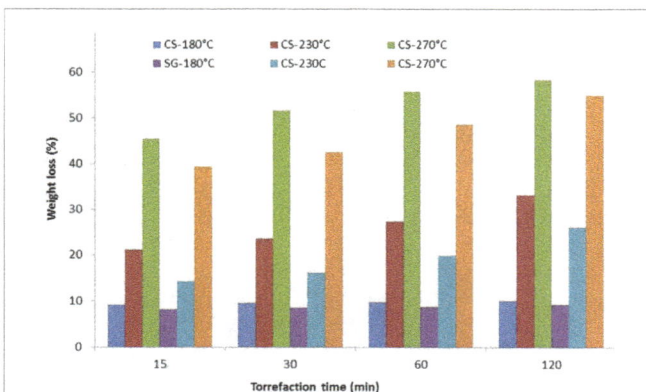

FIGURE 2 | Weight loss in corn stover and switchgrass at different torrefaction temperatures and times.

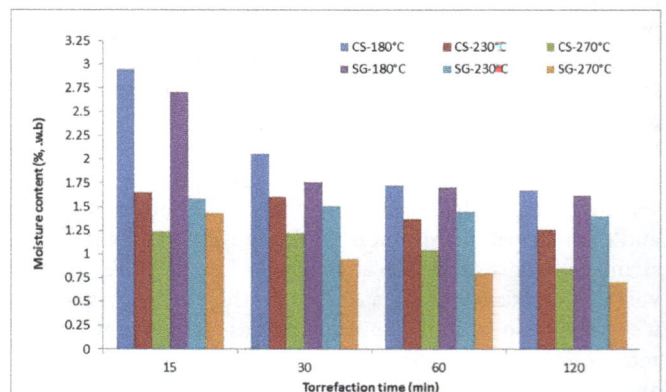

FIGURE 3 | Torrefaction temperature and time effect on moisture content of the switchgrass and corn stover.

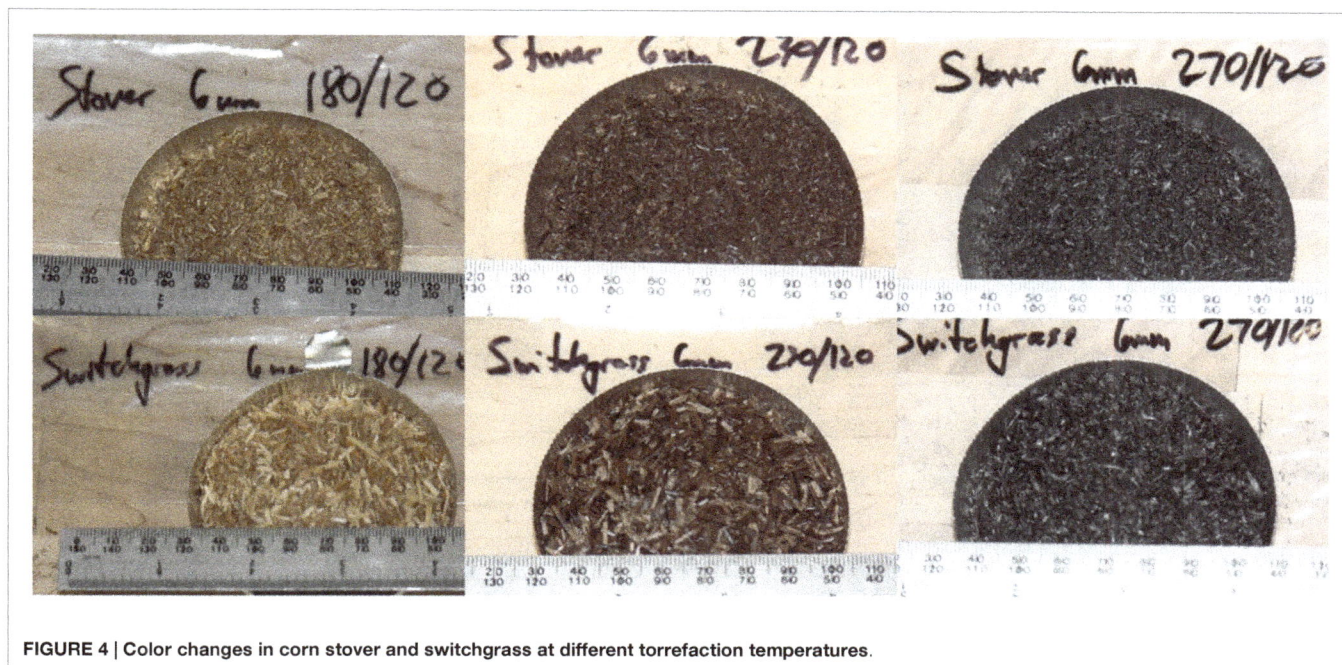

FIGURE 4 | Color changes in corn stover and switchgrass at different torrefaction temperatures.

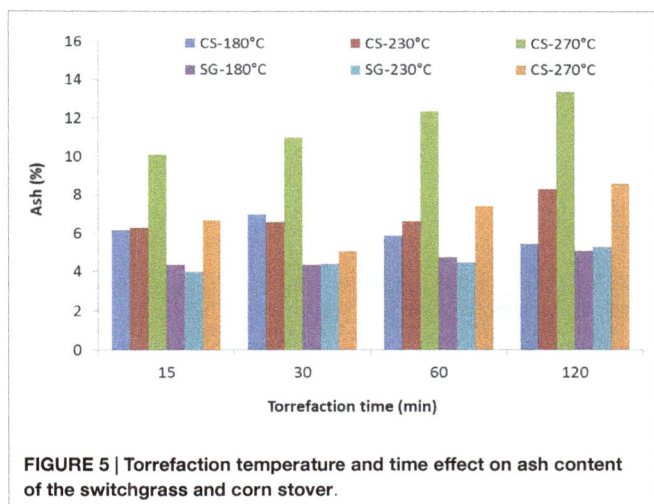

FIGURE 5 | Torrefaction temperature and time effect on ash content of the switchgrass and corn stover.

of corn stover and switchgrass was 5.13 and 4.01%. At 180°C and 15-min residence time, the percent increase in the ash content was about 20.07% for corn stover, and in the case of switchgrass, it increased by about 8.22%. Increasing the residence time to 30 min did increase the ash content to 6.98 and 4.35%. This increase is marginal from 15–30 min for both the feedstocks. **Figure 5** clearly indicates that the ash content in both the biomasses tested increased with the increasing of the torrefaction temperature and residence time. At 230° and 270°C torrefaction temperature and 15-min residence time, the ash content of both the biomasses tested increased. According to Poudel and Oh (2012) and Chen et al. (2014), during torrefaction the volatile content in the biomass decreases leading to an increase in ash content.

Initial volatile content of corn stover was 75.63% and the switchgrass was 73.32%. At 180°C and 15-min residence time, the volatile content decreased to 74.03% (a decrease of about 2% from the original value for corn stover). In the case of switchgrass, the decrease was from 73.32 to 71.39% (a decrease of about 2.6% from the original value). The loss of volatiles increased with the increase in torrefaction temperature and residence time. At 230°C and between 15- and 120-min residence time, the decrease was in the range of 61.23–69.42%; however, at 270°C the decrease in the volatile content was further decreased and was in the range of 50.27–58.78% (**Figure 6**). The study indicated that the volatile losses are higher at higher torrefaction temperature and residence time. At torrefaction temperature of 200–300°C, weight loss in the biomass is mainly due to loss of moisture and hemicellulose and lignin decomposition. Xylan-based hemicellulose generally decomposes around 250–280°C. Lignin decomposition proceeds more slowly, but will gradually increase starting at about 200°C. At these temperatures, the disruption of most inter- and intramolecular hydrogen bonds and C–C and C–O bonds will result in a formation of hydrophilic extractives, carboxylic acids, alcohols, aldehydes, ether, and gases like CO, CO_2, and CH_4 (Tumuluru et al., 2011). The common reactions at these temperatures are limited devolatilization and carbonization of the hemicellulose. At temperatures of >250°C, the hemicellulose decomposes extensively into volatiles and a char-like solid product.

The initial fixed carbon content of corn stover and switchgrass is 15.23 and 16.66%. At lower temperatures the change in the fixed carbon is not significant, where at higher temperature of 270°C the change in the fixed carbon is more significant. At 180°C torrefaction temperature and 15-min residence time, the change in the fixed carbon is marginal for corn stover (16.86%);

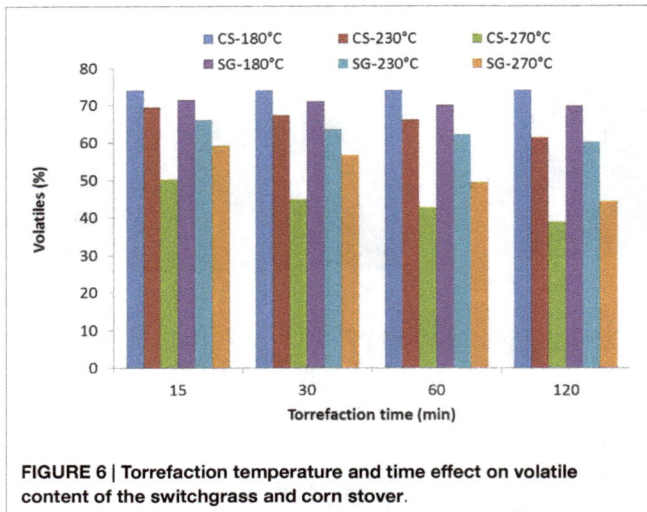

FIGURE 6 | Torrefaction temperature and time effect on volatile content of the switchgrass and corn stover.

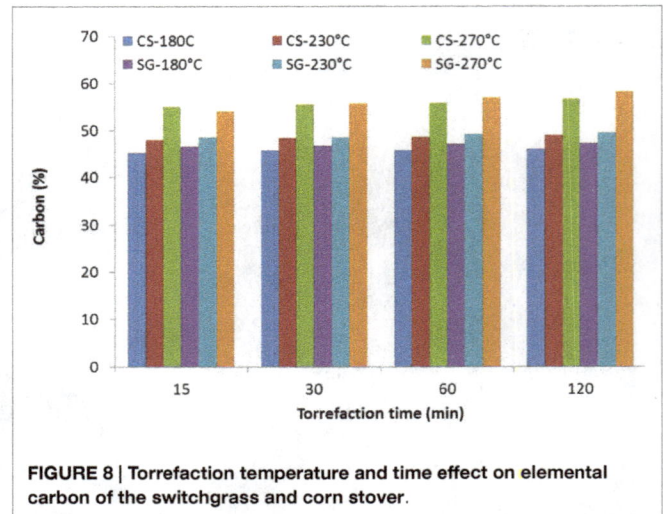

FIGURE 8 | Torrefaction temperature and time effect on elemental carbon of the switchgrass and corn stover.

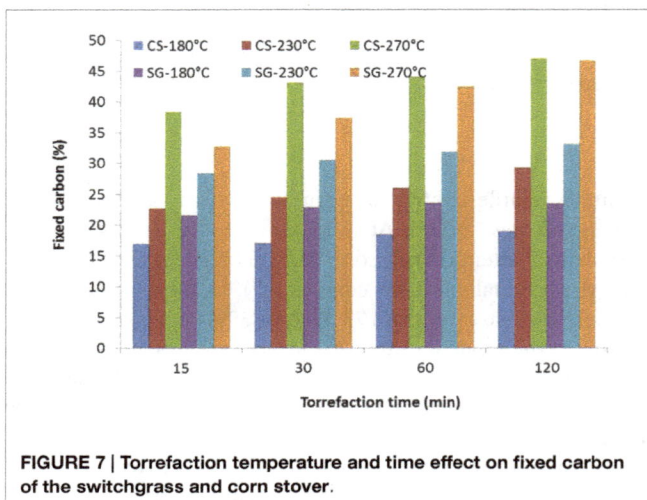

FIGURE 7 | Torrefaction temperature and time effect on fixed carbon of the switchgrass and corn stover.

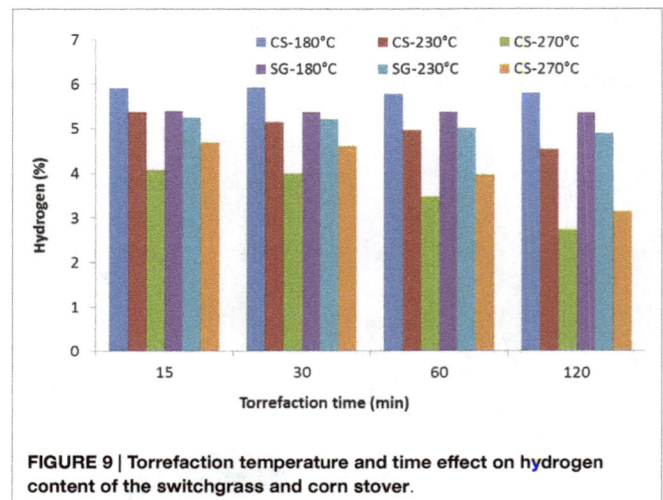

FIGURE 9 | Torrefaction temperature and time effect on hydrogen content of the switchgrass and corn stover.

however, for switchgrass the fixed carbon increased to 21.56%. Increasing the residence time to 30–120 min at 180°C torrefaction temperature increased the fixed carbon to 18.95% for corn stover, and for switchgrass the fixed carbon increased to 23.49% (**Figure 7**). At 230 and 270°C, the increase in fixed carbon content was significant compared with its original value. At 270°C and 15 min, the fixed carbon content of both the biomasses tested almost doubled (28.35 and 32.71%).

Ultimate Composition

The initial carbon content of corn stover and switchgrass is about 43.92 and 42.08%. The increased carbon content at 180°C at 15-min residence time is 45.26% for corn stover and 46.56% for switchgrass. Furthermore, increasing the residence time to 120 min increased the carbon content to a final value of 46.03% for corn stover and 47.28% for switchgrass (**Figure 8**). By increasing the torrefaction temperature to 230 and 270°C at 15-min residence time, the carbon content values observed from both corn stover and switchgrass were 47.92 and 48.53%, and 54.92

and 53.94%. Increasing the residence time further to 120 min also increased the carbon content, but marginally.

The initial hydrogen content of the corn stover and switchgrass is about 6.01 and 5.44%. The hydrogen content of corn stover and switchgrass at 180°C and 15-min residence time decreased to 5.9 and 5.39%. By further increasing the residence to 120 min, the decrease was marginal (5.8 and 5.35%). Increasing the torrefaction temperature to 230°C and 15 min, the hydrogen content of corn stover and switchgrass samples were found to be 5.14 and 5.20%, but increasing the residence time to 120 min decreased the hydrogen content of the samples to 4.53 and 4.89% (**Figure 9**). At 270°C and 15 min, the hydrogen content of corn stover and switchgrass were found to be 4.07 and 4.68%; however, at 30, 60, and 120 min, the hydrogen content of the samples observed was about 3.99 and 4.6; 3.48 and 3.97; and 2.74 and 3.14% for corn stover and switchgrass.

The initial nitrogen content of the corn stover and switchgrass was 0.42 and 0.36%. At 180°C and 120 min, the nitrogen content increased to 0.50% for corn stover; however, for switchgrass, the nitrogen content increased to 0.38%. Increasing the torrefaction temperature to 270°C and 120 min, the nitrogen content observed in the case of corn stover was 0.98%, while in the case of switchgrass it was about 0.8%. The initial sulfur content of corn stover and switchgrass was found to be 0.07 and 0.05%. There is not a significant change in the sulfur content of both of the biomasses tested, but at 270°C and 120 min the observed values were 0.076 and 0.07%.

Oxygen content is determined based on the difference method. The initial oxygen content of the corn stover and switchgrass was about 40.44 and 41.38%. At a lower torrefaction temperature of 180°C and smaller residence time of 15 min, the oxygen content of the corn stover and switchgrass decreased to 39.99 and 41% (**Figure 10**). Additionally, increasing the residence time to 120 min did not bring much change in the oxygen content of the samples. At 230°C and 15-min residence time, the observed oxygen content of the corn stover and switchgrass samples was 38.53 and 39.53%, which is also marginal. Torrefying the switchgrass and corn stover at higher temperatures of 270°C did have an impact on the oxygen content of both corn stover and switchgrass. At 270°C and 15 min residence time, the observed oxygen content of the corn stover and switchgrass samples was 29.67 and 33.29%. Furthermore, increasing the residence time to 120 min still reduced the oxygen content of the samples to 26.41 and 29.5%. The results indicated that the torrefaction temperature had a more-significant effect on the oxygen content compared to residence time.

The H/C and O/C ratios of the corn stover and switchgrass raw samples are calculated and indicated in **Table 1**. Corn stover has higher H/C ratio when compared to switchgrass, and switchgrass has higher O/C ratio when compared to corn stover (H/C and O/C ratios of corn stover showed 1.64 and 1.52 and switchgrass showed 1.38 and 1.45). At a lower torrefaction temperature of 180°C and 15 min, the observed H/C ratio for corn stover was 1.56 and switchgrass was 1.38. Increasing the torrefaction residence time to 120 min reduced the H/C ratio marginally (corn stover 1.51 and switchgrass 1.35). Increasing the torrefaction temperature to 230°C at 15 min residence

time, reduced the H/C ratio to 1.27 for corn stover and 1.28 for switchgrass (**Figure 11**). Further increasing the residence time to 120 min at the same torrefaction temperature of 230°C, the H/C ratio values observed were 1.11 for corn stover and 1.18 for switchgrass. At 270°C torrefaction temperature and 15 min residence time, the observed H/C ratio value for corn stover was 0.88 and switchgrass was 1.04, though at 120 min residence time H/C values reduced to 0.59 and 0.64 for corn stover and switchgrass (**Figure 11**).

With the O/C ratio, a similar trend was observed, where increasing the torrefaction temperature and residence decreased the values for both corn stover and switchgrass. At a torrefaction temperature of 180°C and 15-min residence time, the observed values were 1.32 for both corn stover and switchgrass. Increasing the residence time to 120 min did change the values marginally (1.31 corn stover and 1.29 switchgrass). At the other torrefaction temperatures of 230 and 270°C and a 15-min residence time, the observed O/C ratio values for corn stover and switchgrass observed were 1.20, 1.22 and 0.81, 0.92. At 120 min residence time at 230 and 270°C, O/C values reduced to 1.13, 1.16 and 0.71, 0.76 (**Figure 12**). It is clear from the data that both torrefaction temperature and residence time had significant effect on the H/C ratio and O/C ratios.

The van Krevelen diagram, which is drawn for O/C and H/C ratio, was drawn for corn stover and switchgrass and was compared to different grades of the coal (**Figure 13**). The lower ratio of H/C to O/C in coal is mainly due to lower oxygen content and higher carbon content when compared to biomass. Torrefaction of switchgrass and corn stover helped to lower the oxygen and hydrogen content of the biomass, and made it comparable to different forms of coal. It is clear from this diagram that lower H/C ratio and O/C ratios can be produced at higher torrefaction temperatures and residence times. Torrefying both switchgrass and corn stover at 230°C at different residence times resulted in H/C and O/C ratio closer to North Dakota lignite coal. Additionally, increasing the temperature to 270°C has moved the torrefied switchgrass and corn stover closer to other coal forms like Illinois Basin and Powder River Basin.

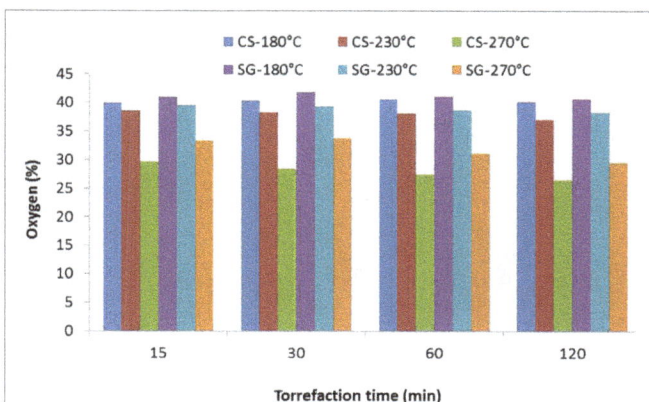

FIGURE 10 | Torrefaction temperature and time effect on oxygen content of the switchgrass and corn stover.

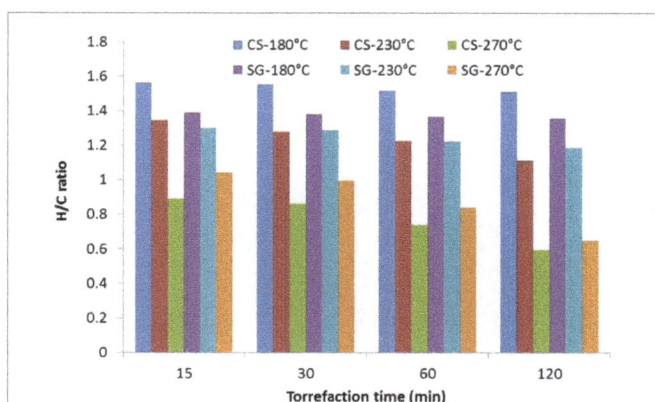

FIGURE 11 | Torrefaction temperature and time effect on H/C ratio of the switchgrass and corn stover.

Higher Heating Value

The calorific value was measured for both raw and torrefied corn stover and switchgrass samples. The higher heating values of raw corn stover and switchgrass observed were 17.31 and 17.36 MJ/kg. At a torrefaction temperature of 180°C and 15-min residence times, the observed higher heating values were 18.43 and 18.06 MJ/kg (**Figure 13**). Increasing the residence to 120 min increased the higher heating values marginally. At 230°C torrefaction temperature and 15-min residence time, the higher heating values observed were 18.77 for corn stover and 18.75 MJ/kg for switchgrass. At 120-min residence time, the higher heating value of corn stover was 19.74 and switchgrass was about 19.34 MJ/kg. At a torrefaction temperature of 270°C and 15-min residence time, the higher heating values observed 20.55 for corn stover and 21.0 MJ/kg for switchgrass. At the same temperature, increasing the residence time to 120 min increased the higher heating values to 21.51 for corn stover and 21.53 MJ/kg for switchgrass (**Figure 14**).

DISCUSSION

In the present study, both the switchgrass and corn stover lost moisture at different torrefaction temperatures and residence times tested. Tumuluru et al. (2011) and other researchers indicated that at temperatures of 150–200°C the loss of moisture is due to dehydration reactions. The dehydration reactions mainly result in loss of unbound moisture. According to Bridgeman et al. (2008) and other researchers, increasing the torrefaction temperature to 230 and 270°C, the loss of moisture can be attributed to drying and depolymerization of hemicellulose. The loss of volatiles at the temperature of 180°C is marginal, also at this temperature thermal devolatilization and depolymerization reactions do not initiate. At 230 and 270°C, the loss of volatiles is more significant, which might be due to depolymerization of hemicellulose, cellulose, and lignin (Tumuluru et al., 2012). Torrefaction temperatures (200–300°C) of biomass will affect the ash content in the biomass. The increase of ash content observed, in the present study, is more relative to a decrease in the biomass components when compared with the original biomass components. In case of ultimate composition, the carbon content of the biomass increased with increases in torrefaction temperature and residence time. In the present study, the increase in elemental carbon is higher at 270°C when compared with 230 and 180°C temperatures. According to Tumuluru et al. (2011) and other researchers, at a temperature of <250°C the decarbonizing reactions are more limited, but at a temperature of >250°C the biomass undergoes extensive devolatilization resulting with more volatiles loss and an increase in carbon content. The decrease in

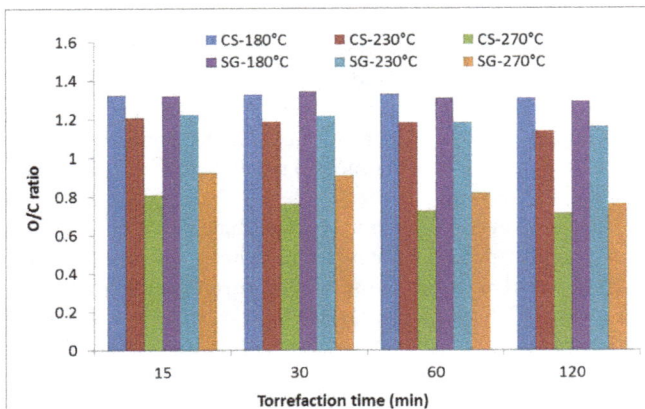

FIGURE 12 | Torrefaction temperature and time effect on O/C ratio of the switchgrass and corn stover.

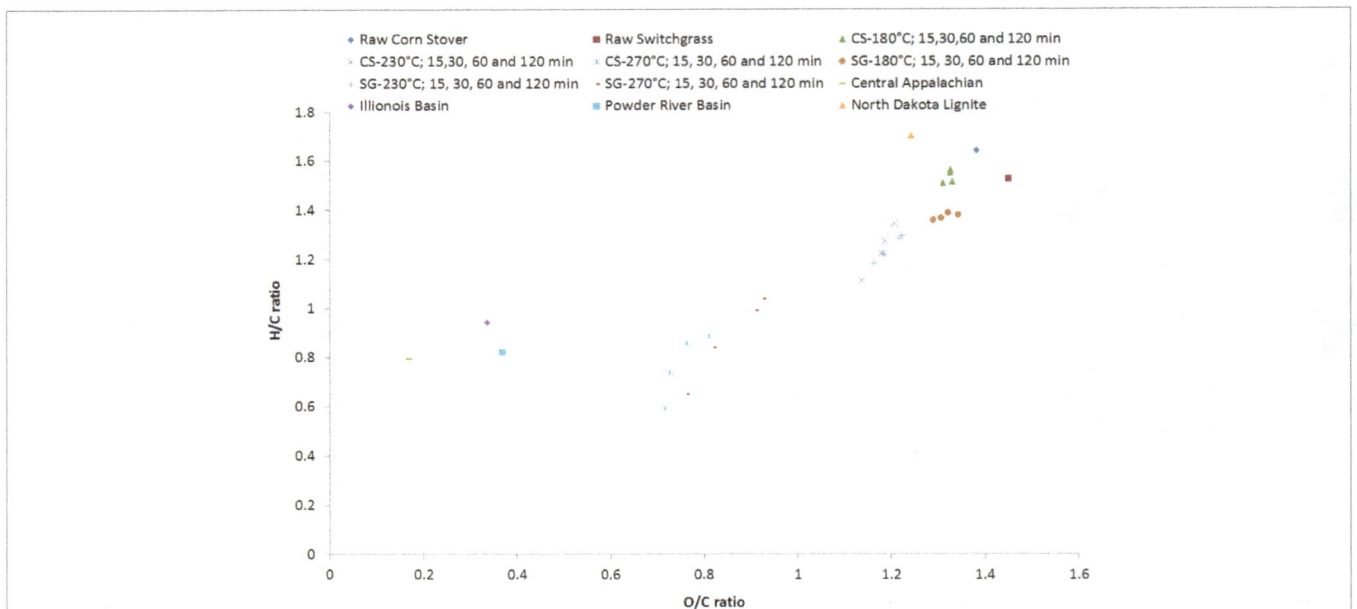

FIGURE 13 | van Krevelen diagram for corn stover, switchgrass, and coals.

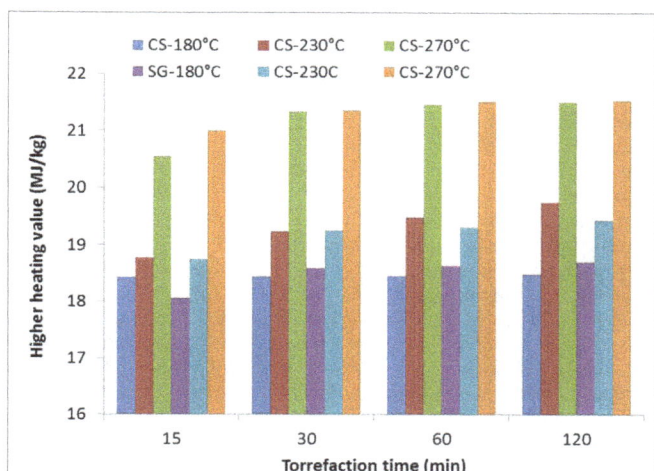

FIGURE 14 | Higher heating value of corn stover and switchgrass at different torrefaction temperature and time.

hydrogen and oxygen content at a lower temperature of 180°C was marginal; whereas, at higher torrefaction temperatures of 230°C and 270°C, the decrease in hydrogen and oxygen is more significant. The major reasons for the decrease in hydrogen and oxygen content at a higher torrefaction temperature are due to formation of water, carbon monoxide, and carbon dioxide. In the present study, it indicates that the nitrogen content increased slightly, while the sulfur content did not change much. The increase in nitrogen and sulfur is more a relative and is due to the decrease of oxygen content. H/C and O/C ratios that were calculated were marginal at the temperature of 180° compared to 230 and 270°C. This observation has corroborated with the findings of the other researchers (Poudel and Oh, 2012; Tumuluru et al., 2012b,c, and Tumuluru et al., 2011). Lower O/C ratio observed at higher torrefaction temperature can be due to the generation of volatiles rich in oxygen, such as CO, CO_2, and H_2O. Lower H/C ratios observed at higher torrefaction temperature can be due to the formation of hydrocarbons, such as CH_4 and C_2H_6. In general, fuels with less H/C and O/C ratios resulted in less smoke and water vapor formation and less energy loss during combustion and gasification processes. The van Krevelen diagram drawn for torrefied corn stover and switchgrass at a torrefaction temperature of 270°C at different torrefaction residence times, indicated that the both corn stover and switchgrass moved closer to higher quality coals like Power River Basin and Illinois Basis. The lower H/C and O/C ratios of these fossil fuels is mainly due low oxygen content and hydrogen content, which makes them suitable for power generation and gasification applications. Torrefaction of switchgrass and corn stove at 270°C and different residence times lowered H/C and O/C ratios to <1 making them more suitable for biopower generation. Also, based on this study it can be concluded that torrefaction residence times of >30 min may not be necessary, as most of the changes in the biomass proximate, ultimate and higher heating value occur at ≤30 min of the residence time. Additionally in the present study, the higher heating value

measured increased with higher torrefaction temperatures. Maximum heating values were observed at 270°C for both corn stover and switchgrass. The increase in higher heating values at higher torrefaction temperature can be due to the loss of lower energy content volatiles, resulting in net increases in the energy content of the torrefied biomass. Nevertheless, torrefaction of corn stover and switchgrass resulted in consistent proximate, ultimate, and energy properties making it more suitable for co-firing applications. The TGA data obtained will be further used to understand the mechanism that results in proximate and ultimate changes in biomass during torrefaction, kinetics of the biomass torrefaction process, and textural changes in torrefied biomass using a scanning electron microscope.

CONCLUSION

Torrefaction of corn stover and switchgrass has resulted in improved chemical and energy properties. Torrefaction temperature had more significant effect on the chemical and energy property compared to residence time. Weight loss observed during torrefaction at 180°C was about 10%, but at 270°C it increased to >45% for both corn stover and switchgrass. At 180°C and 120 min, about 78.8% in corn stover moisture and 88.18% in switchgrass moisture were lost. Loss of moisture at this temperature is mainly due to dehydration reactions. Increasing the temperatures to 230 and 270°C caused significant changes in proximate and ultimate composition. The relative ash content significantly increased to about 10.1 in the corn stover and 6.66% in the switchgrass at 270°C and 15 min. At this temperature, the volatiles decreased to about 50.27 in the corn stover and 52.03% in the switchgrass. In case of ultimate composition, the carbon content increased to about 54.92 in the corn stover and 53.94% in the switchgrass, and hydrogen content decreased to about 4.07 and 4.68%. Furthermore, the nitrogen and sulfur content observed in corn stover and switchgrass were 0.98 and 0.8%, and 0.076 and 0.07%. The oxygen content of the corn stover and switchgrass observed at 270°C and 120 min was about 26.41 and 29.5%. The H/C and O/C ratios also decreased with increasing torrefaction temperature. The van Krevelen diagram drawn for H/C and O/C ratios at 270°C and 15–30 min residence time is closer to some of the coals like Illinois Basis and Powder River Basin. Maximum higher heating values observed for corn stover and switchgrass were 21.51 and 21.53 MJ/kg at 270°C and 120 min. From the present study, it can be concluded that torrefaction resulted in improved and consistent chemical composition and energy properties of corn stover and switchgrass.

ACKNOWLEDGMENTS

This work was supported by the Department of Energy, Office of Energy Efficiency and Renewable Energy under the Department of Energy Idaho Operations Office Contract DE-AC07-05ID14517. Accordingly, the publisher, by accepting the article for publication, acknowledges that the U.S. government retains a non-exclusive, paid-up, irrevocable, worldwide license to publish or reproduce the published form of this manuscript, or allow others to do so, for U.S. government purposes.

REFERENCES

Arias, B., Pevida, C., Fermoso, J., Plaza, M. G., Rubiera, F., and Pis, J. J. (2008). Influence of torrefaction on the grindability and reactivity of woody biomass. *Fuel Process Technol.* 89, 169–175. doi:10.1016/j.fuproc.2007.09.002

ASTM International D3173. (2008). *Standard Test Methods for Moisture in the Analysis Sample of Coal and Coke", Last Modified 2015*. Available from: http://www.astm.org/Standards/D3173.htm

ASTM International D3174. (2002). *Standard Test Methods for Ash Analysis of Coal and Coke", Last Modified 2015*. Available from: https://edis.ifas.ufl.edu/pdffiles/AG/AG29600.pdf.

ASTM International D3175. (2007). *Standard Test Methods for Volatile Matter in the Analysis Sample of Coal and Coke" Last Modified 2015*. Available from: http://www.astm.org/Standards/D3175.htm

ASTM International D3177. (2002). *Standard Test Methods for Total Sulfur in the Analysis Sample of Coal and Coke" Last Modified 2015*. Available from: http://www.astm.org/Standards/D3177.htm

ASTM International D3178. (2002). *Standard Test Methods for Carbon and Hydrogen in the Analysis Sample of Coal and Coke" Last Modified 2015*. Available from: http://www.astm.org/Standards/D5373.htm

ASTM International D3179. (2002). *Standard Test Methods for Nitrogen in the Analysis Sample of Coal and Coke" Last Modified 2015*. Available from: http://www.astm.org/Standards/D3179.htm

ASTM International D5865. (2010). *Standard Test Methods for Gross Calorific Value of Coal and Coke" Last Modified 2015*. Available from: http://www.astm.org/Standards/D5865.htm

Bergman, P. C. A., and Kiel, J. H. A. (2005). "Torrefaction for biomass upgrading," in Proceedings of the 14th European Biomass Conference & Exhibition (Paris), 17–21. Available at: http://www.energy.ca.gov/2009_energypolicy/documents/2009-04-21_workshop/comments/Torrefaction_for_Biomass_Upgrading_TN-51257.PDF

Bridgeman, T. G., Jones, J. M., Shield, I., and Williams, P. T. (2008). Torrefaction of reed canary grass, wheat straw and willow to enhance solid fuel qualities and combustion properties. *Fuel.* 87, 844–856. doi:10.1016/j.fuel.2007.05.041

Chen, D., Zhou, J., Zhang, Q., Zhu, X., and Lu, Q. (2014). Upgrading of rice husk by torrefaction and its influence on the fuel properties. *Bioresour.* 9, 5893–5905.

Chen, W. H., and Kuo, P. C. (2010). A study on torrefaction of various biomass materials and its impact on lignocellulosic structure simulated by a thermogravimetry. *Energy* 35, 2580–2586. doi:10.1016/j.energy.2010.02.054

Lu, J. J., and Chen, W. H. (2014). Product yields and characteristics of corncob waste under various torrefaction atmospheres. *Energies* 7, 13–27. doi:10.3390/en7010013

Medic, D., Darr, M., Shah, A., Potter, B., and Zimmerman, J. (2011). Effects of torrefaction process parameters on biomass feedstock upgrading. *Fuel* 91, 147–154. doi:10.1016/j.fuel.2011.07.019

Newman, Y., Williams, M. J., Helsel, Z., and Vendramini, J. (2014). *Production of Biofuel Crops in Florida: Switchgrass*. Gainesville, FL: IFAS Extension University of Florida, SSAGR291.

Nhuchhen, D. R., Basu, P., and Acharya, B. (2014). A comprehensive review on biomass torrefaction. *Interantinla J. Renew. Energy ad Biofuels* 2014, 1–55. doi:10.5171/2014.506376

Phanphanich, M., and Mani, S. (2011). Impact of torrefaction on the grindability and fuel characteristics of forest biomass. *Bioresour. Technol.* 102, 1246–1253. doi:10.1016/j.biortech.2010.08.028

Poudel, J., and Oh, S. C. (2012). A kinetic analysis of wood degradation in supercritical alcohols. *Ind. Eng. Chem. Res.* 51, 4509–4514. doi:10.1021/ie200496b

Repellin, V., Govin, A., Rolland, M., and Guyonnet, R. (2010). Energy requirement for fine grinding of torrefied wood. *Biomass Bioenergy* 34, 923–930. doi:10.1016/j.biombioe.2010.01.039

Sarkar, M., Kumar, A., Tumuluru, J. S., Patil, K. N., and Bellmer, D. D. (2014). Gasification performance of switchgrass pretreated with torrefaction and densification. *Appl. Energy* 127, 194–201. doi:10.1016/j.apenergy.2014.04.027

Simmons, B. A., Loque, D., and Blanch, H. W. (2008). Next-generation biomass feedstocks for biofuel production. *Gen. Biol.* 9, 242. doi:10.1186/gb-2008-9-12-242

Tillman, D. A., Duong, D. N. B., Miller, B. G., and Bradley, L. C. (2009). "Combustion effects of biomass co-firing in coal-fired boiler," in Proceedings of Power-Gen International (Presentation) (Las Vegas, NV), December 8–10 (2009).

Tumuluru, J. S., Boardman, R. D., and Wright, C. T. (2012c). Response surface analysis of elemental composition and energy properties of corn stover during torrefaction. *J. Biobased Mater. bioenergy.* 6, 25–35. doi:10.1166/jbmb.2012.1187

Tumuluru, J. S., Boardman, R. D., Wright, C. T., and Hess, J. R. (2012b). Some chemical compositional changes in *Miscanthus* and white oak sawdust samples during torrefaction. *Energies* 5, 3928–3947. doi:10.3390/en5103928

Tumuluru, J. S., Hess, J. R., Boardman, R. D., Wright, C. T., and Westover, T. L. (2012). Formulation, pretreatment, and densification options to improve biomass specifications for co-firing high percentages with coal. *Ind. Biotechnol.* 8, 113–132. doi:10.1089/ind.2012.0004

Tumuluru, J. S., Shahab, S., Hess, J. R., Wright, C. T., and Boardman, R. D. (2011). A review on biomass torrefaction process and product properties for energy applications. *Ind. Biotechnol.* 7, 384–401. doi:10.1089/ind.2011.7.384

U.S. Department of Energy. (2011). *U.S. Billion-Ton Update: Biomass Supply for a Bioenergy and Bioproducts Industry*. R. D. Perlack and B. J. Stokes (Leads), ORNL/TM-2011/224. Oak Ridge, TN: Oak ridge National Laboratory. p. 227.

Uemura, Y., Omar, W. N., Tsutsui, T., and Yusup, S. B. (2011). Torrefaction of oil palm wastes. *Fuel* 90, 2585–2591. doi:10.1016/j.fuel.2011.03.021

Wannapeera, J., Fungtammasan, B., and Worasuwannarak, N. (2011). Effects of temperature and holding time during torrefaction on the pyrolysis behaviors of woody biomass. *J. Anal. Appl. Pyrolysis* 92, 99–105. doi:10.1016/j.jaap.2011.04.010

Yang, Z., Sarkar, M., Kumar, A., Tumuluru, J. S., and Huhnke, R. L. (2014). Effects of torrefaction and densification on switchgrass pyrolysis products. *Bioresour. Technol.* 174, 266–273. doi:10.1016/j.biortech.2014.10.032

15

When Push Comes to Shove: Compensating and Opportunistic Strategies in a Collective-Risk Household Energy Dilemma

Anya Skatova[1,2]*, Benjamin Bedwell[1] and Benjamin Kuper-Smith[1,3]

[1] Horizon Digital Economy Research, University of Nottingham, Nottingham, UK, [2] Warwick Business School, University of Warwick, Coventry, UK, [3] Institute of Neurology, University College London, London, UK

Edited by:
Tobias Brosch,
University of Geneva, Switzerland

Reviewed by:
John M. Polimeni,
Albany College of Pharmacy and
Health Sciences, USA
Ulf J. J. Hahnel,
University of Geneva, Switzerland

*Correspondence:
Anya Skatova
anya.skatova@gmail.com

Specialty section:
This article was submitted to Energy
Systems and Policy,
a section of the journal
Frontiers in Energy Research

To solve problems such as climate change, every little push counts. Community energy schemes are a popular policy targeted to reduce a country's carbon emissions but the effect they have on energy use depends on whether people can work together as a community. We often find ourselves caught in a dilemma: if others are not doing their bit, why should I? In our experiment, participants ($N = 118$) were matched in groups of 10 to play in a collective-risk game framed as a community energy purchase scheme. They made only one decision about energy use for their virtual household a day, while a full round of the game lasted 1 week in real time. All decisions were entered via personal phone or a home computer. If in the end of the week the group exceeded a pre-paid threshold of energy use all group members would share a fine. Each day participants received feedback about decisions of their group partners, and in some groups the feedback was manipulated as high (unfair condition) or low (fair condition) use. High average group use created individual risk for participants to be penalized in the end of the week, even if they did not use much themselves. We found that under the risk of having to pay a fine, participants stayed significantly below the fair-share threshold regardless of unfair behavior of others. On the contrary, they significantly decreased their consumption toward the end of the game. Seeing that others are doing their bit – using a fair-share – encouraged people to take advantage of the situation: those who played against fair confederates did not follow the normative behavior but conversely, increased their consumption over the course of the game. These opportunistic strategies were demonstrated by impulsive participants who were also low in punishment sensitivity. We discuss the findings in the light of policy research as well as literature on cooperation and prosocial behavior.

Keywords: cooperation, collective-risk social dilemma, public good, community energy, environmental behaviour, impulsivity, punishment sensitivity, collective purchasing

Citation:
Skatova A, Bedwell B and
Kuper-Smith B (2016) When Push
Comes to Shove: Compensating and
Opportunistic Strategies in a
Collective-Risk Household Energy
Dilemma.
Front. Energy Res. 4:8.

INTRODUCTION

Many environmental choices represent social dilemmas (Irwin and Berigan, 2013), whether they are large-scale decisions about climate change mitigation (Milinski et al., 2008) or everyday choices such as recycling (Lyas et al., 2004) and responsible energy use at home (Leygue et al., 2014). Social dilemmas are scenarios when the communal resources have to be maintained and individuals face a dilemma between either using more than others, while sharing the costs of usage equally – thus, free-riding; or using a fair-share that allows maintaining the consumption of a resource but often with smaller immediate personal benefits. A type of social dilemma – a collective-risk game – is relevant to various social scenarios with repeated interactions and previously has been studied in the context of climate change mitigation (e.g., Milinski et al., 2008). Understanding how people act in dilemmas, such as climate change mitigation, is of high importance. However, realization of policy makers' decisions relies as much on small everyday choices of regular people as on large-scale choices about climate change mitigation by the leaders of policy making. Currently, there is not enough understanding of what people will do given various policy scenarios. We introduce a novel "in-the-wild" design of a social dilemma experiment. It takes an experimental laboratory game to everyday environments while still keeping the structure of the experimental social dilemma game. Through revealing people's behavioral strategies in the situations that resemble real-world scenarios while keeping experimental control, this approach can serve as an alternative or a precursor to expensive field studies helping to understand barriers and enablers of behavior change in the domain of energy use as well as other domains of behavior.

Community Energy Purchase Schemes

Cooperation around environmental resources is vital. For example, to achieve an 80% reduction of carbon emissions target by 2050, UK energy end-users – households, businesses, and third sector – are expected to use energy more efficiently, which among other measures includes better management of supply and demand. The benefits of encouraging communities to engage in managing their energy consumption is outlined in UK government's first Community Energy Strategy (DECC, 2014), which presents a range of initiatives that are to be supported going forward. One of these initiatives is *collective purchasing*, which "can make things cheaper, as buying in larger volumes usually means better deals and lower prices" (www.gov.uk/government/policies). Although examples are few and far between, the personal financial benefits to those who participate in collective purchasing have been demonstrated (Conaty and Mayo, 2012): for example, in the UK, an average saving of £131 was realized by households on the Cheaper Energy Together scheme [DECC, 2013; for similar evidence from Belgium, see Erbmann et al. (2009)]. Importantly, collective purchasing initiatives could also help to achieve carbon emissions targets by increasing engagement of community members in energy issues, and by reducing a variety of related emissions (e.g., the reduction in emissions related to the delivery of fuel, DECC, 2014, p. 6).

However, community energy purchase schemes can introduce interdependence of individual decisions, and so participating households might face a scenario alike a standard social dilemma. This is not accounted for in the policy documents, which focus on the positive outputs of a community purchasing initiative. Researchers have shown that near-future changes in energy infrastructure, e.g., forthcoming smart meter rollouts in the UK, will make it easier to identify which consumers might benefit

by forming collectives (Vinyals et al., 2012) and what the tools might look like that help collectives deal with energy retailers (Ramchurn et al., 2013). However, in reality communities can be transient and marginalizing (Harvey and Braun, 1999), particularly to those not predicated toward collective action (Hoffman and High-Pippert, 2010), and reactions to energy initiatives by different communities will not necessarily remain positive (Walker et al., 2010).

If one of the households in the community, despite an agreement, uses unreasonably high amounts of energy and if there is no opportunities to punish the free-rider (Fehr and Gachter, 2002), will the rest of the community compensate for them by using less? Or will they retaliate and use more themselves, causing a rebound effect (Greening et al., 2000), thus eliminating the benefits of the deal secured by the community? Furthermore, if some households use a fair-share amount, will others follow an establishing norm of cooperation in the group? The present study used a collective-risk game, a type of experimental social dilemma, to model a communal energy purchase scenario "in the wild." We investigated participants' responses to free-riding or fair-share use in their group as they were going about their everyday lives, as well as what consequences the dynamic interactions over communal resources had on cooperation around energy use.

Social Dilemmas to Explain Environmental Decisions

While standard social dilemmas are conceptually applicable to many real-world scenarios, the predominant body of research in the area uses stripped down storylines where participants make decisions about money units (MUs, e.g., Fehr and Gachter, 2002). Building on previous research (Milinski et al., 2008; Jacquet et al., 2013; Leygue et al., 2014), we transformed a laboratory game to investigate whether previous experimental findings apply to more realistic, ecologically valid real-world choices. Such an approach can build a basis for establishing the constraints of current policy strategies on behaviors in schemes such as community energy purchase deals. We improved the design of the laboratory game by introducing a novel "in the wild" aspect, which aimed to enhance the ecological validity of the experiment where the decisions were made on a timeline that is closer to real-world scenarios, as well as in the familiar environment of participants' everyday life.

A collective-risk game is a scenario where a group of players interact over a course of several rounds. They are given individual endowments and have to accumulate (or save, depending on the framing) a certain amount of money in the public pot over the course of the game. If by the end of the game they do not collect enough money (or if they overspend), they are all fined equally. We applied a collective-risk game approach to study household energy decisions by simulating everyday choices in a controlled experimental design. Will a group of households participating in a community energy purchase scheme with a pre-defined limit of energy allowance manage to stay below the threshold, given the benefits of individual use? To study this question, we modified a previously reported design (Milinski et al., 2008) to fit a community energy-buying storyline. In our experiment, if the group went over the threshold, the fine occurred with 100% probability. The fine was distributed among group members regardless of their usage from the communal resource, which is similar to previous research using experimental public goods games (e.g., Croson, 2007). Such a scenario simulates more realistically the case of community energy purchase as, unlike climate change that can happen with a particular probability, energy use in real life can be measured objectively, so its over-usage would be fined with 100% certainty.

In previously reported collective-risk games, participants interacted over a number of turns with the aim to reach a collective target of contributions (Milinski et al., 2008; Santos and Pacheco, 2011; Jacquet et al., 2013). Such give-some scenarios model social decision-making in situations such as climate change mitigation (e.g., everybody needs to contribute enough to prevent a catastrophe of climate change). However, there are many real-world scenarios in for example, community buying schemes or household energy use, which could be better represented by take-some games (Leygue et al., 2014). In these scenarios, a community has to maintain the use of a communal resource under a certain threshold to avoid negative consequences, such as exhausting the resource (Van Dijk et al., 2003) or paying a fine for over-usage.

Importantly, similar to many real-life situations, in the collective-risk game participants receive frequent feedback about the behavior of others in the group throughout the game; however, the outcome for the whole game was only evident in the end. This introduced a dilemma to each individual group member. If the target was not met, the whole group suffered: every individual had to contribute equally to the fine. However, by using more individually, participants received greater private benefit. This could be especially tempting in the short-term given the structure of the game: participants were rewarded through individual usage on each turn, but rewards for cooperation or punishment for not meeting the target were distant and were revealed only in the end of the game. Such a set up gives an opportunity to study how participants react to the behavior of others and adjust their game strategy if necessary in order to reach the target. While achieving the collective target implies some individual sacrifice, it brings benefits to everybody in the group. However, there is always uncertainty for the individual about whether others in the group choose to cooperate or to free ride. Furthermore, collective-risk games allow the study of strategies that are dependent on the behavior of others. For example, one can compensate for free-riding of others (Milinski et al., 2008). Alternatively, one can also be opportunistic and expect others to compensate. Thus, the key feature of this experimental design is to observe how the behavior of others can affect people's choices in the game.

The behavior of others often serves as a cue eliciting certain norms of interaction, which people then can choose to follow (Biel and Thøgersen, 2007), and this is applicable to social dilemmas (Weber and Murnighan, 2008). However, do people always follow the example of the majority? Research on social norms, including energy use domain, suggests that the majority comply with normative usage after seeing the information about others' behavior (Schultz et al., 2007). Feedback about behavior of others is referred to as a descriptive norm of behavior, which in addition to injunctive norms (rules or standards of behavior), is suggested

to affect people's choices. However, the feedback about behavior of others does not always affect decisions in a positive way, especially in the household energy use domain (Leygue et al., 2014). Field studies on household energy use also report "rebound" effects: if people find out that others use more than them, in some circumstances they can increase but not decrease their use (Schultz et al., 2007). One potential explanation for this rebound effect relates to scenarios perceived as social dilemmas where high usage by others could be perceived as unjust. In this case, instead of following the majority and using a fair-share, individuals could increase their usage in retaliation toward free-riders.

Strategies to Deal with Unfair Behavior of Others

Fairness is an important principle of human interactions. It is pervasive throughout human society: we often expect others to behave in a way that is fair to us and others (Binmore, 2014). Strong reciprocity theory suggests that violation of the fairness principle evokes strong negative emotional and behavioral reactions such as altruistic punishment of free-riders (Fehr and Gachter, 2002). Ultimatum game (UG) experiments are specifically designed to study people's reactions to fair or unfair behavior of others. One out of two players is required to divide a pot of money in two parts and the other needs to approve the outcome for both of them to receive the allocation. Around half of participants in UGs refuse the offer, which they consider unfair, even though in this case both parties get nothing (Nowak et al., 2000), which is a way to retaliate in response to free-riding. Leygue et al. (2014) found that when faced with a hypothetical scenario in which one house member overused energy and everybody has to share the bill for their overuse, participants report heightened anger. But would they retaliate and increase their energy use, as strong reciprocity theory suggests? Many social dilemmas in the real world differ from one-shot UGs as we often interact with the same individuals over a number of occasions. Retaliation in such circumstances can have negative effects on the outcome of interactions, especially if there is no opportunity to directly punish free-riders: often retaliation causes complete elimination of cooperation (Fehr and Fischbacher, 2003). This is a highly undesirable outcome for various real-life situations, including community energy purchasing scenarios. Luckily, there are other strategies to deal with free-riding that are also at play in social dilemmas (Axelrod and Hamilton, 1981; Fudenberg et al., 2012).

The literature reports a variety of "nice" strategies in social dilemmas, which under certain circumstances lead to better payoffs for the individuals employing them. In the repeated prisoner's dilemma between one- and two-thirds of participants, depending on conditions, demonstrate lenient strategies by not retaliating to defection straightaway and forgiving strategies by attempting to restore cooperation after inflicting punishment on the free-rider (Fudenberg et al., 2012). Furthermore, in a repeated social dilemma experiment, participants who consistently contribute a high amount to the communal account influence others through establishing a norm of cooperation in the group at no cost to themselves and often with some gain, which subsequently leads to increase in cooperation levels in those groups (Weber and Murnighan, 2008).

Previous research on collective-risk dilemmas has not looked into strategies in response to fair or unfair behavior of others, as well as whether normative behavior presented as feedback about decisions of others influences individual choices. However, Milinski et al. (2008) demonstrated that if the punishment was highly probable, more participants showed altruistic or compensating strategies, while if the punishment was expected with a low probability, a higher proportion of participants were opportunistic or free rode. This is relevant because similar to high versus low probability of punishment, unfair versus fair behavior of others throughout the game in the collective-risk dilemma, respectively, could be perceived by an individual as a higher versus lower chance of having to pay a fine in the end.

We predicted that in the community energy purchase game when others are using a fair-share [similar to Milinski et al. (2008) uncertain punishment condition], participants would realize that the fine is not likely to occur, so they could increase their usage, thus demonstrating opportunistic strategies. An alternative reaction to fair-share behavior of others would be adherence to the social norm of behavior (Biel and Thøgersen, 2007; Schultz et al., 2007) and usage of a fair-share amount.

When others are unfair, in the absence of opportunity to directly punish free-riders, two reactions are possible. First, in accord with retaliation literature, participants in the community energy purchase game could employ an emotionally driven retaliation strategy to punish free-riders or increase their usage. However, this behavior is highly undesirable from the rational point of view as it increases the risk of not meeting the target and, thus, might lead to punishment in the form of a fine for everybody. Thus, similar to Milinski et al. (2008) in certain punishment conditions, participants could use an alternative strategy and decrease their usage or compensate if others were unfair.

Individual Differences in Social Dilemma

While describing behavioral strategies in scenarios that resemble real-world situations – such as communal energy use – can help to explain and predict cooperative and non-cooperative outcomes for the group, it is equally important to understand the individual motivations behind people's decisions. Overall, research shows heterogeneity of behavioral strategies in various types of social dilemmas (e.g., Burlando and Guala, 2005; Zhao and Smillie, 2014); however, this heterogeneity has not been yet explored in the real-world social dilemmas, such as communal energy use. To identify potential mechanisms, we review literature on individual differences in behavior in lab-based social dilemmas.

The heterogeneity in decisions in social dilemmas has been linked to a number of psychological factors, such as personality dispositions, which reflect individual differences in processing rewards and punishments (Scheres and Sanfey, 2006; Skatova and Ferguson, 2011, 2013). Dispositional reward and punishment sensitivities are key to explain individual behavior in domains where reward and punishment processing have been strongly implicated, such as prosociality and cooperation (Gintis et al., 2003; Gneezy and Fessler, 2012; Van Lange et al., 2014; Zhao and Smillie, 2014). A psychological measure that has often been used

to assess individual differences in reward and punishment sensitivity includes behavioral approach system (BAS) and behavioral inhibition system (BIS) scales (Carver and White, 1994).

The BAS scale includes three subscales: two subscales measure reward reactivity aspects of reward sensitivity [BAS-reward responsiveness (BAS-RR) and BAS-drive (BAS-D)], and one measures the impulsivity aspect of reward sensitivity [BAS-Fun Seeking (BAS-FS)]. Impulsivity is associated with the tendency to engage in behaviors which are risky and often require disinhibition, while reward reactivity refers to propensity to be sensitive to opportunities for rewards and rewarding experiences (see Smillie et al., 2006, for discussion of the distinction between reward reactivity and impulsivity). These scales were previously used to explain behavior in the economic games (Scheres and Sanfey, 2006; Skatova and Ferguson, 2011, 2013) and, thus, should be applicable for explaining behavior in a collective-risk game scenario structured around communal energy use. Specifically, participants who self-reported high sensitivity to rewarding experiences (success, social interactions, etc.) in everyday life also demonstrated more strategic behavior in social dilemmas. Skatova and Ferguson (2011) showed that high BAS-RR was associated with lower contributions in a one-shot public goods game after revealing that others in the group contributed a high amount. Scheres and Sanfey (2006) found associations of BAS-RR and BAS-D with lower offers in the Dictator Game (which is similar to the UG except that the respondent does not have an opportunity to reject the offer) but not in the UG. Pothos et al. (2011) showed that participants high in BAS-RR were more likely to defect in the one-shot prisoner's dilemma game. In all cases, participants high in reward reactivity, made a decision to defect while having full control of the situation and no dependency on the decision of other people. Therefore, their decision to defect could be interpreted as strategic and reflect the ability to better learn from reward, which they were getting in this case by defecting.

Previous studies that looked into associations between BAS scales and behavior in one-shot economic games did not find relationships between BAS-FS and individual choices. The BAS-FS scale has strongest conceptual and empirical links with impulsivity and diminished delayed reward gratification (Smillie et al., 2006; Giovanelli et al., 2013). Individual differences in behavior might be associated with differences in reward discounting when each turn of the game introduces a conflict between short-term private benefit and long-term reward by cooperation. Jacquet et al. (2013) demonstrated that discounting mechanisms affected people's decisions in a social dilemma: a greater delay in achieving rewards by cooperation made it less likely for people to cooperate in the short term in a collective-risk dilemma. Specifically, when individuals received benefits from cooperation the day after they played the game, 7 out 10 groups succeeded in reaching a cooperation target. However, when the benefits from cooperation were delayed by 7 weeks, only 4 groups out of 11 succeeded. They also demonstrated variability in individual responses: some groups were able to reach cooperation even when the benefits were delayed by 7 weeks. Previous research using public goods games found a negative association between cooperation and impulsivity but only when the reward from free-riding was tangible (Myrseth et al., 2015). It is plausible that in a game with a longer time span, where it is easier to free ride at a given turn and get away with it, BAS-FS would be associated with more selfish behaviors. That should happen especially when the risk of loss is low, as for impulsive individuals it would be easier to disregard long-term benefits of cooperation. Instead, BAS-RR and BAS-D should be positively associated with strategic behavior, leading to high certain profits in any case.

Differences in decisions in social dilemmas were associated with low self-reported sensitivity to negative experiences in real life (e.g., social disapproval, failure, etc.) measured by the BIS scale: participants with low BIS made smaller contributions in a one-shot social dilemma while facing the risk of punishment (Skatova and Ferguson, 2013). Low BIS was also associated with higher proportion of contributing nothing in a one-shot social dilemma after finding out that others contributed a high amount to the public good (Skatova and Ferguson, 2011). Finally, research suggested that interaction between BIS and BAS traits, or broadly speaking reward and punishment sensitivity systems, is associated with various clinical and behavioral outcomes, including prosocial and antisocial behavior. Specifically, McCabe et al. (2001) demonstrated that cooperation occurs through a neural network, which provides binding joint attention to mutual gains with inhibition of immediate reward: those who cooperate inhibit the dominant response of getting a quick smaller reward in order to gain a larger delayed reward by the means of cooperation.

We predicted that in the situation when others were fair and used a small amount throughout, making the risk of group punishment for overuse low, those who were higher in BAS-FS should take advantage and use more energy to get more private immediate benefits. We predicted that if others were unfair (by using high amounts throughout the game) and the risk of a fine was high, more strategically driven participants (e.g., high in BAS-RR and BAS-D) should use less to avoid paying the fine. In terms of BIS, we predicted that those who were less sensitive to the risk of punishment (e.g., low in BIS) should use more energy when the punishment was uncertain, i.e., in the fair condition. Finally, we predicted that participants high in impulsivity (measured by BAS-FS) and low in punishment sensitivity (measured by BIS) would be more likely to demonstrate opportunistic strategies when the advantage of immediate benefits were high (e.g., in the fair feedback condition).

The Present Study

Our study extended previous research to reveal whether the fair (using the pre-agreed amount of energy) or unfair (using more energy than was pre-agreed) behavior of others influenced individual decisions over the course of a collective-risk social dilemma. Specifically, we employed an experimental game to model household energy use in the context of a community energy deal, where individuals were part of a group of households that collectively pre-paid for a certain amount of energy to use per week. If the group overused energy, a fine was distributed between all group members equally.

We manipulated feedback about the behavior of others as fair or unfair, and investigated how such feedback influenced participants' individual decisions about energy usage in their own households, resulting in a variety of strategies: fair-share (to use as much as established by social convention), opportunistic (use more to gain private benefit even at a risk of a group-level fine), retaliatory (increase the usage after facing unfair behavior of others), or compensatory (decrease the usage in order to compensate for high use of others and so avoid the fine). We further looked at whether the different strategies were associated with individual differences in punishment (measured by BIS) and two distinct aspects of reward sensitivity: reward responsiveness (measured by BAS-RR and BAS-D) and impulsivity (measured by BAS-FS).

Our participants made decisions through a smart-phone or a home PC while going about their everyday lives as opposed to interacting with other group members in laboratory settings. In addition, unlike in laboratory settings, where participants usually make decisions every minute, our participants replied just once a day in the morning, wherever they were at the moment, and by using their mobile phone or computer at home. This is an important feature of the study as it aimed to reveal potential conflict between short-term and long-term benefits: participants were rewarded for their energy use every day, while the bill revealing potential excess would arrive only in the end of the (actual) week. Such features of the game provided a more accurate simulation of real-world decisions. The data presented in this paper are a subset of data collected within the project. Here, we focus on details of the design that are relevant to the aims presented in this paper.

MATERIALS AND METHODS

Participants and Procedure

The study was conducted through Qualtrics software. Overall, 118 UK-based participants volunteered to participate (74 females; age: range 25–66, $M = 35$; 46 were homeowners). Out of 118, 78 participants partook in fair and unfair condition, $N = 39$ for each condition. For all analyses below, we used only data from these 78 participants (see Design for further explanation why only fair and unfair conditions were focus of analyses in this paper).

We aimed to recruit participants who were responsible for paying their own bills as for them the decisions in the game would have greater resemblance to real life. For that reason, we explicitly sought to recruit a non-undergraduate sample of participants. In the UK, students often have their energy bills included as a part of a rental contract. In these circumstances, there is no monetary incentive to use energy responsibly (as they pay the same amount in any case). Participants were recruited in two cities in the Midlands, UK via various university-wide mailing lists and a list of members of an energy trial conducted by a national energy company.

All participants were briefed and debriefed in person. At the briefing, they received full instructions and could try out the game. They also filled in demographic information and a BIS/BAS questionnaire. We used the BIS/BAS questionnaire (Carver and White, 1994) to measure differences in reward and punishment sensitivities. Participants rated various statements on a 4-point

TABLE 1 | Person zero-order correlations for BIS and BAS subscales.

	BIS	BAS-D	BAS-FS
BIS	–		
BAS-D	−0.02	–	
BAS-FS	−0.13	0.44***	–
BAS-RR	0.43***	0.37**	0.17

*** $p < 0.001$, ** $p < 0.01$.

scale ranging from "very true for me" to "very false for me." BIS/BAS questionnaire was scored as four scales: BIS scale ($M = 3.02$, $SD = 0.52$, $\alpha = 0.77$, seven items, example item: "Criticism or scolding hurts me quite a bit") and three BAS scales: BAS-D ($M = 2.67$, $SD = 0.60$, $\alpha = 0.74$, four items, example item: "I go out of my way to get things I want"), BAS-FS ($M = 2.81$, $SD = 0.58$, $\alpha = 0.72$, four items, example item: "I often act on the spur of the moment"), and BAS-RR ($M = 3.38$, $SD = 0.47$, $\alpha = 0.73$, five items, example item: "When I'm doing well at something I love to keep at it"). **Table 1** reports zero-order correlations between BIS and BAS subscales. For the presentation of results, BIS/BAS scores were reversed; so high rating represents high ends of the BAS and the BIS scales. There were no differences on any of BIS/BAS scales between conditions. All scales were z-scored for all analyses.

All participants who completed the study were compensated £40 (~$61) for their time. In addition, they were incentivized by being paid contingent on their choices in the experiment (see The Game). In the end, participants were paid additional £3 (~$5) on average based on their responses. The study was approved by the School of Computer Science Ethics Committee at The University of Nottingham.

Design

Participants were divided into three conditions: fair, unfair, and real. Each condition included 4 groups of 10. As only 118 participants were recruited, two participants were lacking to form 12 full groups of 10. However, for fair and unfair conditions, it did not matter if there was not a full group of 10 as the feedback about group behavior was pre-set. Therefore, we assigned 39 participants for each of manipulated conditions. Thus, we manipulated the feedback about the behavior of others in 8 out of 12 groups in a between-subjects design as fair versus unfair usage. During the game, participants in the "fair" condition received feedback, which indicated that others in their group consumed energy within the pre-agreed norm, i.e., the group's deal allocation. The feedback was generated to represent a plausible distribution with a mean of 3.7 energy units (EUs) and SD of 0.48 EUs. The mean and SD was estimated based on the pilot study data. In the "unfair" condition, feedback indicated that their group partners consumed more than was pre-agreed (simulated in the similar way to fair condition, $M = 4.4$ EUs, $SD = 0.47$ EUs). The exact feedback on each day for each condition can be found in the Supplementary Materials. In the "real" condition, participants received accurate feedback about the consumption of others in their group ($M = 3.87$ EUs, $SD = 0.90$ EUs). To avoid deception, it was explained to participants prior to the study that some groups would receive manipulated feedback but neither experimenter nor

they would know which group they were assigned to (Bardsley, 2000). As the purpose of this paper was to investigate the effect of fair and unfair behavior of others on individual strategies in a collective-risk dilemma, here we only report the results for fair and unfair conditions.

The Game
The Village Energy Deal

Participants had to imagine that they and nine other households in their virtual village were participating in a deal to purchase energy communally. The deal lasted for a week and provided a pre-paid energy amount for the village (the group of 10 households), i.e., 280 EUs shared across the whole group. Each participant received a 62-MU endowment, of which they were deducted 28 MUs for inclusion in the pre-paid deal, leaving a remainder of 34 MUs in their private account. This remainder could be spent on excess energy use (in response to hypothetical situations encountered in the game, as described later), or saved to be converted into pounds in the end of the experiment at a rate of 1 MU = £0.02. Excess energy use, i.e., any energy used over the village's 280 EU allowance, was twice as expensive as energy paid for through the village's deal, costing 2 MUs per 1 EU. The cost of any excess energy that was used had to be paid for communally, divided equally between all group members.

Using Energy

The only way participants used energy during the game was by setting heating in their individual virtual households. The heating was set in heat points (HPs) that reflected a subjective energy scale from very cool (1 HP) to very warm (6 HPs). HPs were introduced (as opposed to degrees Celsius or Fahrenheit) as people have different subjective perceptions of what warm or cold feels like. For example, for somebody 18°C at home might seem as quite "warm," while for somebody else it might seem as "cold" (Li, 2005). The use of 1 HP resulted in expenditure of 1 EU.

Participants received private incentives to heat their homes: 1 HP used added 0.25 MUs to their private account. That meant the more energy they used, the more monetary benefits they would receive after the end of the game. Participants were told that all households in the group were similar in the level of energy efficiency and how much energy they used regularly.

As a result, if all participants kept their use to the norm, as suggested by the rules of the village's energy deal (i.e., up to 4 HPs per day, for 7 days), the group would not consume excess energy and not have to pay extra at the end of the week. If the group overused, all participants had to share a fine (i.e., pay for the excess), regardless of their individual use. Therefore, the scenario represented a social dilemma, where private interest (to use as much as possible in order to gain a monetary incentive) clashed with public concern (to keep the use down in order to avoid a collective fine).

Playing the Game

Participants were instructed that one round of the game lasted a week, with seven turns. One turn took place each day of the week. In the morning of each day of the study, participants received a text message or an email with a link that they had to follow to engage in the game. The link provided the following information:

- A recap of the previous day, including the average energy consumption of other members of the village; a reminder of how much they used themselves; how many MUs they received as a benefit from previous day's use;
- A summary of the week so far, including how much the village had consumed, how much was left in the community deal allocation for the week, and how many days were left in the week.

Following this feedback, participants had to make one decision about temperature in their virtual house for this day. This consisted of choosing a temperature setting from a scale ranging from very cool (1 HP) to very warm (6 HP).

Participants were also provided with a background story to make their hypothetical day-to-day decisions feel more real. Prior to the study, we asked participants to name three close real-life friends and/or family members who might come to visit them at Christmas. We used these names to individualize the reminders sent to participants during the game, telling participants that it was Christmas time and that those friends and/or family members had come to stay with them. Their additional goal in the game was then to make their guests happy by keeping the house warm, while also attempting to save the money in their private account by avoiding the costs incurred by the group exceeding their deal's communal allowance.

On the eighth day of the game, participants received information about energy consumption for the preceding week, learnt whether the group had exceeded its deal's allocation and, thus, whether they needed to contribute a payment towards the fine, if there was a fine, and how much they had to pay. In addition to the decisions about energy use in their virtual house, we also measured a number of psychological variables before and after participants set the temperature every day. As these variables were not the focus of this paper, we are omitting them from any further analyses or discussion. After the first week of the study, participants participated in an extension of the game with the same group partners. Only results from the first week are reported in this paper. A complete design of the project is available from the first author. At the end of the study, participants were rewarded based on the MUs remaining in their private accounts plus the participation fee.

RESULTS

Response Rate

Participants responded on 6.4 days on average and there was no difference in response rate between conditions: $t(68.77) = -0.62$, $p = 0.54$. The rules of the game stated that if participants missed a response on a particular day, the temperature they set for the previous day would be carried over. We excluded from analyses participants' data when they missed more than one response during the week. Eighty-five percent ($N = 66$) of participants responded on at least 6 days of the study. Only these responses were used for all further analyses. Whenever we presented aggregated responses from a specific day of the study, we also excluded participants who did not provide responses on that specific day from all relevant analyses. Response rate for each day of the study for those who responded on at least 6 days for the study was the following: 98% ($N = 65$) participants provided responses responded on day 1, 97% ($N = 64$) on day 2, 98% ($N = 65$) on day 3, 100% ($N = 66$) on day 4, day 5, and day 6, and 74% ($N = 49$) provided responses on day 7. The lower response rate on day 7 can be explained by failure of experimental software on that day. On that specific day, the reminder that went out to participants contained a link with incorrect feedback information. Most participants responded to this incorrect reminder, but we had to disregard those responses. Later in the same day, participants received a correct link with a request to respond again, however, not everybody responded to this second reminder. The response rate to the second reminder on day 7 was similar across conditions: 24 participants responded in the fair and 25 in the unfair condition. We further checked that all our results remained the same if we ran analyses on a restricted sample of those who responded on day 7 (results are presented in **Tables 3** and **4**).

Average Use Compared to the Fair-Share

Mean use across the week was 3.5 HPs (SD = 0.88) (see **Table 2** for means and SDs of energy use per day, overall and per condition and **Figure 1** for graphical representation). Individual use fell in the range of all possible options: participants used from 1 to 6 HPs, with 51% of all responses falling on the choice of 4 HPs, 31%, 3 HPs; 9%, 2 HPs; 3%, 1 HPs; 4%, 5 HPs; and 2%, 6 HPs. As 4 HPs seemed to be a common option for many (which is not surprising as it was suggested as a normative expenditure in the

TABLE 2 | Means and SDs of energy use for each day of the week for the whole sample, fair and unfair condition.

		Day 1	Day 2	Day 3	Day 4	Day 5	Day 6	Day 7
Overall	Mean	3.51	3.41	3.43	3.39	3.61	3.59	3.55
	SD	0.79	0.77	0.85	0.89	0.94	0.91	1.02
	N	65	64	65	66	66	66	49
Fair condition	Mean	3.42	3.37	3.48	3.48	3.81	3.73	4.12
	SD	0.79	0.83	0.91	1.03	0.92	0.94	0.99
	N	33	32	33	33	33	33	24
Unfair condition	Mean	3.59	3.44	3.37	3.30	3.39	3.45	3
	SD	0.80	0.72	0.79	0.73	0.93	0.87	0.71
	N	32	32	32	33	33	33	25

N represents number of participants responded on each day.

TABLE 3 | Model 1: mixed-effects random intercept regression model predicting the usage on day$_i$.

	Full sample ($N = 65$)		Restricted sample ($N = 48$)	
	Fixed effects			
	B (SE)	95% CIs	B (SE)	95% CIs
Intercept	3.66*** (0.09)	3.49; 3.85	3.70*** (0.12)	3.46; 3.94
Usage on Day 1	0.38*** (0.05)	0.27; 0.49	0.40*** (0.06)	0.27; 0.52
Day number	0.03 (0.02)	−0.003; 0.06	0.04 (0.02)	−0.001; 0.07
Condition (0 – fair, 1 – unfair)	−0.32** (0.11)	−0.54; −0.10	−0.39** (0.14)	−0.67; −0.12
Day number × Condition	−0.15*** (0.03)	−0.21; −0.09	−0.18*** (0.04)	−0.25; −0.11
	Random effects			
Intercept σ (participant)	0.38		0.42	
Observations	435		333	

Terms of all interactions were centered to reduce multicollinearity. The table reports unstandardized estimates with SEs in parenthesis and 95% confidence intervals for full (N = 65) and restricted (N = 48) sample.
***p < 0.001, **p < 0.01.

TABLE 4 | Model 2: mixed-effects random intercept regression model predicting the usage on day$_i$.

	Full sample ($N = 65$)		Restricted sample ($N = 48$)		Simulations ($n = 10{,}000$)
	Fixed effects				
	B (SE)	95% CIs	B (SE)	95% CIs	$1 − \beta^a$
Intercept	3.75*** (0.13)	3.53; 3.98	3.82*** (0.16)	3.55; 4.07	1
Use on Day 1	0.42*** (0.07)	0.29; 0.54	0.45*** (0.09)	0.31; 0.60	1
Day number	0.01 (0.02)	−0.02; 0.05	0.01 (0.02)	−0.03; 0.05	0.12
Condition (0 – fair, 1 – unfair)	−0.33* (0.15)	−0.58; −0.07	−0.37* (0.18)	−0.67; −0.08	0.87
Day number × Condition	−0.14*** (0.04)	−0.21; −0.07	−0.14*** (0.04)	−0.21; −0.06	0.98
BAS-RR	0.03 (0.09)	−0.12; 0.18	0.03 (0.11)	−0.15; 0.20	0.07
BAS-RR × Day number	0.02 (0.02)	−0.02; 0.06	0.03 (0.02)	−0.01; 0.07	0.17
BAS-RR × Condition	−0.003 (0.17)	−0.30; 0.29	−0.04 (0.22)	−0.39; 0.30	0.05
BAS-RR × Day number × Condition	−0.01 (0.04)	−0.10; 0.07	−0.01 (0.05)	−0.10; 0.08	0.06
BAS-D	−0.05 (0.09)	−0.21; 0.10	−0.05 (0.12)	−0.24; 0.14	0.12
BAS-D × Day number	−0.04 (0.02)	−0.08; −0.00001	−0.03 (0.20)	−0.08; 0.01	0.48
BAS-D × Condition	0.05 (0.18)	−0.25; 0.35	0.07 (0.22)	−0.29; 0.42	0.07
BAS-D × Day number × Condition	0.07 (0.04)	−0.02; 0.15	0.03 (0.05)	−0.06; 0.12	0.32
BAS-FS	0.05 (0.09)	−0.11; 0.21	0.15 (0.14)	−0.08; 0.37	0.12
BAS-FS × Day number	0.08** (0.02)	0.03; 0.13	0.09** (0.03)	0.03; 0.15	0.95
BAS-FS × Condition	−0.32 (0.18)	−0.64; −0.002	−0.46 (0.28)	−0.92; −0.003	0.66
BAS-FS × Day number × Condition	−0.12* (0.05)	−0.22; −0.03	−0.14* (0.06)	−0.26; −0.03	0.81
BIS	−0.07 (0.08)	−0.20; 0.06	−0.10 (0.10)	−0.26; 0.06	0.19
BIS × Day number	−0.08*** (0.02)	−0.11; −0.04	−0.08*** (0.02)	−0.12; −0.04	0.95
BIS × Condition	0.10 (0.16)	−0.17; 0.38	0.14 (0.20)	−0.20; 0.47	0.14
BIS × Day number × Condition	0.13*** (0.04)	0.06; 0.20	0.13** (0.04)	0.05; 0.21	0.88
BIS × BAS-RR	−0.01 (0.09)	−0.17; 0.15	−0.03 (0.12)	−0.22; 0.17	0.06
BIS × BAS-RR × Day number	0.05 (0.02)	0.001; 0.09	0.04 (0.03)	−0.01; 0.09	0.45
BIS × BAS-RR × Condition	−0.01 (0.19)	−0.33; 0.31	0.10 (0.24)	−0.29; 0.48	0.05
BIS × BAS-RR × Day number × Condition	−0.03 (0.05)	−0.13; 0.06	−0.04 (0.05)	−0.15; 0.05	0.10
BIS × BAS-D	0.02 (0.09)	−0.13; 0.18	0.02 (0.12)	−0.16; 0.21	0.06
BIS × BAS-D × Day number	0.01 (0.02)	−0.03; 0.05	0.001 (0.03)	−0.05; 0.05	0.07
BIS × BAS-D × Condition	−0.11 (0.18)	−0.43; 0.20	−0.24 (0.24)	−0.63; 0.15	0.10
BIS × BAS-D × Day number × Condition	−0.10* (0.05)	−0.18; −0.01	−0.10 (0.05)	−0.20; 0.00003	0.45
BIS × BAS-FS	−0.12 (0.10)	−0.29; 0.05	−0.21 (0.15)	−0.45; 0.04	0.34
BIS × BAS-FS × Day number	−0.04 (0.03)	−0.09; 0.01	−0.04 (0.03)	−0.10; 0.02	0.38
BIS × BAS-FS × Condition	0.20 (0.20)	−0.14; 0.55	0.48 (0.31)	−0.02; 0.98	0.25
BIS × BAS-FS × Day number × Condition	0.15** (0.05)	0.06; 0.25	0.21** (0.07)	0.08; 0.34	0.82
	Random effects				
Intercept σ (participant)	0.42		0.48		−
Observations	435		333		−

Terms of all interactions were centered to reduce multicollinearity.
The table reports unstandardized estimates with SEs in parenthesis and 95% confidence intervals for full (N = 65) and restricted (N = 48) sample. Simulation results represent power calculation for each fixed effect at 0.05-level using 100 random samples and 100 simulations per each sample.
***p < 0.001, **p < 0.01, ${}^a\alpha = 0.05$.

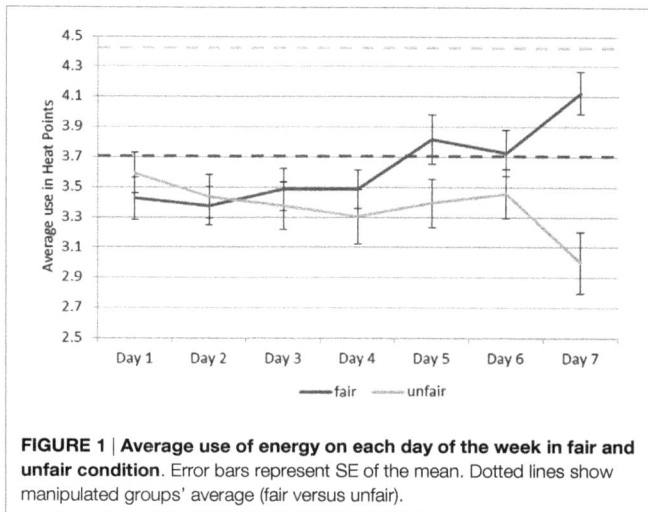

FIGURE 1 | Average use of energy on each day of the week in fair and unfair condition. Error bars represent SE of the mean. Dotted lines show manipulated groups' average (fair versus unfair).

instructions), first we investigated whether the average behavior in the game deviated from a fair-share usage (i.e., a choice of 4 HPs), and whether there were any differences between conditions in this respect. On average, participants in the fair condition used 3.61 HPs during the week (SD = 0.62), while participants in unfair condition used 3.38 HPs (SD = 0.58). Both values were significantly lower than the suggested "norm" of 4 HPs: one-sample t-test comparing a mean usage in each condition to 4: t (32) = 33.48, $p = 2.2e-16$ for fair condition, and t (32) = 33.41, $p = 2.2e-16$ for unfair condition. This suggests that most people did not overuse energy to make private profits – they used a fair-share or up to 4 HPs – even though using as much as possible (up to 6 HPs) would be rational due to the structure of the game.

Differences in Use in the Fair Versus Unfair Condition

Furthermore, we investigated the use for the week day-by-day for fair and unfair conditions separately. There was no difference in day 1 use between fair (M = 3.42, SD = 0.79, SEM = 0.14) and unfair (M = 3.59, SD = 0.80, SEM = 0.14) conditions: t (62.91) = −0.86, p = 0.39. To investigate whether fair or unfair condition had an effect on individual use, we ran a mixed-effects random intercept regression model (Model 1) estimated by maximum likelihood using lme4 package in R (Bates et al., 2014) predicting energy use on each day from condition (fair, coded as "0," versus unfair, coded as "1"). The regression included random intercept for each participant to account for dependency between observations. We controlled for the use on day 1 to account for individual baseline. We also controlled for learning effects through using day number as a predictor: it is possible that participants would change their energy use across the week as they learn about the game and behavior of others. Predictors entered into all interaction terms were mean-centered to reduce multicollinearity.

The results (see **Table 3**) demonstrated that significant predictors of use were the consumption on day 1 (B = 0.38, SE = 0.05, p = 0.00001), condition (B = −0.32, SE = 0.11, p = 0.007), and interaction of the day of response by condition (B = −0.15,

SE = 0.03, p = 0.00001). **Table 2** reports means and SDs per condition per day of response reflecting a steady increase of overall use across the week and differences in the pattern of use for conditions: the use in the fair condition increased toward the end, while there was relative lack of change in use by participants in the unfair condition across all days apart from drastic decrease in use on the last day before the end of the game, day 7. This suggests that there was a general trend of increase in HP use over the week, however, it was reversed for unfair condition: participants in the unfair condition decreased their use toward the end of the week. Specifically, 68% decreased their use on the last day compared to the first day, 20% did not change, and 11% increased their use in the unfair condition, while 63% increased their use on the last day compared to the first day, 20% did not change, and 18% decreased their use in the fair condition. The results remained the same for the restricted sample (see **Table 3**).

Individual Differences and Strategies in the Game

We further investigated whether individual differences in change of strategies in the games can be attributed to personality traits. We predicted energy use on each day and used the same specification for the mixed-effects random intercept regression model as Model 1 to which we added personality traits predictors and interactions of personality with other effects. Model 2 included main effects of BIS and all BAS subscales; two-way interactions between BIS and each BAS subscale; two-way interactions between each personality subscale and day number; two-way interactions between each personality subscale and condition; three-way interactions between each personality subscale, day number, and condition; three-way interactions between BIS, each BAS subscale, and day number; three-way interactions between BIS, each BAS subscale, and condition; as well as four-way interactions between BIS, each BAS subscale, day number, and condition. See **Table 4** for the full specification of the model. All predictors entered into the interaction terms were mean-centered to reduce multicollinearity. In order to assess the posterior power of the results, we ran simulations (Martin et al., 2011). First, we generated 100 samples of simulated data for all independent variables, with each variable randomly drawn from a normal distribution with a mean and SD of the respective variable from our sample (N = 65), restricted to a variable's respective actual minimal and maximum value. As personality traits (BIS, BAS-RR, BAS-FS, and BAS-D) were correlated, their simulated scores were drawn from a normal multivariate distribution which, in addition to means and SDs from the sample and low/high limits of each variable, also accounted for covariance between each variable. Second, for each of 100 samples, we simulated 100 vectors of the dependent variable (energy use on each day) by using simulated dependent variables, as well as fixed and random effects parameters from Model 2. Third, we ran regression models, as specified in **Table 4**, with 10,000 sets of simulated data: 100 samples, each simulated 100 times. Finally, we determined the power of the analysis by looking at the proportion of significant results at 0.05-level for each of the fixed effect in Model 2. The results of power analyses are reported in **Table 4**.

The results confirmed the previous analysis with energy use predicted by the consumption on day 1 ($B = 0.42$, SE $= 0.07$, $p = 0.00001$), condition ($B = -0.33$, SE $= 0.15$, $p = 0.032$), and interactions between day number and condition ($B = -0.14$, SE $= 0.04$, $p = 0.0002$). In addition, there was an effect of personality traits. Specifically, there was a positive effect of the interactions between BAS-FS and day number ($B = 0.08$, SE $= 0.02$, $p = 0.001$), a negative effect of the interactions between BIS and day number ($B = -0.08$, SE $= 0.02$, $p = 0.00001$), a negative effect of the three-way interactions between BAS-FS, day number, and condition ($B = -0.12$, SE $= 0.05$, $p = 0.011$), a positive effect of a three-way interactions of BIS, a day number, and condition ($B = 0.13$, SE $= 0.04$, $p = 0.0001$), and a positive effect of a four-way interactions between BIS, BAS-FS, day number, and condition ($B = 0.15$, SE $= 0.05$, $p = 0.0032$). All results remained significant in the restricted sample. All effects were detected with sufficient power (>80%) at significance level of 0.05. There were no effects of BAS-RR or BAS-D on behavior in the game.

To analyze the results of the interactions, we calculated mean predicted values of consumption for high- and low-end participants of each trait, using results of regression analysis, Model 2. High- and low-end participants were identified as above or below of a respective scale of the sample's average. We then calculated the change for each group of participants (e.g., high versus low BAS-FS group) between days 1 and 7 to illustrate changes in behavior during the week. Analysis of interactions suggests the following interpretation of results: BAS-FS and BIS can explain some variability in individual decisions. Specifically, those who are high in BAS-FS used more toward the end of the week overall, with a predicted average increase between days 1 and 7 of 0.42 EUs, compared to 0.04 EUs increase in those who are low in BAS-FS. Likewise, those who are low in BIS used more toward the end of the week overall, with a predicted average increase of 0.43 EUs compared to high BIS group, who had a predicted decrease of 0.04 EUs. The effects of personality traits were specific to the fair condition: only in the fair condition, participants high in BAS-FS demonstrated a predicted average increase of their use on 1.28 EUs on average between days 1 and 7 (0.33 increase for low BAS-FS participants), while low BIS participants demonstrated a predicted increase of 1.16 EUs (0.16 for high BIS participants). In the unfair condition, both low/high BAS-FS and low/high BIS participants decreased their consumption: high BAS-FS on 0.25, low BAS-FS on 0.31, high BIS on 0.27, and low BIS on 0.29 EUs. Thus, high BAS-FS and low BIS explain some variation in increased use toward the end of the week, but only in the fair condition.

Finally, the investigation of the four-way interaction of BAS-FS, BIS, condition, and day of response suggests that specifically in the fair condition those who are high in BAS-FS and at the same time low in BIS produce the largest increase in use: by 2.77 EUs difference between days 1 and 7, while high BAS-FS and high BIS produced a 0.42 EUs increase, low BAS-FS and low BIS – 0.62 EUs, with low BAS-FS and high BIS participants not changing their use in the fair condition between days 1 and 7. All participants in the unfair condition produced a decrease in use with high BIS/low BAS-FS and low BIS/high BAS-FS decreasing on 0.39 EUs, while high BAS-FS and high BIS on 0.11, low BAS-FS/low BIS on 0.22 EUs. The actual aggregated responses of the groups broken down by high/low BIS and BAS-FS, as well as by condition for each day are depicted on the **Figure 2**.

DISCUSSION

Behavior in social dilemmas often depends on what others do. After we get feedback about others, we adjust our strategy for future interactions. This paper presents the results of a community purchase energy game structured as an inverted collective-risk dilemma with seven turns before the final outcome is revealed. Participants entered decisions simultaneously with nine other players in their group using their own smart phones or computers at home over a course of a week-long game. Each day, before they made a decision, they also received aggregated information about behavior of others in the group for the previous day. In some groups, we manipulated behavior of the group as fair – the followed a pre-defined group norm of use – or unfair, where the rest of the group used more than was pre-defined.

We find that the majority of participants who were in the unfair condition demonstrated generous behavior: 68% decreased their use on the last day compared to the first day to compensate for the high use of others. This indicates that individuals in groups were good at dealing with effects of free-riding of others, especially as the punishment for group-level non-cooperation was certain. Further we find that the majority of participants in the fair condition demonstrated opportunistic behavior: 63% increased their use on the last day compared to the first day. The findings suggest that individuals did not follow either descriptive ("what others do") or injunctive ("what I am supposed to do") social norms of behavior especially under low risk of punishment for free-riding: in the fair condition participants used significantly more than their group on the last day of the study, while in the unfair condition significantly less than their group.

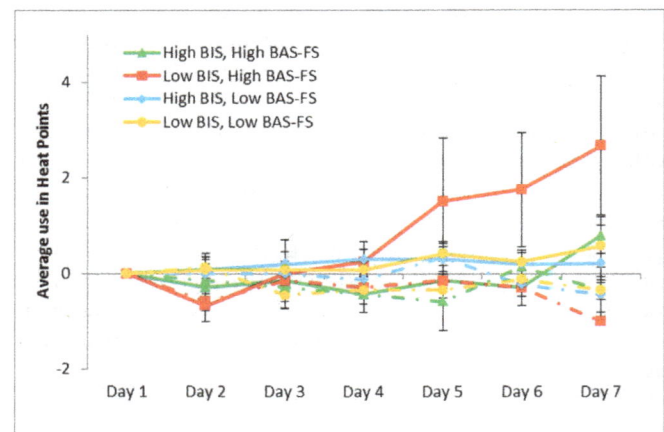

FIGURE 2 | Average use of energy on each day of the week grouped by condition (unfair versus fair) and four different combinations of low and high scores on BIS and BAS-FS: (1) high BAS-FS and high BIS; (2) high BAS-FS and low BIS; (3) low BAS-FS and high BIS; (4) low BAS-FS and low BIS. Error bars represent SE of the mean. The scores are corrected for the baseline use in the first day. High/low groups are identified based on above/below the mean of a respective scale. Dotted lines represent fair, and solid – unfair condition.

We find that individual differences in impulsivity, measured through BAS-FS, and punishment sensitivity, measured through BIS, were associated with opportunistic strategies in the fair condition. In particular, those who were high in BAS-FS and low in BIS used more on the last turn of the game showing over 2 EUs increase on the last day of the game compared to the first day. We further discuss contribution of the results of the paper to the literature on social dilemmas, individual differences, and understanding of energy behaviors.

Compensating and Opportunistic Strategies in Multi-Turn Social Dilemmas

Strong reciprocity theory suggests that if others are free-riding, people will punish free-riders even at a cost to themselves (Fehr and Gachter, 2002). Our results show that only a small proportion of individuals retaliated by increasing their use when others were unfair, with the majority being generous and compensating for others. Thus, we did not find evidence for strong reciprocity theory in our experiment. It is possible that strong reciprocity does not explain cooperation in social dilemmas that involve interactions over multiple turns. This is in line with previous research suggesting that strong reciprocity cannot always explain how people manage free-riding behaviors in social dilemmas (Yamagishi et al., 2012). In the case of our game, from an individual player perspective, the more others used, the higher was the likelihood of the fine for the group. Therefore, the reason why participants were generous can be explained by high risk of punishment. This is in line with Milinski et al. (2008) who found that in a multi-turn collective-risk dilemma in the high punishment risk condition more participants demonstrated compensating strategies toward the end of the game.

In addition to the risk of punishment, it is possible that participants were generous in this condition because of the take-some framing of the game. This is in line with previous research: it is often reported that people cooperate more in take-some dilemmas (Van Lange et al., 2013). Furthermore, the instructions could have had an influence too: participants knew that their task was not to use over the limit as a group. Such instructions could have enhanced a goal to achieve the results that were best for the group. Previous research demonstrated that goal-orientation has an effect on people's behavior in social dilemmas. Specifically, people can assign the importance on self- or other-beneficial outcomes (van Lange, 1999) and depending on the framing of the outcome, this can lead to either selfish or other-regarding behavior (Van Lange et al., 2013).

However, even though in this game the majority of participants did not retaliate, this does not mean they were indifferent to the fact that others in their group were unjust. Psychological factors, such as emotions and cognitive appraisals, are strongly implicated in the way people judge a situation that involves unfairness of others (e.g., Ketelaar and Tung Au, 2003; Sanfey et al., 2003; Nelissen et al., 2007). In our case, even if our participants felt angry, most of them did not act on their emotions as demonstrated by decrease in energy use in the unfair condition. The result also cannot be explained by opportunity to "cool down" (Dickinson and Masclet, 2015) as the decision of how much to use on a particular day had to be made straight after being presented with the feedback about others. It is still not clear what the cost of generosity was for individual participants in circumstances when others were unfair. Being angry and not having an opportunity to deal with the emotion (either by reappraising or acting on it) can have detrimental effects on one's wellbeing (Zammuner and Galli, 2005; Barrett et al., 2013) and have negative consequences for future interactions (e.g., Pillutla and Murnighan, 1996). Future research needs to investigate the role of psychological factors on behavioral strategies in multi-turn games, such as collective-risk social dilemmas with and without opportunities to impose sanctions or act on one's emotions in some way, as well as what consequences psychological factors, such as emotions, could have on people's behavior and interactions beyond an experimental game.

While in the unfair condition, participants were generous, the opposite was observed in the fair condition. In this condition, the majority of participants increased their use toward the end of the game. Contrary to social norms literature (Schultz et al., 2007), which suggests that both descriptive (do as others do, in our case implemented through feedback about behavior of others) and injunctive (do as think is right, in our case implemented through the instruction of normative use of 4 HPs) norms guide people's behavior, in our game, participants did not follow the norm of use demonstrated by their confederates through manipulated feedback, or as it was reinforced in the instructions for the experiment. This suggests the influences of social norms on behavior might be weaker if there are other motivations guiding people's choices, such as getting individual private benefits or maximizing individual profit through avoiding a group-level fine (Charness and Rabin, 2002). However, as the reported results only cover one game, it is not clear what the participants in the unfair condition would do should they have another round of interactions with their group partners.

Participants in the fair condition significantly increased their consumption. The increase throughout the game is in line with rebound effects in energy use domain: when presented with real feedback about energy use of other households people find out that they used less than others, under some circumstances, they can increase their energy use (Leygue et al., 2014). Use of over the fair-share allowance by 30% of participants at some point in the game further indicated a different type of individual behavior in response to feedback about others – a type of weak free-riding (Keser and van Winden, 2000), which has negative consequences especially considering the environmental context of these decisions. While the findings are at odds with social norms literature, they are in line with literature on collective-risk games. Milinski et al. (2008) demonstrated that in low punishment risk condition participants free rode more. In the case of our fair condition, participants presumably also perceived risk of punishment as low; therefore, they chose to take advantage of the situation.

On the one hand, given the structure of the game, it was rational to use more when there was an opportunity, as one could get additional private benefits for usage. What is rational in the current situation could depend on the context and it is possible that our participants just aimed to optimize their profit in the game and disregarded the broader environmental picture. However, the environmental context of the game puts the decision into a different perspective, as the overarching goal of such energy purchase scheme is ultimately to decrease energy use. It is noteworthy that people still increased their energy use for private benefit, despite knowing that doing so can have consequences for the environment. While this result can be also explained by the fact that some participants might have not considered environmental framing of the game, understanding people's motivations behind free-riding is important, as it can explain why people do not make environmentally friendly choices in the real world, including not making links between their own actions and consequences for environment. For example, future energy collectives individuals might employ such opportunistic strategies by accumulating more energy than their fair-share in personal storage. Previous research suggests that people justify free-riding behavior by denying their responsibility for the outcome (Schwartz and Howard, 1982) or by convincing themselves that their behavior would not make a difference to the group outcome (Kerr and Kaufman-Gilliland, 1997). It is possible that in the case of this game, participants felt that increasing usage would not impact the outcome, so they could behave opportunistically. It is also plausible that participants behaved rationally and optimized their profits. However, whichever reason was driving the behavior of the majority, similar choices in real world have negative implications for issues such as climate change mitigation. Our findings can be used by policy makers to develop and model approaches to predict and discourage opportunistic strategies. Based on our findings, policy makers could benefit by building in an incentive structure to encourage cooperation and to prevent opportunistic behavior in future scenarios.

Individual Differences and Opportunistic Strategies

While some participants used an opportunity to get additional profits in the fair condition, about 40% did not demonstrate such behavior: they either did not change or decreased their usage. We showed that individual differences in impulsivity and punishment sensitivity were associated with opportunistic strategies. Specifically, participants with high BAS-FS and low BIS increased their energy use toward the end of the game significantly more than other participants. Results are in line with previous findings that impulsive individuals are more likely to be biased toward an immediate reward in the situations where there is a conflict between immediate and delayed reward (Smillie et al., 2006). In our study, participants knew that using more HPs would increase their profits, so they received immediate gratification from using HPs, while the reward through cooperation was delayed by at least one day (in case of the last turn decision) or more days (in case of all other decisions). This result is in line with findings of Myrseth et al. (2015) who demonstrated negative associations between impulsivity and cooperation when immediate rewards for free-riding were more salient and tangible. Furthermore, in line with predictions, we found that participants who were low in BIS were more likely to free ride when the risk of punishment was low, i.e., in the fair condition. This supports previous research, suggesting that inhibitory mechanisms are implicated in prosocial choices as one needs to withhold an initial impulse to free ride in order

to get better rewards through cooperation in the future (McCabe et al., 2001; Skatova and Ferguson, 2013).

Our findings contribute to the understanding of conditions necessary in order to maintain cooperation in groups especially around environmental issues. For example, Freytag et al. (2014) found that intermediate targets featuring environmental protection as a process helped to improve cooperation in collective-risk dilemmas. It is possible that introduction of intermediate targets and rewards in a community energy purchase scenario, for example, through messages that enhance environmental consequences of various decisions or through opportunities for reputation formation could reduce opportunistic behavior in the collective-risk game among impulsive individuals with low inhibitory control.

We did not find predicted associations between other subscales of BAS (BAS-RR and BAS-D) and behavior in the experiment. In previous studies that demonstrated associations between reward responsiveness component of BAS, namely BAS-RR and BAS-D, and free-riding behavior in economic games, participants had full information about behavior of others or control over the situation while making their decision. Thus, selfish choices of reward responsive participants could have been explained by the fact that they learned better from reward and made a selfish choice to take advantage of a certain increase in profits. In our design even for the last decision, there was some uncertainty about behavior of others. This different structure of the game can explain why there were no associations of BAS-RR and BAS-D with behavior in the unfair condition. While we did not find any associations of individual differences and behavior in the unfair condition, future research could study the motivation behind generous compensating strategies in collective-risk dilemmas.

Implications for Policy Around Energy Use

Many researchers highlight that it is important to extend lab-based paradigms and develop social dilemma research designs that help to mirror important features of real-world behavior in social dilemma-like scenarios (e.g., Van Lange et al., 2013). Such research can help to identify constraints of policies and test out model scenarios in various areas of social decision-making. The results of the study presented here suggest that community energy purchase deals could backfire as we predict that under certain conditions people will increase their energy usage, especially if there is an opportunity to gain private benefits and the risk of punishment is low. Community energy purchase schemes without a system of intermediate rewards and/or risk of punishment might not be as efficient as expected. We further suggest that it is necessary to study the implications of these schemes beyond actual energy use, because opportunistic behavior of others might lead to indirect negative consequences on interpersonal relationships in the community. Future research is needed to understand psychological cost of generous compensatory behavior that we observed in unfair condition, and whether it could spill over to other domains of interactions within community.

While our game modeled one specific case of managing energy supply and demand on a local level, our results have implications to decision-making in other areas of sustainable behavior such as household energy use (Leygue et al., 2014) and climate change mitigation (Milinski et al., 2008). Understanding how people act in dilemmas such as climate change mitigation are of high importance, however, we advocate the approach to employ social dilemmas to study more local decisions, such as community energy purchase schemes. Ultimately, for people, the climate change mitigation dilemma consists of small person-level every day dilemmas, such as the one presented in this paper. Moreover, research suggests that attempts to establish cooperation with large groups is less productive than when small groups are involved (Santos and Pacheco, 2011). Without understanding how to manage free-riding and achieve cooperation on small scale, it will also not be possible to resolve the global climate change mitigation dilemma.

Limitations

Our study had limitations. Failure in experimental software on day 7 meant that all participants whose data were submitted to the final analysis saw the feedback about behavior of others twice, and on the first occasion the feedback was incorrect. This reduced sample size that could have biased the responses and were submitted to the analyses. While our findings are consistent with previous research both in terms of behavioral outcomes (Jacquet et al., 2013) and individual differences (Myrseth et al., 2015; Skatova and Ferguson, 2013), the replication of the main findings can help to affirm the results. The heterogeneity of responses in economic games (which subsequently produces large variation around the mean) is well documented (Burlando and Guala, 2005), however, future research could also help to explain remaining variation that is visible from **Figure 2**'s SE: specifically, there might be other personality or cognitive factors driving variation in behavior in the fair condition.

Our study also did not account for a number of factors that could have impacted cooperation in the collective-risk game scenario: for example, reputation, anonymity, communication between group members, and other factors. Research on social dilemmas suggests that reputation (Milinski et al., 2002) is key in sustaining cooperation in groups. Reputation scenarios assume that players responses could be traced throughout the game, which was not possible in our design. Decreasing anonymity is not directly applicable to energy use at home, as it comes at privacy cost (McKenna et al., 2012; Rouf et al., 2012). However, lower levels of anonymity than we had in our game – where only group-level behavior was shared with others – and some opportunities for reputation building might have improved cooperation in a collective-risk dilemma scenario. Furthermore, our study did not involve any communication between group members, while real-world interactions certainly involve at least some level of communication. Communication provides the group with more opportunities to self-manage cooperation through, for example, imposing social sanctions, such as disapproval (Noussair and Tucker, 2005). Future research could look into whether communication between group partners helps to coordinate the efforts around energy use and reduce the level of opportunistic strategies.

CONCLUSION

We used a social dilemma – a collective-risk game – to model real-world decisions in a community energy purchase scenario. Our study confirms that in order to maintain cooperation the risk of punishment should be high and tangible; otherwise, people take advantage of the situation and free ride. Specifically, individuals high in impulsivity and low in sensitivity to punishment showed higher levels of opportunistic behavior. We also show that when the risk of punishment is high, people compensate for others to avoid the group-level punishment. However, the psychological cost is unclear. Compensating for others could come at an emotional toll and impact negatively on further interactions. Taking advantage at the last moment puts collective good at risk in a way that can lead to a disaster, especially in an environmental context. We suggest that people should have tangible intermittent incentives to save energy and not just be expected to follow what others do as suggested by social norms literature. Taken together, the findings of the study reported here illustrate the benefits of a social dilemma approach to study behaviors around energy use and the constraints of policies in the environmental domain.

AUTHOR CONTRIBUTIONS

AS and BB designed the study, AS and BKS collected and analysed the data, all authors contributed to the writing and revision of the paper, approved the final draft, and agree to be accountable to all aspects of the work presented in the paper.

ACKNOWLEDGMENTS

The study was supported by Horizon Digital Economy Research (Research Councils UK grant EP/G065802/1), from Human Data to Personal Experience (Research Councils UK grant EP/M02315X/1), Creating the Energy for Change (Research Councils UK grant EP/K002589/1), and Network for Integrated Behavioural Science (Research Councils UK grant ES/K002201/1).

REFERENCES

Axelrod, R., and Hamilton, W. D. (1981). The evolution of cooperation. *Science* 211, 1390–1396. doi:10.1126/science.7466396

Bardsley, N. (2000). Control without deception: individual behaviour in free-riding experiments revisited. *Exp. Econ.* 3, 215–240. doi:10.1023/a:1011420500828

Barrett, E. L., Mills, K. L., and Teesson, M. (2013). Mental health correlates of anger in the general population: findings from the 2007 national survey of mental health and wellbeing. *Aust. N. Z. J. Psychiatry* 47, 470–476. doi:10.1177/0004867413476752

Bates, D., Mächler, M., Bolker, B., and Walker, S. (2014). *Fitting Linear Mixed-Effects Models Using LME4. arXiv preprint arXiv:1406.5823.*

Biel, A., and Thøgersen, J. (2007). Activation of social norms in social dilemmas: a review of the evidence and reflections on the implications for environmental behaviour. *J. Econ. Psychol.* 28, 93–112. doi:10.1016/j.joep.2006.03.003

Binmore, K. (2014). Bargaining and fairness. *Proc. Natl. Acad. Sci. U.S.A.* 111(Suppl. 3), 10785–10788. doi:10.1073/pnas.1400819111

Burlando, R. M., and Guala, F. (2005). Heterogeneous agents in public goods experiments. *Exp. Econ.* 8, 35–54. doi:10.1007/s10683-005-0436-4

Carver, C. S., and White, T. L. (1994). Behavioral inhibition, behavioral activation, and affective responses to impending reward and punishment: the BIS/BAS scales. *J. Pers. Soc. Psychol.* 67, 319–333. doi:10.1037/0022-3514.67.2.319

Charness, G., and Rabin, M. (2002). Understanding social preferences with simple tests. *Q. J. Econ.* 117, 817–869. doi:10.1162/003355302760193904

Conaty, P., and Mayo, E. (2012). Towards a co-operative energy service sector. *J. Cooper. Stud.* 45, 46–55.

Croson, R. T. A. (2007). Theories of commitment, altruism and reciprocity: evidence from linear public goods games. *Econ. Inq.* 45, 199–216. doi:10.1111/j.1465-7295.2006.00006.x

DECC. (2013). *Helping Customers Switch: Collective Switching and Beyond*. Available at: https://www.gov.uk/government/uploads/system/uploads/attachment_data/file/253862/Helping_Customers_Switch_Collective_Switching_and_Beyond_final__2_.pdf

DECC. (2014). *Community Energy Strategy*. London: HMSO. Available at: https://www.gov.uk/government/uploads/system/uploads/attachment_data/file/275163/20140126Community_Energy_Strategy.pdf

Dickinson, D. L., and Masclet, D. (2015). Emotion venting and punishment in public good experiments. *J. Public Econ.* 122, 55–67. doi:10.1016/j.jpubeco.2014.10.008

Erbmann, R., Goulbourne, H., and Malik, P. (2009). *Collective Power: Changing the Way We Consume Energy*. London: Co-operative Party.

Fehr, E., and Fischbacher, U. (2003). The nature of human altruism. *Nature* 425, 785–791. doi:10.1038/nature02043

Fehr, E., and Gachter, S. (2002). Altruistic punishment in humans. *Nature* 415, 137–140. doi:10.1038/415137a

Freytag, A., Güth, W., Koppel, H., and Wangler, L. (2014). Is regulation by milestones efficiency enhancing? An experimental study of environmental protection. *Eur. J. Polit. Econ.* 33, 71–84. doi:10.1016/j.ejpoleco.2013.11.005

Fudenberg, D., Rand, D. G., and Dreber, A. (2012). Slow to anger and fast to forgive: cooperation in an uncertain world. *Am. Econ. Rev.* 102, 720–749. doi:10.1257/aer.102.2.720

Gintis, H., Bowles, S., Boyd, R., and Fehr, E. (2003). Explaining altruistic behavior in humans. *Evol. Hum. Behav.* 24, 153–172. doi:10.1016/s1090-5138(02)00157-5

Giovanelli, A., Hoerger, M., Johnson, S. L., and Gruber, J. (2013). Impulsive responses to positive mood and reward are related to mania risk. *Cogn. Emot.* 27, 1091–1104. doi:10.1080/02699931.2013.772048

Gneezy, A., and Fessler, D. M. (2012). Conflict, sticks and carrots: war increases prosocial punishments and rewards. *Proc. Biol. Sci.* 279, 219–223. doi:10.1098/rspb.2011.0805

Greening, L. A., Greene, D. L., and Difiglio, C. (2000). Energy efficiency and consumption–the rebound effect–a survey. *Energy Policy* 28, 389–401. doi:10.1016/S0301-4215(00)00021-5

Harvey, D., and Braun, B. (1999). Justice, nature, and the geography of difference. *Can. Geogr.* 43, 105.

Hoffman, S. M., and High-Pippert, A. (2010). From private lives to collective action: recruitment and participation incentives for a community energy program. *Energy Policy* 38, 7567–7574. doi:10.1016/j.enpol.2009.06.054

Irwin, K., and Berigan, N. (2013). Trust, culture, and cooperation: a social dilemma analysis of pro-environmental behaviors. *Sociol. Q.* 54, 424–449. doi:10.1111/tsq.12029

Jacquet, J., Hagel, K., Hauert, C., Marotzke, J., Röhl, T., and Milinski, M. (2013). Intra-and intergenerational discounting in the climate game. *Nat. Clim. Chang.* 3, 1025–1028. doi:10.1038/nclimate2024

Kerr, N. L., and Kaufman-Gilliland, C. M. (1997). "… and besides, I probably couldn't have made a difference anyway": justification of social dilemma defection via perceived self-inefficacy. *J. Exp. Soc. Psychol.* 33, 211–230. doi:10.1006/jesp.1996.1319

Keser, C., and van Winden, F. (2000). Conditional cooperation and voluntary contributions to public goods. *Scand. J. Econ.* 102, 23–39. doi:10.1111/1467-9442.00182

Ketelaar, T., and Tung Au, W. (2003). The effects of feelings of guilt on the behaviour of uncooperative individuals in repeated social bargaining games: an affect-as-information interpretation of the role of emotion in social interaction. *Cogn. Emot.* 17, 429–453. doi:10.1080/02699930143000662

Leygue, C., Ferguson, E., Skatova, A., and Spence, A. (2014). Energy sharing and energy feedback: affective and behavioral reactions to communal energy displays. *Front. Energy Res.* 2, 29. doi:10.3389/fenrg.2014.00029

Li, Y. (2005). Perceptions of temperature, moisture and comfort in clothing during environmental transients. *Ergonomics* 48, 234–248. doi:10.1080/0014013042000327715

Lyas, J. K., Shaw, P. J., and Van-Vygt, M. (2004). Provision of feedback to promote householders' use of a kerbside recycling scheme: a social dilemma perspective. *J. Solid Waste Technol. Manag.* 30, 7–18.

Martin, J. G., Nussey, D. H., Wilson, A. J., and Reale, D. (2011). Measuring individual differences in reaction norms in field and experimental studies: a power analysis of random regression models. *Methods Ecol. Evol.* 2, 362–374. doi:10.1111/j.2041-210X.2010.00084.x

McCabe, K., Houser, D., Ryan, L., Smith, V., and Trouard, T. (2001). A functional imaging study of cooperation in two-person reciprocal exchange. *Proc. Natl. Acad. Sci. U.S.A.* 98, 11832–11835. doi:10.1073/pnas.211415698

McKenna, E., Richardson, I., and Thomson, M. (2012). Smart meter data: balancing consumer privacy concerns with legitimate applications. *Energy Policy* 41, 807–814. doi:10.1016/j.enpol.2011.11.049

Milinski, M., Semmann, D., and Krambeck, H.-J. (2002). Reputation helps solve the 'tragedy of the commons'. *Nature* 415, 424–426. doi:10.1038/415424a

Milinski, M., Sommerfeld, R. D., Krambeck, H.-J., Reed, F. A., and Marotzke, J. (2008). The collective-risk social dilemma and the prevention of simulated dangerous climate change. *Proc. Natl. Acad. Sci. U.S.A.* 105, 2291–2294. doi:10.1073/pnas.0709546105

Myrseth, K. O. R., Riener, G., and Wollbrant, C. E. (2015). Tangible temptation in the social dilemma: Cash, cooperation, and self-control. *J. Neurosci. Psychol. Econ.* 8, 61–77. doi:10.1037/npe0000035

Nelissen, R. M. A., Dijker, A. J. M., and deVries, N. K. (2007). How to turn a hawk into a dove and vice versa: interactions between emotions and goals in a give-some dilemma game. *J. Exp. Soc. Psychol.* 43, 280–286. doi:10.1016/j.jesp.2006.01.009

Noussair, C., and Tucker, S. (2005). Combining monetary and social sanctions to promote cooperation. *Econ. Inq.* 43, 649–660. doi:10.1093/ei/cbi045

Nowak, M. A., Page, K. M., and Sigmund, K. (2000). Fairness versus reason in the ultimatum game. *Science* 289, 1773–1775. doi:10.1126/science.289.5485.1773

Pillutla, M. M., and Murnighan, J. K. (1996). Unfairness, anger, and spite: emotional rejections of ultimatum offers. *Organ. Behav. Hum. Decis. Process* 68, 208–224. doi:10.1006/obhd.1996.0100

Pothos, E. M., Perry, G., Corr, P. J., Matthew, M. R., and Busemeyer, J. R. (2011). Understanding cooperation in the Prisoner's Dilemma game. *Pers. Individ. Dif.* 51, 210–215. doi:10.1016/j.paid.2010.05.002

Ramchurn, S. D., Osborne, M., Parson, O., Rahwan, T., Maleki, S., Reece, S., et al. (2013). "AgentSwitch: towards smart energy tariff selection," in *Proceedings of the 12th International Conference on Autonomous Agents and Multiagent Systems (AAMAS 2013)*, 2013, St. Paul, MN, eds T. Ito, C. Jonker, M. Gini, and O. Shehory.

Rouf, I., Mustafa, H., Xu, M., Xu, W., Miller, R., and Gruteser, M. (2012). "Neighborhood watch: security and privacy analysis of automatic meter reading systems," in *Proceedings of the 2012 ACM Conference on Computer and Communications Security*. (Raleigh, NC: ACM), 462–473.

Sanfey, A. G., Rilling, J. K., Aronson, J. A., Nystrom, L. E., and Cohen, J. D. (2003). The neural basis of economic decision-making in the ultimatum game. *Science* 300, 1755–1758. doi:10.1126/science.1082976

Santos, F. C., and Pacheco, J. M. (2011). Risk of collective failure provides an escape from the tragedy of the commons. *Proc. Natl. Acad. Sci. U.S.A.* 108, 10421–10425. doi:10.1073/pnas.1015648108

Scheres, A., and Sanfey, A. (2006). Individual differences in decision making: drive and reward responsiveness affect strategic bargaining in economic games. *Behav. Brain Funct.* 2, 35. doi:10.1186/1744-9081-2-35

Schultz, P. W., Nolan, J. M., Cialdini, R. B., Goldstein, N. J., and Griskevicius, V. (2007). The constructive, destructive, and reconstructive power of social norms. *Psychol. Sci.* 18, 429–434. doi:10.1111/j.1467-9280.2007.01917.x

Schwartz, S. H., and Howard, J. A. (1982). "A self-based motivational model of helping," in *Cooperation and Helping Behavior: Theories and Research*, eds V. Derlega and J. Grzelak (New York: Academic Press), 22–35.

Skatova, A., and Ferguson, E. (2011). What makes people cooperate? Individual differences in BAS/BIS predict strategic reciprocation in a public goods game. *Pers. Individ. Dif.* 51, 237–241. doi:10.1016/j.paid.2010.05.013

When Push Comes to Shove: Compensating and Opportunistic Strategies in a Collective-Risk...

159

Skatova, A., and Ferguson, E. (2013). Individual differences in behavioural inhibition explain free riding in public good games when punishment is expected but not implemented. *Behav. Brain Funct.* 9, 3. doi:10.1186/1744-9081-9-3

Smillie, L. D., Jackson, C. J., and Dalgleish, L. I. (2006). Conceptual distinctions among Carver and White's (1994) BAS scales: a reward-reactivity versus trait impulsivity perspective. *Pers. Individ. Dif.* 40, 1039–1050. doi:10.1016/j.paid.2005.10.012

Van Dijk, E., Wilke, H., and Wit, A. (2003). Preferences for leadership in social dilemmas: public good dilemmas versus common resource dilemmas. *J. Exp. Soc. Psychol.* 39, 170–176. doi:10.1016/S0022-1031(02)00518-8

Van Lange, P. A. (1999). The pursuit of joint outcomes and equality in outcomes: an integrative model of social value orientation. *J. Pers. Soc. Psychol.* 77, 337–349. doi:10.1037/0022-3514.77.2.337

Van Lange, P. A., Joireman, J., Parks, C. D., and Van Dijk, E. (2013). The psychology of social dilemmas: a review. *Organ. Behav. Hum. Decis. Process* 120, 125–141. doi:10.1016/j.obhdp.2012.11.003

Van Lange, P. A., Rockenbach, B., and Yamagishi, T. (eds). (2014). *Reward and Punishment in Social Dilemmas*. Oxford: Oxford University Press.

Vinyals, M., Bistaffa, F., Farinelli, A., and Rogers, A. (2012). "Coalitional energy purchasing in the smart grid," in *Energy Conference and Exhibition (ENERGYCON), 2012 IEEE International.* (Florence, IT: IEEE), 848–853.

Walker, G., Devine-Wright, P., Hunter, S., High, H., and Evans, B. (2010). Trust and community: exploring the meanings, contexts and dynamics of community renewable energy. *Energy Policy* 38, 2655–2663. doi:10.1016/j.enpol.2009.05.055

Weber, J. M., and Murnighan, J. K. (2008). Suckers or saviors? Consistent contributors in social dilemmas. *J. Pers. Soc. Psychol.* 95, 1340. doi:10.1037/a0012454

Yamagishi, T., Horita, Y., Mifune, N., Hashimoto, H., Li, Y., Shinada, M., et al. (2012). Rejection of unfair offers in the ultimatum game is no evidence of strong reciprocity. *Proc. Natl. Acad. Sci. U.S.A.* 109, 20364–20368. doi:10.1073/pnas.1212126109

Zammuner, V. L., and Galli, C. (2005). Wellbeing: causes and consequences of emotion regulation in work settings. *Int. Rev. Psychiatry* 17, 355–364. doi:10.1080/09540260500238348

Zhao, K., and Smillie, L. D. (2014). The role of interpersonal traits in social decision making exploring sources of behavioral heterogeneity in economic games. *Pers. Soc. Psychol. Rev.* 19, 277–302. doi:10.1177/1088868314553709

Conflict of Interest Statement: The authors declare that the research was conducted in the absence of any commercial or financial relationships that could be construed as a potential conflict of interest.

16

Low-Concentration Solar-Power Systems Based on Organic Rankine Cycles for Distributed-Scale Applications: Overview and Further Developments

Christos N. Markides*

Clean Energy Processes (CEP) Laboratory, Department of Chemical Engineering, Imperial College London, London, UK

Edited by:
*Michael Folsom Toney,
SLAC National Accelerator
Laboratory, USA*

Reviewed by:
*Boon Han Lim,
Universiti Teknologi Malaysia,
Malaysia
Zhibin Yu,
University of Glasgow, UK
George Kosmadakis,
Agricultural University of
Athens, Greece*

*Correspondence:
*Christos N. Markides
c.markides@imperial.ac.uk*

Specialty section:
*This article was submitted to
Solar Energy, a section of the
journal Frontiers in Energy Research*

Citation:
*Markides CN (2015) Low-
Concentration Solar-Power Systems
Based on Organic Rankine Cycles for
Distributed-Scale Applications:
Overview and Further Developments.
Front. Energy Res. 3:47.*

This paper is concerned with the emergence and development of low-to-medium-grade thermal-energy-conversion systems for distributed power generation based on thermo-dynamic vapor-phase heat-engine cycles undergone by organic working fluids, namely organic Rankine cycles (ORCs). ORC power systems are, to some extent, a relatively established and mature technology that is well-suited to converting low/medium-grade heat (at temperatures up to ~300–400°C) to useful work, at an output power scale from a few kilowatts to 10s of megawatts. Thermal efficiencies in excess of 25% are achievable at higher temperatures and larger scales, and efforts are currently in progress to improve the overall economic viability and thus uptake of ORC power systems, by focusing on advanced architectures, working-fluid selection, heat exchangers and expansion machines. Solar-power systems based on ORC technology have a significant potential to be used for distributed power generation, by converting thermal energy from simple and low-cost non-concentrated or low-concentration collectors to mechanical, hydraulic, or electrical energy. Current fields of use include mainly geothermal and biomass/biogas, as well as the recovery and conversion of waste heat, leading to improved energy efficiency, primary energy (i.e., fuel) use and emission minimization, yet the technology is highly transferable to solar-power generation as an affordable alternative to small-to-medium-scale photovoltaic systems. Solar-ORC systems offer naturally the advantages of providing a simultaneous thermal-energy output for hot water provision and/or space heating, and the particularly interesting possibility of relatively straightforward onsite (thermal) energy storage. Key performance characteristics are presented, and important heat transfer effects that act to limit performance are identified as noteworthy directions of future research for the further development of this technology.

Keywords: solar power, distributed power, thermal-energy conversion, heat engine, organic Rankine cycle, heat transfer

INTRODUCTION AND MOTIVATION

The recently heightened interest in issues relating to energy and the environment has given rise to an intensified debate in the public domain, the scientific community, industry, government and policy circles, and even the financial and investment sectors, concerning the role that a wide variety of energy (fuel-to-power and heat-to-power) technologies can play within competing visions of both global and national energy futures. Rapid developments have been observed in the areas of energy generation, management (transportation, conversion, storage and supply) and consumption in all their facets, spanning a range of scales and diverse applications. An important aspect of the energy challenge concerns the harnessing of renewable and sustainable energy sources, such as the solar resource, for the provision of power, heating, and also, depending on the need, cooling (Markides, 2013). The IEA projects that solar energy has the potential to cover one-third of the world's energy consumption by 2060 under favorable conditions. This represents a significant displacement in the utilization of fossil fuels, and thus both in the consumption of this finite resource and in the consequent release of harmful emissions to the atmosphere.

Although fossil fuels will remain the most important (and dominant) primary energy resource over the next decades, renewable energy technologies have received particular attention in both developed and developing countries, yet for different reasons. In the case of the former, the attention has been strongly supported by public opinion, driven by a desire for energy diversification and decarbonization in an effort to move away from an existing reliance on fossil fuels and toward a more secure, clean, and sustainable energy portfolio. In the latter, including in China, India, and the rest of the BRICKS, this has arisen in response to a need to drive the strong economic growth that is being experienced and a desire to raise living standards, which are known to correlate with higher energy use (MacKay, 2009; Markides, 2013), while addressing health (e.g., clean environment) concerns that have emerged from rapid industrialization. In both cases, these trends have established renewable technologies as an indispensable contributor to energy generation, with exponential growth experienced in the sector in recent decades.

Solar-based renewable energy systems can be deployed to deliver electricity and also to provide hot water, generate space heating and even cooling at different application scales, depending on the specific requirements and the technologies employed. It is generally accepted that photovoltaic (PV) panels and solar-thermal (ST) collector systems are highly suitable options for onsite renewable energy generation. The former can provide an electrical-energy output to cover end-users' electricity needs, while the latter can provide a thermal-energy output for water or space heating at much higher efficiency [50–80% (Freeman et al., 2015a)]. A UK-based analysis by McKinsey in 2007 identified solar hot water as the leading solution in terms of greenhouse gas (GHG) abatement potential (Confederation of British Industry, 2007), even in this northern climate with its restricted solar resource compared to other regions.

The PV market has been experiencing a well-documented exponential growth for over two decades, driven strongly by installations in Europe, and more recently Asia, assisted by a broad range of incentivization programs. Global installed capacity has been increasing on average by at least 40% year-on-year, and in some cases close to 50%, with a doubling time of 1.8–2.0 years over the last 20 years. At the same time, costs have been decreasing strongly, driven by increased production mainly in China, to the point where solar PV is now close to achieving "grid parity" (i.e., at a price competing with the purchase of conventional power from the electricity grid). The price of crystalline silicon (c-Si) solar cells has been falling by 10–15% per year with a 4.5–5.0-year halving time over the last 30 years. Currently, the largest PV power station in the world (the Topaz Solar Farm in the US) stretches over 25 km^2 and has a ~0.5 GW$_p$ peak capacity; given actual load factors and performance, this corresponds to an output of up to ~1 TWh/year (~0.1 GW or 3.6 PJ/year).

When considering *electricity* generation from solar energy, and with exception of large/utility-scale (defined as >4–5 MW, Wolfe, 2015) ST power-generation plants, the focus is often placed exclusively on PV, typically flat-panel non-concentrated PV based on a range of semiconductor materials. Arguably, this practice is done to an extent that, for the most part, PV technologies are often mentioned synonymously with solar energy, thus largely displacing ST power from the discussion. Nevertheless, large-scale ST power plants based invariably on *concentrated* solar collectors (mainly parabolic trough concentrators with single-axis solar tracking, but also solar tower receivers with two-axis tracking heliostats), and referred to as concentrated solar-power (CSP) plants, are also a mature technology which is suitable for solar-power generation. The largest CSP plant in the world (Ivanpah Solar Power Facility, USA) covers 14 km^2 and has a rating of ~0.4 GW (or 13 PJ/year).

The global overall operational (full-load) capacity of utility-scale PV power stations amounted to ~36 GW$_p$ (or, ~7 GW) at the end of 2014 (Wolfe, 2015) (at the same time, total cumulative capacity across all scales was closer to 180 GW$_p$, or 30 GW), while that of utility-scale CSP electricity production reached ~4 GW at the end of 2013 and was projected to approach ~5 GW at the end of the same period as above (i.e., 2014) (Hashem, 2015). The levelized electricity cost (LEC) of CSP is similar to that of PV (both fall in the range ~\$100–200/MWh depending on the scale of application), although CSP has an edge in terms of efficiency (Lazard, 2013; Markides, 2013). Cradle-to-grave GHG emissions from the solar options are also comparable. A meta-analysis reported in Markides (2013) gave for PV: 18–67 g for thin-film CdTe and 32–104 g for Si and for CSP: 14–90 g for trough, 21–60 g for receiver, and 22–58 gCO$_2$/kWh for dish collectors. Importantly, up to 15 h of thermal storage can be provided in the case for demand matching and load factor improvement at a fraction of the cost of equivalent-scale electricity storage.

Yet beyond conventional solar-power from PV and CSP, hybrid PV-ST (PVT) systems and also solar combined heat and power (S-CHP) systems based on non-concentrated or low-concentration ST collectors in conjunction with thermodynamic power cycles are alternative solar-energy options in smaller/distributed-scale applications, which offer the distinct advantage of providing from a single system both a thermal-energy (e.g., for water heating) and an electrical-energy output. These systems are unlikely to be considered for large-scale use, since the thermal output, although equally (if not more, arguably, in certain northern regions) important, is less fungible compared to electricity. Still, distributed ST power systems based on these technologies can and should play a role in a future with increased penetration of renewable technologies into the energy landscape.

SOLAR HYBRID AND COMBINED HEAT AND POWER SYSTEMS

In hybrid PVT systems, the synergistic combination of ST and PV technology allows for the electrical and thermal outputs to be obtained simultaneously, while reducing the losses in the electrical efficiency of the PV module caused by the increase in the operating temperature of the cell due to solar heating. The loss reduction is achieved in practice by using a cooling flow of either air or water through or over the unit. If designed correctly, this allows improved efficiencies compared to stand-alone PV modules (Herrando et al., 2014), while also making available a hot stream of air or water as a thermal output from the system that can be used for hot water provision, space heating, or cooling.

Current applications of hybrid PVT systems typically prioritize the electrical output, which requires the panels and thus the cooling fluid (air or water) to be kept at a low temperature. This allows the PV cells to achieve high electrical efficiencies but also decreases the usefulness of the thermal output. On the other hand, if a PVT system is designed to provide higher cooling-fluid delivery temperatures, then the PV cell efficiency will deteriorate to some extent relative to the optimal electrical power-output setting. Hence a trade-off between the two outputs is sought that depends on the end-user needs.

It is necessary to consider that PVT panels are associated with significantly higher (approximately ×2) capital costs per unit area compared to PV-only equivalents (Herrando et al., 2014). This introduces competition from ST-based alternative solar-energy options that are also capable of providing combined thermal and electrical-energy outputs, such as S-CHP systems (Freeman et al., 2015b). S-CHP systems employ ST collectors to convert solar radiation into a hot fluid-stream (i.e., an enthalpy flow) and a power-generation component to convert this (partially) to a mechanical or hydraulic output. The focus here is on simple and affordable non-concentrated and low-concentration collectors, and power generation based on thermodynamic power cycles (i.e., heat engines). From this mechanical or hydraulic output, electrical power can be generated with the use of generators. S-CHP systems can, therefore, be used to deliver heating and a mechanical, hydraulic or electrical power output depending on the requirements of the application. In addition, as is the case with PVT technology, the thermal output can be used for hot-water provision, space heating, or cooling. Since these systems are based fundamentally on the use of thermal energy, including for the generation of the electrical output, they benefit naturally from thermal-energy storage (TES) as a part of their operation.

One aspect of S-CHP technology that is of importance concerns suitable ST collector designs for such systems. Given the coupling of the performance of the ST collector component(s) and that of the power-generating component(s), optimized S-CHP systems require collectors that are designed to operate efficiently at temperatures higher than those typically associated with solar hot-water provision. Although conventional collectors can be utilized in such systems, this is an area of particular interest.

Recent simulations relevant to Northern European climatic conditions (specifically, the UK) have shown that a simple domestic S-CHP system-design operating with a 15 m^2 rooftop collector array (specifically, conventional non-tracking evacuated-tube collectors) can produce power in the region of 700–780 kW$_e$h/year (continuous power of 80–90 W$_e$) and displace 310–350 kgCO$_2$(e) in emissions at a capital cost of $6,800–8,500, of which $4,200–6,000 is attributed to electrical power generation and the rest to solar hot-water heating. This corresponds to an installed total cost per unit power generation of $85–95/W$_e$, $53–68/W$_e$ of which is associated with electricity generation alone. This system also demonstrated a potential for producing up to 86% of the household's hot-water requirement, corresponding to an additional 470 kgCO$_2$(e) in emission reductions (Freeman et al., 2015b). With an advanced system architecture incorporating a two-stage solar collector/evaporator configuration and a more suitable working fluid, a maximum net annual electrical-work output of 1,070 kW$_e$h/year (122 W$_e$) and a solar-to-electrical efficiency of 6.3% have been reported (Freeman et al., 2015c). This would cover ~32% of the electricity demand of a typical, average UK home, and represents an improvement of more than 50% over the previous effort by the same authors. By comparison, a similarly sized (15 m^2) c-Si PV system costing around $11,600 can be expected to output 200 W$_e$ in the same climate, at an installed cost of ~$59/W$_e$; a value in the middle of the $53–68/W$_e$ range given above for the simple S-CSHP system. The purchase and installation cost of equivalent side-by-side PV and ST systems (covering the same area) would range from $14,700 to $15,500, giving a cost per unit generating capacity of ~$113/W$_e$. This figure can be directly compared to the $85–95/W$_e$ range that was given above for the simple S-CHP system. For a PVT system, the purchase and installation cost is in the region $12,400–13,200, with a cost per unit capacity of ~$59/W$_e$ based on an output of ~215 W$_e$. This value may appear lower than the total capital cost per unit delivered power by the simple S-CHP system ($85–95/W$_e$); however, the PVT alternative has a significantly reduced capacity for hot water provision, amounting to 35% of household consumption at best (as predicted in Herrando et al., 2014).

The present author stated in Markides (2013) that "In summary, dwindling resources and rising energy prices, together with a growing public acceptance and even demand for government regulation to address sustainable development, environmental and health concerns, fuel economy and energy security issues are acting, and will continue to act, as major and intensifying drivers for the widespread application of energy efficiency schemes and the utilization of alternative energy sources to fossil fuels." In this present paper we are concerned with non-concentrated and low-concentration combined solar systems (i.e., S-CHP) for the supply of electricity, and if necessary also hot water and/or space heating or cooling. Furthermore, the interest is in distributed applications in the domestic (1–10 kW$_e$) and commercial/industrial (10–100 s of kW$_e$) sectors, i.e., individual households, whole residential and commercial buildings, and industrial plants. Therefore, the conversion technologies considered here cover a range of (electrical) power-output scales from 1 kW$_e$ to 1 MW$_e$. As a rough guide, scales of the order of ~100 kW$_{th}$ (thermal) and ~10 kW$_e$ (electric) would correspond to approximate collection areas of the order of 1,000 m^3 (or, ~30 m × 30 m). In addition, the focus is on heat-source temperatures below 400°C.

The following sections attempt to justify rationally the interest in suitable solar-energy technologies based on thermodynamic power cycles focusing in particular on power-generation performance, costs, and other important characteristics. Aspects of scale and the use of distributed versus centralized energy systems will be discussed, and suitable technologies that can contribute in the short-to-medium term toward a high efficiency and sustainable energy future will be identified.

TECHNOLOGY APPRAISAL

Efficiency Considerations of Common Systems

Figure 1 is a performance map that shows the thermal efficiencies, η_{th}, of common power systems over a range of heat-source temperatures, T_{hot}, from 100 to 1,400°C. Included in this figure is the relative performance of thermoelectric generators (TEGs), a competing technology for thermal-energy conversion directly to electricity based on the Seebeck effect. The maximum thermodynamic limit imposed by the Carnot efficiency, $\eta_C = \eta_{th} = 1 - T_{cold}/T_{hot}$, is indicated by the blue line, and the Novikov and Curzon–Ahlborn efficiency results from endoreversible analyses, $\eta_{th} = 1 - (T_{cold}/T_{hot})^{0.5}$, is indicated by the green line. A heat sink is selected with a fixed temperature $T_{cold} = 25°C$.

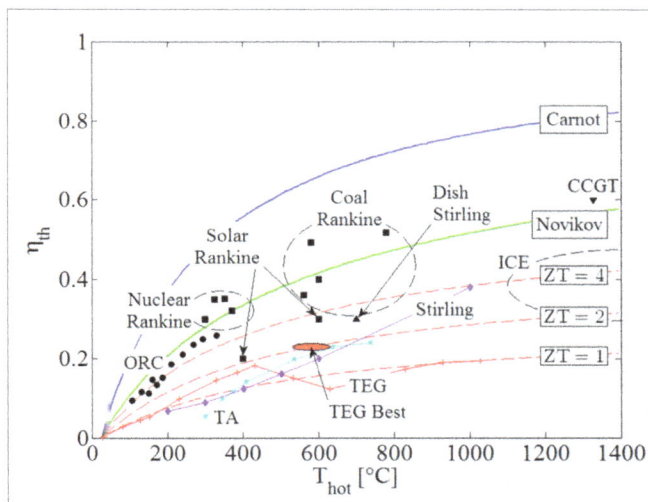

FIGURE 1 | Thermal efficiency, η_{th}, of common thermodynamic heat engines and TEGs over a range of heat-source temperatures, T_{hot}.
The circles represent actual ORC and Kalina cycles' applications; squares are for various Rankine cycles; triangles for solar dish Stirling and CCGT cycles; diamonds for conventional Stirling; and stars for TA engines. The solid red line represents the current performance of TEGs, with the three dashed red lines indicating TEG figure-of-merits ZT = 1, 2, and 4.

The circular points in **Figure 1** represent systems based on organic Rankine cycle (ORC) and Kalina (ammonia–water) cycles in actual solar, geothermal and waste-heat plants up to $T_{hot} \approx 350°C$ (Bianchi and Pascale, 2011). The square points represent, in order of increasing heat-source temperature:

- $T_{hot} \approx 300–400°C$: large-scale nuclear-powered steam-Rankine cycles;
- $T_{hot} \approx 400–600°C$: large-scale CSP Rankine cycles; and
- $T_{hot} \approx 550–800°C$: large-scale conventional coal-fired Rankine and advanced supercritical coal Rankine cycles.

In addition, the diamond points in the figure are taken from Nightingale (1986) and Bianchi and Pascale (2011), and represent the performance of Stirling-engine cycles, while the triangle at $T_{hot} \approx 700°C$ represents a highly concentrated solar-dish Stirling cycle. The stars are from the high-performance (traveling-wave) thermoacoustic (TA) engine reported in Backhaus and Swift (2000). Internal combustion engines (ICEs) based on Diesel and Otto cycles are also shown on the far left of the figure, along with a combined-cycle gas turbine (CCGT), i.e., Joule/Brayton top cycle plus Rankine bottoming cycle, at $T_{hot} > 1,300°C$.

The solid red line in **Figure 1** indicates current performance of TEGs, and the three dashed red lines indicate theoretical efficiencies, $\eta_{th} = \eta_C \times [(1 + ZT)^{0.5} - 1]/[(1 + ZT)^{0.5} + T_{cold}/T_{hot}]$, attained by TEGs given figure-of-merits: ZT = 1, 2, and 4. As above, a heat sink temperature of $T_{cold} = 25°C$ is used. In this expression, the modifying ratio multiplied by the Carnot efficiency accounts for Joule losses (i.e., losses due to parasitic electrical power dissipation and conversion to heat) and other inherent irreversible processes in TEGs (Snyder and Toberer, 2008). In evaluating the performance of actual TEG systems, $ZT(T_{hot})$ values for different materials were taken (Li et al., 2010; Szczech et al., 2011) and used in this expression at the corresponding heat-source temperature, T_{hot}, at which they are mentioned in the stated references. The dashed lines where generated with this same expression, assuming that the value ZT is maintained constant over the range of investigated heat-source temperatures, T_{hot}. It is noted that current "best" performance in terms of ZT (also indicated in **Figure 1**), is around 2.1 at 800 K/530°C (Hsu et al., 2004) and 2.2 at 900 K/630°C (Biswas et al., 2012) attained under laboratory conditions, while commercially available systems can be found with ZT values of unity (ZT ≈ 1). Vining (2009) also mentions a material with a ZT value of 3.5, but this does not yet lend itself to being produced in bulk quantities as would be required in practical applications.

The value of ZT = 4 is referred to in Vining (2009) as being "ambitious," yet possibly feasible. The opinion of the present author is that a significant breakthrough will be required to attain a working, commercially available and economically competitive TEG system operating at an average value of ZT = 4, and even then it is unlikely to emerge in the range of power-output scales and temperatures that are of interest here, i.e., >1 kW and <400°C. Therefore, the inescapable conclusion from **Figure 1**, which is reached also by Vining (2009), is that although TEGs may yet become appropriate for small-scale applications which require power outputs <100 W, it is unlikely that they will play a role in the type and range of applications that are considered in the present work, i.e., >1 kW and <400°C.

It can also be seen from **Figure 1**, as would be expected from simple thermodynamic principles, that plants based on vapor-cycle heat engines (i.e., involving phase change) outperform gas-phase heat engines for heat-source temperatures <700–800°C. This is due to the significant penalty paid for the compression of the gas relative to the power produced during expansion.

Moreover, the figure suggests that ORCs are the preferred heat-conversion technology at temperatures <400°C. It is both interesting and important to consider the reasons for which ORCs have the potential to outperform equivalent power-generation systems such as conventional steam-Rankine cycles at these lower temperatures, especially in the power range of our focus applications, i.e., 1 kW to 1 MW. This is done in Section "Rankine Cycle Ideal Maximum Power," but before we proceed to these considerations, a brief mention ought to be made of alternative technologies that have been attracting increased interest recently owing to their suitability for use with low-grade (i.e., tempera-ture) heat sources, such as ST energy, in the same applications and, therefore, range of temperatures and scales.

Alternative Technologies

Thermofluidic oscillators are unsteady thermodynamic heat-engine devices in which persistent and reliable thermodynamic property (pressure, temperature, etc.) oscillations are generated and sustained by constant temperature differences imposed by static external heat sources and sinks. A defining characteristic of these unsteady heat engines is that the working fluid contained within the device undergoes a thermodynamic cycle by virtue of the oscillatory time-varying flow of the fluid through the various connections (i.e., pipes and tubes) and compartments of the device. Oscillatory working-fluid motion is thus a neces-sary condition for operation, in direct contrast to conventional systems in which the cycle is undergone as the working fluid flows steadily from one individual component to the next, with each component responsible for a specific and well-defined process of the cycle. Common examples of such systems are TA engines, and Fluidyne and Stirling engines. As shown in **Figure 1**, however, the efficiency of these devices is limited at lower temperatures, a characteristic which is in some applications can be offset by their simple designs, long lifetimes, and low costs.

Two-phase thermofluidic oscillators (TFOs) (Smith, 2006) share a common feature with these types of devices in that reciprocating, positive-displacement work is produced by sustained flow and pressure oscillations of the working fluid. TFOs are *vapor-phase* heat engines, in that a cyclic (periodic) *two-phase* thermal interaction with two heat exchangers (hot and cold) contained within the device is established. The hot heat exchanger (HHX) introduces a high-temperature region inside the device (hotter than the saturation temperature of the working fluid), while the cold heat exchanger (CHX) introduces a cold-temperature region. The alternating thermal interaction of the working fluid with the hot and cold regions results in periodic evaporation and condensation processes that induce the forcing necessary to sustain the thermodynamic cycle and to drive the positive-displacement work done by the fluid in a suitable load.

Therefore, it can be noted that the key, defining characteristic of TFOs compared to (single-phase) TA engines, Fluidyne and Stirling engine variants is their inherent reliance on phase change during operation. This choice carries a set of important advantages and also inevitable disadvantages. One key advantage arises from the high heat transfer coefficients that are associated with phase change, which can be an order of magnitude (or more) higher than those associated with single-phase forced convection. This allows significant heat transfer over relatively small temperature differences, which is important when dealing with low-grade heat sources, and also over smaller areas. In turn, it implies smaller, more compact, and simpler heat exchangers, which has a direct implication on the eventual capital costs of these systems; an important consideration especially for the conversion of solar and other renewable energy streams (e.g., geothermal), as well as waste heat (see Section "Cost Considerations").

Example TFO devices are the "Non-Inertive-Feedback Thermofluidic Engine" (NIFTE) (Markides and Smith, 2011; Solanki et al., 2012, 2013a,b; Markides and Gupta, 2013; Markides et al., 2013, 2014; Palanisamy et al., 2015) that is being developed as a solar-powered fluid pump, and the even more recent "liq-uid Stirling engine" (also known as the "Up-THERM" engine) (Glushenkov et al., 2012) that is currently under development as a combined heat and power (CHP) prime-mover. In both cases, these technologies are suited also to the conversion of ST energy, and a key strength is their simple construction with a reduced number of moving parts and dynamic seals, thus leading to low capital and maintenance costs and long lifetimes, as mentioned earlier. However, although the thermodynamic performance of these TFOs has improved significantly in recent years, it is expected to remain considerably lower than that of equivalent ORC power systems, at least for heat-source temperatures upwards of 80°C. Specifically for the NIFTE, thermal, and exergy efficiency values at low temperatures (<100–200°C) are expected to remain within the range originally predicted in Markides (2013) for this technology, i.e., 1–5% and 5–20%, respectively, depending on the characteristics of the device configuration, the application and the mode of operation. Given these low values, TFOs are not considered further in this paper.

Furthermore, a particularly interesting alternative technology, known as the trilateral cycle or trilateral flash cycle (Fischer, 2011; Ajimotokan and Sher, 2015), can offer potential performance benefits at the lower temperature ranges (typically below ~150°C), even compared to ORCs. This option was not discussed in Section "Efficiency Considerations of Common Systems" and is absent from **Figure 1**, since the focus here is on solar applications where higher temperatures can be attained from low-cost collectors, in which case ORCs should hold a performance edge.

Rankine Cycle Ideal Maximum Power

We return to consider the performance of Rankine cycles, and in particular the comparative performance of organic fluids relative to water/steam as the working fluid in such cycles. It is commonly perceived that the employment of organic compounds as working fluids with lower boiling points compared to water (at the same pressure) is necessitated when low heat-source temperatures are involved, since these cycles are associated also with low evaporation temperatures. A typical argument suggests an inability to employ water at heat-source temperatures <100°C, given that water will not evaporate below this temperature. However, this is not strictly speaking correct since water can be evaporated at these temperatures at sub-atmospheric pressures.

Yet the use of organic working fluids *does* offer certain advantages, both theoretical (thermodynamic and thermal) and practical, over conventional steam-Rankine systems, which arise from the properties of the available fluids (or mixtures thereof) that act to replace water as the working fluid. Briefly, these include a reduced reliance on superheating to avoid problematic condensation in the case where turbomachines are used for expansion and work extraction, as well as simpler and more affordable evaporator and condenser designs with reduced exergetic losses owing to the more flexible selection of the thermodynamic and transport fluid properties and conditions, including pressures, heat transfer rates, and a greater degree of freedom in designing the single- and two-phase processes in key components. This section is aimed at a high-level exploration of the reasons underpinning the potential advantages displayed by organic working fluids relative to water/steam. A more direct study into the relative performance of these working fluids is presented in Section "Direct Performance Comparison with Steam-Rankine Cycles"; the discussion here is restricted to simple thermodynamic arguments with idealized cycles.

Figure 2 shows a contour plot of the normalized power output from an infinite series of infinitesimal thermodynamically ideal (Carnot cycle) engines operating between (varying) heat-source *and* heat-sink temperatures, effectively modeling an idealized energy integration application in which power is generated between a heat source and sink. We consider a heat-source fluid stream entering a hot heat exchanger (HHX) within which an infinitesimal amount of heat, $d\dot{Q}_{hot}$ (>0), is added from the fluid stream to the working fluid in each successive cycle. During this

process, the enthalpy of the hot fluid stream decreases according to $d\dot{H}_{hot} = -d\dot{Q}_{hot}$. Similarly, a heat-sink fluid stream enters a cold heat exchanger (CHX) within which heat is rejected from the cycles to the fluid stream. The heat-source stream enters the HHX at a temperature $T_{hot,in}$, and experiences a total temperature drop through that heat exchanger ΔT_{hot}, such that $\Delta \dot{H}_{hot} = (\dot{m}c_p)_{hot} \Delta T_{hot}$, while the heat-sink stream enters the CHX at a temperature $T_{cold,in}$, and experiences a temperature rise ΔT_{cold}, such that $\Delta \dot{H}_{cold} = (\dot{m}c_p)_{cold} \Delta T_{cold}$. For simplicity, but without loss of generality, we consider the case when the two streams have equal heat-capacity rates, $(\dot{m}c_p) = (\dot{m}c_p)_{hot} = (\dot{m}c_p)_{cold}$.

Assuming that the two heat exchangers (HHX and CHX) are ideal with no losses to the surroundings, the enthalpy difference across the HHX is equal to the heat transferred to the working fluid in all of the cycles, $\Delta \dot{H}_{hot} = -\Delta \dot{Q}_{hot}$. Note that this analysis is subtly different from a conventional maximum work (exergy) analysis, in which heat is rejected from a similar arrangement to a constant "dead" state temperature, rather than to a varying cooling stream temperature as is done here.

The horizontal axis in **Figure 2** is the inlet temperature of the heat-source fluid stream to the HHX, $T_{hot,in}$, while the vertical axis is the (fractional) temperature drop of the same stream through the HHX normalized by the inlet temperature, $\Delta T_{hot}/T_{hot,in}$. The inlet temperature of the heat-sink fluid stream to the CHX is set to $T_{cold,in} = 20°C$. Also superimposed on this plot are two lines. The white line is a locus of the maximum power output at each value of $T_{hot,in}$, which corresponds to a monotonically increasing value of $\Delta T_{hot}/T_{hot,in}$ that obeys the relationship $1-(T_{hot,in}/T_{cold,in})^{0.5}$. The red line traces the output of this ideal arrangement for a given application with a fixed heat flow rate, $d\dot{Q}_{hot}$, and a fixed heat-capacity rate, $(\dot{m}c_p)$. This case has a fixed heat flow per unit heat capacity, $\dot{Q}_{hot}/(\dot{m}c_p)$, chosen here arbitrarily to be equal to 100.

Two important interpretations emerge from **Figure 2**. First, the *ideal* conversion of heat at *higher* temperatures such that maximum power is extracted requires that the normalized heat-source temperature drop is high. For instance, consider an ideal system to be used for converting heat at $T_{hot,in} = 100°C$. For this system, maximum power is attained for a normalized heat-source temperature drop of $\Delta T_{hot}/T_{hot,in} = 0.11$. This corresponds to a temperature drop of $\Delta T_{hot} = 42°C$, from 100 to 58°C through the HHX. At the same time (not shown in the figure) the cold temperature through the CHX increases by 38°C from 20 to 58°C. Conversely, consider a second system to be used for converting heat at $T_{hot,in} = 400°C$. Maximum power for this second system is attained for a normalized heat-source temperature drop of $\Delta T_{hot}/T_{hot,in} = 0.34$, which corresponds to a temperature drop of $\Delta T_{hot} = 230°C$, from 400 to 170°C. The cold temperature though the CHX increases by 150°C from 20 to by 170°C.

Now, for a given heat-capacity rate of the heat-source and sink fluid-streams the second system will be more than 5 times larger in terms of the heat input to the cycle and, thanks to its higher efficiency, almost 10 times larger in terms of the power output. In other words, *for the same heat-source fluid-stream heat capacity rate*, larger-scale (centralized) plants are better suited to the effective utilization of higher temperature heat-sources, whereas smaller-scale (distributed) systems are more appropriate in optimally converting lower-temperature sources. The

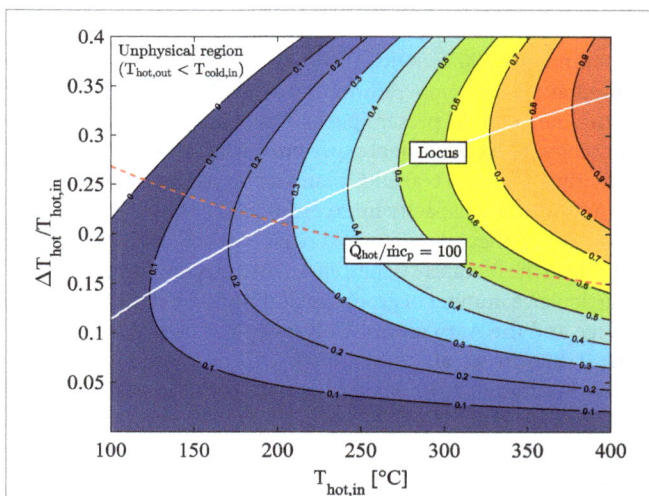

FIGURE 2 | Effect of heat-source cooling and heat-sink heating on ideal maximum cycle power output per unit heat-capacity rate $(\dot{m}c_p)_{hot} = (\dot{m}c_p)_{cold}$, showing locus of maximum net output power, with $T_{cold,in} = 20°C$.

increased stream heat capacity rates expected in the larger-scale systems will only act to amplify this distinction.

Secondly, an important difference between the employment of steam/water and organic compounds as working fluids (i.e., between conventional steam-Rankine cycles and ORCs) is the much greater specific enthalpy associated with phase change of the former. An increased specific enthalpy associated with heat addition will lead to an increased heat (per unit mass of working fluid) intake into the cycle, shifting the ideal operation of this cycle toward higher temperatures and larger systems as a consequence of the discussion in the previous paragraphs. It is also true that the *specific* power output from Rankine cycles, even at low temperatures, can be higher than ORC equivalents. However, this advantage is overcome and negated by the need to use much lower working fluid mass flow rates in Rankine cycles operating at low temperatures compared to ORCs, owing to the large differences in specific enthalpy of heat addition.

Cost Considerations

An acceptable performance from a technical standpoint can be judged based on indicators, such as primary energy/fuel efficiency, emissions, flexibility of operation, and ability to match variable demand. Yet, beyond these purely technical considerations, the widespread deployment of any successful solution to the energy challenge must be associated with, either a cost benefit or at the very least a cost level that is affordable and economically justifiable to the end-user or investor (Markides, 2013).

In conventional power generation, fuel costs are the single largest contributor toward the total cost of electricity. Consider, for example, a typical coal-fired steam power plant with a typical efficiency of 38–40%, a capital cost in the region of ~$1,300/kW, an additional operating and maintenance cost of $30–45/kW, and an economic life expectancy of 30 years (Royal Academy of Engineering, 2004). This plant has a total LEC of $50/MWh produced over the lifetime of the plant. Moreover, the largest single contributor toward this cost is the cost of fuel, which amounts to 35%. The case is even stronger for gas-fired power plants. A typical simple open-cycle gas turbine power plant with a typical efficiency of 39–43%, a capital cost of $510/kW, an operating and maintenance cost of $55/kW, and an economic life expectancy of 20 years has a total LEC of $55/MWh of which >60% is attributed to the fuel. Similarly, closed-cycle combined gas turbine power plant with a typical efficiency of 58–60%, a capital cost of $470/kW, an operating and maintenance cost of $40/kW, and an economic life expectancy of 25 years has a total LEC of $55/MWh of which 60% is again attributed to the purchase of gas (Royal Academy of Engineering, 2004).

Hence, beyond its formal definition, it is reasonable to argue that, for the case of conventional power generation, the thermal efficiency is also a figure-of-merit that is a reasonable measure of the electrical-energy output (and thus profit) *per unit* total cost. This cannot be said, however, for power-generation systems whose energy input is not associated with a significant *operating* cost, such as solar heat (as well as geothermal energy and waste heat). In this case, the total cost is dominated by the up-front initial investment required for the necessary capital expenditure, and consequently, the figure-of-merit that is the electrical-energy output (i.e., profit) per unit total cost must be evaluated directly as the electrical-energy output per unit installation cost, or at least per unit capital cost. In both cases, this figure-of-merit goes some way toward reflecting the true economic viability of such systems, in a way that thermal efficiency alone cannot.

Distributed Energy-Conversion Systems

In the previous sections, a brief overview was given of the considerations that are acting to motivate an interest in technologies that are capable of converting solar heat to useful work (either mechanical, hydraulic, or electrical). Additionally, some of these technologies are diverse enough to be suitable for the conversion of other renewable energy sources such as geothermal heat and biomass/biogas, as well as waste heat in a variety of settings.

These technologies are being proposed, in particular, for distributed power generation (and/or simultaneous heat/cooling provision). Benefits from an increased deployment of distributed power-generation solutions include enhanced reliability and security, reduced losses from energy transmission and distribution, as well as reduced infrastructure and maintenance costs for transmission and distribution, and easier plant sizing (Gullì, 2006; Strachan and Farrell, 2006).

It is implied that distributed systems will be smaller in scale than centralized equivalents and will not benefit from the economies of scale the latter enjoy. One must also remain aware of the fact that centralized, larger-scale systems will retain an edge in plant efficiency, but that this efficiency will be compromised by increased transmission/distribution losses from the plant to the consumer/end-user. In many cases, these losses are not negligible, amounting to an efficiency reduction of 5–10% points.

ORGANIC RANKINE CYCLES

Technology Overview

ORC systems have been indicated in the previous sections as a highly appropriate technology for the conversion of heat at temperatures <400°C. ORCs with suitable working fluids can be used at higher temperatures, but we will focus on this temperature range in the present paper. ORCs are a relatively mature technology, with operational experience available since the 1960s. Currently, more than 600 ORC plants are in operation worldwide, with a cumulative capacity in excess of 2,000 MW.

A number of excellent reviews of all types of ORC systems are available in the literature (Chen et al., 2010; Lecompte et al., 2015). In particular, a number of groups including those at the University of Liege and the Agricultural University of Athens (Orosz et al., 2009; Tchanche et al., 2009, 2010, 2011; Delgado-Torres and García-Rodríguez, 2010; Malavolta et al., 2010; Kosmadakis et al., 2011; Declaye et al., 2012) have considered ORC systems specifically in solar applications in a series of noteworthy studies that include installing and testing solar-ORC systems. The present paper focuses on the most advanced in terms of development/readiness and lowest-cost system, the sub-critical ORC without regeneration. A simple sub-critical, non-regenerative ORC system is shown in **Figure 3**, along with a cycle on a *T–S* diagram with R-245fa as the working fluid.

The main components of the system in **Figure 3** are a feed pump (this can be multistage), evaporator (this can comprise a number of components, including a superheater), expander/turbine (again this can comprise a number of stages) and condenser (including a desuperheater). A regenerator can also be used to recover some of the heat-rejected downstream of the expander (Point 4 in the diagram) and to use this to perform part of the heating downstream of the pump (Point 2).

As mentioned in Section "Rankine Cycle Ideal Maximum Power," ORCs are associated with a number of specific advantageous features compared to water/steam-Rankine cycles. First, unlike wet fluids, such as water, dry and isentropic organic fluids (see **Figure 4**) have positively sloped or vertical dry saturation curves. Therefore, they do not require a significant degree of superheating to avoid condensation and droplet formation in turbines/expanders. Such a scenario can cause mechanical damage to the turbine blades and also degrade the thermodynamic performance of this component. The former would not apply to more structurally robust expander designs (e.g., reciprocating piston expanders), but the latter would remain.

In the case of wet working fluids, the desire to keep the flow through the expander/turbine outside of the saturation region over the entire expansion process, and hence for the exit state from this process to also be outside the saturation region, translates to a requirement for significant superheating prior to entry into the turbine. The absence of adequate superheating leads to an intersection of the dry saturation curve during expansion, and thus, expansion into the saturation (two-phase) region.

Second, it is advantageous thermodynamically to expand the working fluid to the lowest possible pressure, which corresponds to condensation and heat rejection at a temperature as close as possible to the cooling stream temperature. Assuming this is at ambient conditions (20–25°C), the condensation temperature would be a few degrees higher than this, as determined by the pinch temperature difference in the condenser. For water, a saturation temperature of 30°C corresponds to an absolute saturation pressure close to 0.04 bar. The large pressure difference between the surrounding atmosphere and any components that

operate at such low pressures can lead to ingress of air into the cycle with significant detrimental effects on system performance. The design of components that can operate reliably at such a degree of sub-atmospheric pressures is difficult and expensive. Conversely, for R-245fa, the saturated condensation pressure at a saturated temperature of 30°C is 1.8 bar, which is above atmospheric.

Furthermore, it can be said that, in general, the large choice of currently available (and possible future) organic compounds that can be used as working fluids, and mixtures thereof, allows ORCs to be "tuned" to specific applications. Therefore, ORCs comprise a more flexible solution by allowing some degree of control over the phase behavior of the working fluid, the design of the processes that comprise the cycle, and in matching the cycle to available heat sources and heat sinks.

Direct Performance Comparison with Steam-Rankine Cycles

When discussing **Figure 2**, a rudimentary analysis was used to indicate the underlying reasons for which ORCs may outperform conventional Rankine cycles when converting low-grade heat in small-scale systems. The current section proceeds to compare these two cycles directly and to offer further insight into their relative performance, and also, the approximate cost of related power-generation systems. Specifically, we focus here on case-study application where it is desired to generate electrical power from a fluid stream at an initial temperature of $T_{hot,in} = 200°C$. The heat-source fluid stream is allowed to interact thermally with the heat engine, such that its enthalpy (and thus temperature) will decrease progressively as heat is taken in the cycle. This is similar to the rudimentary analysis that led to the result in **Figure 2**, only that analysis considered a series of multiple ideal cycles, i.e., fully reversible, Carnot cycles, whereas here the cycle is a single theoretical Rankine cycle with either a water or an organic compound as the working fluid. The heat-source fluid stream is taken to have a mass flow rate $\dot{m}_{hot} = 500 \, kg/s$ and a specific heat capacity $c_{p,hot} = 1 \, kJ/kg \, K$, such that the stream heat capacity duty is $(\dot{m}c_p)_{hot} = 5 \times 10^5 \, W/K$. The heat-sink (cooling) fluid stream is assumed to be (constant) at an ambient temperature of 20°C,

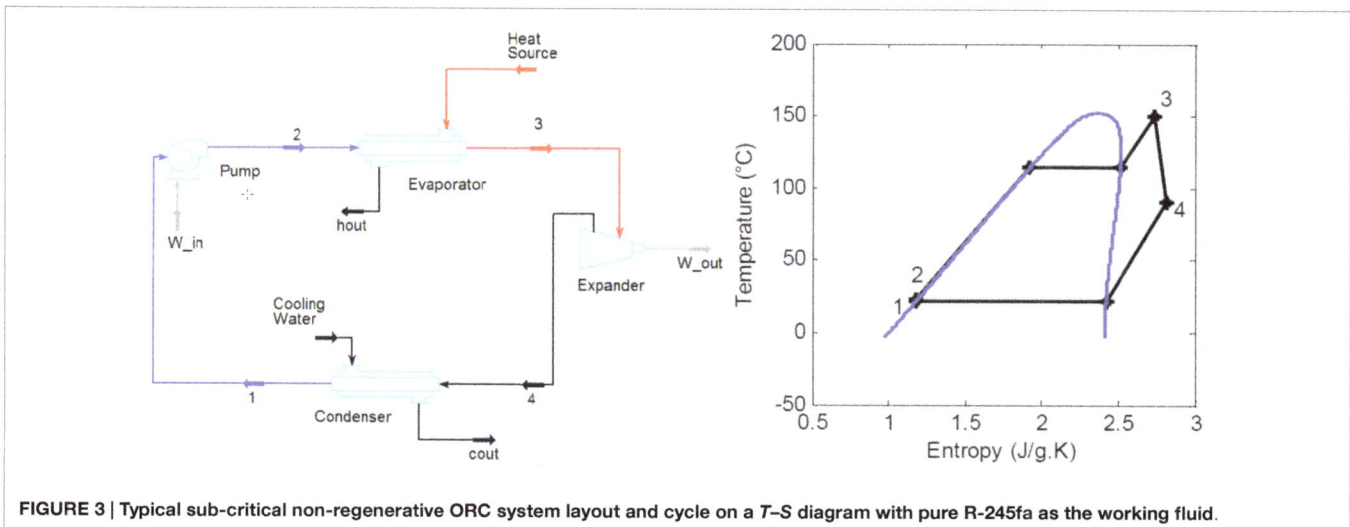

FIGURE 3 | Typical sub-critical non-regenerative ORC system layout and cycle on a T–S diagram with pure R-245fa as the working fluid.

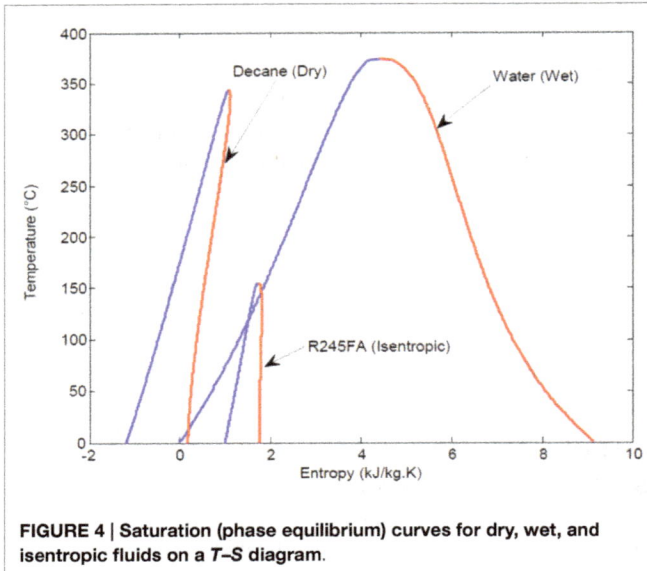

FIGURE 4 | Saturation (phase equilibrium) curves for dry, wet, and isentropic fluids on a *T–S* diagram.

while the condensation temperature and temperature at the inlet of the pump, T_1, is taken to be 10°C higher than this, $T_1 = 30$°C.

Figures 5 and **6** show the specific work output and thermal efficiency, respectively, for a number of Rankine cycles operating with water and fluid R-245fa. Three lines are shown on each plot. Each one of these corresponds to a different saturation temperature (and thus also a different saturation pressure) during evaporation, as per the legend. Results for water/steam-Rankine cycles are shown for the case of expansion to and condensation at: (i) 1 bar and 100°C, and (ii) 0.04 bar and 30°C.

Clearly, it is thermodynamically beneficial to expand to a low temperature that is as close as possible to atmospheric *temperature*, which in this case is 30°C. However, this may come at a severe cost especially for small-scale systems, as discussed previously. Hence, expansion to near atmospheric *pressure* is also shown. Expansion to the lower temperature leads to a 2.5–4-fold increase in both specific work output and thermal efficiency.

In **Figures 5** and **6**, better performance (specific work *and* efficiency) can be observed at higher evaporation pressures. The extent of superheating does not strongly affect water-based cycle performance but has a significant effect on work output from the ORCs, even though this does not appear in the ORC efficiencies. Essentially this is due to a near-proportional increase in both the heat input to the cycle along with the specific work output, as the degree of superheating is increased.

When comparing the working fluids, it is found that water outperforms the organic fluids with respect to specific work output by a factor of between two and five at the higher condensation pressure and temperature (for water). This increases to a factor of 10 or more at the lower condensation pressure and temperature. Although the specific work potential of the steam-cycles is clearly higher than the equivalent potential of the organic-fluid cycles, the performance in terms of efficiency presents a more mixed picture. In fact, at the higher condensation pressure and temperature for water, the organic fluids outperform water by a factor of 2–3, while at its lower condensation pressure and temperature

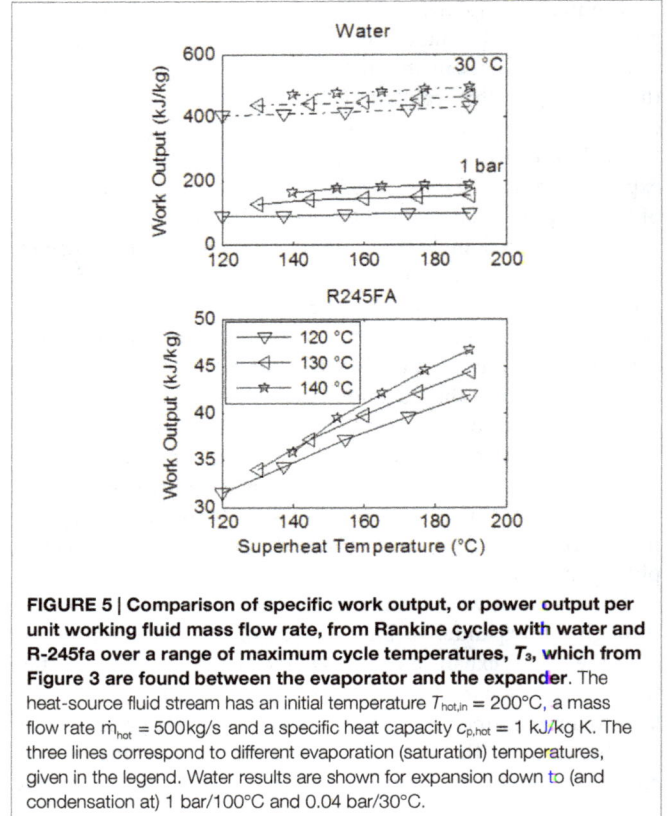

FIGURE 5 | Comparison of specific work output, or power output per unit working fluid mass flow rate, from Rankine cycles with water and R-245fa over a range of maximum cycle temperatures, *T₃*, which from Figure 3 are found between the evaporator and the expander. The heat-source fluid stream has an initial temperature $T_{hot,in} = 200$°C, a mass flow rate $\dot{m}_{hot} = 500$ kg/s and a specific heat capacity $c_{p,hot} = 1$ kJ/kg K. The three lines correspond to different evaporation (saturation) temperatures, given in the legend. Water results are shown for expansion down to (and condensation at) 1 bar/100°C and 0.04 bar/30°C.

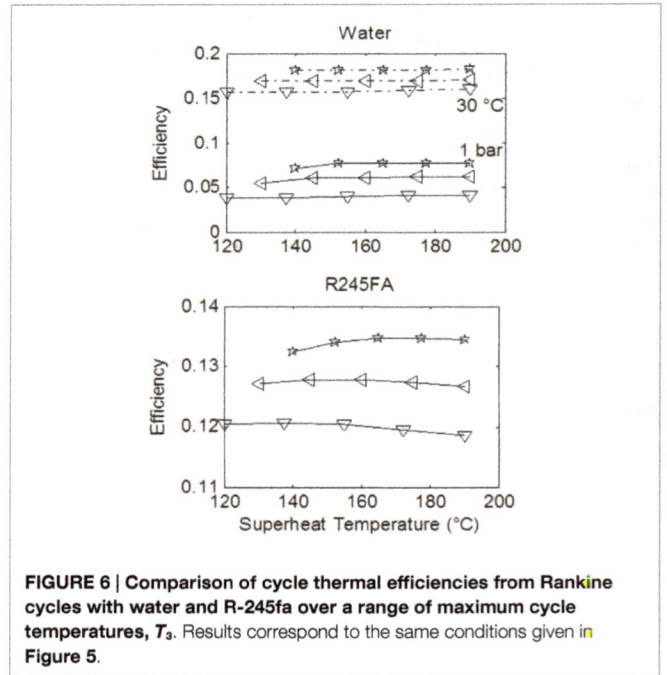

FIGURE 6 | Comparison of cycle thermal efficiencies from Rankine cycles with water and R-245fa over a range of maximum cycle temperatures, *T₃*. Results correspond to the same conditions given in Figure 5.

water outperforms the organic fluids only marginally, by 3–4 absolute points, or 25–30% in relative terms.

Therefore, if one is to accept that it is not economically desirable to design a system in which steam is expanded down to and condenses at pressures of 0.04 bar, which is a reasonable point of view for affordable, distributed, and small-scale power generation, ORCs show a potential for improved performance compared to conventional (steam) Rankine cycles in terms of efficiency. Furthermore, it is important to consider not only the specific work output of these cycles, but the actual power output once the mass flow rate of the working fluid is evaluated based on the thermal interaction between the heat-engine cycle and the external heat-source fluid stream. The result from such a consideration is shown in **Figure 7**, where we include data from three organic fluids: Butane, R-245fa, and Perflenapent.

The results in **Figure 7** were generated by progressively increasing the mass flow rate of the working fluid in each cycle (i.e., each point on this plot) until a pinch temperature difference of 10°C was reached in the evaporator between the heat-source stream and the working fluid for that cycle. This is the maximum working fluid mass flow rate. Interestingly, superheating is detrimental to ORC power output, but not to water. This figure demonstrates that, at least theoretically, organic fluids have the potential to outperform water by a considerable extent, also when considering power output in the chosen case study. In particular, power output for R-245fa is higher than that for water by a factor of 4–5. It is emphasized that this figure does not show water data at the low condensation conditions (0.04 bar and 30°C). Nevertheless, the underlying conclusion remains unchanged, even when this data are considered. The organic fluids in this case outperform water by a factor between 1.5 and 2.

The observation that organic working fluids have (desirably) higher power outputs than water, even when compared to water condensing at the lower pressures and temperatures that showed much higher specific work outputs (per unit mass flow rate of working fluid) by more than an order of magnitude (refer back to **Figure 5**), can be understood by the much higher mass flow rates permitted in the ORCs before any pinch violation is reached. This can also be seen directly in **Figure 7** and arises from the significantly higher specific enthalpy change during heat addition for water/steam compared to that of the organic fluids, as indicated in **Figure 4**.

In **Figure 8**, a basic attempt is made at estimating approximate system costs. Here, we show the sum of costs associated with the purchase of the four basic components that form the Rankine heat-engine systems. Each data point corresponds to the same systems contained in **Figure 7**. Heat exchanger costs were evaluated by using the C-value method, while costs for the pumps and expanders were obtained by compiling price information from a market study and establishing a correlation with component power, pressure ratio, and flow rate (Oyewunmi et al., 2014, 2015). The C-value method is an approximate approach for the costing of heat exchangers, described in Hewitt et al. (1994).

Figure 8 shows that, due to their larger heat exchangers (allowing higher power outputs) ORCs are more expensive when considering the total system costs compared to steam-Rankine cycles. However, when the cost of the system is normalized by the power output capacity of the system, thus providing the all-important indicator of cost per unit useful output, the ORCs are shown to be a more affordable solution.

Finally, it is possible and instructive to condense the information contained in **Figures 6–8** into a single performance-cost map. This is attempted in **Figure 9**. For simplicity and clarity, we do not show all data corresponding to each working fluid in this figure, which can be done by drawing an area for each working fluid. Instead, we select a single degree of superheating that corresponds to the maximum total power output. We recall, from **Figure 7**, that for organic fluids this is attained with little or

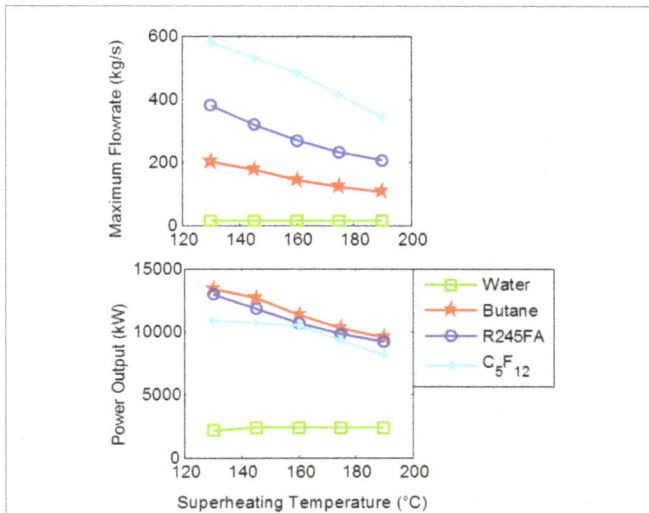

FIGURE 7 | Maximum working fluid mass flow rate and total power output from Rankine cycles with water and indicated organic fluids over a range of maximum cycle temperatures, T_3. Results correspond to the same conditions given in **Figures 5** and **6**. Water results are shown only for expansion down to (and condensation at) 1 bar/100°C.

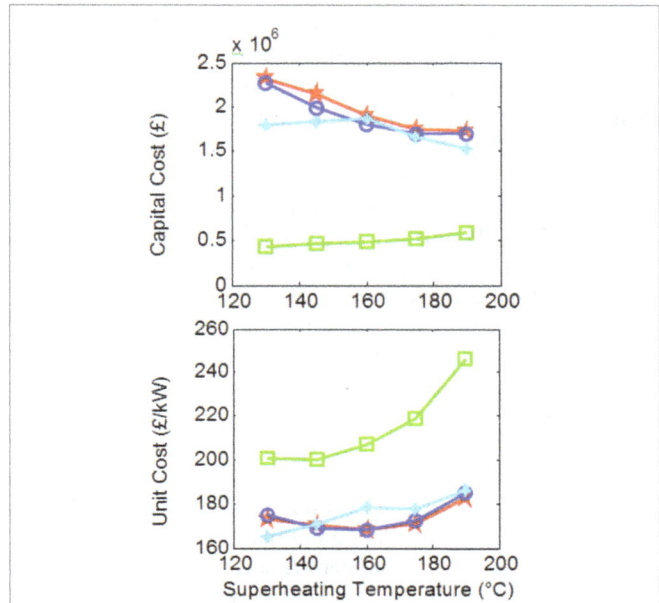

FIGURE 8 | System costs corresponding to Figure 7, over a range of maximum cycle temperatures, T_3.

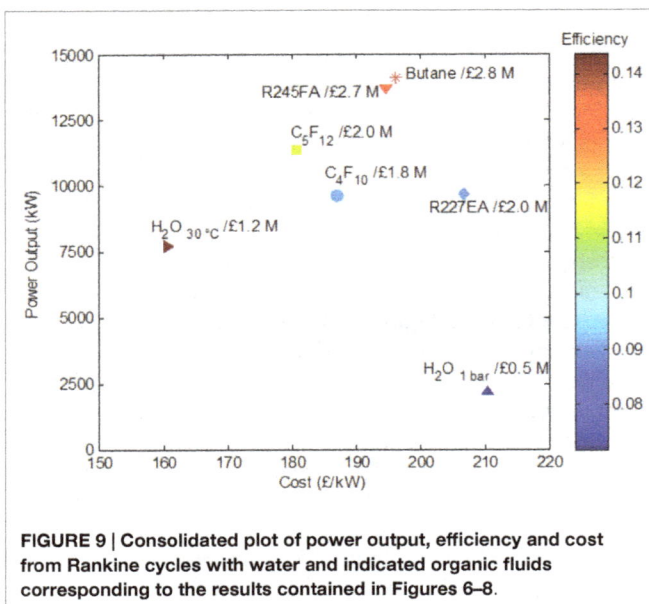

FIGURE 9 | Consolidated plot of power output, efficiency and cost from Rankine cycles with water and indicated organic fluids corresponding to the results contained in Figures 6–8.

no superheating. So, for example, for R-245fa, this would be at an evaporation temperature of 120°C, when the power output is ~14 MW and the cost per unit power ~£195/kW (or $300/kW).

Opportunities for Improvement and Future Developments

It is known from second law (exergy) analyses that about 75–80% of the ultimate potential to do useful work in a sub-critical ORC is lost in the heat exchangers (evaporator, condenser and regenerator) and about 20–25% in the expansion machine. The lost work (exergy destruction) in the evaporator amounts to ~1.5:1–2:1 times that lost in the condenser (Freeman et al., 2015b). Hence, significant performance improvements can come from advances in these areas or components. Additionally, there is a great interest in the identification of optimal working fluids for specific applications, and attention has turned to the possible employment of binary and even tertiary mixtures of organic compounds as working fluids in ORCs (as mentioned in Section "Heat Exchanger Thermodynamic Performance Versus Size/Cost"). This is an area of active research, and a challenge arises in predicting reliably the properties of these fluids.

Advanced ORC system models that include a computer-aided molecular design (CAMD) framework with explicit information on the role of molecular size and structure on thermodynamic and thermal properties of working fluids are also currently in development (Lampe et al., 2012), based on thermodynamic theories such as the statistical-associated fluid theory (Lampe et al., 2012; Oyewunmi et al., 2014, 2015). Such models will play an important role in identifying optimal compromises between thermodynamic and thermal performance, which controls efficiency, power output, system size, and cost.

Finally, at the scales of operation of interest, the selection of the expansion machine is an open question, with positive-displacement expanders presenting a real challenge to turbomachines. A

significant effort is being made in the modeling and development of ORC systems featuring positive-displacement expanders, which promise higher efficiencies, for the applications identified in the present paper.

FINITE HEAT TRANSFER EFFECTS

Heat Exchanger Thermodynamic Performance Versus Size/Cost

Recently, the selection of working fluids for ORC systems has received close attention from the ORC community, including a particular interest in multicomponent fluid mixtures, due to the opportunities they offer in improving thermodynamic performance. Various authors have carried out investigations to demonstrate and quantify these benefits, which have shown that working-fluid mixtures can exhibit an improved thermal match with the heat source compared to the isothermal profile of (isobaric) evaporation of pure-component fluids, therefore reducing exergy losses due to heat transfer, and increasing thermal and exergy efficiencies (Angelino and di Paliano, 1998; Wang et al., 2010; Garg et al., 2013).

Investigators have carried out both experimental and theoretical studies across a range of heat-source temperatures into the benefits of employing working-fluid mixtures based on refrigerants (Sami, 2010; Chen et al., 2011; Aghahosseini and Dincer, 2013), hydrocarbons (Heberle et al., 2012; Shu et al., 2014), and siloxanes (Dong et al., 2014). Compared to pure fluids, binary mixtures showed increased power outputs by up to 30% and thermal efficiencies by >15% in some cases. Excellent second law analyses have also shown significant potential benefits (Lecompte et al., 2014). [Some exceptions to these general trends have also been reported (Li et al., 2014).]

Additionally, fluid mixtures can be used to adjust the environmental and safety-related properties of ORC working fluids or to improve design parameters of system components; this is increasingly of importance. At the same time, some investigators have begun to develop and apply advanced CAMD methodologies (Papadopoulos et al., 2010; Lampe et al., 2014) with a view toward identifying or designing optimal fluids for ORC systems.

While these efforts have demonstrated the potential *thermodynamic* advantages of working-fluid mixtures, notably in terms of power output and efficiency, many of the associated conclusions have been derived strictly based on the thermodynamic cycle analyses that do not fully consider the expected heat transfer performance between the heat-source/sink and working-fluid streams in the heat exchangers of ORC engines. In particular, the heat transfer and cost implications of using working-fluid mixtures have not been properly addressed. Essentially this arises from the minimization of the temperature differences between the working fluid stream (cycle) and the heat-source/sink streams on the other side of the heat exchange components. This practice is thermodynamically beneficial, but detrimental in terms of heat transfer, and it opens up an important area of research aiming to enhance heat transfer with low-cost modifications across small temperature differences. Moreover, refrigerant mixtures are known to exhibit reduced heat transfer coefficients compared to their pure counterparts (Jung et al., 1989). Specifically, heat transfer coefficients for refrigerants mixtures are usually lower than the ideal values, linearly interpolated between the mixture components. This may invariably lead to larger and more expensive heat exchangers in an ORC system that employs a working-fluid mixture. Therefore, although working-fluid mixtures may allow a thermodynamic advantage over single-component working fluids, they may also lead to higher system costs owing to a deterioration in their thermal performance.

Recent analyses (Oyewunmi and Markides, 2015 ; Oyewunmi et al., 2015) have revealed that the temperature glides of the working-fluid mixtures during evaporation and condensation can result in higher power output and thermal/exergy efficiencies for fluid mixtures (at least for the two sets of mixtures in the specific cases studied). The pure fluids did however result in smaller expanders due to their lower volumetric flow rates and expansion ratios and also smaller evaporators and condensers, requiring less expensive components than the fluid mixtures. Therefore, although the mixtures were found to have the highest power output, they also had the highest rated costs (equipment cost per kW of power generated), which resulted from larger equipment/component sizes compared to the constituent pure fluids.

These observations imply that the thermodynamic benefits derived from using working-fluid mixtures may be outweighed by the increased costs incurred, although this is in need of confirmation and generalization. The fact that these insights were only possible from a direct consideration of thermal and cost factors as exemplified here, underlines the importance of employing a combined thermodynamic, thermal, and cost approach in the selection of optimal working-fluid (mixtures) for ORC systems.

Thermally Induced Thermodynamic Losses

Time-mean heat transfer can act to affect heat-engine performance detrimentally by giving rise a direct loss of the available heat from the heat source to the surroundings, which does not then take part in the thermodynamic cycle. This can be alleviated by careful design of the relevant components, for example, by insulating the components and/or by separating hot and cold sections in order to force thermal-energy transport into the working fluid cycle. Beyond these losses, situations arise in which *unsteady* heat transfer (even in the case that the time-averaged heat transfer is zero) plays a significant role in affecting the performance of the energy-conversion systems under consideration, as well as of similar systems. This is the case in positive-displacement expansion machines that are being envisioned as high efficiency alternatives to turbomachines when used in small-scale ORC systems (but also, as an side, in the heat exchangers and in the nominally adiabatic vapor volumes of TFOs, such as the NIFTE).

Some peculiarities arise with respect to unsteady heat transfer in these systems owing to the fact that, unlike time-mean heat transfer, it is not possible to arbitrarily minimize this component of heat transfer with increasing levels of insulation due to a thin solid region (known as the thermal diffusion length or "penetration depth") that is in thermal contact with and experiencing time-varying heat exchange with the fluid domain. This region

will interact thermally with the fluid in a time-varying manner and affect the magnitude and phase of the heat transfer process. This unsteady thermal process and its detrimental effect on thermodynamic performance (also in the absence of time-mean heat transfer) are dealt with in the following sections.

Unsteady and Conjugate Heat Transfer

Unsteady and conjugate heat transfer is defined as a time-varying thermal-energy transport process in which a solid is in thermal contact with a fluid, with both domains exhibiting a time-varying temperature and heat flux at their common boundary, i.e., the solid–fluid interface. **Figure 10** shows a conjugation map for a one-dimensional thermal interaction between a solid of finite thickness a and a fluid within which a flow imposes a constant convective heat transfer coefficient h. This map is plotted as a function of the unsteady Biot number, $Bu = h\delta/k_s$, where the relevant length scale in the solid is the thermal penetration depth or diffusion length $\delta = (\alpha_s \tau/\pi)^{0.5}$ that takes part in the unsteady thermal process rather than the full extent/thickness of the solid domain a, and the Fourier number, $Fo = \alpha_s \tau/a^2$. It is noted that the unsteady Biot number Bu is related to the conventional steady Biot number Bi via the dimensionless length scale $a^* = a/\delta_s$, such that $Bu = Bi/a^*$. In these definitions, k_s is the thermal conductivity of the solid, α_s is the thermal diffusivity of the solid, and τ is the period of the temperature oscillations in the fluid domain due to some time-varying (periodic) thermodynamic process.

The red region in **Figure 10** indicates large temperature fluctuations and small heat-flux fluctuations (i.e., an isoflux boundary condition) on the (inner) solid–fluid interface, whereas the blue region indicates large heat-flux fluctuations and small temperature fluctuations on the same interface (i.e., an isothermal

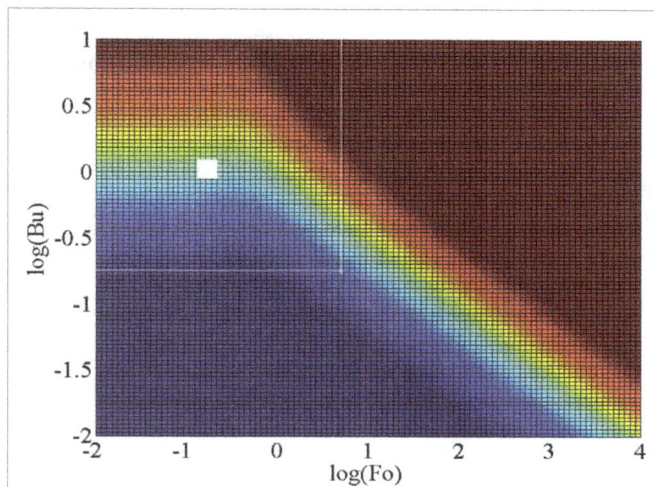

FIGURE 10 | Map of conjugation in Fo–Bu space, showing the extent of conjugation in unsteady 1D solid–fluid systems, with an isoflux outer wall boundary condition. Red indicates large temperature fluctuations and small heat-flux fluctuations (approaching an isoflux boundary condition) on the inner wall surface at the (wetted) solid–fluid interface, and blue indicates large heat-flux fluctuations and small temperature fluctuations on the inner wall surface (approaching an isothermal boundary condition).

boundary condition). Since data from ORC expanders are not available, the large white square is an approximate narrow area occupied by the NIFTE prototype water-pump TFO as reported in Smith (2006), and the extended white space is the estimated design area within which the TFO technology is expected to be found given reasonable variations from the initial design (Markides and Smith, 2011; Glushenkov et al., 2012; Solanki et al., 2012, 2013a,b; Markides and Gupta, 2013; Markides et al., 2013, 2014; Palanisamy et al., 2015).

It is evident that the region of interest straddles the two extreme cases defined above. This implies that the boundary condition on the working fluid is neither isothermal nor isoflux, and that the solid and fluid are thermally coupled in such a way that in order to predict the temperature and heat flux at the solid–fluid interface the heat transfer problem must be solved in both domains and the solutions matched at this interface. This observation is important and has serious implications because it suggests that any effort to understand and predict the unsteady thermal losses in such a device must contain explicit information not just on the thermal processes in the fluid (i.e., heat transfer coefficients), but also in the solid which actively takes part in determining the thermal solution.

Thermodynamic Losses in Gas Springs

Mathie et al. (2014) also considered the thermodynamic losses that result from cyclic, unsteady conjugate heat transfer in reciprocating components termed "gas springs." A gas spring is simplified model of a reciprocating compressor or expander, in which a fixed mass of gas is trapped in a cylinder, with a piston acting to impose volumetric variations. In the case considered in Mathie et al. (2014), the variations where sinusoidal, $V(t) = V_\mathrm{o} + V_\mathrm{a} \sin \omega t$, with a varying frequency, ω, whose dimensionless description is the Péclet number, $\mathrm{Pe} = \omega D^2/\alpha_\mathrm{f}$, where D is the diameter of the cylinder, and α_f is the thermal diffusivity of the fluid (gas). This arrangement is a convenient way to isolate the thermodynamic irreversibility due to thermal processes and remove those due to valve (pressure) losses. In addition to the frequency of the reciprocating motion, the framework allowed variations to the thickness and thermal properties of the solid walls of the cylinder, which are captured by the normalized cylinder wall thicknesses, $a^* = a/\delta$, where $\delta = (2\alpha_\mathrm{s}/\omega)^{0.5}$ is the thermal penetration depth.

A result from the investigation in Mathie et al. (2014) is shown in **Figure 11**, which indicates the ability of the solid domain variables to affect the thermodynamic loss. It is clear that the effect of the solid can be significant, depending on the solid and gas material properties (thermal conductivity, density, specific heat capacity), and also that mid-speed (intermediate Pe) constitutes the worse-case scenario for this type of loss mechanism, which can be close to 20% of the work exchanged with the gas spring for these conditions.

Non-Linear Heat Transfer Augmentation

Section "Thermodynamic Losses in Gas Springs" considered the effects of unsteady and conjugate heat transfer for the case that the heat transfer coefficient, h, is set to a constant (but complex) value. This is the conventional approach taken when dealing with gas-spring problems, in order to account for the observed phase shift between the heat flux at the wall, that arises from the thermal

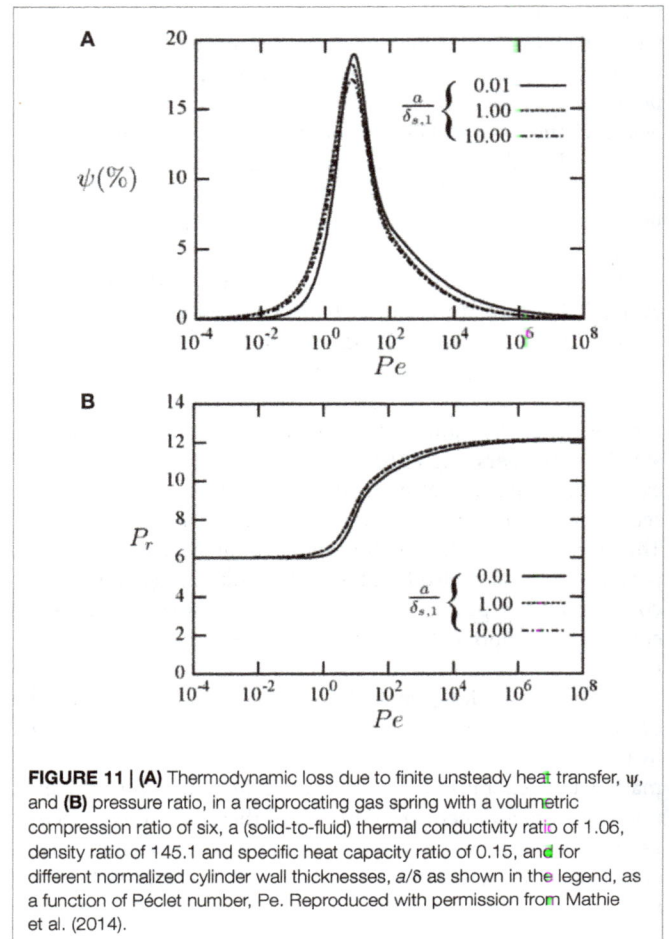

FIGURE 11 | (A) Thermodynamic loss due to finite unsteady heat transfer, ψ, and **(B)** pressure ratio, in a reciprocating gas spring with a volumetric compression ratio of six, a (solid-to-fluid) thermal conductivity ratio of 1.06, density ratio of 145.1 and specific heat capacity ratio of 0.15, and for different normalized cylinder wall thicknesses, a/δ as shown in the legend, as a function of Péclet number, Pe. Reproduced with permission from Mathie et al. (2014).

boundary layer there, and the temperature difference across the (bulk) fluid (Mathie and Markides, 2013; Mathie et al., 2013, 2014). One additional phenomenon is suspected to take place in the systems of interest, which is due to a non-linear interaction between the (time-varying) heat transfer coefficient, h, and the (time-varying) temperature difference across the fluid domain, ΔT. This phenomenon is referred to as "heat transfer augmentation" (Mathie and Markides, 2013; Mathie et al., 2013).

Mathematically, heat transfer augmentation can be described as follows; consider a fluid undergoing a time-varying thermal and fluid-flow process, such that $\Delta T(t) = \overline{(\Delta T)} + \Delta T'(t)$ and $h(t) = \bar{h} + h'(t)$ using Reynolds decompositions of a fluctuation around a mean. Then, the time-mean heat flux is given by the expression $\bar{\dot{q}} = \overline{(h \Delta T)} = \bar{h}\overline{(\Delta T)} + \overline{(h' \Delta T')} = A\bar{h}\overline{\Delta T}$. Essentially, this equation for the time-mean heat flux states that the fluctuations of the heat transfer coefficient, $h'(t)$, and those of the temperature difference, $\Delta T'(t)$, can become non-linearly coupled. Physically, we would expect an instantaneous increase in the heat transfer coefficient to give rise to a decrease in the instantaneous temperature difference, and *vice versa*.

Figures 12 and **13** show results from a semi-analytical study on the augmentation ratio, A, as a function of the: (i) heat transfer coefficient fluctuation intensity, $h^* = h_\mathrm{a}/h_\mathrm{o}$, where the sinusoidally

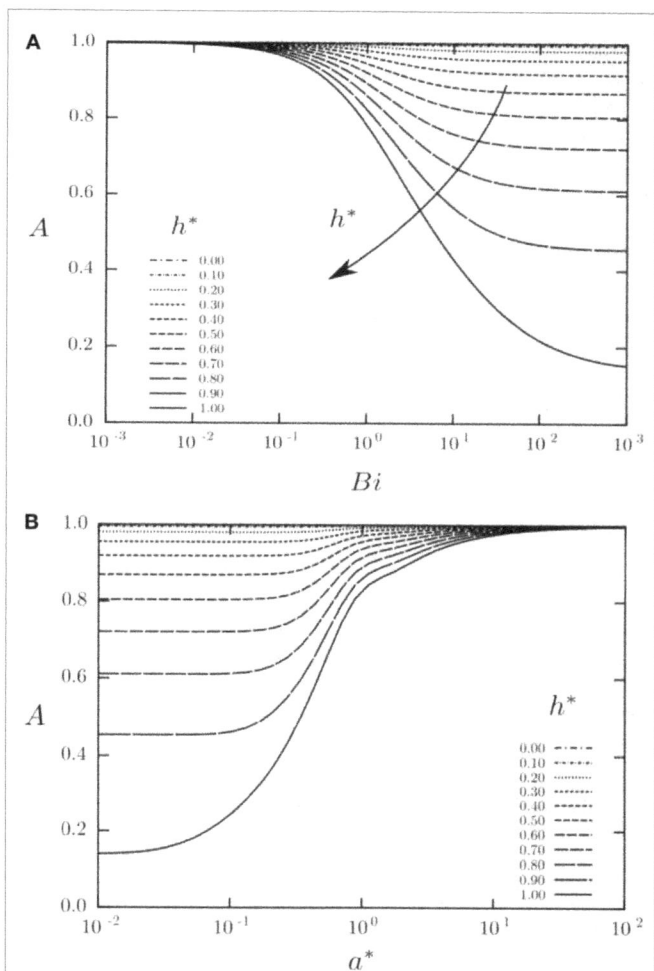

FIGURE 12 | Augmentation ratio, A = $\overline{(h\Delta T)}$ / $\overline{h}\overline{\Delta T}$, in a planar 1-D convective-conductive domain with an isoflux outer boundary condition (q̇' = 0) as a function of heat transfer coefficient fluctuation intensity ($h^* = h_a/h_o$; where the sinusoidally varying heat transfer coefficient is: $h(t) = h_o + h_a \sin \omega t$), Biot number (Bi = $h_o a/k$) and Fourier number ($a^* = a/\delta = \pi^{0.5}/Fo^{0.5}$). Reproduced with permission from Mathie and Markides (2013).

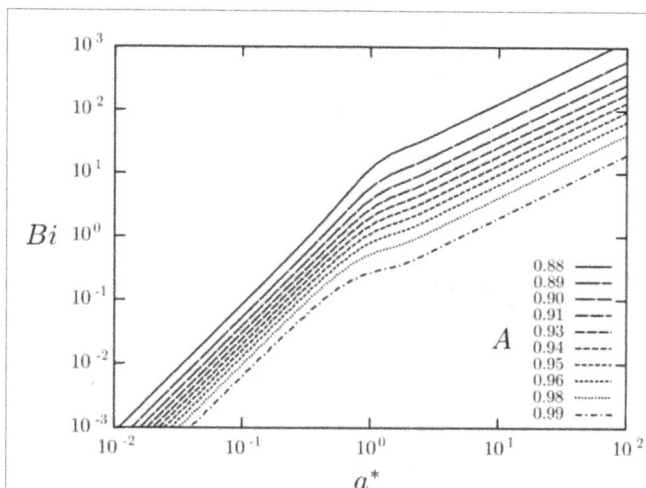

FIGURE 13 | Combination of the augmentation plots in Figures 12 into a single map. Reproduced with permission from Mathie and Markides (2013).

varying heat transfer coefficient is: $h(t) = h_o + h_a \sin \omega t$, (ii) Biot number, Bi = $h_o a/k$, and (iii) the normalized solid wall thicknesses, $a^* = a/\delta$, where $\delta = (2\alpha_s/\omega)^{0.5}$ is the thermal penetration depth (Mathie and Markides, 2013). Note that in **Figure 12A** the normalized solid wall thickness is kept constant at $a^* = 1$, whereas in **Figure 12B** the Biot number is kept constant at Bi = 1. It can be observed that the augmentation ratio is always $A \leq 1$, suggesting a _reduction_ in time-mean heat transfer relative to expectations from $\bar{q} = A\bar{h}\overline{\Delta T}$. Importantly, at small a^*, large h^*, and/or large Bi, this effect can become very significant.

The role of this phenomenon has not yet being considered in the energy-conversion systems under consideration, but it has been identified and measured in a number of flows, such as in Mathie and Markides (2013) and Mathie et al. (2013) which deal with two such cases: (i) the unsteady heat transfer between low-dimensional falling films and heated substrates and (ii) the

unsteady heat transfer between downstream of a broadband (turbulent) backwards-facing step. This forms an interesting and important avenue for further work.

FURTHER DISCUSSION AND CONCLUSION

This paper has been concerned with energy technologies capable of converting solar heat from non-concentrated or low-concentration solar collectors at temperatures <400°C to useful power, aimed at the domestic (1–10 kWe) and commercial/industrial (10–100 s of kW) sectors, thus covering a range of power output scales from a 1 kW to 1 MW. The availability of solar energy is strongly dependent on geography and also exhibits strong diurnal and possibly also seasonal variations. These latter variations of solar intensity depend on geography and in some cases are mild but in others quite strong. The solar resource, and therefore the potential availability of heat and power from combined solar-energy systems, such as PVT and S-CHP technologies, must also be considered in conjunction with end-user needs in energy consumption. Northern European climates, the UK for example, are less favorable for these technologies, even excluding climatic conditions (cloud coverage). The daily load factor varies strongly between summer and winter, with the peak demand at 6 p.m. in January when solar intensity is almost zero. Therefore, solar electricity will require significant storage to be effective and for this reason it will be considerably more expensive than predictions in generic LEC studies that are based in more favorable geographic locations. This plays to the strengths of small-scale thermodynamic power-generation (S-CHP) systems, which can benefit directly from TES as a part of their operation, relative to PV.

Such systems were considered based on thermodynamic vapor-phase heat-engine cycles undergone by organic working fluids, namely ORCs. ORCs are a relatively well-established and mature technology compared to some of the aforementioned alternative technologies, such as the TFOs, which have not yet found technological maturity and commercial application to the extent that ORC systems have, with high efficiencies especially when used with higher temperatures heat sources and at larger scales. Specifically, ORC systems are particularly well-suited to the conversion of low-to-medium-grade heat to mechanical or electrical work, and at an output power scale from kilowatts up to a few 10s of megawatts. Thermal efficiencies in excess of 25% are achievable at higher temperatures (i.e., with heat-source temperatures up to about 300–400°C), and efforts are currently in progress to develop improved systems by focusing on working fluid selection, the heat exchangers, and expansion machines at the scale of interest.

Models capable of accurate and reliable predictions of system performance were used to provide insight on operational characteristics and performance. Challenges and opportunities were identified, and recommendations made for further improvements, in particular with regards to the minimization of thermodynamic losses inflicted by finite heat transfer effects. It was noted, beyond conventional loss mechanisms, that losses can arise from inherently unsteady, conjugate and non-linear thermal processes and interactions between the working fluids within the systems of interest and the solid walls of key system components.

ACKNOWLEDGMENTS

The author would like to thank O. A. Oyewunmi for generating some of the results contained in the paper, as well as the following members of the CEP group: R. Mathie, I. Zadrazil, A. Charogiannis, R. Edwards, J. Freeman, A. I. Taleb, I. Guarracino, and C. Kirmse, without whom this work would not have been possible; it has been a pleasure and a privilege to work with them.

REFERENCES

Aghahosseini, S., and Dincer, I. (2013). Comparative performance analysis of low-temperature organic Rankine cycle (ORC) using pure and zeotropic working fluids. *Appl. Therm. Eng.* 54, 35–42. doi:10.1016/j.applthermaleng.2013.01.028

Ajimotokan, H. A., and Sher, I. (2015). Thermodynamic performance simulation and design optimisation of trilateral-cycle engines for waste heat recovery-to-power generation. *Appl. Energy* 154, 26–34. doi:10.1016/j.apenergy.2015.04.095

Angelino, G., and di Paliano, P. C. (1998). Multicomponent working fluids for ORCs. *Energy* 23, 449–463. doi:10.1016/S0360-5442(98)00009-7

Backhaus, S., and Swift, G. W. (2000). A thermoacoustic-stirling heat engine: Detailed study. *J. Acoust. Soc. Am.* 107, 3148–3166. doi:10.1121/1.429343

Bianchi, M., and Pascale, A. D. (2011). Bottoming cycles for electric energy generation: Parametric investigation of available and innovative solutions for the exploitation of low and medium temperature heat sources. *Appl. Energy* 88, 1500–1509. doi:10.1016/j.apenergy.2010.11.013

Biswas, K., He, J., Blum, I. D., Wu, C.-I., Hogan, T. P., Seidman, D. N., et al. (2012). High-performance bulk thermoelectrics with all-scale hierarchical architectures. *Nature* 489, 414–418. doi:10.1038/nature11439

Chen, H., Goswami, D. Y., Rahman, M. M., and Stefanakos, E. K. (2011). A supercritical Rankine cycle using zeotropic mixture working fluids for the conversion of low-grade heat into power. *Energy* 36, 549–555. doi:10.1016/j.energy.2010.10.006

Chen, H., Yogi Goswami, D., and Stefanakos, E. K. (2010). A review of thermodynamic cycles and working fluids for the conversion of low-grade heat. *Renew. Sustain. Energy Rev.* 14, 3059–3067. doi:10.1016/j.rser.2010.07.006

Confederation of British Industry. (2007). *Climate Change: Everyone's Business.* Available at: http://www.cbi.org.uk/media/1058204/climatereport2007full.pdf

Declaye, S., Georges, E., Bauduin, M., Quoilin, S., and Lemort, V. (2012). "Design of a small-scale organic Rankine cycle engine used in a solar power plant," in *Heat Powered Cycles Conference*, Alkmaar.

Delgado-Torres, A. M., and García-Rodríguez, L. (2010). Analysis and optimization of the low-temperature solar organic Rankine cycle (ORC). *Energy Convert. Manag.* 51, 2846–2856. doi:10.1016/j.enconman.2010.06.022

Dong, B., Xu, G., Cai, Y., and Li, H. (2014). Analysis of zeotropic mixtures used in high-temperature organic Rankine cycle. *Energy Convert. Manag.* 84, 253–260. doi:10.1016/j.enconman.2014.04.026

Fischer, J. (2011). Comparison of trilateral cycles and organic Rankine cycles. *Energy* 36, 6208–6219. doi:10.1016/j.energy.2011.07.041

Freeman, J., Hellgardt, K., and Markides, C. N. (2015a). An assessment of solar-thermal collector designs for small-scale combined heating and power applications in the United Kingdom. *Heat Transf. Eng.* 36, 1332–1347. doi:10.1080/01457632.2015.995037

Freeman, J., Hellgardt, K., and Markides, C. N. (2015b). An assessment of solar-powered organic Rankine cycle systems for combined heating and power in UK domestic applications. *Appl. Energy* 138, 605–620. doi:10.1016/j.apenergy.2014.10.035

Freeman, J., Hellgardt, K., and Markides, C. N. (2015c). Optimisaton of a domestic-scale solar-ORC heating and power system for maximum power output in the UK. *Paper Presented at ASME-ORC2015. Proceedings of the 3rd International Seminar on ORC Power Systems*, Brussels, Belgium.

Garg, P., Kumar, P., Srinivasan, K., and Dutta, P. (2013). Evaluation of isopentane, R-245fa and their mixtures as working fluids for organic Rankine cycles. *Appl. Therm. Eng.* 51, 292–300. doi:10.1016/j.applthermaleng.2012.08.056

Glushenkov, M., Sprenkeler, M., Kronberg, A., and Kirillov, V. (2012). Single-piston alternative to Stirling engines. *Appl. Energy* 97, 743–748. doi:10.1016/j.apenergy.2011.12.050

Gullì, F. (2006). Small distributed generation versus centralised supply: A social cost–benefit analysis in the residential and service sectors. *Energy Policy* 34, 804–832. doi:10.1016/j.enpol.2004.08.008

Hashem, H. (2015). *Shifting Sands in the Global CSP Market*. Available from: http://social.csptoday.com/markets/shifting-sands-global-csp-market

Heberle, F., Preißinger, M., and Brüggemann, D. (2012). Zeotropic mixtures as working fluids in organic Rankine cycles for low-enthalpy geothermal resources. *Renew. Energy* 37, 364–370. doi:10.1016/j.renene.2011.06.044

Herrando, M., Markides, C. N., and Hellgardt, K. (2014). A UK-based assessment of hybrid PV and solar-thermal systems for domestic heating and power: System performance. *Appl. Energy* 122, 288–309. doi:10.1016/j.apenergy.2014.01.061

Hewitt, G. F., Shires, G. L., and Bott, T. R. (1994). *Process Heat Transfer*. London: CRC Press.

Hsu, K. F., Loo, S., Guo, F., Chen, W., Dyck, J. S., Uher, C., et al. (2004). Cubic AgPbmSbTe2+m: Bulk thermoelectric materials with high figure of merit. *Science* 303, 818–821. doi:10.1126/science.1092963

Jung, D. S., McLinden, M., Radermacher, R., and Didion, D. (1989). Horizontal flow boiling heat transfer experiments with a mixture of R22/R114. *Int. J. Heat Mass Transf.* 32, 131–145. doi:10.1016/0017-9310(89)90097-5

Kosmadakis, G., Manolakos, D., and Papadakis, G. (2011). "Investigating the double-stage expansion in a solar ORC," in *Proc. 1st Int. Seminar ORC Power Syst.*, Delft.

Lampe, M., Gross, J., and Bardow, A. (2012). Simultaneous process and working fluid optimisation for organic Rankine cycles (ORC) using PC-SAFT. *Comput.-Aided Chem. Eng.* 30, 572–576. doi:10.1016/B978-0-444-59519-5.50115-5

Lampe, M., Kirmse, C., Sauer, E., Stavrou, M., Gross, J., and Bardow, A. (2014). Computer-aided molecular design of ORC working fluids using PC-SAFT. *Comput.-Aided Chem. Eng.* 34, 357–362. doi:10.1016/B978-0-444-63433-7.50044-4

Lazard. (2013). *Levelized Cost of Energy – Version 7.0.*

Lecompte, S., Ameel, B., Ziviani, D., van den Broek, M., and Paepe, M. D. (2014). Exergy analysis of zeotropic mixtures as working fluids in organic Rankine cycles. *Energy Convert. Manag.* 85, 727–739. doi:10.1016/j.enconman.2014.02.028

Lecompte, S., Huisseune, H., van den Broek, M., and De Paepe, M. (2015). Methodical thermodynamic analysis and regression models of organic Rankine cycle architectures for waste heat recovery. *Energy* 87, 60–76. doi:10.1016/j.energy.2015.04.094

Li, J.-F., Liu, W.-S., Zhao, L.-D., and Zhou, M. (2010). High-performance nano-structured thermoelectric materials. *NPG Asia Mater* 2, 152–158. doi:10.1038/asiamat.2010.138

Li, Y.-R., Du, M.-T., Wu, C.-M., Wu, S.-Y., and Liu, C. (2014). Potential of organic Rankine cycle using zeotropic mixtures as working fluids for waste heat recovery. *Energy* 77, 509–519. doi:10.1016/j.energy.2014.09.035

MacKay, J. C. D. (2009). *Sustainable Energy – Without the Hot Air*. Cambridge: UIT Cambridge.

Malavolta, M., Beyene, A., and Venturini, M. (2010). "Experimental implementation of a micro-scale ORC-based CHP energy system for domestic applications," in *Proc. ASME 2010 Int. Mechanical Eng. Congress Exposition*, Vancouver.

Markides, C. N. (2013). The role of pumped and waste heat technologies in a high-efficiency sustainable energy future for the UK. *Appl. Therm. Eng.* 53, 197–209. doi:10.1016/j.applthermaleng.2012.02.037

Markides, C. N., and Gupta, A. (2013). Experimental investigation of a thermally powered central heating circulator: Pumping characteristics. *Appl. Energy* 110, 132–146. doi:10.1016/j.apenergy.2013.03.030

Markides, C. N., Osuolale, A., Solanki, R., and Stan, G.-B. V. (2013). Nonlinear heat transfer processes in a two-phase thermofluidic oscillator. *Appl. Energy* 104, 958–977. doi:10.1016/j.apenergy.2012.11.056

Markides, C. N., and Smith, T. C. B. (2011). A dynamic model for the efficiency optimization of an oscillatory low grade heat engine. *Energy* 36, 6967–6980. doi:10.1016/j.energy.2011.08.051

Markides, C. N., Solanki, R., and Galindo, A. (2014). Working fluid selection for a two-phase thermofluidic oscillator: Effect of thermodynamic properties. *Appl. Energy* 124, 167–185. doi:10.1016/j.apenergy.2014.02.042

Mathie, R., and Markides, C. N. (2013). Heat transfer augmentation in unsteady conjugate thermal systems – Part I: Semi-analytical 1-D framework. *Int. J. Heat Mass Transf.* 56, 802–818. doi:10.1016/j.ijheatmasstransfer.2012.08.023

Mathie, R., Markides, C. N., and White, A. J. (2014). A framework for the analysis of thermal losses in reciprocating compressors and expanders. *Heat Transf. Eng.* 35, 1435–1449. doi:10.1080/01457632.2014.889460

Mathie, R., Nakamura, H., and Markides, C. N. (2013). Heat transfer augmentation in unsteady conjugate thermal systems – Part II: Applications. *Int. J. Heat Mass Transf.* 56, 819–833. doi:10.1016/j.ijheatmasstransfer.2012.09.017

Nightingale, N. P. (1986). "Automotive Stirling engine; Mod II design report," in *Report DOE/NASA/0032-28; NASA CR-175106; MT186ASE58SRI* (Cleveland, OH: National Aeronautics and Space Administration) Available at: http://ntrs.nasa.gov/archive/nasa/casi.ntrs.nasa.gov/19880002196.pdf.

Orosz, M., Mueller, A., Quoilin, S., and Hemond, H. (2009). "Small scale solar ORC system for distributed power," in *Conf. SolarPaces. 15th SolarPACES Conference, 15–18 September 2009, Berlin, Germany*.

Oyewunmi, O. A., and Markides, C. N. (2015). Effect of working-fluid mixtures on organic Rankine cycle systems: Heat transfer and cost analysis. *Paper Presented at ASME-ORC2015. Proceedings of the 3rd International Seminar on ORC Power Systems*, Brussels, Belgium.

Oyewunmi, O. A., Taleb, A. I., Haslam, A. J., and Markides, C. N. (2014). An assessment of working-fluid mixtures using SAFT-VR Mie for use in organic Rankine cycle systems for waste-heat recovery. *Comput. Therm. Sci.* 4, 301–316. doi:10.1615/.2014011116

Oyewunmi, O. A., Taleb, A. I., Haslam, A. J., and Markides, C. N. (2015). On the use of SAFT-VR Mie for assessing large-glide fluorocarbon working-fluid mixtures in organic Rankine cycles. *Appl. Energy*. doi:10.1016/j.apenergy.2015.10.040

Wolfe, P. (2015). Capacity rating for solar generating stations, *Wiki-Solar*.

Palanisamy, K., Taleb, A. I., and Markides, C. N. (2015). Optimising the non-inertive-feedback thermofluidic engine for the conversion of low-grade heat to pumping work. *Heat Transf. Eng.* 36, 1303–1320. doi:10.1080/01457632.2015.995014

Papadopoulos, A. I., Stijepovic, M., and Linke, P. (2010). On the systematic design and selection of optimal working fluids for organic Rankine cycles. *Appl. Therm. Eng.* 30, 760–769. doi:10.1016/j.applthermaleng.2009.12.006

Royal Academy of Engineering. (2004). *"The Costs of Generating Electricity: A Study Carried out by PB Power for the Royal Academy of Engineering"*. Available from: http://www.countryguardian.net/generation_costs_report2.pdf

Sami, S. M. (2010). Energy and exergy analysis of new refrigerant mixtures in an organic Rankine cycle for low temperature power generation. *Int. J. Ambient Energy* 31, 23–32. doi:10.1080/01430750.2010.9675805

Shu, G., Gao, Y., Tian, H., Wei, H., and Liang, X. (2014). Study of mixtures based on hydrocarbons used in ORC (organic Rankine cycle) for engine waste heat recovery. *Energy* 74, 428–438. doi:10.1016/j.energy.2014.07.007

Snyder, G. J., and Toberer, E. S. (2008). Complex thermoelectric materials. *Nat. Mater.* 7, 105–114. doi:10.1038/nmat2090

Solanki, R., Galindo, A., and Markides, C. N. (2012). Dynamic modelling of a two-phase thermofluidic oscillator for efficient low grade heat utilization: Effect of fluid inertia. *Appl. Energy* 89, 156–163. doi:10.1016/j.apenergy.2011.01.007

Solanki, R., Galindo, A., and Markides, C. N. (2013a). The role of heat exchange on the behaviour of an oscillatory two-phase low-grade heat engine. *Appl. Therm. Eng.* 53, 177–187. doi:10.1016/j.applthermaleng.2012.04.019

Solanki, R., Mathie, R., Galindo, A., and Markides, C. N. (2013b). Modelling of a two-phase thermofluidic oscillator for low-grade heat utilisation: Accounting for irreversible thermal losses. *Appl. Energy* 106, 337–354. doi:10.1016/j.apenergy.2012.12.069

Strachan, N., and Farrell, A. (2006). Emissions from distributed vs. centralized generation: The importance of system performance. *Energy Policy* 34, 2677–2689. doi:10.1016/j.enpol.2005.03.015

Szczech, J. R., Higgins, J. M., and Jin, S. (2011). Enhancement of the thermoelectric properties in nanoscale and nanostructured materials. *J. Mater. Chem.* 21, 4037–4055. doi:10.1039/C0JM02755C

Smith, T. C. B. (2006). *Thermally Driven Oscillations in Dynamic Applications. Ph.D. thesis.* Cambridge: University of Cambridge.

Tchanche, B. F., Lambrinos, G., Frangoudakis, A., and Papadakis, G. (2011). Low-grade heat conversion into power using organic Rankine cycles–a review of various applications. *Renew. Sustain. Energy Rev.* 15, 3963–3979. doi:10.1016/j.rser.2011.07.024

Tchanche, B. F., Papadakis, G., Lambrinos, G., and Frangoudakis, A. (2009). Fluid selection for a low-temperature solar organic Rankine cycle. *Appl. Therm. Eng.* 29, 2468–2476. doi:10.1016/j.applthermaleng.2008.12.025

Tchanche, B. F., Quoilin, S., Declaye, S., Papadakis, G., and Lemort, V. (2010). "Economic optimization of small scale organic Rankine cycles," in *23rd Int. Conf. Efficiency, Cost, Optimization, Simulation Environmental Impact Energy Syst*.

Vining, C. B. (2009). An inconvenient truth about thermoelectrics. *Nat. Mater.* 8, 83–85. doi:10.1038/nmat2361

Wang, J. L., Zhao, L., and Wang, X. D. (2010). A comparative study of pure and zeotropic mixtures in low temperature solar Rankine cycle. *Appl. Energy* 87, 3366–3373. doi:10.1016/j.apenergy.2010.05.016

Conflict of Interest Statement: The author declares that the research was conducted in the absence of any commercial or financial relationships that could be construed as a potential conflict of interest.

17

The Potential for Electrofuels Production in Sweden Utilizing Fossil and Biogenic CO₂ Point Sources

Julia Hansson[1,2]*, Roman Hackl[1], Maria Taljegard[3], Selma Brynolf[2] and Maria Grahn[2]

[1] *Climate and Sustainable Cities, IVL Swedish Environmental Research Institute, Stockholm, Sweden,* [2] *Division of Physical Resource Theory, Department of Energy and Environment, Chalmers University of Technology, Göteborg, Sweden,* [3] *Division of Energy Technology, Department of Energy and Environment, Chalmers University of Technology, Göteborg, Sweden*

Edited by:
Katy Armstrong,
University of Sheffield, UK

Reviewed by:
Hyungwoong Ahn,
University of Edinburgh, UK
Adam Hughmanick Berger,
Electric Power Research Institute,
USA

*Correspondence:
Julia Hansson
julia.hansson@ivl.se

Specialty section:
This article was submitted to Carbon Capture, Storage, and Utilization, a section of the journal Frontiers in Energy Research

Citation:
Hansson J, Hackl R, Taljegard M, Brynolf S and Grahn M (2017) The Potential for Electrofuels Production in Sweden Utilizing Fossil and Biogenic CO₂ Point Sources. Front. Energy Res. 5:4.

This paper maps, categorizes, and quantifies all major point sources of carbon dioxide (CO_2) emissions from industrial and combustion processes in Sweden. The paper also estimates the Swedish technical potential for electrofuels (power-to-gas/fuels) based on carbon capture and utilization. With our bottom-up approach using European data-bases, we find that Sweden emits approximately 50 million metric tons of CO_2 per year from different types of point sources, with 65% (or about 32 million tons) from biogenic sources. The major sources are the pulp and paper industry (46%), heat and power production (23%), and waste treatment and incineration (8%). Most of the CO_2 is emitted at low concentrations (<15%) from sources in the southern part of Sweden where power demand generally exceeds in-region supply. The potentially recoverable emissions from all the included point sources amount to 45 million tons. If all the recoverable CO_2 were used to produce electrofuels, the yield would correspond to 2–3 times the current Swedish demand for transportation fuels. The electricity required would correspond to about 3 times the current Swedish electricity supply. The current relatively few emission sources with high concentrations of CO_2 (>90%, biofuel operations) would yield elec-trofuels corresponding to approximately 2% of the current demand for transportation fuels (corresponding to 1.5–2 TWh/year). In a 2030 scenario with large-scale biofuels operations based on lignocellulosic feedstocks, the potential for electrofuels production from high-concentration sources increases to 8–11 TWh/year. Finally, renewable elec-tricity and production costs, rather than CO_2 supply, limit the potential for production of electrofuels in Sweden.

Keywords: carbon dioxide, CO₂ recovering, carbon capture and utilization, carbon recycling, power-to-gas, alternative transportation fuels

HIGHLIGHTS

- Sweden emits 50 million metric tons of CO_2 per year from different types of point sources, the vast majority of which is emitted at low concentrations.
- Of this, 65% is from biogenic sources, most of which are located in southern Sweden.
- Currently, the high-concentration sources of CO_2 in Sweden can provide a potential 1.5–2 TWh electrofuels/year (2% of current transportation demand).

- The Swedish potential for electrofuels is currently limited by the electricity required and production costs rather than the amount of recoverable CO_2.

INTRODUCTION

Anthropogenic greenhouse gas (GHG) emissions need to be reduced in order to limit global climate change and reach ambitious climate targets (Pachauri et al., 2014). Carbon dioxide (CO_2) emissions can be reduced by using less fossil fuels or by using fossil fuels in combination with carbon capture and storage (CCS) or carbon capture and utilization (CCU) [e.g., Cuéllar-Franca and Azapagic (2015), Wismans et al. (2016)]. In Sweden, the overall national vision is for zero net emissions of GHG to the atmosphere by 2050 (likely to be changed to 2045), along with a fossil fuel-independent vehicle fleet by 2030 (Government offices of Sweden, 2009; Swedish Government Official Reports, 2016). An extensive official investigation commissioned by the Swedish government has concluded that a range of options are needed to reduce CO_2 emissions from the transport sector, including biomass-based liquid and gaseous fuels (biofuels) along with hydrogen and electricity produced from renewable energy sources (Swedish Government Official Reports, 2013).

However, neither government nor academia have explored electrofuels (i.e., power-to-gas/fuels or synthetic hydrocarbons produced from CO_2 and water using electricity), extensively. Interest in electrofuels is on the rise, both in the literature (Graves et al., 2011; Mohseni, 2012; Nikoleris and Nilsson, 2013; Taljegård et al., 2015)[1] and in terms of demonstration plants in the EU, in some cases, including CO_2 capture (Gahleitner, 2013). Studies mainly investigate electrofuels as a (i) technology for storing intermittent electricity [e.g., Streibel et al. (2013), de Boer et al. (2014), Vandewalle et al. (2014), König et al. (2015), Qadrdan et al. (2015), Varone and Ferrari (2015), Zakeri and Syri (2015), Zhang et al. (2015), and Kötter et al. (2016)], (ii) fuel for transport [e.g., Connolly et al. (2014), Ridjan et al. (2014), Larsson et al. (2015)], or (iii) means of producing chemicals [e.g., Ganesh (2013), Perathoner and Centi (2014), and Chen et al. (2016)]. Different types of energy carriers [e.g., methane, methanol, DME (dimethyl ether), gasoline, and diesel] can be produced, which makes electrofuels a potentially interesting option for all transport modes, especially shipping, aviation, and long distance road transport, where the potential for other renewable fuel options, such as electricity and hydrogen, may be limited. Electrofuels may allow increased use of biofuels, if the CO_2 associated with their production is used for production of electrofuels instead of being emitted to the atmosphere (Mignard and Pritchard, 2008; Mohseni, 2012; Hannula, 2015, 2016).

CO_2 emissions can be captured from various point sources, including industrial processes that produce CO_2, such as biofuel production (including anaerobic digestion and fermentation), natural gas processing, steel plants, and oil refineries, fossil and biomass combustion in heat and power plants, or directly from the air.

Many studies have estimated CO_2 emissions from point sources in China [e.g., Chen and Chen (2010), Liu et al. (2010), Zhang and Chen (2014)]. Zhang and Chen (2014) used a bottom-up approach to estimate CO_2 emissions from fuel combustion and the main industrial processes at 7.7 Gt CO_2 per year in 2008, with coal as the main source. The potential global supply of CO_2 from point sources is estimated in Naims (2016). The total estimated global capturable CO_2 supply from point sources amount to approximately 12.7 Gton of CO_2 (Naims, 2016). High purity point sources (e.g., fermentation of biomass and ammonia production) and other low cost sources (e.g., bioenergy, natural gas, and hydrogen production) represent in total approximately 0.3 Gton of CO_2. Naims (2016) further indicates that there is enough CO_2 to meet the estimated global CO_2 demand in the near and long term.

In Austria, the iron and steel, cement industry, and power and heat industries are the largest point sources of CO_2 emissions (Reiter and Lindorfer, 2015). Biofuel production, a relatively modest point source at about 113 kton in 2013, is considered the most suitable Austrian source for power-to-gas application by Reiter and Lindorfer (2015). A German feasibility study by Trost et al. (2012) identifies a large potential for biogenic CO_2 sources, including biogas upgrading, bioethanol plants, and sewage treatment plants. Trost et al. (2012) also found a substantial electrofuels potential of over 130 TWh fuel per year in the form of methane produced using CO_2 from industrial processes and biogenic sources. Reiter and Lindorfer (2015) and Trost et al. (2012), both conclude that availability of CO_2 will not be a limiting factor for using power-to-gas as a balancing strategy for intermittent renewable power sources (wind power and photovoltaics) in Austria or Germany.

In Sweden, carbon capture is currently implemented at, for instance, Agroetanol in Norrköping. Agroetanol produces grain-based ethanol; the resulting CO_2 is purified and sold to the AGA Gas AB. Detailed quantification of current and/or future Swedish CO_2 emissions from point sources is, however, lacking in the scientific literature, and there are no assessments of the technical potential for Swedish production of electrofuels. Electrofuels may represent an interesting option in Sweden, that is a forest-rich country, due to the ambitious GHG emission reduction targets in general and specifically in the transport sector. Assessing the Swedish potential for CCS and CCU requires detailed knowledge of the stationary CO_2 emissions. The overall impact on CO_2 emissions of the production and use of electrofuels mainly depends on the electricity-related CO_2 emissions. The Swedish electricity production consists mainly of hydro power and nuclear power implying relatively low GHG emissions.

The overall aim of this paper is to map and quantify stationary Swedish CO_2 emissions by concentration, origin, and geographical distribution, as well as investigate the potential for CCU. Specifically, we aim to (i) map and quantify the major point sources of CO_2 emissions from industrial and combustion processes in Sweden with a bottom-up approach and estimate the technical potential for CO_2 capture or recovery and (ii) estimate the technical potential for production of electrofuels

---

[1]Brynolf, S., Taljegård, M., Grahn, M., and Hansson, J. (2017). Electrofuels for the transport sector: a review of production costs. *Renew. Sust. Energ. Rev.* (Submitted).

in Sweden, as an example of CCU. We analyze the potential for biofuels-related CO_2 in the future (a 2030 scenario), since the use of biomass and biofuels is expected to increase and use of fossil fuels decrease. Additionally, we estimate the potential demand for CO_2 and electricity corresponding to the use of electrofuels for road transport, heavy trucks, and shipping, at scale, in order to give a first indication of the potential role for electrofuels in transportation in Sweden.

MATERIALS AND METHODS

This section describes the methodology for estimating both CO_2 emissions from major point sources and the potential for capturing and using the emissions.

Assumptions about the CO_2 Sources Included

CO_2 emission sources can be divided into diffuse sources (e.g., transport and agriculture) and point sources (e.g., factories and power production). This study uses a bottom-up approach to estimate CO_2 emissions from the following point sources in Sweden:

- Industrial process plants (including iron and steel, non-ferrous metal, oil and gas refineries, lime and cement, pulp and paper, chemical, metal, and other similar plants)
- Heat and power production (including biomass, waste, and fossil fuel-fired plants)
- Biofuels production facilities (including ethanol, biogas, and more advanced biofuels).

Emissions data for year 2013 from the European Environment Agency's "European Pollutant Release and Transfer Register" (European Environment Agency, 2015) was used to estimate (i) the available amount of CO_2 and (ii) the share of fossil and biogenic CO_2, for Swedish point sources, including all sources emitting 0.1 million metric tons of CO_2 per year or more. Other CO_2 sources are assumed to be negligible (except in the case of biofuels production). The concentration of CO_2 for each type of

sources was estimated using (Chapel et al., 1999; Bosoaga et al., 2009) (see **Table 1**). For the purposes of analysis, the concentrations were divided in three ranges: low (<15 vol%), medium (15–90 vol%), and high (>90 vol%).

For biofuels plants, the CO_2 estimates are based on data gathered by Swedish Energy Agency and Energigas Sverige (2015) and Grahn and Hansson (2015) in 2012–2013. Also, the sources emitting less than 0.1 million metric tons of CO_2 per year are included in the case of biofuels since these are relatively pure and, therefore, well suited for electrofuels production. In most biofuels production processes, there is a surplus of CO_2 and the CO_2 is of high purity (Xu et al., 2010). When biogas is upgraded to transport fuel quality, a cleaning step to remove CO_2 is included, resulting in a relatively pure stream of CO_2. The CO_2 emissions from domestic biofuel production in a 2030 scenario are estimated based on biofuels production scenarios from Grahn and Hansson (2015) and on scenarios for anaerobic digestion and gasification-based biogas production from Dahlgren et al. (2013). Grahn and Hansson (2015) assessed the potential contribution of domestically produced biofuels for transport in Sweden in 2030 based on a mapping of the prospects for current and potential Swedish biofuel producers. Some of the planned biofuels production plants included in the scenario for 2030 have been canceled or put on hold and are, therefore, excluded in this study.

The 2030 scenario was constructed exclusively for biofuel plants because these represent a relatively pure stream of CO_2 of particular interest in electrofuels production, and because the use of biofuels is expected to increase in the future. For many biofuels, no extra major purification step is needed in the capture process, which leads to a relatively low capture cost. This can also be assumed for the case of biogas since CO_2 is already removed when biogas is upgraded to transport fuel quality. This can be compared to the CO_2 capture cost linked to processes requiring an extra purification step like steel and iron, ammonia, refinery, cement, and fossil or biomass combustion plants estimated at $20€_{2015}$–$170€_{2015}$/ ton CO_2 in the short term (10–15 years) and $10€_{2015}$–$100€_{2015}$/ton CO_2 in the more long term (Damen et al., 2007; Finkenrath, 2011; Kuramochi et al., 2012, 2013; IEA, 2013). Even though it has been

TABLE 1 | The type of CO_2 stream, CO_2-concentration range, range of CO_2 emissions per unit, and share of recoverable CO_2, for different point sources in Sweden based on European Environment Agency (2015).

Production facility and location	Type of CO_2 stream	Typical concentration	Process CO_2 emissions (kton/year) for smallest and largest plant	Recoverable share (%)
Oil and gas refineries	Flue gases, by-product	3–13 vol%[a]	122–1,573	90
Power and heat production	Flue gases	3–13 vol%	104–1,990	90
Iron and steel production	Flue gases	Approx. 15 vol%	102–1,540	90
Non-ferrous metal production	Flue gases	Approx. 15 vol%	101–256	90
Cement and lime production	Flue gases, by-product	Approx. 14–33 vol%	110–1,940	90
Production of chemicals	Flue gases, by-product	3–13 vol%[a]	13–620	90
Pulp and paper production	Flue gases	Approx. 15 vol%	165–1,740	90
Waste treatment or incineration	Flue gas	Approx. 10 vol%	105–837	90
Fermentation-based biofuels	By-product	Pure stream	0.11–154	100
Anaerobic digestion-based biofuels	By-product	>90 vol-%	0.14–21	54
Gasification-based biofuels	By-product	>90 vol-%	1.84–37	100
Other	Flue gas	3–13 vol%	134	90

For CO_2 concentration and recoverability references, see Section "Availability of CO_2 for Carbon Capture and Utilization."
[a]*Minor amounts of CO_2 are available at higher concentrations (up to 100 vol%).*

indicated that the cost for carbon capture represents a relatively modest share (a few percent) of the total electrofuel-production cost unless air capture is assumed (Graves et al., 2011; Tremel et al., 2015; Varone and Ferrari, 2015; see text footnote 1), using CO_2 from biofuel production represent an attractive source for electrofuel production since more pure streams will likely be used first for economic reasons and the domestic biofuel actors, representing a considerable biofuel production capacity, in order to comply with sustainability requirements need to improve their production processes in terms of CO_2 emissions.

Table 1 presents the type of CO_2 stream, typical concentration of CO_2, the range of CO_2 emissions per unit, and the amount of recoverable CO_2, for different point sources. **Table 2** includes a list of all the biofuel production facilities in operation in 2015, their production capacity and associated CO_2 emissions, and the corresponding information for the biofuels plants planned by 2030. **Table 3** summarizes the main assumptions used in estimating the amount of CO_2 that is available for recovery from current and future biofuels plants.

Availability of CO_2 for CCU

In order for CO_2 to be used to produce electrofuels, the gas needs to be separated from other substances in emissions from industrial and combustion processes, such as sulfur dioxide. The concentration of CO_2 in power plant flue gases is relatively low (<15 vol%) (Chapel et al., 1999); for process-related emissions, e.g., in the lime and cement industry, CO_2 concentrations are somewhat higher (14–33 vol%) (Bosoaga et al., 2009) (see **Table 1**). In this study, we assume that 90% of the CO_2 from medium- (15–90 vol%) and low- (<15 vol%) concentration CO_2 sources is recoverable (Chapel et al., 1999). Current CO_2 capture technologies do not usually capture all the CO_2 as this is too expensive and requires too much energy.

In biofuels production processes (fermentation, anaerobic digestion, gasification), relatively pure streams (>90 vol%) of CO_2 are available in latter cases due to the demand for high fuel purity in the transport sector. We assume that 100% of the CO_2 from biofuel plants is recoverable and could be converted into fuel. Approximately 54% of the biogas produced in Sweden is upgraded for the transportation sector (Swedish Energy Agency and Energigas Sverige, 2016), which means that CO_2 capturing technology already exist on several Swedish anaerobic digestion facilities. Another opportunity for anaerobic digestion-based biogas plants is to feed raw biogas to a methanation reactor, thereby combining biogas upgrading and electrofuels production (Johannesson, 2016). Biogas plants that currently do not upgrade their gas are generally small implying high costs for upgrading and currently supplying other markets than the transport sector, making them less suitable as a source of CO_2 for electrofuels production. Therefore, only CO_2 from biogas-upgrading plants is considered in this study. For simplicity, we assume that the share of upgraded biogas of total biogas production by 2030 remains at 54%.

Geographic Distribution of CO_2 Emissions

The CO_2 emission sources have been mapped and categorized by concentration and geographical area. The geographical areas are those used for the Swedish electricity market, i.e., four price areas (SE1, SE2, SE3, and SE4) (Swedish Energy Markets Inspectorate, 2014) (see **Figure 1**). The electricity price areas were implemented in Sweden in order to control the transmission of electricity between regions and to promote the construction of power generation and transmission capacity in and to areas with electricity deficits. On average, the northern parts of the country (SE1 and SE2) are characterized by an excess of electricity production due to the available hydropower resources and relatively

TABLE 2 | Biofuels production facilities and associated CO_2 emissions.

Production facility and location	Biofuel	Biofuel production (GWh/year)	Process CO_2 emissions (ton/year)	Reference[a]
Facilities operational in 2015				
Agroetanol, Line 1, Norrköping	Ethanol	391	53,466[b]	Axelsson et al. (2014) and Grahn and Hansson (2015)
Agroetanol, Line 2, Norrköping	Ethanol	1,126	154,014[b]	Axelsson et al. (2014) and Grahn and Hansson (2015)
ST1, Göteborg	Ethanol	34	4,617	Axelsson et al. (2014) and ST1 (2016)
SEKAB, Örnsköldsvik	Ethanol	64	7,807	Arvidsson and Lundin (2011) and Grahn and Hansson (2015)
SP, pilot plant, Örnsköldsvik	Ethanol	0.9	109	Arvidsson and Lundin (2011) and Grahn and Hansson (2015)
LTU Green Fuels, pilot plant, Piteå[c]	DME	6	1,836	Pettersson and Harvey (2012) and Grahn and Hansson (2015)
GoBiGas, Göteborg Energi, Göteborg	Gasification-based biogas	180	36,900	Heyne (2013) and Grahn and Hansson (2015)
Swedish anaerobic digestion-based biogas production (277 plants)	Biogas	1,686	245,680	SGC (2012) and Swedish Energy Agency and Energigas Sverige (2016)
Additional production capacity until 2030				
Fermentation	Ethanol	3,300	402,033	Hansson and Grahn (2013)
Anaerobic digestion	Biogas	4,600	672,342	SGC (2012), Dahlgren et al. (2013), and Hansson and Grahn (2013)
Gasification	Biogas, methanol, DME	4,050	1,023,260	Dahlgren et al. (2013) and Hansson and Grahn (2013)

[a]References for the amount of biofuels produced and the estimated CO_2 emissions per unit of fuel are provided here.
[b]CO_2 produced at Agroetanol in Norrköping is currently purified and sold to the AGA Gas AB.
[c]The closure of this pilot plant was announced in April 2016.

TABLE 3 | Main assumptions for assessing CO₂ availability from current and future biofuels plants in Sweden.

Production technology	Assumed amount of available CO_2 per GWh biofuel
Fermentation	Cereal based: 136.8 ton CO_2/GWh (Axelsson et al., 2014)
	Lignocellulose based: 121.7 ton CO_2/GWh (Arvidsson and Lundin, 2011)
Anaerobic digestion	Upgraded biogas: 145.7 ton CO_2/GWh (SGC, 2012)
Gasification	Black liquor gasification: 305 ton CO_2/GWh (Pettersson and Harvey, 2012)
	Indirect gasification: 206 ton CO_2/GWh (Heyne, 2013)

low overall power consumption. In the southern parts (SE3 and SE4), electricity consumption often exceeds production, which leads to relatively higher electricity prices in these areas (Nord Pool, 2016).

Electrofuel-Production Efficiency and Cost

The focus in this study is on electrofuels in the form of methane, methanol, and DME since these are the most discussed electrofuels in the literature (see text footnote 1), are of interest for the relevant transport sector (shipping and trucks), and include fuels in liquid and gaseous form. The amounts of CO_2 and electricity necessary for the types of electrofuels included in this study are given in **Table 4** and are based on lower heating value (LHV).

Table 4 also presents cost ranges for 2015 and 2030 estimated in the base case reference scenario in Brynolf et al. (see text footnote 1). The electricity-to-fuel efficiency of the electrofuel-production process strongly depends on the type of electrolyzer and the future development of production technologies. Alkaline electrolysers have efficiencies in the range of 43–69% today, while the most efficient electrolysers are expected to reach efficiencies above 80% based on LHV (Smolinka et al., 2011; Benjaminsson et al., 2013; Grond et al., 2013; Mathiesen et al., 2013; Bertuccioli et al., 2014; Hannula, 2015; Schiebahn et al., 2015). Combining this with the efficiency for fuel synthesis yields electricity-to-fuel efficiencies in the 30–75% range for methane, methanol, and DME, this corresponds to an electricity demand of 1.33–3.33 MWh electricity/MWh electrofuel.

Brynolf et al. (see text footnote 1) suggest costs for different electrofuels (methane, methanol, DME, gasoline, and diesel) in the span of 120€$_{2015}$–1,050€$_{2015}$/MWh$_{fuel}$ and 100€$_{2015}$–430€$_{2015}$/MWh$_{fuel}$ in 2015 and 2030, respectively. However, in the base case of the reference scenario representing average data, the same costs are 200€$_{2015}$–280€$_{2015}$/MWh$_{fuel}$ and 160€$_{2015}$–210€$_{2015}$/MWh$_{fuel}$ in 2015 and 2030, respectively. The most important factors affecting the production cost of electrofuels are the capital cost of the electrolyzer, the electricity price, the capacity factor of the unit, and the lifetime of the electrolyzer. The base case reference scenario assumes alkaline electrolyzer with a capital cost of 600€$_{2015}$/kW$_{el}$, capacity factor of 80%, lifetime of the electrolyzer at 25 years, carbon capture cost at 30€$_{2015}$/ton, and electricity price of 50€$_{2015}$/MWh. A capacity factor at 80% implies that the plant is run the major part of the year. However, if electrofuels are used to balance intermittent renewable power production (i.e., there is production only when there

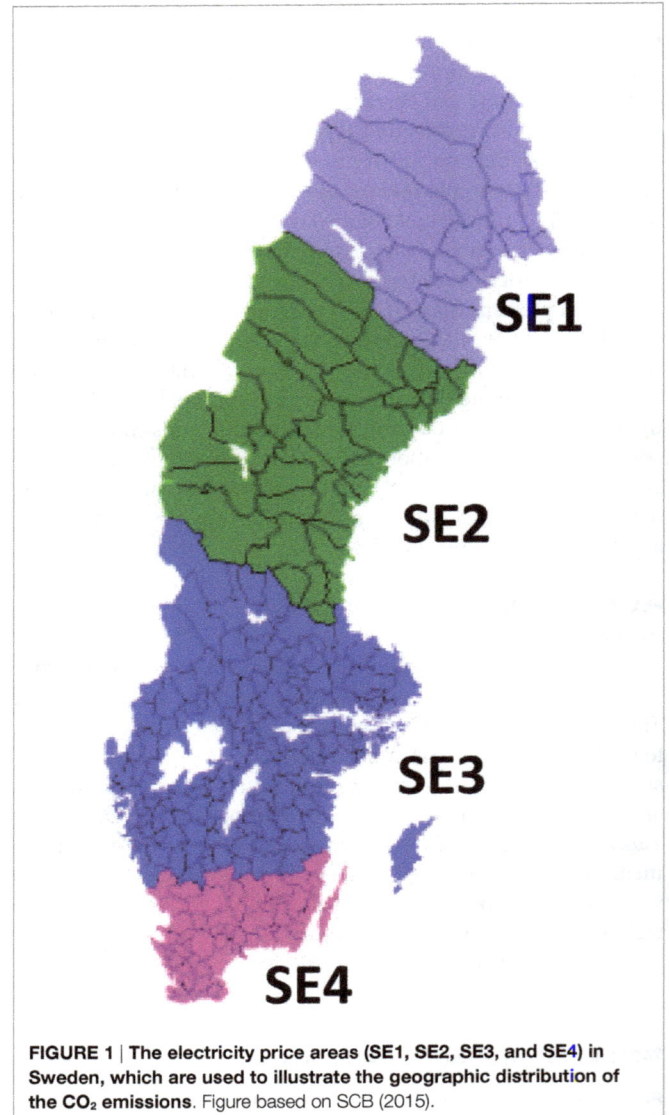

FIGURE 1 | The electricity price areas (SE1, SE2, SE3, and SE4) in Sweden, which are used to illustrate the geographic distribution of the CO_2 emissions. Figure based on SCB (2015).

is a surplus of power from these sources), the capacity factor will be reduced. This will not influence the estimated technical potential for production of electrofuels in Sweden in this study, but it will lead to increased electrofuel-production costs [which is further assessed in Brynolf et al. (see text footnote 1)]. In the case of a carbon capture cost at 10€$_{2015}$/ton representing more pure streams like biofuels operation, the production cost of electrofuels is reduced by approximately 3%. In their review of the literature, Brynolf et al. (see text footnote 1) also found that the cost of capturing CO_2 generally is a minor factor in the total production cost of electrofuels representing less than 10% (when not considering CO_2 capturing from air). CO_2 can be captured from various industrial sources with costs ranging from about 10€$_{2015}$ to 170€$_{2015}$/ton CO_2, depending on the CO_2 concentration (Damen et al., 2006, 2007; Finkenrath, 2011; Goeppert et al., 2012; Kuramochi et al., 2012, 2013; IEA, 2013; see text footnote 1). This indicates that from an economic point

TABLE 4 | Estimated values for CO₂ and electricity demand per unit of electrofuel and production cost for 2015 and 2030 (based on literature review and base case reference scenario by Brynolf et al. (see text footnote 1) representing average data and based on lower heating value, for assumptions see the text).

Electrofuel	Fuel synthesis efficiency (%)	CO₂ per unit of fuel (t/MWh_{fuel})	Electricity per unit of fuel (MWh_{el}/MWh_{fuel})	Production cost 2015 (€_{2015}/ MWh_{fuel})	Production cost 2030 (€_{2015}/MWh_{fuel})
Methane	77[a]	0.21	2.00	200	160
Methanol	79[b]	0.28	1.93	210	160
DME	80[b]	0.27	1.95	210	160

[a]Mohseni (2012), Grond et al. (2013), Schiebahn et al. (2015), and Tremel et al. (2015).
[b]Hannula and Kurkela (2013) and Tremel et al. (2015).

of view, all CO_2 sources (except from pure air) might be of interest for electrofuel production in the future.

RESULTS

CO₂ Emissions in Sweden

In Sweden, major stationary point sources currently emit approximately 50 Mton CO_2 per year. Of this, about 45 Mton CO_2 is recoverable (see **Figure 2**). Our analysis includes 148 facilities, with 14 U emitting more than 1 Mton CO_2/year, 88 U emitting between 1 Mton and 100 kton CO_2/year, and 47 U emitting less than 100 kton/year.

Figure 2 shows the distribution of CO_2 emissions among different types of point sources. Pulp and paper plants and heat and power plants are the two major types of point sources, corresponding to 23 Mton CO_2 (45% of the total) and 11.5 Mton CO_2 (23% of the total) per year, respectively. In total, biogenic sources account for 65% or 32 Mton of CO_2 emissions per year. The high share of biogenic CO_2 is mainly due to the extensive use of biomass in producing pulp, paper, heat, and power and from waste treatment and incineration. Emissions from biofuel production represent a small share of the current total amount of available CO_2, with approximately 0.5 Mton of recoverable CO_2 per year. According to Andreas Gundberg, Innovation manager at Lantmännen Agroetanol, CCU has already been implemented at the main Swedish ethanol producer representing approximately 90% of the total Swedish ethanol production capacity. The emissions from this ethanol production (about 100 kton/year) are included in the analysis.

Figure 3 shows the amount of CO_2 available and the corresponding potential production of electrofuels in the form of methanol at different CO_2 concentrations in Sweden in 2013 and in 2030. The majority of the CO_2 is available at low and medium concentrations, equally spread between the categories low and medium but mainly below 20 vol%. A small share of the CO_2, mainly from the biofuels industry, is available at higher, significantly more accessible, concentrations.

About 90% of the high-concentration emissions come from sources in geographic region SE3, along with about 60% of the rest of the CO_2 emission sources (see **Figure 4**). Anaerobic digestion and ethanol production from agricultural crops currently dominate biofuels production, and these are mostly located in densely populated areas (producing biogas from digestion of sewage sludge and food waste) or in proximity to agricultural

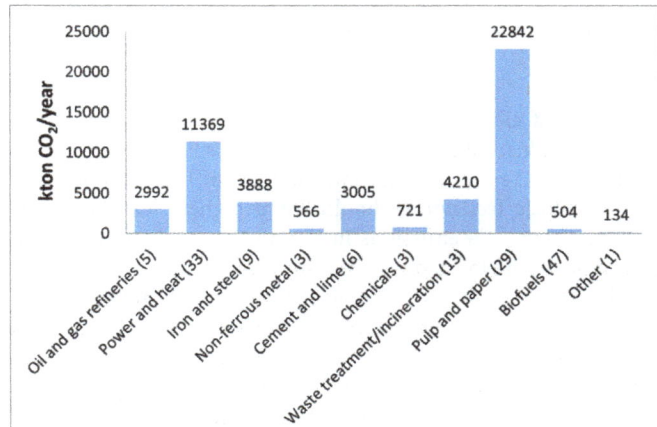

FIGURE 2 | Current recoverable CO₂ from major point sources in Sweden, based on European Environment Agency (2015), Grahn and Hansson (2015), and Dahlgren et al. (2013). In total, 149 point sources are included; the number of plants in each category is given in parenthesis.

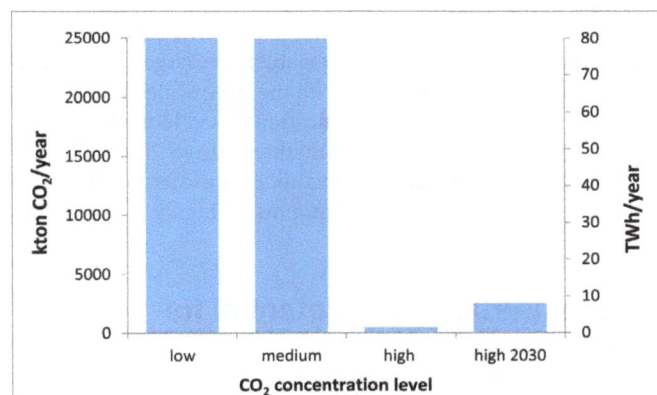

FIGURE 3 | Recoverable CO₂ and potential for production of electrofuels in the form of methanol at three different concentration levels (low: <15 vol%, medium: 15–90 vol% and, high: >90 vol%) in 2013 and at high concentration in 2030.

operations (farm-based ethanol and biogas production), which are mainly found in southern Sweden. However, electricity prices in the southern parts are currently less favorable than further north where hydropower resources and lower demand create

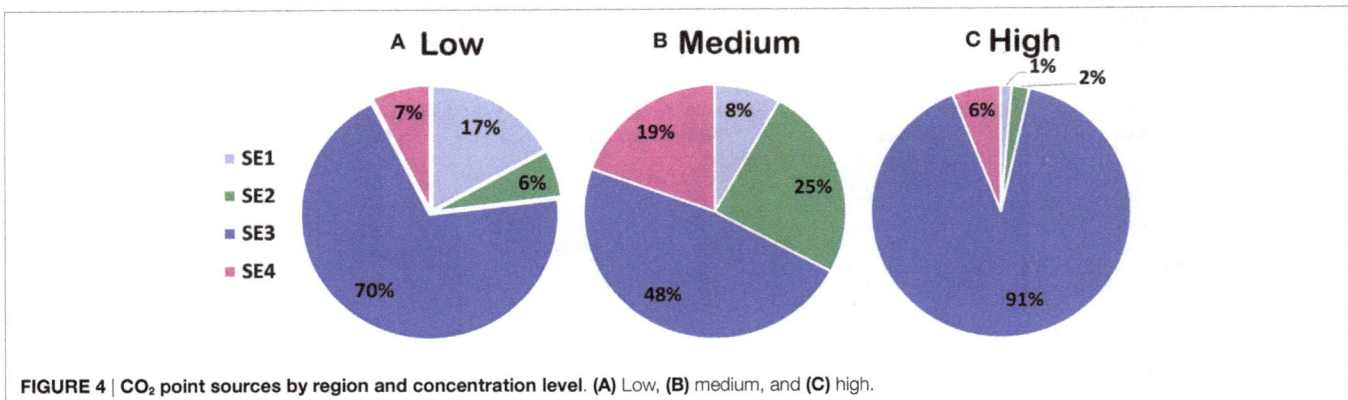

FIGURE 4 | CO₂ point sources by region and concentration level. (A) Low, **(B)** medium, and **(C)** high.

an excess of electricity while the transmission capacity to the southern industrial and population centers is limited.

The projected large-scale introduction of biofuels based on lignocellulosic feedstocks should entail higher shares of high-concentration CO_2 emissions in the northern regions, SE1 and SE2, if plants are located near feedstock resources.

The biofuels sector is expected to grow significantly in Sweden during the coming years in order to achieve national climate and transport targets. **Figure 5** illustrates the current and estimated amount of CO_2 available for electrofuels production from different biofuel production technologies and a minor share of others sources available by 2030 in Sweden based on Dahlgren et al. (2013) and Hansson and Grahn (2013). Only CO_2 from the production of upgraded biogas is included. In 2030, the CO_2 originates mainly from gasification, anaerobic digestion, and fermentation-based biofuels production (utilizing both cereals and lignocellulosic biomass and considering recent implementation plans). In 2030, these sources could potentially yield 2.2 Mton CO_2 for electrofuels production (approximately 5.5 times the amount currently available). The largest increase in production capacity is expected with the large-scale implementation of a variety of biomass-gasification-based biofuels, such as synthetic natural gas, DME, or methanol from lignocellulosic biomass. Ethanol produced from lignocellulosic feedstocks could also potentially generate large amounts of highly concentrated biogenic CO_2.

Swedish Production Potential for Electrofuels

Using all the currently recoverable CO_2 from the point sources identified in this study to produce electrofuel in the form of methane would yield approximately 224 TWh per year. This corresponds to approximately 2.5 times the current Swedish demand for transportation fuels [approximately 85 TWh per year in 2014 (Swedish Energy Agency, 2015b)]. For electrofuels with lower conversion efficiencies (e.g., methanol and DME), production could instead cover about twice the current demand. Producing 224 TWh per year of electro-methane requires about 448 TWh of electricity (assuming 2 MWh_{el}/MWh_{fuel}), which corresponds to three times the current Swedish electricity generation [149 TWh (Swedish Energy Agency, 2015a)].

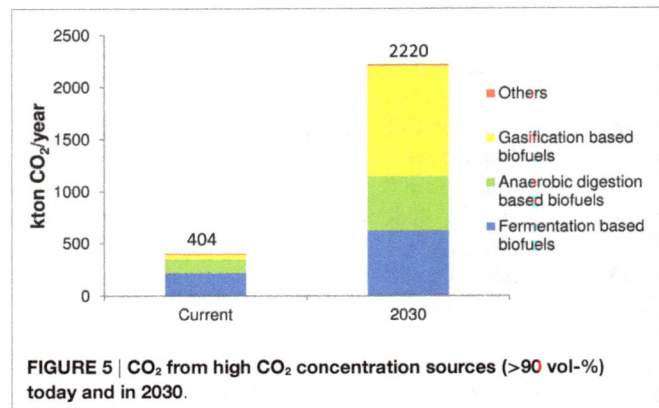

FIGURE 5 | CO₂ from high CO₂ concentration sources (>90 vol-%) today and in 2030.

The high-concentration sources, represented mainly by biofuel plants, suffice to provide only about 2% of the current demand for transportation fuels (corresponding to 1.5–2/year, see **Figure 6**). Converting the high-concentration emissions to electrofuels would require about 3–4 TWh of electricity (2–3% of the current national production). In 2030, the potential production of electrofuels in the form of methane, methanol, and DME from high-CO_2 sources is 8–11 TWh (see **Figure 6**). This corresponds to approximately 9–13% of the current demand for transportation fuels and would require about 15–21 TWh of electricity (10–14% of current electricity production).

Table 5 shows the requirements for meeting the current Swedish fuel demand for (non-air) transport with electrofuels in the form of methanol. As seen in **Table 5**, about half of the recoverable CO_2 (23 Mton) would be needed to supply the entire current Swedish road transport demand with electrofuels in the form of methanol (assuming a conversion factor of 0.275 ton CO_2/MWh methanol). The corresponding amount of CO_2 needed to satisfy the entire fuel demand from heavy trucks and all domestic and international shipping currently bunkering in Sweden is estimated to be about 5 and 6 Mton CO_2, respectively. This implies that in the case of large-scale introduction of electrofuels for road transport (including heavy trucks), heavy trucks only, or shipping in Sweden, the supply of CO_2 is not a limiting factor.

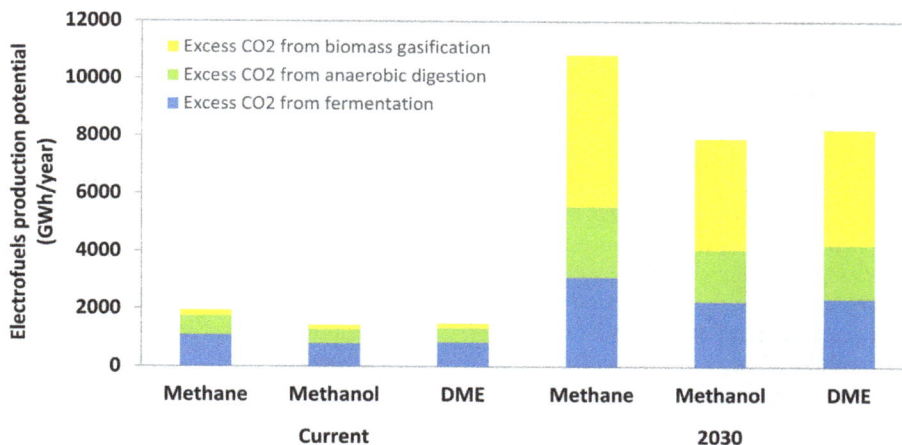

FIGURE 6 | Production potential for electrofuels in the form of methane, methanol and DME from current and future biofuel plants with high CO_2 concentrations.

TABLE 5 | Outputs and inputs to electrofuels production if fulfilling the fuel demand with electrofuels in the form of methanol in three different transport modes.

	Road transport	Heavy trucks	Shipping
Fuel demand 2014 (TWh)	85 (Swedish Energy Agency, 2015b)	18 (Swedish Government Official Reports, 2013)[a]	21 (Swedish Energy Agency, 2015b)[b]
Electrofuel replacement (%)	100	100	100
Electrofuel production			
Methanol (TWh)	85	18	21
Electrofuel requirements			
Electricity (TWh)	164	35	41
Carbon dioxide (Mton)	23	5	6

For electricity and CO_2 demand per unit of electrofuel see **Table 4**.
[a]Expected to increase to approximately 25 TWh by 2050.
[b]Represents the total Swedish use of bunker fuels in 2014 of which 96% was used for international sea transport.

However, meeting the entire current road transport demand with electrofuels would require about 164 TWh$_{el}$ of electricity (with methanol at 1.93 MWh$_{el}$/MWh$_{fuel}$). This would more than double the current demand for electricity. To meet the current Swedish fuel demand for passenger cars (at about 41 TWh) (Swedish Government Official Reports, 2013) with electrofuels in the form of methanol would require approximately 11 ton CO_2 and 79 TWh$_{el}$ of electricity. For comparison, if the entire passenger car fleet were replaced by electric vehicles, the increased demand for electricity would be approximately 10 TWh (based on Swedish Government Official Reports, 2013).

Using electrofuels for the heavy truck sector and for shipping bunker fuel sold in Sweden would require about 35 and 41 TWh$_{el}$, respectively. For comparison, in 2014, domestic power generation was 150 TWh (SCB, 2016). Further, the goal is to increase domestic generation from renewable sources by about 30 TWh by

2020, compared to 2002 figures and current production of renewable electricity is approximately 85 TWh (SCB, 2016). Large-scale introduction of electrofuels would require a major increase in the supply of electricity from renewable energy sources.

DISCUSSION AND CONCLUSION

This study shows that Swedish point sources emit approximately 50 million metric tons of CO_2 per year, 65% of which is biogenic in origin. The potentially recoverable emissions amount to 45 Mton. The main point sources are in the pulp and paper industry along with heat and power, while emissions from biofuel production (with relatively high concentrations of recoverable CO_2) amounted to 0.5 Mton CO_2 in 2015, with an estimated potential for 2.2 Mton CO_2 in 2030. Thus, the potential streams of relatively pure CO_2 are modest, at least in the near term. Currently, the potential yield from these sources is 1.5–2 TWh of electrofuels per year, corresponding to approximately 2% of the current Swedish demand for transportation fuels.

However, in Sweden, all types of CO_2 emissions, whether fossil or biogenic, and whether low-concentration or high, are of interest in terms of CCU (although carbon capture can be expected to first be applied to systems with higher concentrations of CO_2 because capture costs are somewhat lower for these, generally speaking). In the case of electrofuels, as mentioned earlier, it has been indicated that the cost for carbon capture represents a relatively modest share of the total electrofuel-production cost which makes the purity of the CO_2 sources less important. However, CO_2 from biofuel operations seem like an attractive source since biofuel actors strive to reduce their CO_2 emissions due to sustainability requirements. Further, biomass-related CO_2 emissions are expected to increase in the future, since the use of biomass for energy is expected to increase while fossil CO_2 emissions are expected to decrease.

We conclude that the Swedish supply of CO_2 does not have to be a limiting factor for the potential future production of electrofuels for the Swedish transport sector, even if the current supply of pure CO_2 streams is limited. However, there might be other limiting factors such as the associated electricity demand.

As indicated in the introduction, electrofuels represent a potential long-term energy storage option and could, therefore, be of interest in terms of managing grid-integration of more intermittent renewable energy sources (e.g., wind and solar power). But large-scale introduction of electrofuels in the transport sector would in turn represent a huge new demand for electricity. The direct use of electricity needed to supply the entire current transport demand for passenger cars would increase current electricity demand by 10%, while using electrofuels would require increasing the Swedish electricity generation by about 60% to meet the same transport demand (Swedish Energy Agency, 2015b). The electrofuels production process and combustion engine are simply that much less efficient than electric motors. Therefore, large-scale introduction of electrofuels might potentially increase the challenge of balancing intermittent renewable generation, rather than help solve it with long-term energy storage, since an increased demand for power would most likely be met with new wind power installations in Sweden. Producing electrofuels only part of the year is one option to limit this problem. However, according to Brynolf et al. (see text footnote 1), the production cost of electrofuels increases drastically per megawatt hours fuel when the capacity factor (i.e., actual production as share of total production capacity) of the wind turbines is decreased. Thus, the benefit of using electrofuels for balancing renewable energy need to be further assessed.

The production cost of different electrofuels is also a limiting factor for the potential future production of electrofuels in Sweden. The literature contains a fairly broad range of estimates, but the most important factors in the production cost of electrofuels are the capital cost of the electrolyzer, the electricity price, the capacity factor of the unit, and the lifetime of the electrolyzer (see text footnote 1).

The majority of the current CO_2 sources are located in southern Sweden, which is also the case for the current CO_2 sources with relatively pure CO_2 emissions. However, from the perspective of the electric-grid, electrofuels production may be more suitable in the northern parts of Sweden where there is generally a surplus of power generation and lower electricity prices. An increasing demand for electricity in southern Sweden might put additional pressure on the transmission capacity from north to south. Future biofuel plants based on forest biomass (as included in the 2030 scenario) are expected to be located mostly in northern Sweden and, therefore, represent an interesting source of CO_2 for production of electrofuels.

From a climate perspective, it might be preferable to capture and store CO_2 underground, using CCS technology, and not convert CO_2 into a fuel that after combustion will be released to the atmosphere again (van der Giesen et al., 2014; Sternberg and Bardow, 2015). If the CO_2 has been captured from burning fossil fuels, CCS will avoid increased CO_2 concentration, and if the CO_2 is captured from burning biomass (or from air), CCS will decrease the atmospheric CO_2 concentration, *ceteris paribus*. Today, however, there are several obstacles that have to be overcome before CCS could be available at a large scale, including public acceptance (Oltra et al., 2010; Dütschke, 2011). CCS is also only applicable for relatively large CO_2 sources and storage possibilities depend on geological prerequisites.

The overall impact on CO_2 emissions of the production and use of electrofuels mainly depends on the electricity-related CO_2 emissions and what the fuels replace (van der Giesen et al., 2014; Sternberg and Bardow, 2015). van der Giesen et al. (2014) conclude that for some production paths, the climate impact is worse than for fossil fuels, and achieving a net climate benefit requires using renewable electricity and renewable CO_2 sources. Sternberg and Bardow (2015) evaluate electrofuels relative to the case in which the same amount of CO_2 is instead either emitted or stored. They find that electrofuels can at best only make a small contribution to mitigation compared to other available solutions and that using CO_2 emissions for electrofuels is worse from a climate perspective compared to storing them. It would be interesting to more thoroughly study the environmental impact of electrofuels compared to other CCU technologies with a lifecycle perspective. For example, the amount of CO_2 emissions from electricity production will depend on (i) the time perspective (for example using a marginal or average electricity mix) and (ii) the geographical boundaries of the electricity supply. However, GHG emissions from electricity production are expected to decrease significantly as a consequence of stringent energy and climate policies changing the mix of energy sources.

To summarize, electrofuels are limited by electricity demand rather than the demand for CO_2 and, at scale, require a substantial amount of renewable electricity at relatively low cost. The GHG impact of electrofuels compared to other options, in particular CCS, needs to be further assessed.

AUTHOR CONTRIBUTIONS

JH is the main author; planned the work and led the writing. RH was responsible for the mapping and quantification of the major Swedish point sources of CO_2 emissions and contributed to further assessments and paper writing. SB, MT, and MG contributed with the electrofuel-production characteristics, participated in the assessment, and contributed to paper writing.

ACKNOWLEDGMENTS

Financial support from the Swedish Research Council Formas, Nordic Energy Research through the Nordic flagship project Shift (Sustainable Horizons for Transport), the Swedish Energy Agency, and the Swedish Knowledge Centre for Renewable Transportation Fuels (f3) is acknowledged. This publication is partly the result of a project within the Renewable Fuels and Systems Program (Samverkansprogrammet Förnybara drivmedel och system), financed by the Swedish Energy Agency and the Swedish Knowledge Centre for Renewable Transportation Fuels (f3). The f3 Centre contributes, through knowledge based on science, to the development of environmentally, economically, and socially sustainable and renewable transportation fuels, as part of a future sustainable society (see www.f3centre.se). The authors also thank Magnus Fröberg, Scania CV AB, and Paulina Essunger for valuable input.

FUNDING

This work has received financial support from (i) the Swedish Research Council Formas *via* the project titled "Cost-effective choices of marine fuels under stringent carbon dioxide reduction targets," (ii) Nordic Energy Research through the Nordic flagship project Shift (Sustainable Horizons for Transport), and (iii) the Swedish Knowledge Centre for Renewable Transportation Fuels and the Swedish Energy Agency *via* the project titled "The role of electrofuels: a cost-effective solution for future transport?"

REFERENCES

Arvidsson, M., and Lundin, B. (2011). *Process Integration Study of a Biorefinery Producing Ethylene from Lignocellulosic Feedstock for a Chemical Cluster*. Master thesis, Chalmers University of Technology, Gothenburg, Sweden.

Axelsson, P., Cederlöf, F., and Forslöf, S. (2014). *Carbon Dioxide Capture from the Ethanol Industry and Use in Industrial Applications*. B.Sc. thesis, Linköping University, Linköping, Sweden.

Benjaminsson, G., Benjaminsson, J., and Rudberg, R. B. (2013). *Power to Gas – A Technical Review (El till gas – system, ekonomi och teknik)*. Malmö, Sweden: Svenskt Gastekniskt Center AB.

Bertuccioli, L., Chan, A., Hart, D., Lehner, F., Madden, B., and Standen, E. (2014). *Development of Water Electrolysis in the European Union, Fuel Cells and Hydrogen Joint Undertaking*. Lausanne, Switzerland. Available at: http://www.fch.europa.eu/sites/default/files/study%20electrolyser_0-Logos_0.pdf

Bosoaga, A., Masek, O., and Oakey, J. E. (2009). CO_2 capture technologies for cement industry. *Energy Procedia* 1, 133–140. doi:10.1016/j.egypro.2009.01.020

Chapel, D. G., Mariz, C. L., and Ernest, J. (1999). *Recovery of CO_2 from Flue Gases: Commercial Trends*. Available at: http://citeseerx.ist.psu.edu/viewdoc/download?doi=10.1.1.204.8298&rep=rep1&type=pdf

Chen, G., and Chen, Z. (2010). Carbon emissions and resources use by Chinese economy 2007: a 135-sector inventory and input-output embodiment. *Commun. Nonlinear Sci. Numer. Simulat.* 15, 3647–3732. doi:10.1016/j.cnsns.2009.12.024

Chen, Q., Lv, M., Tang, Z., Wang, H., Wei, W., and Sun, Y. (2016). Opportunities of integrated systems with CO_2 utilization technologies for green fuel & chemicals production in a carbon-constrained society. *J. CO_2 Utilization* 14, 1–9. doi:10.1016/j.jcou.2016.01.004

Connolly, D., Mathiesen, B. V., and Ridjan, I. (2014). A comparison between renewable transport fuels that can supplement or replace biofuels in a 100% renewable energy system. *Energy* 73, 110–125. doi:10.1016/j.energy.2014.05.104

Cuéllar-Franca, R. M., and Azapagic, A. (2015). Carbon capture, storage and utilisation technologies: a critical analysis and comparison of their life cycle environmental impacts. *J. CO_2 Utilization* 9, 82–102. doi:10.1016/j.jcou.2014.12.001

Dahlgren, S., Liljeblad, A., Cerruto, J., Nohlgren, I., and Starberg, K. (2013). *Realiserbar biogaspotential i Sverige år 2030 genom rötning och förgasning*. Stockholm, Sweden: WSP.

Damen, K., van Troost, M., Faaij, A., and Turkenburg, W. (2006). A comparison of electricity and hydrogen production systems with CO_2 capture and storage. Part A: review and selection of promising conversion and capture technologies. *Prog. Energy Combust. Sci.* 32, 215–246. doi:10.1016/j.pecs.2005.11.005

Damen, K., van Troost, M., Faaij, A., and Turkenburg, W. (2007). A comparison of electricity and hydrogen production systems with CO_2 capture and storage-Part B: chain analysis of promising CCS options. *Prog. Energy Combust. Sci.* 33, 580–609. doi:10.1016/j.pecs.2007.02.002

de Boer, H. S., Grond, L., Moll, H., and Benders, R. (2014). The application of power-to-gas, pumped hydro storage and compressed air energy storage in an electricity system at different wind power penetration levels. *Energy* 72, 360–370. doi:10.1016/j.energy.2014.05.047

Dütschke, E. (2011). What drives local public acceptance – comparing two cases from Germany. *Energy Procedia* 4, 6234–6240. doi:10.1016/j.egypro.2011.02.636

European Environment Agency. (2015). *European Pollutant Release and Transfer Register*. Available at: http://prtr.ec.europa.eu/#/home

Finkenrath, M. (2011). *Cost and Performance of Carbon Dioxide Capture from Power Generation*. Paris, France: International Energy Agency (IEA).

Gahleitner, G. (2013). Hydrogen from renewable electricity: an international review of power-to-gas pilot plants for stationary applications. *Int. J. Hydrogen Energy* 38, 2039–2061. doi:10.1016/j.ijhydene.2012.12.010

Ganesh, I. (2013). Conversion of carbon dioxide into several potential chemical commodities following different pathways-a review. *Mater. Sci. Forum* 764, 1–82. doi:10.4028/www.scientific.net/MSF.764.1

Goeppert, A., Czaun, M., Surya Prakash, G. K., and Olah, G. A. (2012). Air as the renewable carbon source of the future: an overview of CO_2 capture from the atmosphere. *Energy Environ. Sci.* 5, 7833–7853. doi:10.1039/C2EE21586A

Government offices of Sweden. (2009). *Regeringens Proposition (Government Bill) 2008/2009:163: En sammanhållen Klimat och Energipolitik – Energi respektive Klimat (A Coherent Climate and Energy Policy – Energy and Climate Respectively)*. Available at: www.regeringen.se/rattsdokument/proposition/2009/03/prop.-200809163/ and www.regeringen.se/rattsdokument/proposition/2009/03/prop.-200809162/

Grahn, M., and Hansson, J. (2015). Prospects for domestic biofuels for transport in Sweden 2030 based on current production and future plans. *Wiley Interdiscip. Rev. Energy Environ.* 4, 290–306. doi:10.1002/wene.138

Graves, C., Ebbesen, S. D., Mogensen, M., and Lackner, K. S. (2011). Sustainable hydrocarbon fuels by recycling CO_2 and H_2O with renewable or nuclear energy. *Renew. Sustain. Energ. Rev.* 15, 1–23. doi:10.1016/j.rser.2010.07.014

Grond, L., Schulze, P., and Holstein, S. (2013). *Systems Analyses Power to Gas: Deliverable 1: Technology Review*. Groningen, the Netherlands: DNV KEMA Energy & Sustainability.

Hannula, I. (2015). Co-production of synthetic fuels and district heat from biomass residues, carbon dioxide and electricity: performance and cost analysis. *Biomass Bioeng.* 74, 26–46. doi:10.1016/j.biombioe.2015.01.006

Hannula, I. (2016). Hydrogen enhancement potential of synthetic biofuels manufacture in the European context: a techno-economic assessment. *Energy* 104, 199–212. doi:10.1016/j.energy.2016.03.119

Hannula, I., and Kurkela, E. (2013). *Liquid Transportation Fuels via Large-Scale Fluidised-Bed Gasification of Lignocellulosic Biomass*. VTT Technical Research Centre of Finland Report. Available at: http://www.ieatask33.org/download.php?file=files/file/publications/new/Techno-economics.pdf

Hansson, J., and Grahn, M. (2013). *Utsikt för förnybara drivmedel i Sverige, IVL report B 2083*. Available at: http://spbi.se/wp-content/uploads/2013/03/IVL_B2083_2013_final.pdf

Heyne, S. (2013). *Bio-SNG from Thermal Gasification-Process Synthesis, Integration and Performance*. Doctoral thesis, Linköping University, Gothenburg, Sweden.

IEA. (2013). *Technology Roadmap – Carbon Capture and Storage*. Paris: International Energy Agency.

Johannesson, T. (2016). *Implementation of Electrofuel Production at a Biogas Plant – Case Study at Borås*. Master thesis, Chalmers University of Technology, Gothenburg, Sweden.

König, D. H., Baucks, N., Dietrich, R. U., and Wörner, A. (2015). Simulation and evaluation of a process concept for the generation of synthetic fuel from CO_2 and H_2. *Energy* 91, 833–841. doi:10.1016/j.energy.2015.08.099

Kötter, E., Schneider, L., Sehnke, F., Ohnmeiss, K., and Schröer, R. (2016). The future electric power system: impact of power-to-gas by interacting with other renewable energy components. *J. Energy Storage* 5, 113–119. doi:10.1016/j.est.2015.11.012

Kuramochi, T., Ramírez, A., Turkenburg, W., and Faaij, A. (2012). Comparative assessment of CO_2 capture technologies for carbon-intensive industrial processes. *Prog. Energy Combust. Sci.* 38, 87–112. doi:10.1016/j.pecs.2011.05.001

Kuramochi, T., Ramírez, A., Turkenburg, W., and Faaij, A. (2013). Techno-economic prospects for CO_2 capture from distributed energy systems. *Renew. Sustain. Energy Rev.* 19, 328–347. doi:10.1016/j.rser.2012.10.051

Larsson, M., Grönkvist, S., and Alvfors, P. (2015). Synthetic fuels from electricity for the Swedish transport sector: comparison of well to wheel energy efficiencies and costs. *Energy Procedia* 75, 1875–1880. doi:10.1016/j.egypro.2015.07.169

Liu, L. C., Wang, J. N., Wu, G., and Wei, Y. M. (2010). China's regional carbon emissions change over 1997–2007. *Int. J. Energy Environ.* 1, 161–176.

Mathiesen, B. V., Ridjan, I., Connolly, D., Nielsen, M. P., Vang Hendriksen, P., Bjerg Mogensen, M., et al. (2013). *Technology Data for High Temperature Solid Oxide Electrolyser Cells, Alkali and PEM Electrolysers*. Aalborg University, Denmark: Department of Development and Planning.

Mignard, D., and Pritchard, C. (2008). On the use of electrolytic hydrogen from variable renewable energies for the enhanced conversion of biomass to fuels. *Chem. Eng. Res. Des.* 86, 473–487. doi:10.1016/j.cherd.2007.12.008

Mohseni, F. (2012). *Power to Gas – Bridging Renewable Electricity to the Transport Sector*. Licentitate thesis, KTH Royal Institute of Technology, Stockholm, Sweden.

Naims, H. (2016). Economics of carbon dioxide capture and utilization – a supply and demand perspective. *Environ. Sci. Pollut. Res.* 23, 22226–22241. doi:10.1007/s11356-016-6810-2

Nikoleris, A., and Nilsson, L. (2013). *Elektrobränslen en kunskapsöversikt (Electrofuels An Overview)*. IMES Report Series, Vol. 85. Lund, Sweden: Lund University. Available at: http://portal.research.lu.se/portal/files/3697397/3738230.pdf

Nord Pool. (2016). *Historical Market Data*. Available at: http://www.nordpoolspot.com/historical-market-data/

Oltra, C., Sala, R., Solà, R., Di Masso, M., and Rowe, G. (2010). Lay perceptions of carbon capture and storage technology. *Int. J. Greenhouse Gas Control* 4, 698–706. doi:10.1016/j.ijggc.2010.02.001

Pachauri, R. K., Allen, M. R., Barros, V. R., Broome, J., Cramer, W., Christ, R., et al. (2014). *Climate Change 2014: Synthesis Report. Contribution of Working Groups I, II and III to the Fifth Assessment Report of the Intergovernmental Panel on Climate Change*, eds R. Pachauri, and L. Meyer (Geneva, Switzerland: IPCC), 151.

Perathoner, S., and Centi, G. (2014). CO_2 recycling: a key strategy to introduce green energy in the chemical production chain. *ChemSusChem* 7, 1274–1282. doi:10.1002/cssc.201300926

Pettersson, K., and Harvey, S. (2012). Comparison of black liquor gasification with other pulping biorefinery concepts – systems analysis of economic performance and CO_2 emissions. *Energy* 37, 136–153. doi:10.1016/j.energy.2011.10.020

Qadrdan, M., Abeysekera, M., Chaudry, M., Wu, J., and Jenkins, N. (2015). Role of power-to-gas in an integrated gas and electricity system in Great Britain. *Int. J. Hydrogen Energy* 40, 5763–5775. doi:10.1016/j.ijhydene.2015.03.004

Reiter, G., and Lindorfer, J. (2015). Evaluating CO_2 sources for power-to-gas applications – a case study for Austria. *J. CO_2 Utilization* 10, 40–49. doi:10.1016/j.jcou.2015.03.003

Ridjan, I., Mathiesen, B. V., and Connolly, D. (2014). Synthetic fuel production costs by means of solid oxide electrolysis cells. *Energy* 76, 104–113. doi:10.1016/j.energy.2014.04.002

SCB. (2015). *Energy Prices and Switching of Suppliers, 3rd Quarter 2015 (Prisutveckling på energi samt leverantörsbyten, tredje kvartalet 2015)*. Stockholm, Sweden: Statistics Sweden. Available at: www.scb.se/Statistik/EN/EN0304/2015K03/EN0304_2015K03_SM_EN24SM1504.pdf

SCB. (2016). *Tillförsel och användning av el 2001–2015 (GWh) (Production and Use of Electricity 2001-2015)*. Available at: http://www.scb.se/hitta-statistik/statistik-efter-amne/energi/tillforsel-och-anvandning-av-energi/arlig-energistatistik-el-gas-och-fjarrvarme/tillforsel-och-anvandning-av-el-20012015-gwh/

Schiebahn, S., Grube, T., Robinius, M., Tietze, V., Kumar, B., and Stolten, D. (2015). Power to gas: technological overview, systems analysis and economic assessment for a case study in Germany. *Int. J. Hydrogen Energy* 40, 4285–4294. doi:10.1016/j.ijhydene.2015.01.123

SGC. (2012). *Basic Data on Biogas. Report*. Malmö, Sweden: Svenskt gastekniskt centrum. Available at: http://www.sgc.se/ckfinder/userfiles/files/BasicDataonBiogas2012.pdf

Smolinka, T., Günther, M., and Garche, J. (2011). *Stand und Entwicklungspotenzial der Wasserelektrolyse zur Herstellung von Wasserstoff aus regenerativen Energien, Kurzfassung des Abschlussberichtes NOW-Studie*. Freiburg im Breisgau: Fraunhofer ISE and FCBAT. Available at: http://www.hs-ansbach.de/uploads/tx_nxlinks/NOW-Studie-Wasserelektrolyse-2011.pdf

ST1. (2016). *St1 Built a Waste-Based Etanolix® Ethanol Production Plant in Gothenburg*. Available at: http://www.st1.eu/news/st1-built-a-waste-based-etanolix-ethanol-production-plant-in-gothenburg

Sternberg, A., and Bardow, A. (2015). Power-to-what? – environmental assessment of energy storage systems. *Energy Environ. Sci.* 8, 389–400. doi:10.1039/C4EE03051F

Streibel, M., Nakaten, N., Kempka, T., and Kühn, M. (2013). Analysis of an integrated carbon cycle for storage of renewables. *Energy Procedia* 40, 202–211. doi:10.1016/j.egypro.2013.08.024

Swedish Energy Agency. (2015a). *Energy in Sweden 2015*. Eskilstuna, Sweden: Swedish Energy Agency. Available at: https://www.energimyndigheten.se/globalassets/statistik/overgripande-rapporter/energy-in-sweden-till-webben.pdf

Swedish Energy Agency. (2015b). *Transportsektorns energianvändning 2014 (Energy Use in the Transport Sector 2014)*. Eskilstuna, Sweden: Swedish Energy Agency.

Swedish Energy Agency and Energigas Sverige. (2015). *Produktion och användning av biogas och rötrester år 2014, ES 2015:03*. Available at: https://www.energimyndigheten.se/globalassets/nyheter/2015/produktion-och-anvandning-av-biogas-och-rotrester-ar-2014.pdf

Swedish Energy Agency and Energigas Sverige. (2016). *Produktion och användning av biogas och rötrester år 2015, ES 2016:04*. Available at: https://www.energimyndigheten.se/globalassets/nyheter/2016/es-2016-04-produktion-och-anvandning-av-biogas-och-rotrester-ar-2015.pdf

Swedish Energy Markets Inspectorate. (2014). *Sverige är indelat i fyra elområden. Fact Sheet*. Available at: http://ei.se/Documents/Publikationer/fakta_och_informationsmaterial/Elomraden.pdf

Swedish Government Official Reports. (2013). *Fossilfrihet på väg (On Its Way to Fossil Fuel Independence). SOU 2013:84*. Available at: http://www.regeringen.se/rattsdokument/statens-offentliga-utredningar/2013/12/sou-201384/

Swedish Government Official Reports. (2016). *En klimat – och luftvårdsstrategi för Sverige – Del 1 (A Climate and Air Pollution Treatment Strategy for Sweden – Part 1). Delbetänkande av Miljömålsberedningen, Swedish Government Official Reports SOU 2016:47*. Available at: http://www.regeringen.se/rattsdokument/statens-offentligautredningar/2016/06/en-klimat-och-luftvardsstrategi-for-sverige/

Taljegård, M., Brynolf, S., Hansson, J., Hackl, R., Grahn, M., and Andersson, K. (2015). "Electrofuels – a possibility for shipping in a low carbon future?" in *Proceedings of International Conference on Shipping in Changing Climates*. Glasgow, 405–418.

Tremel, A., Wasserscheid, P., Baldauf, M., and Hammer, T. (2015). Techno-economic analysis for the synthesis of liquid and gaseous fuels based on hydrogen production via electrolysis. *Int. J. Hydrogen Energy* 40, 11457–11464. doi:10.1016/j.ijhydene.2015.01.097

Trost, T., Jentsch, M., and Sterner, M. (2012). Erneuerbares Methan: Analyse der CO_2-Potenziale für Power-to-Gas Anlagen in Deutschland. *Zeitschrift für Energiewirtschaft* 36, 173–190. doi:10.1007/s12398-012-0080-6

van der Giesen, C., Kleijn, R., and Kramer, G. J. (2014). Energy and climate impacts of producing synthetic hydrocarbon fuels from CO_2. *Environ. Sci. Technol.* 48, 7111–7121. doi:10.1021/es500191g

Vandewalle, J., Bruninx, K., and D'haeseleer, W. (2014). *The Interaction of a High Renewable Energy/Low Carbon Power System with the Gas System through Power to Gas*. Rome, 28–31.

Varone, A., and Ferrari, M. (2015). Power to liquid and power to gas: an option for the German Energiewende. *Renew. Sustain. Energ. Rev.* 45, 207–218. doi:10.1016/j.rser.2015.01.049

Wismans, J., Grahn, M., and Denbratt, I. (2016). *Low-Carbon Transport: Health and Climate Benefits. Background Report to Intergovernmental Ninth Regional Environmentally Sustainable Transport (EST) Forum in Asia*. Gothenburg: Chalmers University of Technology.

Xu, Y., Isom, L., and Hanna, M. A. (2010). Adding value to carbon dioxide from ethanol fermentations. *Bioresour. Technol.* 101, 3311–3319. doi:10.1016/j.biortech.2010.01.006

Zakeri, B., and Syri, S. (2015). Electrical energy storage systems: a comparative life cycle cost analysis. *Renew. Sustain. Energ. Rev.* 42, 569–596. doi:10.1016/j.rser.2014.10.011

Zhang, B., and Chen, G. (2014). China's CH_4 and CO_2 emissions: bottom-up estimation and comparative analysis. *Ecol. Indicators* 47, 112–122. doi:10.1016/j.ecolind.2014.01.022

Zhang, X., Chan, S. H., Ho, H. K., Tan, S. C., Li, M., Li, G., et al. (2015). Towards a smart energy network: the roles of fuel/electrolysis cells and technological perspectives. *Int. J. Hydrogen Energy* 40, 6866–6919. doi:10.1016/j.ijhydene.2015.03.133

Conflict of Interest Statement: The authors declare that the research was conducted in the absence of any commercial or financial relationships that could be construed as a potential conflict of interest.

18

Vinasse from Sugarcane Ethanol Production: Better Treatment or Better Utilization?

Cristiano E. Rodrigues Reis and Bo Hu*

Department of Bioproducts and Biosystems Engineering, University of Minnesota, Saint Paul, MN, USA

Ethanol production from sugarcane in Brazil is a well-established industry, with relatively simple operations and high yield. The ethanol primarily serves as a renewable fuel blending with gasoline and diesel to increase the energy security in Brazil. Several environmental concerns are emerged around the by-products from this industry. Vinasse, the liquid fraction generated from the rectification and distillation operations of ethanol, is a sulfur-rich, low pH, dark-colored, and odorous effluent, produced at volumes as high as 20-fold of ethanol. Traditional wastewater treatments, such as bioprocessing, advanced oxidative processes, anaerobic digestion (AD), and chemical-based processes, have been applied to vinasse management. Despite most of its utilization being in fertirrigation practices, vinasse may represent a key factor in enhancing profitability and environmental outcomes of a sugarcane-to-ethanol plant. The application of some upgrade solutions to sugarcane-derived vinasse may represent additional sources of energy, production of animal feed components, and reduction in water consumption within a plant. The use of mature technologies, yet not widespread in the sugarcane-to-ethanol industry, could help attenuate environmental concerns. Oxidation and chemical processes, AD, and microbial fermentation have been presented as alternative impactful alternatives to (i) reduce its organic and mineral load, converting it to a feedstock with fewer environmental applications when applied as fertilizer and (ii) to convert organic matter and nutrients to a nutritious biomass, simultaneously increasing water reclamation potential by plants. This mini-review article provides a critical and comprehensive summary of the alternatives developed or under development to vinasse management.

Keywords: vinasse, sugarcane, anaerobic digestion, fertirrigation, ethanol

Edited by:
Rongxin Su,
Tianjin University, China

Reviewed by:
Chao Ma,
Bristol Myers Squibb, USA
Renliang Huang,
Tianjin University, China

*Correspondence:
Bo Hu
bhu@umn.edu

Specialty section:
This article was submitted to
Bioenergy and Biofuels,
a section of the journal
Frontiers in Energy Research

Citation:
Rodrigues Reis CE and Hu B (2017)
Vinasse from Sugarcane Ethanol
Production: Better Treatment
or Better Utilization?
Front. Energy Res. 5:7.

INTRODUCTION

Brazil is home to over 300 active sugarcane biorefineries (Filoso et al., 2015), with an ethanol-rich history dating back to the 1970s (Goldemberg et al., 2008). The oil embargo crises over 40 years ago forced the Brazilian government to find alternative solutions for energy generation. The most successful strategy was the National Ethanol Program (Proálcool), which increased Brazil's energy security and posed it as the largest ethanol producer for decades (Goldemberg et al., 2008). Even with recent global booming of ethanol and biofuel industry, Brazil is still the second largest ethanol producer in the world, summing values of 25 billion L per year (Walter et al., 2011). Sugarcane-to-ethanol productivity reaches about 5.6 m³ ha⁻¹ over the course of a year (Badger, 2002).

Over 75% of the Brazilian distilleries operate using the Melle-Boinot process, a fed-batch system using yeast cell recycling. Yeast cells are collected at the end of the fermentation cycle and are either centrifuged or filtered, and reinoculated to the next fermentation cycle (Brethauer and Wyman, 2010). The high-density cell culture and simple composition of sugarcane juice allow a quick fermentation cycle (6–10 h) and low cell growth rates (Della-Bianca et al., 2013). The upstream steps include a sulfitation process, which enrich the downstream products with sulfur compounds, especially sulfate species (Della-Bianca et al., 2013). After fermentation, the fermented juice is processed into an ethanol stream and a liquid-rich by-product—vinasse.

Sugarcane vinasse is a residue from the sugar-ethanol industry, characterized as being an acidic suspension, high COD values, unpleasant odors, and dark brown color (Gómez and Rodríguez, 2000; Jiang et al., 2012; Christofoletti et al., 2013). The characteristics of vinasse are largely dependent on the feedstock, and on the fermentation and distillation conditions applied (España-Gamboa et al., 2011). A summary of vinasse composition is presented in **Table 1**. Sugarcane vinasse is reported to be a nitrogen-deficient medium, which is most composed as acid-insoluble nitrogen (Parnaudeau et al., 2008). Vinasse is also characterized as a feedstock rich in phenolic compounds and melanoidins (FitzGibbon et al., 1998). Using NMR and FTIR, Benke et al. (1998) detected levels of cellulose and hemicellulose in vinasse, which are derived from the grinding conditions of sugarcane. Recent development on lignocellulosic ethanol production from sugarcane bagasse may further increase the fiber content of vinasse, as that the various methods applied to sugarcane bagasse pretreatment will likely affect the production of liquid byproducts (Moraes et al., 2015). Until now, there is a lack of information regarding the composition of liquid streams generated as by-products, except for a patent application (Cammarota et al., 2012). The composition of second-generation vinasse contains greater organic matter content and similar BOD/COD ratios (**Table 1**). Mineral composition, especially for potassium, is significantly lower.

Vinasse has been mostly used on fertirrigation practices, i.e., utilizing it as a liquid fertilizer for crops, reducing the water input for plant growth (Walter et al., 2011). Fertirrigation usually has negative effects on soil and ground waters in the long term (Rocha et al., 2007). A few adequate uses for vinasse management have been identified and used in large-scale operations, such as vinasse recycling to fermentation streams (Fadel et al., 2014; Yang et al., 2016), fertirrigation (Christofoletti et al., 2013; Filoso et al., 2015), energy production (Cortez et al., 1992; Walter et al., 2011), and animal feed production (Cortez et al., 1992). Recently, studies

TABLE 1 | Composition of sugarcane vinasse from first and second generation sugarcane-to-ethanol production.

Mineral analysis			Organic analysis		
Component	Value	Reference	Component	Value	Reference
Cl^- (mg L^{-1})	59.4[a]	dos Santos et al. (2013)	Organic matter (%)	3.96[a]	Mariano et al. (2009)
SO_4^{2-} (mg L^{-1})	1,680[a], 44–366[b]	dos Santos et al. (2013), Cammarota et al. (2012)	C:N ratio	10[a], 49.2–124.9[b]	Mariano et al. (2009), Cammarota et al. (2012)
Na^+ (mg L^{-1})	8.6[a]	dos Santos et al. (2013)	COD (mg L^{-1})	32,000–92,800[a], 75,800–109,700[b]	Mariano et al. (2009), Paz-Pino et al. (2014), Cammarota et al. (2012)
K^+ (mg L^{-1})	1,620[a]	dos Santos et al. (2013)	BOD_5 (mgO_2 L^{-1})	13,514–36,847[a], 31,500–87,700[b]	Paz-Pino et al. (2014)
Ca^{2+} (mg L^{-1})	3,160[a]	dos Santos et al. (2013)	BOD_5/COD	0.18–0.34[a], 0.39–0.80[b]	Paz-Pino et al. (2014), Cammarota et al. (2012)
Mg^{2+} (mg L^{-1})	162.4[a]	dos Santos et al. (2013)	Phenols (mg L^{-1})	230–390[a], 0.4–12.4[b]	Paz-Pino et al. (2014), Cammarota et al. (2012)
$PO4^{3-}$ (mg L^{-1})	560[a], 33.26[b]	dos Santos et al. (2013), Cammarota et al. (2012)	NH_4^+ (mg L^{-1})	23.9[a]	dos Santos et al. (2013)
NO_3^- (mg L^{-1})	823.7[a]	dos Santos et al. (2013)	Protein (%)	2.92[a]	Dowd et al. (1994)
Fe (mg L^{-1})	44.9[a]	dos Santos et al. (2013)	Fiber (%)	0.2[a]	Dowd et al. (1994)
Mn (mg L^{-1})	4.9[a]	dos Santos et al. (2013)	Fat (%)	0.41[a]	Dowd et al. (1994)
Zn (mg L^{-1})	1.2[a]	dos Santos et al. (2013)	Ash (%)	3.61[a]	Dowd et al. (1994)
BO_3^{3-} (mg L^{-1})	1.94[a]	dos Santos et al. (2013)	Carbohydrate (%)	3.42[a]	Dowd et al. (1994)
Ba (mg L^{-1})	0.54[a]	Mariano et al. (2009)	Acetaldehyde (g L^{-1})	0.697[a]	Dowd et al. (1994)
Cd (mg L^{-1})	1.06[a]	Mariano et al. (2009)	Ethanol (g L^{-1})	3.83[a]	Dowd et al. (1994)
Cr (mg L^{-1})	0.15[a]	Mariano et al. (2009)	Propylene glycol (g L^{-1})	0.084[a]	Dowd et al. (1994)
Ni (mg L^{-1})	0.26[a]	Mariano et al. (2009)	2,3-butanediols (g L^{-1})	0.568[a]	Dowd et al. (1994)
Al (mg L^{-1})	72.5[a]	Mariano et al. (2009)	Glycerol (g L^{-1})	5.86[a]	Dowd et al. (1994)
MoO_4^{2-} (mg L^{-1})	0.17[a]	dos Santos et al. (2013)	Erythritol (g L^{-1})	0.088[a]	Dowd et al. (1994)
Cu (mg L^{-1})	0.06[a]	Mariano et al. (2009)	Arabinitol (g L^{-1})	0.064[a]	Dowd et al. (1994)
Physicochemical analysis			Chiro-inositol (g L^{-1})	0.114[a]	Dowd et al. (1994)
Density (g mL^{-1})	1[a]	Mariano et al. (2009)	Sucrose (g L^{-1})	0.222[a]	Dowd et al. (1994)
Ph	4.84[a], 4.0–4.9[b]	Dowd et al. (1994)	Acetic acid (g L^{-1})	1.56[a]	Dowd et al. (1994)
DO (mg L^{-1})	4.3[a]	Mariano et al. (2009)	Formic acid (g L^{-1})	0.582[a]	Dowd et al. (1994)
Moisture (%)	89.64[a]	Dowd et al. (1994)	Lactic acid (g L^{-1})	7.74[a]	Dowd et al. (1994)
Eh (Mv)	260[a]	Mariano et al. (2009)	Quinic acid (g L^{-1})	0.508[a]	Dowd et al. (1994)
Conductivity (mS cm^{-1})	8.52[a]	Mariano et al. (2009)			

[a]First generation.
[b]Second generation.

on transforming vinasse into a high value-added feedstock have also been performed (Nitayavardhana and Khanal, 2010).

Sugarcane and ethanol production in Brazil has been largely criticized due to several ecological factors (Sparovek et al., 2009), and negative environmental impacts, especially to the increase of contaminants in soil and surface water (Jiang et al., 2012). This review article describes some of the alternative uses to diminish the environmental impact by vinasse management practices, as shown in **Figure 1**.

FERTIRRIGATION PRACTICES WITH VINASSE

The utilization of vinasse in fertirrigation practices started in the 1950s (Valsechi and Gomes, 1954), and by the 1980s, it was a common practice for sugarcane refineries to utilize the liquid residual as fertilizer (Walter et al., 2011). The concept behind fertirrigation consists on a sum of irrigation to sugarcane fields, by the percolation of vinasse liquid to the soil, with the simultaneous fertilization, transferring its nutrients to the plants (Christofoletti et al., 2013). Besides decreasing the costs involved with chemical fertilizers (Jiang et al., 2012), completely supplying phosphorus (Moran-Salazar et al., 2016) and being of low capital cost, vinasse utilization in fertirrigation practices could be considered of certain level of environmental concern (Sparovek et al., 2009). Fertirrigation practices have been linked with increase in eutrophication of waterbodies and the formation of dead aquatic bodies in Brazil and in other countries (Eykelbosh et al., 2015). The correct application of fertirrigation has proven not to impact the physical, chemical, and biological properties of the soil to which vinasse is applied (Christofoletti et al., 2013), such as, and levels up to 300 m^3 vinasse ha^{-1} with potassium levels of 3–4 kg m^{-3} do not impact negatively the soil (Penatti et al., 1988). However, conditions with increase in crop losses, soil pH change, increase in phytotoxicity, and release of sulfurous odors are not uncommon (Christofoletti et al., 2013).

Sugarcane crops occupy nearly three million hectares in the Brazilian state of São Paulo (Có Júnior et al., 2008). Current production of ethanol could supplement up to 80% of sugarcane plantation by fertirrigation (Christofoletti et al., 2013). A study conducted at the Pirapama basin river, home to three ethanol plants in Brazil, producing over 500,000 L of ethanol per day during the peak season, has estimated biochemical oxygen demand disposal rates of 226,335 kg on a daily basis, correspondent to a city of 4.2 million people (Alcoforado de Moraes et al., 2009). The high toxicity potential of vinasse being utilized in fertirrigation practices may lead to hydrologic, agronomic, and social problems. Since there are no pollution charges applied to sugarcane farmers and ethanol producers, fertirrigation still stands as the predominantly application of vinasse.

VINASSE AS A FEEDSTOCK FOR BIOLOGICAL TREATMENT

Vinasse typically has a moisture content of about 93%, and the present organic solids and minerals, such as potassium, calcium, and magnesium (Christofoletti et al., 2013), may provide a rich culture medium for biological cultivation. Detoxification from soils contaminated by excessive vinasse utilization has been evaluated (Abioye, 2011), indicating that the organic

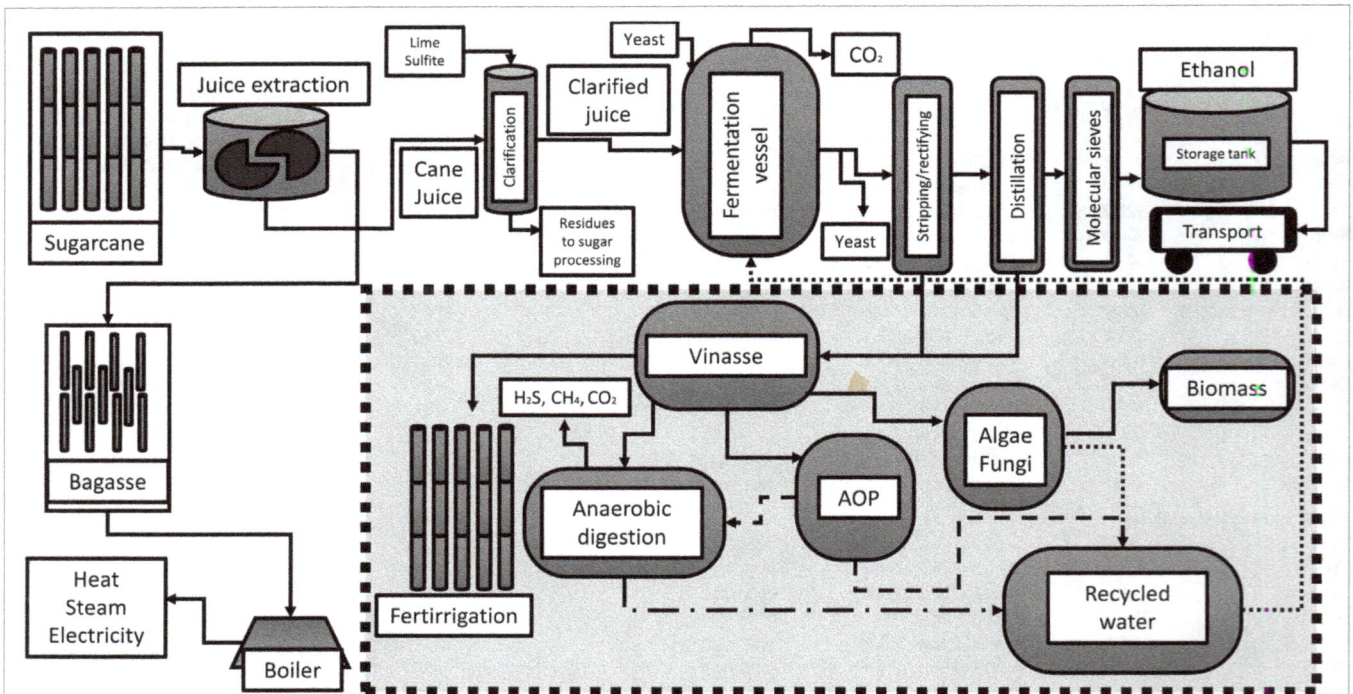

FIGURE 1 | Simplified sugarcane-to-ethanol process with potential modifications in vinasse management.

compounds present in the effluent may serve as nutrient sources for microorganisms, despite the low concentration of nitrogen and phosphorus. Prata et al. (2001) also indicated that the readily available carbon sources in vinasse, such as glycerol, accelerated the degradation of the herbicide ametryn in vinasse-contaminated soils.

In situ biological treatment of vinasse could be a potentially solution. Among different fungi that can utilize vinasse as a substrate for growth, *Rhizopus oligosporus* was grown on a vinasse-rich medium (75% v/v) using an airlift bioreactor (Nitayavardhana et al., 2013). The 2.5-L reactor used for the study had supplementation of nitrogen and phosphorus, and aeration rates ranging from 0.5 to 2.0 vvm, achieving a maximum biomass accumulation of 8.04 g of increase compared to the initial mycelium inoculated. Nitayavardhana et al. (2013) reported a decrease in 80% of COD and observed that the fungal biomass achieved a high accumulation of protein (around 50%), which could be redirected to livestock production, especially since *R. oligosporus* cultivations usually yield amino acid profiles comparable to those of soybean meal (Lim and Akiyama, 1992), being only deficient in methionine and phenylalanine. *R. oligosporus* is a commonly used starter culture for Indonesian tempeh production and are known to fully utilize carbon sources rich in sucrose, as the one in vinasse, to grow (Egounlety and Aworh, 2003). Fungal treatment of with white-rot *Trametes versicolor* was performed in order to evaluate the fungus potential to produce laccase and decrease the concentration of phenol and chromophoric compounds in vinasse (España-Gamboa et al., 2015, 2017). Achieving 60% removal of COD, and over 80% of total phenol, with a decrease in almost 20% in color, *T. versicolor* has proven to be excellent laccase-producing microorganisms, achieving production of 1,630 laccase units per liter of medium (España-Gamboa et al., 2015). Since the presence of melanoidins, phenols, and polyphenols have been described to potentially have negative effects on crop productions (Constabel and Ryan, 1998), bioprocessing technologies, such as the study using *T. versicolor*, can decrease the harmful effects of applying vinasse in fertirrigation.

Utilizing microorganisms to treat vinasse does not only serve for the purpose of reducing COD and toxic compounds. The utilization of fungi cultivated in vinasse potentially can bring additional revenue to sugarcane-based ethanol plants by providing feed and feed supplements to livestock production. Nair and Taherzadeh (2016) cultivated *Neurospora intermedia* and *Aspergillus oryzae* in vinasse and observed that an integration to a medium-sized facility, producing 100,000 m^3 of ethanol a year, could reach up to 250,000 tons of protein-rich (45 weight %) dry fungal biomass per year. *N. intermedia* and *A. oryzae* are characterized as generally regarded as safe materials (Ferreira et al., 2015; Todokoro et al., 2015) and have been traditionally used in the preparation of traditional dishes in Southeast and East Asia, being recently used in starch-based ethanol waste streams (Ferreira et al., 2015). Heterotrophic algae have also been reported as potential microorganisms to utilize the nutrients in vinasse to grow. The dark color characteristic to vinasse may be comparable to other dark effluents, such as municipal leachate, and may significantly hinder photoautotrophic growth of algae (Reis et al., 2014b). Therefore, heterotrophic growth may be the most appropriate growth mode for algae in vinasse. An example with green algae *Desmodesmus*

sp. indicated slight elevation of pH, low oxygen, and low carbon dioxide removal, with a decrease of 52.1% in nitrogen and 36.2% of COD. *Desmodesmus* also achieved high yield of COD to biomass in the first hour of growth (0.5 g g^{-1}) and specific growth rate of 0.15 h^{-1}. *Scenedesmus* sp. was cultivated on a Guillard-modified medium supplemented with 40% of vinasse and has been reported as able to grow in rates comparable to the control experiments (Ramirez et al., 2014).

Several opportunities used in starch-based ethanol research could be utilized with vinasse as bioprocess medium, such as those producing ethanol, malic acid, butanol, and many other commodity chemicals (Reis et al., 2017). The production of secondary ethanol using *N. intermedia* and *A. oryzae* on vinasse, for instance, would potentially provide extra 12.6% ethanol produced annually (Nair and Taherzadeh, 2016). The direct use of vinasse as feed material, in a similar fashion as the conditions applied to the U.S. dry-grind corn-to-ethanol industry, may not be suitable with the current composition of vinasse, especially due to the low protein content, and surplus of sulfur. An opportunity is to utilize protein-rich biomass grown in vinasse as animal feed, which may represent a significant change in the utilization of downstream processing of sugarcane-to-ethanol products. The utilization of fungal protein grown on sugarcane vinasse could provide feedstock for a market similar to soybean meal, generating at least USD 9.5 million annually to ethanol plants (Nitayavardhana and Khanal, 2010).

ADDITION OF VALUE TO VINASSE *VIA* CHEMICAL AND ADVANCED OXIDATION PROCESSES (AOPs)

Advanced oxidation process methods have been extensively used in wastewater treatment facilities, which principles lies behind the high reactivity of HO• radicals, driving oxidation processes to all sorts of recalcitrant pollutants (Andreozzi et al., 1999). Utilizing AOP to treat vinasse is an opportunity to recycle the water back to the fermentation process, decreasing operational costs and environmental footprint of sugarcane plants. The effectiveness of ozone-based AOP (O_3, O_3/UV, and O_3/UV/H_2O_2) has been tested on a vinasse-like effluent, achieving a fast kinetic degradation profile ($k = 6.5 \times 10^{-3}$ min^{-1}) with O_3/UV/H_2O_2 and being reported as an economical process (1.31€ m^{-3} g^{-1} of TOC mineralized under optimized conditions) (Lucas et al., 2010). The process behind recovering high-value phenolic compounds in vinasse, present at a concentration within the region of 600 mg L^{-1}, is likely to be economically not feasible (Santos et al., 2003).

The combination of AOP with other forms of value addition to vinasse is also a research and commercial opportunity. A study conducted by Siles et al. (2011) evaluated the serialization of short-retention time ozonation, which was responsible to a decrease in over 50% of phenols, with anaerobic digestion (AD). Pretreated vinasse had anaerobic degradability of around 80% of the total COD, with enhanced methane yield coefficients and methane production rates enhanced by 13.6 and 41.2% when compared to raw vinasse (Siles et al., 2011). Potential studies with phototrophic microorganisms for production of value-added compounds could be coupled with AOP focused on color removal. Fagier et al. (2016) evaluated Fe^{2+}-activated persulfate and peroxymonosulfate oxidation on vinasse to which an addition of 15 g L^{-1} of coagulant provided over 70% total organic carbon removal, and near 100% of UV_{254} and color removal, which lowers the toxicity levels for phototrophic microorganisms. AOP consist on a series of mature and well-understood steps that can help detoxify vinasse to further utilization and value addition, such as production of protein-rich microbial biomass.

AD OF VINASSE: OPERATION, ENERGY GENERATION, AND DIGESTATE USE

Anaerobic digestion is a common practice in the current dry-grind corn-to-ethanol industry, being responsible to generate energy and degrade complex organic matter present in the corn stillage (Reis et al., 2017). Despite the composition of sugarcane vinasse differing significantly from wastewaters with high carbon load to which AD processes are ubiquitous and profitable, the possibility of decreasing the negative impacts to vinasse application in soil usage accelerated the research and development of AD in sugarcane vinasse. The first industrial application in Brazil was built in the 1990s and consisted of an upflow anaerobic sludge blanket (UASB) reactor in São Martinho mill with a capacity of 5,000 m^3 (Souza et al., 1992). The biogas generated from the UASB reactor was used on the drying process of yeast.

The use of AD in sugarcane vinasse is characterized according the number of steps, process temperature, and reactor design (Rajeshwari et al., 2000). The use of digestate vinasse was used to cultivate the microalgae *Chlorella vulgaris* (Marques et al., 2013; Candido and Lombardi, 2017) and *Neochloris oleoabundans* (Olguín et al., 2015). Initially, research shows that vinasse was highly toxic to *C. vulgaris* at concentrations greater than 4%, reaching allowable concentrations of about 8.6% after treatment (Marques et al., 2013). *C. vulgaris*, a widely used species in accumulation of valuable microbial lipids (Reis et al., 2014b), was able to achieve specific growth rates of 0.76 day^{-1}, higher than the control experiments in nutrient sufficient medium (0.53 day^{-1}) (Marques et al., 2013). The use of AD can represent a feasible opportunity for making vinasse an appropriate cultivation medium for microbial cultivation, such as in the use of fungi with resilience to unfavorable conditions of growth (Reis et al., 2014a). *N. oleoabundans* was able to grow on supplementation of vinasse up to 8%, with addition of sodium bicarbonate, achieving lipid concentrations up to 38.5%, high $N-NH_4^+$ removal (85.2%), and high flocculation efficiency (42% after 30 min) (Olguín et al., 2015).

Due to the high sulfur composition of vinasse, AD produces a sulfur-rich biogas (Barrera et al., 2013), which could be highly corrosive to the burners by the production of SO_2. Removal of SO_x from biogas can be accomplished, among other technologies, through physical processes, such as sulfur-specific membrane filters, or through biological processes (Barrera et al., 2014). Lebrero et al. (2016) evaluated biotrickling filters (BTFs) and algal–bacterial photobioreactors (PBRs) as alternative to membrane removal. Both BTF and PBR constitute removal efficiencies greater than 98%, with elimination capacities as high as 26 g $S-H_2S$ m^{-3} h^{-1}. BTF showed impressive robustness as it was completed revived after a 15-day shut down, and it was able to utilize the nutrients from vinasse, with exogenous nitrate addition (Lebrero et al., 2016). Lebrero et al. (2016) also reported that PBR supported CO_2 removal of $23 \pm 11.8\%$, increasing to 62% at pH of 8.1, with an overall fixation rate of 285 mg CO_2 L^{-1} day^{-1}, thus, it could be coupled with a BTF to upgrade AD-generated biogas, removing inert gases and sulfur-rich compounds. Overall, AD of vinasse could generate approximately 4.5 MW yearly, which would correspond to over 14.5 million m^3 of biogas, with concentrations of 60% CH_4, replacing up to 12% of the bagasse from burning to combined heat and power operations (Moraes et al., 2014). AD in a large-scale sugarcane ethanol plant has a potential of supplying electricity to a city of 130,000 inhabitants or a replacement up to 40% of the annual diesel supply in the agricultural energy requirements of a sugarcane biorefinery (Moraes et al., 2014). Therefore, optimization and widespread use of AD in sugarcane ethanol plants still remain as opportunities.

CONCLUSION AND FUTURE DIRECTIONS

While for many waste sources, a broad literature may be available (e.g., municipal wastewater), for others, references are scarcer and many opportunities are still under-performed—such is the case for sugarcane vinasse. The use of traditional practices, such as fertirrigation, may cause environmental issues in water and soil quality, and the use of more robust approaches must be a practical solution for vinasse management. Fertirrigation is often a practice that provides a false impression of solving the problem of vinasse disposal. The use of AD is an underperformed process in sugarcane plants and could significantly increase the energy output, while reducing the amount of water used within a plant. AOP can help decrease the toxicity of vinasse and can be used as a pretreatment for microbial growth, which could generate high throughput of value-added chemicals. Treatment of vinasse prior to recycling it as fertilizer and irrigation water would potentially lower the environmental impact of applying a nutrient-rich suspension to soils. Thus, vinasse management is an issue in ethanol plants in Brazil, home to the second largest ethanol production in the world, which represents a realm of hidden opportunities for mature technologies.

AUTHOR CONTRIBUTIONS

CR is the first author of the paper. He conceptualized the idea, analyzed the references, and drafted the paper. BH serves as the Ph.D. advisor for CR and worked with him together on designing and finalizing the paper.

FUNDING

Authors are grateful for the support of MnDRIVE-Global Food Ventures for the funding support. CR acknowledges MnDRIVE-UMII and CAPES, Ministry of Education of Brazil (BEX 13252/13-5) for his fellowship.

REFERENCES

Abioye, O. P. (2011). "Biological remediation of hydrocarbon and heavy metals contaminated soil," in *Soil Contamination*, ed. S. Pascucci, Rijeka: INTECH. Available from: https://www.intechopen.com/books/soil-contamination/biological-remediation-of-hydrocarbon-and-heavy-metals-contaminated-soil

Alcoforado de Moraes, M. M. G., Cai, X., Ringler, C., Albuquerque, B. E., Vieira da Rocha, S. P., and Amorim, C. A. (2009). Joint water quantity-quality management in a biofuel production area—integrated economic-hydrologic modeling analysis. *J. Water Resour. Plann. Manag.* 136, 502–511. doi:10.1061/(ASCE)WR.1943-5452.0000049

Andreozzi, R., Caprio, V., Insola, A., and Marotta, R. (1999). Advanced oxidation processes (AOP) for water purification and recovery. *Catal. Today* 53, 51–59. doi:10.1016/S0920-5861(99)00102-9

Badger, P. C. (2002). "Ethanol from cellulose: a general review," in *Trends New Crops and New Uses*, eds J. Janick and A. Whipkey, Alexandria, VA: ASHS Press.

Barrera, E. L., Spanjers, H., Dewulf, J., Romero, O., and Rosa, E. (2013). The sulfur chain in biogas production from sulfate-rich liquid substrates: a review on dynamic modeling with vinasse as model substrate. *J. Chem. Technol. Biotechnol.* 88, 1405–1420. doi:10.1002/jctb.4071

Barrera, E. L., Spanjers, H., Romero, O., Rosa, E., and Dewulf, J. (2014). Characterization of the sulfate reduction process in the anaerobic digestion of a very high strength and sulfate rich vinasse. *Chem. Eng. J.* 248, 383–393. doi:10.1016/j.cej.2014.03.057

Benke, M., Mermut, A., and Chatson, B. (1998). Carbon-13 CP/MAS NMR and DR-FTIR spectroscopic studies of sugarcane distillery waste. *Can. J. Soil Sci.* 78, 227–236. doi:10.4141/S97-036

Brethauer, S., and Wyman, C. E. (2010). Review: continuous hydrolysis and fermentation for cellulosic ethanol production. *Bioresour. Technol.* 101, 4862–4874. doi:10.1016/j.biortech.2009.11.009

Cammarota, M. C., Camporese, S. E. F., Absai, D. A. C. G., Larissa De, C. A., Groposo, S. C. J., De Castro, A. M., et al. (2012). *Method for Producing Energy-Rich Gases from Lignocellulosic Material Streams*. Google patents.

Candido, C., and Lombardi, A. T. (2017). Growth of *Chlorella vulgaris* in treated conventional and biodigested vinasses. *J. Appl. Phycol.* 29, 45–53. doi:10.1007/s10811-016-0940-2

Christofoletti, C. A., Escher, J. P., Correia, J. E., Marinho, J. F. U., and Fontanetti, C. S. (2013). Sugarcane vinasse: environmental implications of its use. *Waste Manag.* 33, 2752–2761. doi:10.1016/j.wasman.2013.09.005

Có Júnior, C., Marques, M. O., and Tasso Júnior, L. C. (2008). Sugarcane technological parameters affected by sewage sludge and vinasse added in soil for four consecutive years. *Eng. Agric.* 28, 196–203. doi:10.1590/S0100-69162008000100020

Constabel, C. P., and Ryan, C. A. (1998). A survey of wound-and methyl jasmonate-induced leaf polyphenol oxidase in crop plants. *Phytochemistry* 47, 507–511. doi:10.1016/S0031-9422(97)00539-6

Cortez, L., Magalhães, P., and Happi, J. (1992). Principais subprodutos da agroindústria canavieira e sua valorização. *Rev. Bras. Energ.* 2, 111.

Della-Bianca, B. E., Basso, T. O., Stambuk, B. U., Basso, L. C., and Gombert, A. K. (2013). What do we know about the yeast strains from the Brazilian fuel ethanol industry? *Appl. Microbiol. Biotechnol.* 97, 979–991. doi:10.1007/s00253-012-4631-x

dos Santos, J. D., da Silva, A. L. L., da Luz Costa, J., Scheidt, G. N., Novak, A. C., Sydney, E. B., et al. (2013). Development of a vinasse nutritive solution for hydroponics. *J. Environ. Manage.* 114, 8–12. doi:10.1016/j.jenvman.2012.10.045

Dowd, M. K., Johansen, S. L., Cantarella, L., and Reilly, P. J. (1994). Low molecular weight organic composition of ethanol stillage from sugarcane molasses, citrus waste, and sweet whey. *J. Agric. Food Chem.* 42, 283–288. doi:10.1021/jf00038a011

Egounlety, M., and Aworh, O. (2003). Effect of soaking, dehulling, cooking and fermentation with *Rhizopus oligosporus* on the oligosaccharides, trypsin inhibitor, phytic acid and tannins of soybean (*Glycine max Merr.*), cowpea (*Vigna unguiculata L. Walp*) and groundbean (*Macrotyloma geocarpa Harms*). *J. Food Eng.* 56, 249–254. doi:10.1016/S0260-8774(02)00262-5

España-Gamboa, E., Mijangos-Cortes, J., Barahona-Perez, L., Dominguez-Maldonado, J., Hernández-Zarate, G., and Alzate-Gaviria, L. (2011). Vinasses: characterization and treatments. *Waste Manag. Res.* 29, 1235–1250. doi:10.1177/0734242X10387313

España-Gamboa, E., Vicent, T., Font, X., Dominguez-Maldonado, J., Canto-Canché, B., and Alzate-Gaviria, L. (2017). Pretreatment of vinasse from the sugar refinery industry under non-sterile conditions by *Trametes versicolor* in a fluidized bed bioreactor and its effect when coupled to an UASB reactor. *J. Biol. Eng.* 11, 6. doi:10.1186/s13036-016-0042-3

España-Gamboa, E., Vicent, T., Font, X., Mijangos-Cortés, J., Canto-Canché, B., and Alzate, L. (2015). Phenol and color removal in hydrous ethanol vinasse in an air-pulsed bioreactor using *Trametes versicolor*. *J. Biochem. Technol.* 6, 982–986.

Eykelbosh, A. J., Johnson, M. S., and Couto, E. G. (2015). Biochar decreases dissolved organic carbon but not nitrate leaching in relation to vinasse application in a Brazilian sugarcane soil. *J. Environ. Manage.* 149, 9–16. doi:10.1016/j.jenvman.2014.09.033

Fadel, M., Zohri, A.-N. A., Makawy, M., Hsona, M., and Abdel-Aziz, A. (2014). Recycling of vinasse in ethanol fermentation and application in Egyptian distillery factories. *Afr. J. Biotechnol.* 13, 4390–4398. doi:10.5897/AJB2014.14083

Fagier, M., Ali, E., Tay, K., and Abas, M. (2016). Mineralization of organic matter from vinasse using physicochemical treatment coupled with Fe^{2+}-activated persulfate and peroxymonosulfate oxidation. *Int. J. Environ. Sci. Technol.* 13, 1189–1194. doi:10.1007/s13762-016-0963-x

Ferreira, J. A., Lennartsson, P. R., and Taherzadeh, M. J. (2015). Production of ethanol and biomass from thin stillage by *Neurospora intermedia*: a pilot study for process diversification. *Eng. Life Sci.* 15, 751–759. doi:10.1002/elsc.201400213

Filoso, S., do Carmo, J. B., Mardegan, S. F., Lins, S. R. M., Gomes, T. F., and Martinelli, L. A. (2015). Reassessing the environmental impacts of sugarcane ethanol production in Brazil to help meet sustainability goals. *Renew. Sustain. Energ. Rev.* 52, 1847–1856. doi:10.1016/j.rser.2015.08.012

FitzGibbon, F., Singh, D., McMullan, G., and Marchant, R. (1998). The effect of phenolic acids and molasses spent wash concentration on distillery wastewater remediation by fungi. *Process Biochem.* 33, 799–803. doi:10.1016/S0032-9592(98)00050-8

Goldemberg, J., Coelho, S. T., and Guardabassi, P. (2008). The sustainability of ethanol production from sugarcane. *Energy Policy* 36, 2086–2097. doi:10.1016/j.enpol.2008.02.028

Gómez, J., and Rodríguez, O. (2000). Effects of vinasse on sugarcane (*Saccharum officinarum*) productivity. *Rev. Fac. Agron.* 17, 318–326.

Jiang, Z.-P., Li, Y.-R., Wei, G.-P., Liao, Q., Su, T.-M., Meng, Y.-C., et al. (2012). Effect of long-term vinasse application on physico-chemical properties of sugarcane field soils. *Sugar Tech* 14, 412–417. doi:10.1007/s12355-012-0174-9

Lebrero, R., Toledo-Cervantes, A., Muñoz, R., del Nery, V., and Foresti, E. (2016). Biogas upgrading from vinasse digesters: a comparison between an anoxic biotrickling filter and an algal-bacterial photobioreactor. *J. Chem. Technol. Biotechnol.* 91, 2488–2495. doi:10.1002/jctb.4843

Lim, C., and Akiyama, D. (1992). Full-fat soybean meal utilization by fish. *Asian Fish. Sci.* 5, 181–197.

Lucas, M. S., Peres, J. A., and Puma, G. L. (2010). Treatment of winery wastewater by ozone-based advanced oxidation processes (O_3, O_3/UV and O_3/UV/H_2O_2) in a pilot-scale bubble column reactor and process economics. *Sep. Purif. Technol.* 72, 235–241. doi:10.1016/j.seppur.2010.01.016

Mariano, A. P., Crivelaro, S. H. R., Angelis, D. D. F. D., and Bonotto, D. M. (2009). The use of vinasse as an amendment to ex-situ bioremediation of soil and groundwater contaminated with diesel oil. *Braz. Arch. Biol. Technol.* 52, 1043–1055. doi:10.1590/S1516-89132009000400030

Marques, S. S. I., Nascimento, I. A., de Almeida, P. F., and Chinalia, F. A. (2013). Growth of *Chlorella vulgaris* on sugarcane vinasse: the effect of anaerobic digestion pretreatment. *Appl. Biochem. Biotechnol.* 171, 1933–1943. doi:10.1007/s12010-013-0481-y

Moraes, B. S., Junqueira, T. L., Pavanello, L. G., Cavalett, O., Mantelatto, P. E., Bonomi, A., et al. (2014). Anaerobic digestion of vinasse from sugarcane biorefineries in Brazil from energy, environmental, and economic perspectives: profit or expense? *Appl. Energy* 113, 825–835. doi:10.1016/j.apenergy.2013.07.018

Moraes, B. S., Zaiat, M., and Bonomi, A. (2015). Anaerobic digestion of vinasse from sugarcane ethanol production in Brazil: challenges and perspectives. *Renew. Sustain. Energ. Rev.* 44, 888–903. doi:10.1016/j.rser.2015.01.023

Moran-Salazar, R., Sanchez-Lizarraga, A., Rodriguez-Campos, J., Davila-Vazquez, G., Marino-Marmolejo, E., Dendooven, L., et al. (2016). Utilization of vinasses as soil amendment: consequences and perspectives. *Springerplus* 5, 1–11. doi:10.1186/s40064-016-2410-3

Nair, R. B., and Taherzadeh, M. J. (2016). Valorization of sugar-to-ethanol process waste vinasse: a novel biorefinery approach using edible ascomycetes filamentous fungi. *Bioresour. Technol.* 221, 469–476. doi:10.1016/j.biortech.2016.09.074

Nitayavardhana, S., Issarapayup, K., Pavasant, P., and Khanal, S. K. (2013). Production of protein-rich fungal biomass in an airlift bioreactor using vinasse as substrate. *Bioresour. Technol.* 133, 301–306. doi:10.1016/j.biortech.2013.01.073

Nitayavardhana, S., and Khanal, S. K. (2010). Innovative biorefinery concept for sugar-based ethanol industries: production of protein-rich fungal biomass on vinasse as an aquaculture feed ingredient. *Bioresour. Technol.* 101, 9078–9085. doi:10.1016/j.biortech.2010.07.048

Olguín, E. J., Dorantes, E., Castillo, O. S., and Hernández-Landa, V. J. (2015). Anaerobic digestates from vinasse promote growth and lipid enrichment in *Neochloris oleoabundans* cultures. *J. Appl. Phycol.* 27, 1813–1822. doi:10.1007/s10811-015-0540-6

Parnaudeau, V., Condom, N., Oliver, R., Cazevieille, P., and Recous, S. (2008). Vinasse organic matter quality and mineralization potential, as influenced by raw material, fermentation and concentration processes. *Bioresour. Technol.* 99, 1553–1562. doi:10.1016/j.biortech.2007.04.012

Paz-Pino, O. L., Barba-Ho, L. E., and Marriaga-Cabrales, N. (2014). Vinasse treatment by coupling of electro-dissolution, hetero-coagulation and anaerobic digestion. *Dyna* 81, 102–107. doi:10.15446/dyna.v81n187.38922

Penatti, C., Cambria, S., Boni, P., Arruda, F., and Manoel, L. (1988). Efeitos da aplicação de vinhaça e nitrogênio na soqueira da cana-de-açúcar. *Bol. Tec. Copersucar* 44, 32–38.

Prata, F., Lavorenti, A., Regitano, J. B., and Tornisielo, V. L. (2001). Degradação e sorção de ametrina em dois solos com aplicação de vinhaça. *Pesq. Agropec. Bras.* 36, 975–981. doi:10.1590/S0100-204X2001000700007

Rajeshwari, K., Balakrishnan, M., Kansal, A., Lata, K., and Kishore, V. (2000). State-of-the-art of anaerobic digestion technology for industrial wastewater treatment. *Renew. Sustain. Energ. Rev.* 4, 135–156. doi:10.1016/S1364-0321(99)00014-3

Ramirez, N. N. V., Farenzena, M., and Trierweiler, J. O. (2014). Growth of microalgae *Scenedesmus* sp in ethanol vinasse. *Braz. Arch. Biol. Technol.* 57, 630–635. doi:10.1590/S1516-8913201401791

Reis, C. E., Zhang, J., and Hu, B. (2014a). Lipid accumulation by pelletized culture of *Mucor circinelloides* on corn stover hydrolysate. *Appl. Biochem. Biotechnol.* 174, 411–423. doi:10.1007/s12010-014-1112-y

Reis, C. E. R., de Souza Amaral, M., Loures, C. C. A., da Rós, P. C. M., Hu, B., Izário Filho, H. J., et al. (2014b). "Microalgal feedstock for bioenergy: opportunities and challenges," in *Biofuels in Brazil*, eds S. S. Silva and A. K. Chandel (Cham: Springer), 367–392.

Reis, C. E. R., Rajendran, A., and Hu, B. (2017). New technologies in value addition to the thin stillage from corn-to-ethanol process. *Rev. Environ. Sci. Biotechnol.* 16, 175–206. doi:10.1007/s11157-017-9421-6

Rocha, M. H., Lora, E. E. S., and Venturini, O. J. (2007). Life cycle analysis of different alternatives for the treatment and disposal of ethanol vinasse. *Proc. Int. Soc. Sugar Cane Technol.* 26, 108–114.

Santos, M., Fernández Bocanegra, J., Martín Martín, A., and García García, I. (2003). Ozonation of vinasse in acid and alkaline media. *J. Chem. Technol. Biotechnol.* 78, 1121–1127. doi:10.1002/jctb.908

Siles, J., García-García, I., Martín, A., and Martín, M. (2011). Integrated ozonation and biomethanization treatments of vinasse derived from ethanol manufacturing. *J. Hazard. Mater.* 188, 247–253. doi:10.1016/j.jhazmat.2011.01.096

Souza, M., Fuzaro, G., and Polegato, A. (1992). Thermophilic anaerobic digestion of vinasse in pilot plant UASB reactor. *Water Sci. Technol.* 25, 213–222.

Sparovek, G., Barretto, A., Berndes, G., Martins, S., and Maule, R. (2009). Environmental, land-use and economic implications of Brazilian sugarcane expansion 1996-2006. *Mitigation Adapt. Strateg. Global Change* 14, 285–298. doi:10.1007/s11027-008-9164-3

Todokoro, T., Fukuda, K., Matsumura, K., Irie, M., and Hata, Y. (2015). Production of the natural iron chelator deferriferrichrysin from *Aspergillus oryzae* and evaluation as a novel food-grade antioxidant. *J. Sci. Food Agric.* 96:2998–3006. doi:10.1002/jsfa.7469

Valsechi, O., and Gomes, F. P. (1954). Solos incorporados de vinhaça e seu teor de bases. *An. Esc. Super. Agric. Luiz Queiroz* 11, 136–158. doi:10.1590/S0071-12761954000100012

Walter, A., Dolzan, P., Quilodrán, O., de Oliveira, J. G., da Silva, C., Piacente, F., et al. (2011). Sustainability assessment of bio-ethanol production in Brazil considering land use change, GHG emissions and socio-economic aspects. *Energy Policy* 39, 5703–5716. doi:10.1016/j.enpol.2010.07.043

Yang, X., Wang, K., Wang, H., Zhang, J., and Mao, Z. (2016). Ethanol fermentation characteristics of recycled water by *Saccharomyces cerevisiae* in an integrated ethanol-methane fermentation process. *Bioresour. Technol.* 220, 609–614. doi:10.1016/j.biortech.2016.08.040

Conflict of Interest Statement: The authors declare that the research was conducted in the absence of any commercial or financial relationships that could be construed as a potential conflict of interest.

The reviewer, RH, and handling editor declared their shared affiliation, and the handling editor states that the process nevertheless met the standards of a fair and objective review.

19

Carbon Capture and Storage and Carbon Capture and Utilization: What Do They Offer to Indonesia?

Didi Adisaputro[1,2]* and Bastian Saputra[1]

[1] Department of Chemical and Biological Engineering, University of Sheffield, Sheffield, UK, [2] Department of Energy Security, Indonesian Defence University, Bogor, Indonesia

Keywords: CO_2 emission, energy, carbon capture and storage, CCU, Indonesia

Edited by:
Yuhan Sun,
Shanghai Advanced Research
Institute, China

Reviewed by:
Mingquan Wang,
Shanghai Advanced Research
Institute, China

*Correspondence:
Didi Adisaputro
didiadisaputro@gmail.com

Specialty section:
This article was submitted to Carbon
Capture, Storage, and Utilization,
a section of the journal
Frontiers in Energy Research

Citation:
Adisaputro D and Saputra B (2017)
Carbon Capture and Storage and
Carbon Capture and Utilization:
What Do They Offer to Indonesia?
Front. Energy Res. 5:6.

INTRODUCTION

Indonesia is a developing country with abundance resource of fossil fuel in the world, and this fossil fuel will remain as the main source of energy over the next few decades. However, the Indonesian Government has committed to reducing greenhouse gas emissions from fossil fuel consumption as an effort to mitigate climate change. In view of this, two possible energy scenarios are envisioned to honor this commitment: "business as usual" (BaU) and the National Energy Policy (NEP) scenario (National Energy Council, 2014). The NEP scenario reduces CO_2 emissions by up to 26% through an improved energy mix, less reliance on carbon-based fuels, and the deployment of renewable energy sources from 2020 to 2050. However, these actions are considered insufficient to further reduce the CO_2 emission target, leading to an initiative to implement carbon capture and storage (CCS) technology.

Although Indonesia is a developing country, it has a mature manufacturing industry, which means that carbon-capture utilization (CCU) technology, which converts CO_2 into useful products and intermediates, could provide an additional method to complement CCS technology and help decouple economic growth from excessive CO_2 emission (Styring et al., 2014).

This report reviews how CCS and CCU are applied in Indonesia, particularly in relation to the energy sector. In addition, CO_2 emissions from future fossil fuel consumption are predicted together with the amount of CO_2 emitted that is suitable for CCS. The barriers hindering the application of CCS are also reviewed together with the alternatives to CCU technology. Because of some drawbacks in the development of CCS, the potential of CCU deployment is assessed, leading to several recommendations. Finally, we discuss what the application of both CCS and CCU can offer Indonesia.

CARBON CAPTURE AND STORAGE

Implementing CCS involves the application of CO_2 capture technology, transportation and injection into storage facilities, and monitoring (ADB, 2013). The Global CCS Institute (2016) reports that, in 2016, most ongoing CCS projects worldwide are applied in oil and gas, power, and the industrial sector. Some examples include the Uthmaniyah CO_2-enhanced oil recovery (EOR) project in Saudi Arabia and the Boundary Dam Power Plant CCS Project in Canada.

The Indonesian Government has also considered CCS as a national strategy, as demonstrated by the existing CCS pilot project in Merbau Gas Gathering Station. The captured CO_2 is injected into the depleted oil and gas reservoir surrounding the point of source. This project will store 50–100 t of CO_2 per day for several months starting in 2016 and will be shut-in for assessment in 2018 [LEMIGAS (R&D Centre for Oil and Gas Technology), 2012].

However, some barriers remain that hinder the deployment of CCS. The retrofitting of CO_2 capture technology to power plants often confronts technical barriers that increase the cost of producing electricity. Other technical barriers come from site selection, storage capacity, transportation, and the degree of confidence in terms of geological modeling and monitoring. Social barriers may also arise because of the safety issues connected with leakage, long-term liability, and public acceptance (Styring et al., 2011).

CARBON CAPTURE AND UTILIZATION

Indonesia has initiated several research programs targeting climate change and CO_2 reduction. However, most of these programs examine only the role of natural CO_2 sequestration rather than CO_2 conversion into commercial products. The idea of CCU technology is to convert CO_2 into various valuable products. Aresta (2010) has researched CO_2 synthesis into carboxylates, carbonates, carbamates, isocyanates, and polymeric materials that have been examined for use in electrochemistry and photo electrochemistry. CO_2 and its co-reactant can be activated and further transformed into a useful chemical either by creating high-energy salts or by using catalysts (Styring et al., 2014).

However, other than for producing urea, the use of CCU on an industrial scale remains insignificant, especially for direct capture from flue gas. This paper considers the CCU concept only in conjunction with the conversion. Thus, CO_2 use other than for conversion, such as the use of CO_2 as a solvent, a working fluid for heat transfer, is not considered herein.

Carbon capture and utilization might also bring significant benefits to Indonesia because the wide-ranging products have increasing market potential, which is of interest because of Indonesia's large population coupled with its economic growth. The population of Indonesia is currently around 257 million and is predicted to rise to 285 million in 2030 and 322 million in 2050 (United Nations, 2015).

CCS AND CARBON CAPTURE AND UTILIZATION FOR INDONESIA

Figure 1 shows the trend of fossil fuel consumption and the related CO_2 emission in Indonesia. Fossil fuel consumption has increased from 53.4 Mtoe in 1990 to 154.93 Mtoe in 2013, when the oil share was around 50–60%. The NEP scenario would reduce the oil share to 25–30%, but the overall fossil fuel consumption should increase to 690 Mtoe in 2050. This increase in fossil fuel consumption will be accompanied by a rise in CO_2 emission. In 1990, CO_2 emission was 133.9 Mtoe, which increased to 133.9 Mtoe in 2013. In 2030 and 2050, it is predicted to reach 1000.6 and 2065.98 Mtoe, respectively.

Although the NEP scenario aims to increase the share of renewable energy sources, it will be insufficient to achieve an overall 26% reduction in CO_2 emission in 2020. As a result, CCS must be applied as a tool to significantly mitigate CO_2 emission, particularly CO_2 emission from the oil and gas, power, and industrial sectors.

In 2012, through LEMIGAS, the Indonesian Government estimates a storage capacity of 640 Mtoe of CO_2 in depleted oil and gas fields, most of which are in Kutai, Tarakan, and the South Sumatra region. Considering the level of CO_2 emission given in **Figure 1**, this CO_2 storage capacity will not cover the predicted CO_2 emissions from fossil fuel in 2030 or 2050. Therefore, CCS might be more feasible if it focuses on the existing power plants and oil and gas facilities, particularly for locations close to the potential CO_2 storage sites.

Currently, the most attractive option for CCS in Indonesia is CO_2 storage in conjunction with EORs, because oil production has the potential to generate additional revenue that can offset the cost of CCS (Syahrial et al., 2010). However, for long-term CCS, this option is not expected to achieve the requisite CO_2 reduction target because, when consumed, oil produced from EOR releases CO_2. Furthermore, deploying CCS in the power sector may not only mitigate CO_2 emissions but may also speed up industrialization by increasing the number of new power plants to achieve the national electrification target. The new construction of "CO_2 capture ready" power plants may avoid the retrofit cost of CO_2 capture (Syahrial et al., 2010).

The predicted fossil fuel consumption, the amount of CO_2 emission, and the future population growth in Indonesia should increase the effort to explore and deploy CCS. A promising first step has been taken by the government in the form of the CO_2 Storage Mapping Program run by the Coordinating Committee for Geoscience Programs in East and Southeast Asia (CCOP), whose goal is to investigate potential CO_2 storage sites.

Carbon Capture and Utilization

The three largest-scale CO_2-derived products are urea with ~157 Mtoe worldwide, followed by salicylic acid and cyclic carbonate with, respectively, ~90 and ~80 thousand tonnes worldwide (Styring et al., 2014). Indonesia contributed almost 5% to the global urea production in 2015 and is predicted to increase its production over the coming decades. According to the Ministry of Industry of the Republic of Indonesia (2010), after the world recession of 2008, Indonesia experienced no economic instability. Instead, all economic sectors enjoyed growth during 2009, including manufacturing. The fertilizer and chemical sectors alone had stable growth, averaging about 5% from 2005 to 2008. These industries are promising for spurring the development of CCU in Indonesia.

For CCU applications, significant energy is required to satisfy specific capturing and conversion technology. Therefore, the energy must come from an excess of renewable power rather than from fossil fuel, otherwise a net reduction in CO_2 would be impossible (Styring et al., 2014). Thus, CCU is more applicable in the NEP scenario than in the BaU scenario because, in the former, sufficient energy for CCU comes from renewables and fossil fuel-based power plants are a source of CO_2.

The transformation of energy systems involves significant investment and regulation from the related institutions. For regulations, the stabilization of political conditions in Indonesia is highly likely to affect the deployment of CCU technology. In addition to transforming energy systems, qualified human resources, mature carbon capture, and a transportation infrastructure

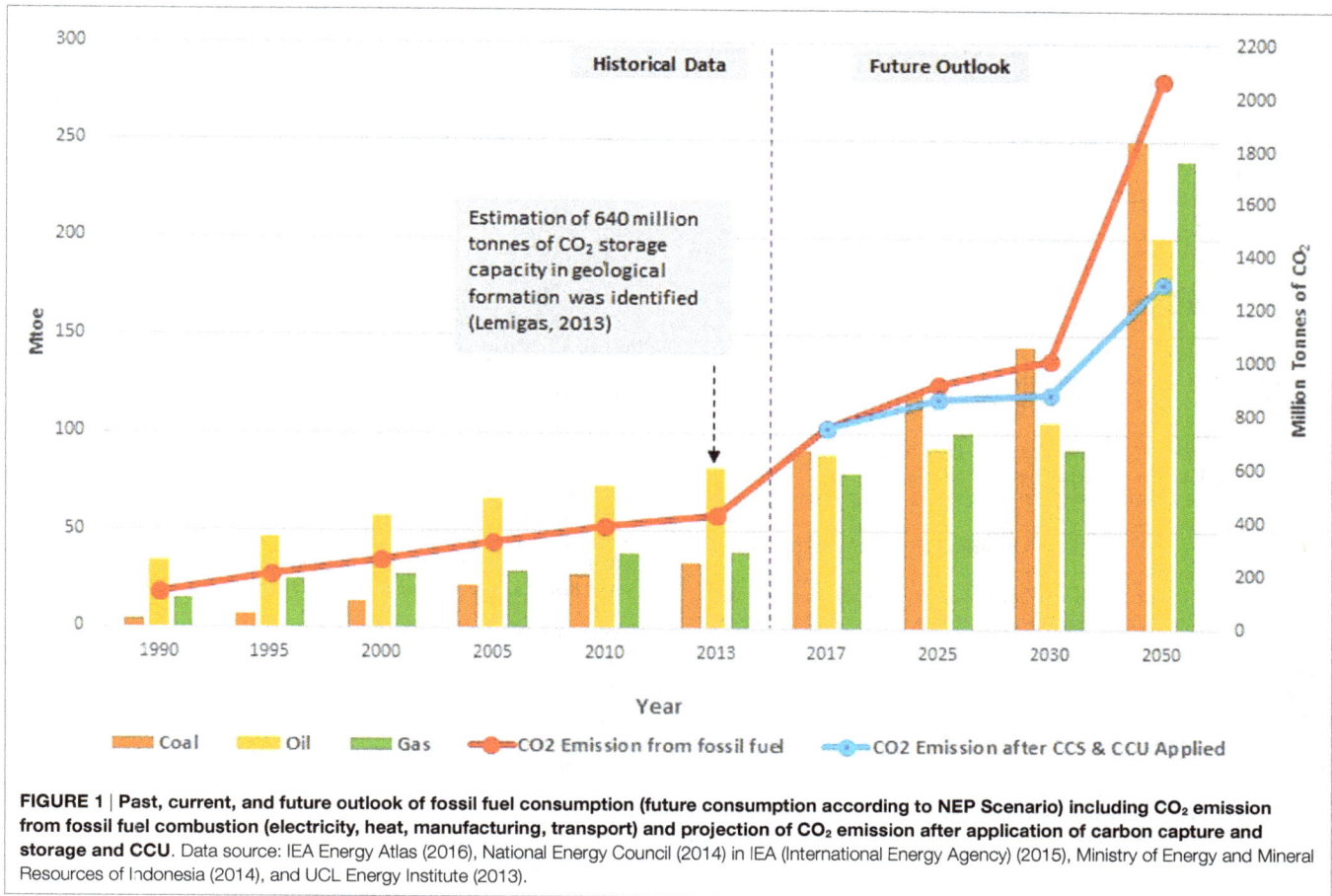

FIGURE 1 | Past, current, and future outlook of fossil fuel consumption (future consumption according to NEP Scenario) including CO_2 emission from fossil fuel combustion (electricity, heat, manufacturing, transport) and projection of CO_2 emission after application of carbon capture and storage and CCU. Data source: IEA Energy Atlas (2016), National Energy Council (2014) in IEA (International Energy Agency) (2015), Ministry of Energy and Mineral Resources of Indonesia (2014), and UCL Energy Institute (2013).

should also be available for CO_2 conversion products to compete with conventional products. However, despite the advance of low-cost energy from renewables, market intervention by policy is still likely to be required to create market space for CCU products.

The Indonesian Government has proposed the National Industrial Development Master Plan (RIPIN) for the years 2015–2035. The CCU concept fits very well into this master plan because the plan regulates some important visions to develop green industries, including the use of eco-products, renewables, and low hazard substances (Presiden Republik Indonesia, 2015). RIPIN also mentions that the Indonesian Government might adopt a mature approach to green industry from another country. The European policy SCOT (a vision for smart CO_2 transformation in Europe) is one recommended approach to support RIPIN. In SCOT, the initial plans for CCU include research into life-cycle assessment, policy frameworks, and carbon capture installations (Wilson et al., 2016). To successfully implement CCU, the technical and non-technical challenges of CCU must be addressed by Indonesia through sufficient funding and support from both the public and private sectors.

Figure 1 shows the prediction of how CO_2 emission can change if the CCS and CCU are applied in the future. The prediction of CO_2 emission would change significantly with the increase of emission reduction from 6% in 2025 to 37% in 2050 in case the CCS and CCU application has been applied. This estimation

is based on the assumption of carbon capture technology will be applied to point CO_2 sources (oil and gas facilities, power plants, and manufacturing) and considers the CCS capacity and efficiency (UCL Energy Institute, 2013) including the potential application of CCU in Indonesia.

CONCLUSION

Indonesia plans to reduce future CO_2 emissions in spite of its energy scenario remaining reliant on fossil fuels. The potential of CCS in Indonesia would reduce CO_2 emission from oil and gas, power plants, and industry. Because CCS alone is insufficient to attain the CO_2 emission goals, CCU technology offers an alternative in which CO_2 is put to use rather than simply sequestered. Carbon capture and utilization will be suitable if the Indonesian Government accepts the NEP scenario, which foresees a significant share of the energy mix coming from renewables. Carbon capture and utilization might be more attractive to Indonesia than CCS alone because the industrial products from CO_2 have a significant potential market due to Indonesia's increasing population and concomitant economic growth. However, in Indonesia, the research and development of CCU is not well developed. Therefore, a framework within which CCU applications can be produced would lead to more efficient and sustainable use of Indonesia's resources and would result in an eco-competitive industry. In conclusion, applying CCS with CCU offers not only a reduction in CO_2 emission but also the potential to decouple economic growth from CO_2 emission.

AUTHOR CONTRIBUTIONS

DA and BS contributed equally to this work.

REFERENCES

ADB. (2013). *Prospects for Carbon Capture and Storage in Southeast Asia.* Philippines: Asian Development Bank.

Aresta, M. (2010). *Carbon Dioxide as Chemical Feedstock.* Weinheim: John Wiley & Sons.

Global CCS Institute. (2016). *Large Scale CCS Projects.* Available at: https://www.globalccsinstitute.com/projects/large-scale-ccs-projects

IEA (International Energy Agency). (2015). *Indonesia 2015.* France: International Energy Agency.

IEA Energy Atlas. (2016). *IEA Energy Atlas Statistics.* Available at: http://www.iea.org/statistics/ieaenergyatlas/

LEMIGAS (R&D Centre for Oil and Gas Technology). (2012). *The Latest Status of Carbon Capture and Storage (CCS) in Indonesia.* Republic of Indonesia: Ministry of Energy and Mineral Resources.

Ministry of Energy and Mineral Resources of Indonesia. (2014). *Handbook of Energy and Economic Statistics of Indonesia.* Indonesia: Pusdatin ESDM.

Ministry of Industry of the Republic of Indonesia. (2010). *Industry for a Better Life.* Jakarta: Kemenprin.

National Energy Council. (2014). *Indonesia Energy Outlook 2014.* Republic of Indonesia: National Energy Council.

Presiden Republik Indonesia. (2015). *Rencana Induk Pembangunan Industry Nasional (RIPIN) Tahun 2015-2035.* Jakarta: Peraturan Pemerintah Republik Indonesia Nomor 14 tahun 2015.

Styring, P., Jansen, D., de Coninck, H., Reith, H., and Armstrong, K. (2011). *Carbon Capture and Utilisation in the Green Economy.* New York: Centre for Low Carbon Futures, 60.

ACKNOWLEDGMENTS

We would like to say thank you to our sponsor, this paper would not have been completed without support from LPDP (Indonesian Endowment Fund for Education). We are grateful to Dr. Grant Wilson for his supervision and insightful suggestion during the preparation of this paper.

Styring, P., Quadrelli, E. A., and Armstrong, K., editors. (2014). *Carbon Dioxide Utilisation: Closing the Carbon Cycle.* Elsevier.

Syahrial, B. E., Pasarai, U., and Iskandar, U. P. (2010). CCS potential in Indonesia. *Understanding Carbon Capture and Storage (CCS) Potential in Indonesia.* 33, 129–134.

UCL Energy Institute. (2013). *Modelling of Global Energy Scenarios under CO_2 Emissions Pathways with TIAM-UCL.* London: UCL Energy Institute.

United Nations, Department of Economic and Social Affairs, Population Division. (2015). *World Population Prospects: The 2015 Revision, Key Findings and Advance Tables. Working Paper No. ESA/P/WP.241.* New York.

Wilson, G., Travaly, Y., Brun, T., Knipples, H., Armstrong, K., Styring, P., et al. (2016). *A VISION for Smart CO_2 Transformation in Europe (SCOT). Using CO_2 ASA Resource.* Seventh Framework Programme and European Union.

Conflict of Interest Statement: The authors declare that the research was conducted in the absence of any commercial or financial relationships that could be construed as a potential conflict of interest.

The reviewer MW and handling editor declared their shared affiliation, and the handling editor states that the process nevertheless met the standards of a fair and objective review.

20

Synthetic Biology and Metabolic Engineering Approaches and Its Impact on Non-Conventional Yeast and Biofuel Production

Aravind Madhavan[1,2], Anju Alphonsa Jose[1], Parameswaran Binod[1], Raveendran Sindhu[1]*, Rajeev K. Sukumaran[1], Ashok Pandey[1,3] and Galliano Eulogio Castro[4]

[1] Biotechnology Division, National Institute for Interdisciplinary Science and Technology, Council of Scientific and Industrial Research, Trivandrum, India, [2] Rajiv Gandhi Centre for Biotechnology, Trivandrum, India, [3] Center for Innovative and Applied Bioprocessing, Mohali, Punjab, India, [4] Dpt. Ingeniería Química, Ambiental y de los Materiales Edificio, Universidad de Jaén, Jaén, Spain

Edited by:
Rajesh K. Sani,
South Dakota School of Mines and
Technology, USA

Reviewed by:
Neha Srivastava,
Banaras Hindu University, India
Sen Li,
Chinese Academy of Sciences, China
Navanietha Krishnaraj,
National Institute of Technology
Durgapur, India

*Correspondence:
Raveendran Sindhu
sindhurgcb@gmail.com,
sindhufax@yahoo.co.in

Specialty section:
This article was submitted to
Bioenergy and Biofuels,
a section of the journal
Frontiers in Energy Research

Citation:
Madhavan A, Jose AA, Binod P,
Sindhu R, Sukumaran RK, Pandey A
and Castro GE (2017) Synthetic
Biology and Metabolic Engineering
Approaches and Its Impact on
Non-Conventional Yeast and
Biofuel Production.
Front. Energy Res. 5:8.

The increasing fossil fuel scarcity has led to an urgent need to develop alternative fuels. Currently microorganisms have been extensively used for the production of first-generation biofuels from lignocellulosic biomass. Yeast is the efficient producer of bioethanol among all existing biofuels option. Tools of synthetic biology have revolutionized the field of microbial cell factories especially in the case of ethanol and fatty acid production. Most of the synthetic biology tools have been developed for the industrial workhorse *Saccharomyces cerevisiae*. The non-conventional yeast systems have several beneficial traits like ethanol tolerance, thermotolerance, inhibitor tolerance, genetic diversity, etc., and synthetic biology have the power to expand these traits. Currently, synthetic biology is slowly widening to the non-conventional yeasts like *Hansenula polymorpha*, *Kluyveromyces lactis*, *Pichia pastoris*, and *Yarrowia lipolytica*. Herein, we review the basic synthetic biology tools that can apply to non-conventional yeasts. Furthermore, we discuss the recent advances employed to develop efficient biofuel-producing non-conventional yeast strains by metabolic engineering and synthetic biology with recent examples. Looking forward, future synthetic engineering tools' development and application should focus on unexplored non-conventional yeast species.

Keywords: synthetic biology, yeast, biofuel, metabolic engineering, ethanol

INTRODUCTION

The price for non-renewable fuels as well as the level of CO_2 in the atmosphere is increasing constantly. Hence, the production of bio-based fuels has attracted the attention of researchers as an alternative source of energy. Biofuels can be used as fuel for combustion engines which should be compatible. Lignocellulosic biomass is the major source of biofuel since it is economical and easily available. The yeast cell factory *Saccharomyces cerevisiae* is a well-known producer of ethanol and fatty acids (Tsai et al., 2015).

The topmost challenge in the marketing of microbial fuels and chemicals is the inability to cross the gap between the laboratory and the commercial market. This is primarily due to the fact that engineered strains do not meet the standard of commercialization. Synthetic biology is the fusion of biological parts and designs which evolved from the huge data of transcriptomics, proteomics, metabolomics, and fluxomics that lead to the design of novel synthetic circuits (Peralta-Yahya et al., 2012; Nielsen, 2015). The challenge in synthetic biology is to design synthetic circuits from genetic constructs toward the development of host systems that can accommodate several complex metabolic regulatory circuits. Rapid advent of DNA sequencing technologies makes the design and construction of synthetic biological devices into a reality (Unkles et al., 2014). Synthetic biology, its application in the production of fuels and chemicals, is a rapidly growing field, and synthetic microbial host cell design is the most complex area in microbial genetic engineering. Various technical platforms for assembling a library of DNA segments into synthetic pathways or circuits, which in turn inserted into the host genome while manipulating existing genes in the host, have recently been developed. Synthetic biology and metabolic engineering techniques have been widely used in *Escherichia coli*, *S. cerevisiae*, and *Zymomonas mobilis* to enhance ethanol production (Chubukov et al., 2016). So investigation and findings in synthetic biology hold extreme importance and need to accelerate the new design and optimization of pathways.

Saccharomyces cerevisiae have been considered as a model system for studying cell and molecular biology and is a well-studied organism (Blazeck et al., 2012; Blount et al., 2012a,b; Bao et al., 2015). Many synthetic biology and metabolic engineering efforts have been already established in *S. cerevisiae* for improving biofuel production (Tsai et al., 2015). An overview of synthetic biology design is depicted in the **Figure 1**. Thus, yeast is a good platform for developing synthetic biology techniques, and this system can be used for improving the status of biofuel production. The main disadvantage of *S. cerevisiae* is its inability to consume a wide range of substrates (e.g., xylose, arabinose) (Garcia Sanchez et al., 2010), and glycerol (Swinnen et al., 2013). *S. cerevisiae* is also not good for use in biofuel application that requires high temperatures (>34°C) (Caspeta et al., 2014).

Non-conventional yeast species possess several advantages over *S. cerevisiae* in terms of its physiology and metabolic pathways and regulation (Wolf, 2012). Non-conventional yeasts, like *Yarrowia lipolytica*, *Hanensula polymorpha*, *Pichia pastoris*, and *Kluyveromyces lactis*, are extensively studied yeast species and attractive production platforms. Most of these yeast systems developed extremely efficient mechanisms to withstand under harsh environmental conditions (Wagner and Alper, 2016). Several yeast species are diverged by evolution from *S. cerevisiae* and possess several unique genes and growth characteristics to withstand different stress conditions (Souciet et al., 2009). Still, several non-conventional yeast species are yet to be characterized and the advent of next-generation sequencing strategies and the other genomics and proteomics tools delineates the mechanism behind this stress tolerance strategies.

Constructing microorganisms toward desired fuel production should take into account several considerations, including increasing the yield, utilization of a wide range of substrates, and should be economical and highly efficient and simple downstream processes. In addition to these, fuel tolerance, inhibitor tolerance, thermotolerance, etc., deserve attention for enhanced fuel production. Control of redox balance is also one of the most important obstacles in strain development for the production

of fuels at high yields. Non-conventional yeasts with these advantageous characters than *S. cerevisiae* have been utilized as industrial microorganisms for biofuel production (Ruyters et al., 2015).

The present review addresses the current status of essential synthetic biology tools being applied to construct synthetic circuits in yeast and furthermore, how these techniques can be applicable to non-conventional yeast system in order to efficiently engineer them for improving biofuel production.

NON-CONVENTIONAL YEAST SYSTEMS

There are several non-conventional yeast species (e.g., *Y. lipolytica*) which offer many potential advantages over *S. cerevisiae* with respect to metabolic pathway requirements, recombinant product profile, and physiological responses (Gellissen and Hollenberg, 1997). The availability of superior quality genome sequences in public domain (Ramezani-Rad et al., 2003; Sherman et al., 2009), development of transformation vectors, gene transformation strategies (Faber et al., 1994), and metabolic engineering tools may change this scenario. Each of these organisms presents with different advantages, similarities, and differences when compared with *S. cerevisiae*. Most of the non-conventional yeast, like *K. lactis*, *P. pastoris*, etc., possess a broad range of substrates which is superior to *S. cerevisiae* and reduce the cost of industrial biofuel production (Gellissen et al., 2005). *Y. lipolytica* and *K. lactis* secrete high titers of secretory proteins extracellularly which is better than the model yeast *S. cerevisiae* (Dominguez et al., 1998).

The main obstacle of non-conventional yeast genetic modification is its non-homologous end-joining pathway compared to *S. cerevisiae* which favors homologous recombination. This results in the ectopic integration of targeted constructs which hampers synthetic biology applications (Vogl et al., 2013). Disruption of the genes involved in the NHEJ pathway Ku70 (Näätsaari et al., 2012; Verbeke et al., 2013) or Ku80 (Kooistra et al., 2004; Saraya et al., 2012) enhanced the efficiency of homologous recombination. The frequency of homologous recombination in *S. cerevisiae* is high achieved with 40 bp, while in the case of non-conventional yeast 500–3,000 bp of flanking sequence is required (Blazeck et al., 2014; Horwitz et al., 2015). Construction of expression cassettes for such long adaptor sequences requires a large number of PCR reactions, cloning steps, etc. Advanced synthetic biology tools may alleviate these drawbacks. In summary, non-conventional yeasts have several unique growth characteristics and serve as a potent cell factory for biofuel production with the help of synthetic biology tools.

BASIC TOOLS FOR METABOLIC ENGINEERING OF NON-CONVENTIONAL YEAST

This section focuses on the characterized promoters from different non-conventional and previously attempted metabolic engineering strategies and its future implications in biofuel production using synthetic biology tools.

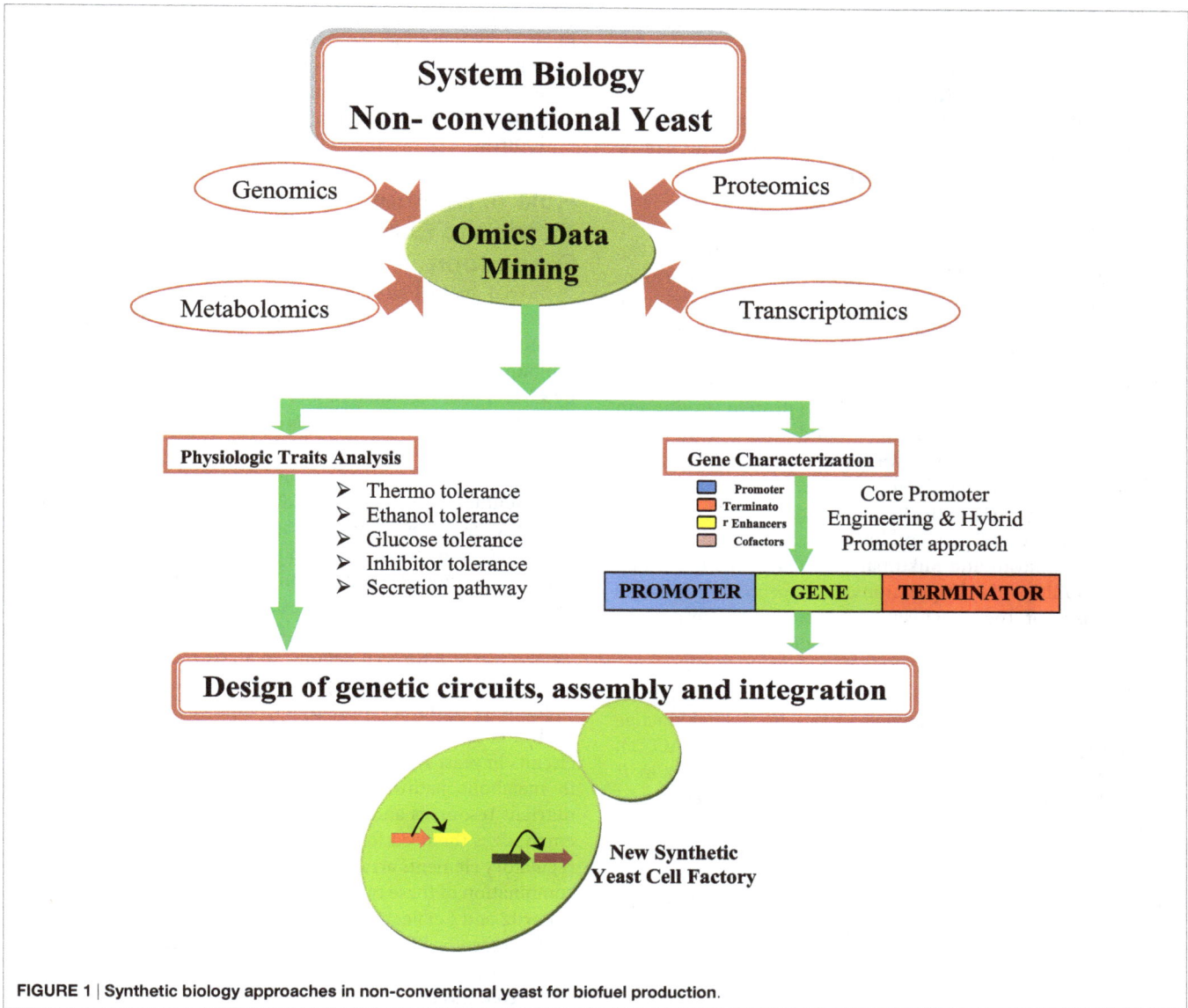

FIGURE 1 | Synthetic biology approaches in non-conventional yeast for biofuel production.

Available Promoters in Non-Conventional Yeast and Promoter Engineering

To develop a microbial cell factory, the selection of the promoter elements required for driving heterologous protein expression is crucial. At present, only very few promoters have been identified in non-conventional yeast and their metabolic regulation is not fully elucidated. Engineered promoter elements are an integral part of synthetic biology and metabolic engineering and are used to enhance the titers of homologous and heterologous proteins. Eukaryotic core promoter elements are the crucial region for the binding of transcription factors and other regulatory factors involved in transcriptional control. So the development of suitably engineered promoters in organisms with poorly defined genetic tools is extremely important for making these organisms as industrial production hosts for fuels and chemicals. The synthetic promoter approach has been developed quickly in *S. cerevisiae*

compared to other non-conventional yeast. However, recently a large number of genome sequences of non-conventional yeasts are available (e.g., *Y. lipolytica*) and have boosted both basic and applied research to understand the genotypic and phenotypic features and further development of metabolic engineering tools for biofuel production. Most of the promoter engineering work has been concentrated on upstream regulatory sequences (Hartner et al., 2008; Xuan et al., 2009), engineering core promoter elements (Blazeck et al., 2012), and/or by creating random mutations in core promoter elements (Berg et al., 2013), and this will lead to tight and tuneable control over a metabolic regulatory pathways (Teo and Chang, 2014).

Along with the core promoter elements, upstream elements such as binding sequences for transcription factors can also be modified to increase the strength of the promoters. For example, TetR protein, which is a well-studied and extensively used protein in bacterial molecular biology, retains its binding to DNA control elements in yeast and sensitive to synthetic inducers, such as anhydrotetracycline or doxycycline. So promoter with TetR-binding sites has been used to control different genes in different yeasts (Blount et al., 2012b). Still, non-conventional yeast promoters are poorly characterised, and findings from *S. cerevisiae* can be applicable to the most the highly efficient non-conventional yeast systems also because most of the yeast promoters are capable of cross recognition (Van Ooyen et al., 2006).

The commercial *K. lactis* uses inducible promoter PLAC4, which is a lactose inducible promoter for recombinant protein production, and the proteins are produced into the culture fluid, which makes the recombinant protein purification easy. Highly efficient protein secretion machinery and the high biomass attained in submerged culture condition makes *K. lactis* an attractive cell factory for heterologous protein production (Van den Berg et al., 1990; Gellissen and Hollenberg, 1997). A hybrid promoter approach was established in *K. lactis* by combining core promoter elements of *Trichoderma reesei* cellobiohydrolase1 (cbh1) to the β-galactosidase (lac4) promoter of *K. lactis* for increasing recombinant protein production in the yeast (Madhavan and Sukumaran, 2014). This may open up the possibilities of manipulation of core promoter elements across classes of the organisms by introducing important core promoter sequences to create highly active synthetic promoters.

Pichia pastoris is a widely used protein production platform for enzymes and pharmaceuticals (Vogl et al., 2013). Highly active methanol-inducible promoter, alcohol oxidase 1 gene (pAOX1), is the most widely used and tightly controlled promoter in *P. pastoris*. This has been well studied in terms of promoter control elements and different transcription factor. Random mutated library of pAOX1 has been constructed and resulted in less glucose repression and increased activity compared to wild pAOX1 (Berg et al., 2013). The constitutive pGAP (glyceraldehyde-3-phosphate dehydrogenase promoter) served as a scaffold, and this resulted in the development of a series of promoters with a range of activity from 0.01- to 19.6-fold of wild-type pGAP activity (Qin et al., 2011). Blazeck et al. (2012) reported that the combination of core promoter elements with synthetic upstream activator sequence could increase the expression level.

CRISPR-Cas

The field of synthetic biology witnessed a transformation by the advent of CRISPR–Cas (Clustered Regularly Interspaced Short Palindromic Repeat) system in several host organisms (Cong et al., 2013). CRISPR-mediated RNA-guided control of gene using nuclease-deficient Cas9 (dCas9) was reported to control and regulate the expression of genes inside the yeast, with synthetic single stranded-guide RNAs (sgRNAs) targeting the genes to silence on a genome. This technique helped to create a series of simultaneous targeted integrations and double- stranded breaks in *S. cerevisiae* rapidly and efficiently (Bao et al., 2015). The dCas9 was fused to Mxi1, a protein that involved in the attraction of histone deacetylase Sin3p homolog, a component of gene-silencing complex in yeast. Further, the dCas9 along with sgRNAs repressed the *TEF* promoter of *S. cerevisiae* about 10-fold, Mxi1 (fusion protein) could downregulate the promoter by 53-fold

(Gilbert et al., 2013). This technique is also been applied to an industrially important non-conventional *K. lactis* strain, where muconic acid production was improved by incorporating six genes at three targeted loci and helped in the introduction of a pathway for the production of muconic acid precursor (Horwitz et al., 2015).

Role of Synthetic Biology in Yeast Metabolic Engineering and Fuel Production

Industrial biofuel production often requires strains that can withstand several stress conditions like extreme pH, high temperature, osmotolerance, shearing forces, organic acids, and other inhibitors. All these properties are complex traits and are encoded by several genes at different loci. Due to the complexity of yeast genome, normal genetic analysis methods, therefore, fail to characterize the underlying genetic regulatory network, and efforts to optimize one of these traits classically depend on adaptive laboratory evolution or other controlled breeding strategies. The phenotypic and genotypic characterization of non-conventional yeast species deserve special attention and will help to identify strains and species with novel and/or improved industrially important properties, and the advent of next-generation sequencing at low cost and in a comparably short time have now opened the way for the use of advanced genome analysis tools like quantitative trait loci analysis to identify, at least under some conditions, all causative genes for a certain trait (Bloom et al., 2013).

Tsai et al. (2015) reviewed about the requirement of genetic circuits in yeast synthetic biology. The timely expression of genes in metabolic pathway could minimize the extra energy and nutrient resources and maximize the production capacity. This emphasizes the need for genetic circuits which include multiple regulatory elements arranged to create different logical gates. The combination of these modules resulted in a complicated network (Elowitz and Leibler, 2000) and genetic switches (Gardner et al., 2000). Recently, Blount et al. (2012a) reviewed the status of Yeast genetic circuits. The genetic circuits in *S. cerevisiae* are far behind the *E. coli*. Blount et al. (2012b) also proposed three important principles to create genetic circuits in yeast: (1) the circuit elements should not be related to the host physiologic elements to retain orthogonality, (2) the circuit elements needs to be tuneable and inducible, and (3) self-contained groups should be built to obtain modularity. In order to construct genetic circuits, technologies that create fast prototyping, evaluation, and optimization of metabolic pathways in host species are essential. The current key research area of synthetic biology is how to reduce the time required for making genetic circuits. The finding of new metabolic pathways and its combination can be efficiently addressed by better pathway assembly techniques, like ligation-free assembly (Li and Elledge, 2007; Vroom and Wang, 2008) and BioBricks (Shetty et al., 2008). Another key area of research of synthetic biology is the synthesis of reusable constructs with predictable behavior. The construction of a different array of controllable and tunable expression system is very useful in metabolic engineering. A genetic switch that responds to environmental cues and leads to further downstream regulation is also advantageous for biofuel production (Dueber et al., 2007).

For example, incorporation of FadR-binding sites in GAL1 promoter region, after RNA polymerase binding region and before the transcription start site, represses the promoter activity. But the presence of fatty acid, like myristic acid, changes the conformation of FadR, reduces the affinity to DNA-binding sequences, and deregulates the expression of genes (Blount et al., 2012a). Teo and Chang (2014) incorporated several upstream activator elements which in turn detect copper and phosphate starvation, and then downstream gene will only be activated by the depletion of both fatty acid and copper or fatty acid and phosphate. This synthetic device is useful for sensing the presence of fatty acid and can be included into pathways with the production or depletion of fatty acids. The regulatory responses of yeast to external stimuli have also been engineered for synthetic biology applications. For example, mitogen-activated kinase cascade have genetically engineered to rewire the novel regulatory networks (Kiel and Serrano, 2011).

NON-CONVENTIONAL YEAST AND BIOFUEL PRODUCTION

Lignocellulosic biomass is the richest source of sugars for the fermentation to ethanol and other valuable chemicals. For the efficient conversion of lignocellulosic biomass, it is necessary that all the sugars in hydrolyzates should be fermentable to ethanol and other value-added chemicals. The main problem associated with this bioconversion is lack of efficient ethanol-producing yeast, the inefficiency to convert pentoses remains to be topmost challenge in bioconversion of lignocellulosic biomass. Pentose-utilizing yeasts can utilize both glucose and xylose from biomass to ethanol. Their conversion rate is very low due to the presence of inhibitors in the hydrolyzate, less ethanol tolerance, and glucose repression.

Saccharomyces cerevisiae is the best characterized currently used industrial workhorse. In addition to the efforts attempted to genetic engineer *S. cerevisiae* to overcome a variety of stresses during fermentation and subsequent ethanol production, *S. cerevisiae* has limitations. Less explored non-conventional yeast species possess better ethanol-producing capabilities, better tolerance to most of these stresses, and could be a better candidate than *S. cerevisiae* to study the molecular mechanism behind inhibitor tolerance and substrate utilization for second-generation biofuel production.

Saccharomyces cerevisae utilizes glucose by fermentative pathway (Crabtree positive) and some non-conventional yeasts, like *K. lactis*, *P. pastoris*, and *Y. lipolytica*, are predominantly oxidative (Crabtree negative). Several attempts were initiated to convert Crabtree-negative yeasts to Crabtree-positive for improving ethanol fermentation efficiency. Metabolic engineering of *K. lactis* has been reported by construction of a null mutant in the single gene encoding for a mitochondrial alternative internal dehydrogenase. The mutant showed unaffected rate of oxidation of exogenous NADH, and the mutation also shifted the metabolism to fermentative instead of respiratory. This increased the rate of ethanol production in *K. lactis* (Gonzalez-Siso et al., 2015).

Dekkera bruxellensis is a crab-positive yeast producing high yield of ethanol. Schifferdecker et al. (2016) developed a metabolically engineered strain by increasing its capability of fermentative metabolism. The encoding for alcohol dehydrogenase was homologously overexpressed under the control of highly active TEF1 promoter, resulted in 1.4–1.7 times more ethanol than the wild-type strain.

The tolerance of yeasts to several stresses like osmolarity and temperature is extremely important for first- and second-generation biofuel production. *Zygosaccharomyces rouxii* is capable of growing in osmolarity of 3 M NaCl and glucose concentrations of 90% due to the presence of unique transporters in plasma membrane which is higher than *S. cerevisiae* (Leandro et al., 2011). The temperature tolerance is also essential during the industrial process, which should be up to 50°C. This temperature is different from the normal fermentation temperature of *S. cerevisiae*, which is between 25 and 37°C (Abdel-Banat et al., 2010). Some non-conventional yeast species-like *Ogataea polymorpha* have been found to ferment xylose at 45°C (Kurylenko et al., 2014).

Two thermotolerant isolates (withstand above 45°C), NIRE-K1 and NIRE-K3, were screened for fermenting both glucose and xylose and identified as *Kluyveromyces marxianus* NIRE-K1 and *K. marxianus* NIRE-K3. The final ethanol yield for both strains was found to be 39.12 and 43.25 g/L (Arora et al., 2015a,b). Another study by Kumar et al. (2014) reported the fermentation of xylose and glucose-rich bagasse hydrolyzates by the thermotolerant (50°C) yeast *Kluyveromyces* sp. IIPE453 and yielded 0.43 g/g ethanol. The genome sequence information of most of these non-conventional yeasts is available. However, the molecular mechanisms underlying the tolerance of these species to these stress conditions remain poorly investigated for all of them.

Substrate Utilization of Non-Conventional Yeasts

Extensive attempts have been done to increase the range of substrate utilization capabilities of *S. cerevisiae* through metabolic engineering. A schematic diagram for different routes of substrate assimilation is illustrated in the **Figure 2**. Efficient conversion of carbon substrates, such as arabinose, xylose, and cellobiose, has been materialized through the introduction of hydrolytic enzymes and transporter genes in yeast. Engineered strains of *S. cerevisiae* capable of utilizing starch and xylose have been reported (Wei et al., 2013). But all these recombinant strains perform well at a fermentation temperature of 30°C but not at the higher temperatures. But most of the hydrolytic enzymes for lignocellulosic biomass function optimally at 50°C. Majority of the non-conventional yeasts are thermotolerant and ferment sugars at high temperature, but the range of substrate utilization is less (Katahira et al., 2004). Metabolic engineering of these non-conventional yeasts is essential to widen the range of substrate utilization. However, certain non-conventional yeasts utilize a variety of carbon substrates, but they do not possess other important characteristics (ethanol tolerance, inhibitor tolerance, etc.).

Pichia stipitis are the ascomycetous yeast that can ferment xylose to ethanol at nearly maximum yield without any by-product formation. Few metabolic engineering approaches have been established in *P. stipitis* to resolve its crabtree effect, low ethanol tolerance, etc. Although *P. stipitis* is an efficient biofuel producer due to its natural capability for xylose fermentation, further improvement in production host is hampered by a lack of genetic

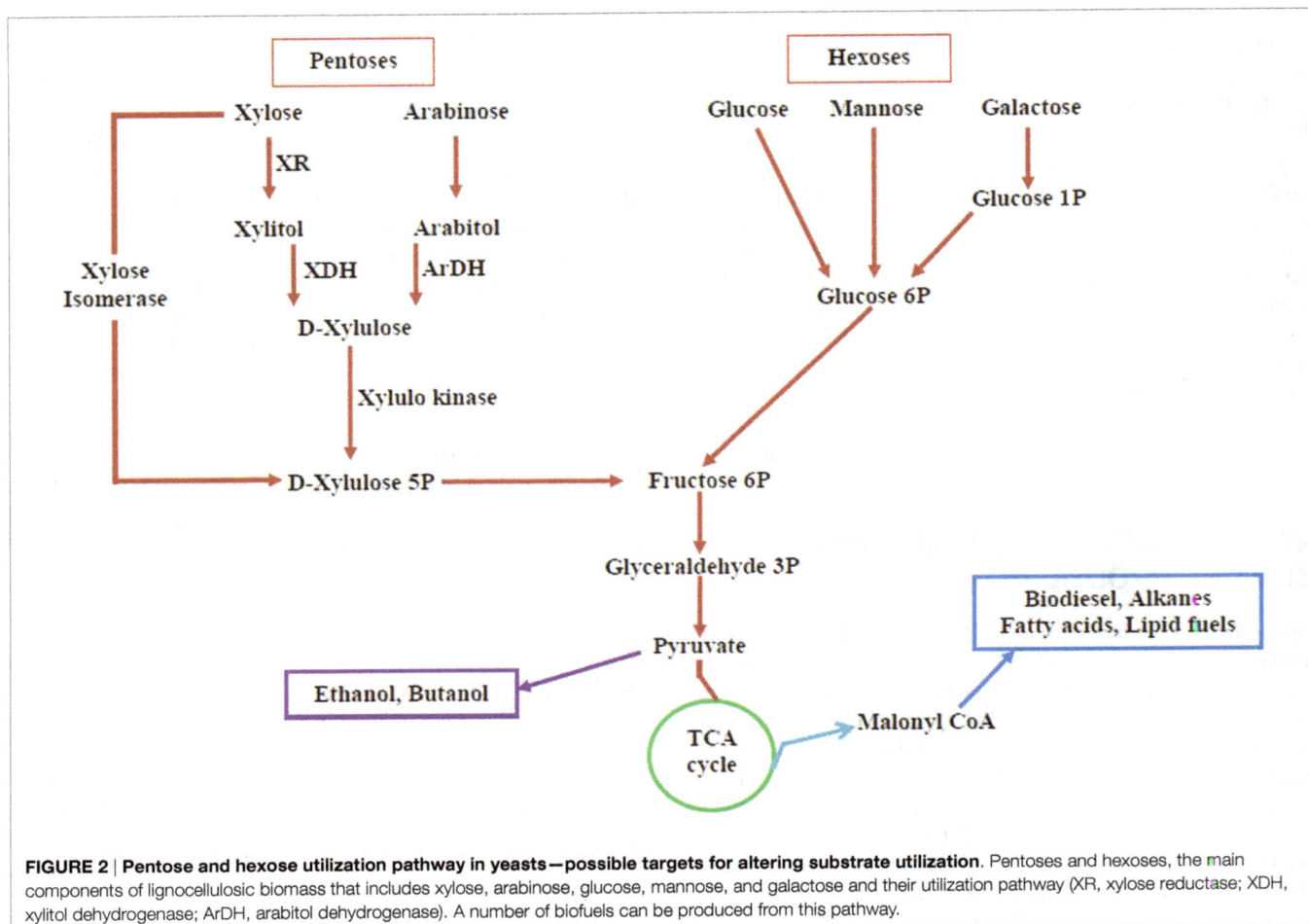

FIGURE 2 | Pentose and hexose utilization pathway in yeasts—possible targets for altering substrate utilization. Pentoses and hexoses, the main components of lignocellulosic biomass that includes xylose, arabinose, glucose, mannose, and galactose and their utilization pathway (XR, xylose reductase; XDH, xylitol dehydrogenase; ArDH, arabitol dehydrogenase). A number of biofuels can be produced from this pathway.

engineering tools. For further improvement of this strain, RNA sequencing was done in search of highly active constitutive promoters. Several promoters like TEF1, Xy11p, ADH1p, and ADH2p have been exploited in S. stipitis for efficient gene expression (Cho and Jeffries, 1999). Few transcriptomic approaches were carried out in S. stipitis under glucose/xylose inducing conditions. Recent transcriptomic study under lignocellulosic biomass inducing condition and oxygen limited condition resulted in the mapping of several important genes (Bullard et al., 2010). Synthetic biology applications in S. stipitis are very less at this moment, and most of them involve the genetic transfer of its efficient biosynthetic genes to S. cerevisiae to make it able to utilize pentose sugars, although the full genome sequencing (Jeffries and Van Vleet, 2009) and the more efficient genetic transformation systems (e.g., plasmid vectors and a loxP/Cre recombination system) and drug resistance markers (Laplaza et al., 2006) might enable better room for genetic modification of this industrial strain.

The thermotolerant yeast H. polymorpha cannot utilize starch and xylose and its ferment glucose and xylose to ethanol at high temperature. For the efficient utilization of xylose and starch, amylolytic and xylanolytic enzymes were heterologously expressed. Genes encoding α amylase and glucoamylase, SWA2 and GAM1, from the yeast Schwanniomyces

occidentalis, encoding α-amylase and glucoamylase, were transferred in to H. polymorpha under the well characterized constitutive promoter of the H. polymorpha glyceraldehyde-3-phosphate dehydrogenase gene (Voronovsky et al., 2009). Since the organism is highly thermotolerant, engineering the substrate utilization strategies will help to improve the industrial production.

Kluyveromyces marxianus KY3 is a highly efficient hexose-fermenting yeast and has many other advantages like thermotolerance, high cell density, temperature and pH tolerance, high secretion of heterologous proteins, and efficient substrate utilization (Yanase et al., 2010). The xylose utilization of K. marxianus is weak. To alleviate the xylose utilization, this strain was engineered for the conversion of xylose to xylitol and ethanol at the higher temperature. This was done by replacing native xylose reductase gene of the K. marxianus strain with xylose reductase gene from P. stipitis. This modified strain with both NADPH and NADH as the coenzyme could produce 55 g/L ethanol and 32 g/L xylitol (Zhang et al., 2013). Other engineering approaches include the protein engineering for altering the cofactor requirement of xylose reductase of Candida tenuis (Petschacher and Nidetzky, 2008) and optimized expression levels of Candida shehatae xylose reductase, Candida tropicalis xylose dehydrogenase, and

P. pastoris xylose kinase by combinatorial transcriptional engineering (Tsai et al., 2015). Recently, evolutionary adaptation technique has been used to enhance the utilization of xylose in *K. marxianus* (Sharma et al., 2016).

Development of Consolidated Bioprocessing for Non-Conventional Yeast Strains

Many microorganisms possess biomass-degrading enzymes that can efficiently degrade lignocellulose materials, but their fermentative production of ethanol is less. Recently, Arora et al. (2015a,b) reviewed the importance of highly efficient cellulosomes for the production of biofuels from lignocellulosic biomass. Several attempts have been done to convert enzymatically efficient hosts into ethanol producing host to serve as the cell factory for ethanol production. But still, an efficient ethanol fermenter needs to be developed. To develop yeast for CBP bioethanol production, a synthetic biology technique, called "promoter-based gene assembly and simultaneous overexpression" (PGASO), that can simultaneously insert and express multiple genes into yeast. *K. marxianus* has a number of advantages, such as heat and toxin tolerance, over the model organisms *K. lactis* and *S. cerevisiae*. To formulate an efficient cellulose cocktail, a filter-paper-activity assay for selecting heterologous cellulolytic enzymes was developed in this study and used to select five cellulase degrading genes (two cellobiohydrolases, two endo-β-1,4-glucanases and one beta-glucosidase genes) from different fungi. A fungal cellodextrin transporter gene was selected to transport cellodextrin into the cytoplasm. These six multiple genes were assembled into the genome of the host using PGASO technology. Experimental results indicated that the developed strain KR7 contains five recombinantly expressed heterologous cellulase genes and that strain could transform crystalline cellulose into ethanol (Chang et al., 2013).

Improvement of Tolerance against Inhibitors and Biofuel Products

Despite the engineering of metabolic networks in yeast for biofuel production, yeast could not achieve the high level of biofuel production because of the toxic nature of products. The toxicity is mainly due to the hydrophobicity of the accumulated products inside the membrane, and this will lead to membrane disruption by the inhibition of ATP-generating pumps and conformation changes in the proteins that maintain the fluidity (Jeffries and Jin, 2000; Dunlop, 2011). Several genetic manipulations elicit enhanced tolerance against several advanced biofuels. The success of all these attempts was not up to the mark.

Even though most of the non-conventional yeast species can naturally fight with most of the inhibitors, such as nitroaromatics, aromatics, halogenated organo- phosphates, metals, and alkanes (Zinjarde et al., 2014), still, there is an urgent need for engineering some non-conventional, like *Y. lipolytica*, against fermentation inhibitors present in biomass hydrolyzates, like acids and phenolics. The inhibitor tolerance is a quantitative trait and is determined by several complex genes. Hence large-scale analysis of gene expression, different expression cassettes construction, and optimization will be essential to impart inhibitor tolerance to yeast. *Y. lipolytica* has been engineered to

express recombinant laccases (Madzak et al., 2005), which could improve its detoxification capacities. Several studies reported the strains of non-conventional yeast species, such as *Schizo saccharomycespombe*, *Pichia kudriavzevii*, *D. bruxellensis*, *Torulaspora delbrueckii*, and *Wickerhamomyces anomala*, with promising fermentative features and superior ethanol tolerance levels than that of *S. cerevisiae* (Mukherjee et al., 2014; Ruyters et al., 2015). *D. bruxellensis* has been reported as one of the excellent yeast in terms of product tolerance and production. *D. bruxellensis and S. cerevisiae* possesses an almost similar molecular mechanism for this trait (Piskur et al., 2006). Kwon et al. (2011) reported that *P. kudriavzevii* is extremely tolerant to 3 g/L furfural and also tolerate acetic acid of up to 10 g/L (Oberoi et al., 2012) and formic acid up to 2 g/L (Dandi et al., 2013). The genetic engineering tools for this yeast are limited, and genome sequence has recently been reported by Chan et al. (2012).

Y. lipolytica Cell Factory for Biofuel Production

Yarrowia lipolytica is a well-studied oleaginous organism and extensively used for industrial biofuel production, and it has been served as a model organism for biofuel research, especially for fatty acid-derived fuels (Beopoulos et al., 2009; Tai and Stephanopoulos, 2013; Blazeck et al., 2014; Zhou et al., 2016). Several metabolic engineering tools are available for *Y. lipolytica* (Juretzek et al., 2001; Madzak, 2015). The completely annotated genome is available (Dujon et al., 2004), and its metabolism and regulation are also studied in detail (Pan and Hua, 2012). System and synthetic biology approaches have been established in this organism (Morin et al., 2011; Pomraning et al., 2015). Several metabolic engineering studies strengthen the lipid production in this organism. Different target genes were found for overexpression from the metabolic pathway and manipulated to increase the fatty acid accumulation. For example, inhibiting beta-oxidation, by targeted deletion of the six *POX* genes or the *MFE* gene (Dulermo and Nicaud, 2011) and overexpression of enzymes leading to TAG production (*DGA2*) (Beopoulos et al., 2012) and *GPD1* (Dulermo and Nicaud, 2011), increased the lipid content. An improved *Y. lipolytica* strain could produce a very high lipid yield and lipid titers of ~55 g/L under optimized conditions. This further proved the economic feasibility of *Y. lipolytica* for biofuel production (Qiao et al., 2015). *Y. lipolytica* is not able to utilize cellulose and starch. Wei et al. (2014) modified the strain by incorporating cellulases for the conversion of cellulosic substrates. In addition, two alpha amylases—starch-degrading enzyme—have been expressed in this host (Park et al., 1997; Celinska et al., 2015).

Studies revealed that the use of intron-containing translation elongation factor-1a (TEF) promoter is capable of increasing the oil production *Y. lipolytica* compared to intron-less TEF promoter. They have exploited this expression system for the overexpression of diacylglycerol acyl transferase (DGA1), the final key enzyme of the triglyceride (TAG) synthesis pathway, which resulted in a fourfold enhancement in lipid yield compared to wild-type, to a lipid yield of 33.8% of DCW. They also proved that the overexpression of acetyl-CoA carboxylase (ACC1), the first key enzyme of fatty acid biosynthesis, increased lipid content twofold over control, or 17.9% lipid content. The co-expression of ACC1 and DGA1 also improved the production of oil content (Tai and Stephanopoulos, 2013). Ledesma-Amaro et al. (2015) performed a strain engineering approach to obtain a consolidated bioprocess for the direct production of lipids from raw starch. Further, they proved that lipid production from starch can be enhanced by both metabolic engineering and culture condition optimization. Blazeck et al. (2014) rewired *Y. lipolytica* native metabolism for increased lipid titers, by coupling combinatorial multiplexing of lipogenesis targets with phenotypic induction. The tri-level metabolic control results in accumulation of 90% lipid content and 60-fold improvement over wild-type strain. Sheng and Feng (2015) reviewed the metabolic engineering of non-conventional yeast strains to improve the production of fatty acid-derived biofuels, and they discussed the bottlenecks that limit the productivity of biofuels and suggested the appropriate strategies to overcome the current bottlenecks. Some of the non-conventional yeast system currently employed for biofuel production is listed in **Table 1**.

OPTIMIZATION OF PRODUCTION PATHWAYS AND HOST

High yield is very important for biofuel production. The system biology data-driven approaches to synthetic biology for improving microorganisms for producing fuels decreases the cost by reducing the number of expression constructs required for different optimization experiments. The omics data help to identify the rate-limiting steps in biosynthetic pathways. Targeted proteomics approach identified two poorly expressed enzymes in heterologous mevalonate pathway. Codon optimization of the target gene and introduction of a promoter at the 5′ region of the most poorly expressed gene also lead to sesquiterpene production (Redding-Johanson et al., 2011). Some computational prediction tools are involved in changing the metabolic pathway by using knockouts or by adding new catalytic enzymes, and this may help in increasing biofuel production. One important tool for the prediction of metabolic pathway is From Metabolite to Metabolite (FMM)—freely available software (Medema et al., 2012). It compiles the KEGG map and KEGG ligand data to obtain a combined pathway. Recently, a metabolic model with the help of computational tools deleted the NADPH-dependent glutamate dehydrogenase that would result in a 10-fold increase in the production of sesquiterpenes in *S. cerevisiae* (Asadollahi et al., 2009). These tools can be applied to non-conventional yeasts to expand its desirable traits for biofuel production.

CURRENT BOTTLENECKS OF SYNTHETIC BIOLOGY APPLICATIONS IN NON-CONVENTIONAL YEAST AND FUTURE PERSPECTIVES

Several non-conventional yeast systems with highly desirable characteristics and genome sequences for biofuel production are available now. The current requirement is to expand the tool box

TABLE 1 | Non-conventional Yeasts and their characterestics and currently available genetic tools.

Non-conventional yeast species	Glucose	Ethanol	Temperature (°C)	Salt	Acetic acid	Available gene manipulation tools	Reference
Kluyveromyces marxianus			52			NHEJ mediated Transformation Complete genome is available	Hoshida et al. (2014)
Zygosaccharomyces bailii	60% (W/V)				24 g/L	Homologous recombination Complete genome is available	Branduardi et al. (2014)
Dekkera bruxellensis		10–16%				Well-established transformation protocols Complete genome is available	Miklenic et al. (2013)
Ogataea polymorpha			50			Recombinant vectors are available with promoters	Saraya et al. (2012)
Hansenula polymorpha		9.8 g/L (xylose as carbon source)	50			Tightly regulated native promoters Complete genome is available	Wagner and Alper (2016); Kurylenko et al. (2014)
Kluyveromyces lactis		4 mol of ethanol from 1 mol of lactose				Tightly regulated native promoters More similar to *S. cerevisiae* Complete genome is available	Wagner and Alper (2016); Gonzalez-Siso et al. (2015)
Pichia pastoris						Tightly regulated native promoters Complete genome is available	Wagner and Alper (2016)
Yarrowia lipolytica						Strong constitutive hybrid promoters	Wagner and Alper (2016)
Zygosaccharomyces rouxii	90% (W/V)			3M		Homologous recombination Modified plasmids	Leandro et al. (2011)

Synthetic Biology and Metabolic Engineering Approaches and Its Impact on Non-Conventional Yeast...

207

to engineer the non-conventional to make it suitable to digest lignocellulosic biomass and ferment to ethanol. The availability of the complete genome sequences for several non-conventional yeasts will open new opportunity to the system biology and for the enhancement of native production capabilities of these yeasts. Genome-scale metabolic models (GEMs) is tool which can be used to assess the metabolic nature of a cell, its metabolite production capability, species to species relation, identification of genes and transcription factors for metabolic engineering, metabolic flux design, etc. (Zhang and Hua, 2015).

RNA interference is still not implemented in non-conventional yeast despite its success in *S. cerevisiae*. Optimized RNAi technique in *S. cerevisiae* has greatly helped for developing tools of metabolic engineering and synthetic biology. Likewise, CRISPR-Cas technology has already been established in *S. cerevisiae* for biofuel production (Papapetridis et al., 2016). Extension of these technologies to non-conventional yeast strains would enable fastest methods to target many genes and thus by the alteration of metabolic pathway for enhanced biofuel production. CRISPR–Cas technology has been initiated in the non-conventional yeast system *K. lactis* and *S. pombe* (Horwitz et al., 2015). The main challenge associated with CRIPR-Cas technology has been the production of guide RNA. In *S. cerevisiae*, this is achieved by use of RNA polymerase III promoters. But RNA polymerase III promoters are not well in non-conventional yeast. This hinders the progress of CRISPR-Cas technology in non-conventional yeast. Once the CRISPR–Cas technique is a success in the non-conventional yeasts, we will be able to establish transcriptional regulation, genetic circuits, and other metabolic networks to establish a fuel production system and heterologous expression host (Zalatan et al., 2015). The targeted generation of double-stranded breaks also alleviates the challenges associated with homologous recombination in non-conventional yeast.

CONCLUDING REMARKS

Current increases in fuels costs have resulted in increased demand in finding an alternative to fossil fuels. This leads to the concept of biofuels by manipulating microbial cellular metabolism for the production of fuels and chemicals. Computational tools are essential for the analysis of high-throughput data generated by gene-sequencing programmes and gene-expression analysis. Analysis of complex metabolic pathway is another challenge in synthetic and metabolic engineering concepts. The field of synthetic biology and metabolic engineering has been growing rapidly in the case of *S. cerevisiae*. This led to the construction of highly engineered metabolic pathways, strongly regulated metabolic networks, efficiently engineered native and synthetic control elements, and efficient recombinant vectors. Development of mathematical models for studying the complex metabolic pathways is also extremely important. Well-established synthetic biology approaches in *S. cerevisiae* can be a boost to other non-conventional yeast species by rewiring the metabolic pathways related to biofuel production.

In summary, biofuel production by engineered yeast through synthetic biology requires several synthetic biology pipelines and models. This can be used for predicting metabolically optimal pathways, gene constructs, and metabolite screening. The developments in different "omics" technologies can serve as tool for future synthetic biology of non-conventional yeast.

AUTHOR CONTRIBUTIONS

AM wrote and reviewed the article. AJ cross checked and modified article and references. PB, RKS, AP, and GC reviewed and modified the article. RS, corresponding author, reviewed and modified the article.

FUNDING

The authors are grateful to the Ministry of New and Renewable Energy, Government of India, New Delhi; Department of Science and Technology, Government of India, New Delhi; and Technology Information, Forecasting and Assessment Council, New Delhi, for the financial support provided to the Centre for Biofuels R&D, CSIR-NIIST, Trivandrum. One of the authors RS acknowledges Department of Biotechnology for financial support under DBT Bio-CARe scheme. AM, GC, RS, and PB acknowledge European Commission Seventh Framework Programme, Marie Curie Actions-International Research Staff Exchange Scheme—Contact Number 318931. AM acknowledges Department of Biotechnology for financial support under DBT Research Associateship programme.

REFERENCES

Abdel-Banat, B. M., Hoshida, H., Ano, A., Nonklang, S., and Akada, R. (2010). High-temperature fermentation: how can processes for ethanol production at high temperatures become superior to the traditional process using mesophilic yeast? *Appl. Microbiol. Biotechnol.* 85, 861–867. doi:10.1007/s00253-009-2248-5

Arora, R., Behera, S., Sharma, N. K., and Kumar, S. (2015a). Bioprospecting thermostable cellulosomes for efficient biofuel production from lignocellulosic biomass. *Bioresour. Bioprocess* 2, 38. doi:10.1186/s40643-015-0066-4

Arora, R., Behera, S., Sharma, N. K., and Kumar, S. (2015b). A new search for thermotolerant yeasts, its characterization and optimization using response surface methodology for ethanol production. *Front. Microbiol.* 6:889. doi:10.3389/fmicb.2015.00889

Asadollahi, M. A., Maury, J., Patil, K. R., Schalk, M., Clark, A., and Nielsen, J. (2009). Enhancing sesquiterpene production in *Saccharomyces cerevisiae* through *in silico* driven metabolic engineering. *Metab. Eng.* 11, 328–334. doi:10.1016/j.ymben.2009.07.001

Bao, Z., Xiao, H., Liang, J., Zhang, L., Xiong, X., Sun, N., et al. (2015). Homology-integrated CRISPR-Cas (HI-CRISPR) system for one-step multigene disruption in *Saccharomyces cerevisiae*. *ACS Synth. Biol.* 4, 585–594. doi:10.1021/sb500255k

Beopoulos, A., Cescut, J., Haddouche, R., Uribelarrea, J. L., Molina-Jouve, C., and Nicaud, J. M. (2009). *Yarrowia lipolytica* as a model for bio-oil production. *Prog. Lipid Res.* 48, 375–387. doi:10.1016/j.plipres.2009.08.005

Beopoulos, A., Haddouche, R., Kabran, P., Dulermo, T., Chardot, T., and Nicaud, J. M. (2012). Identification and characterization of DGA2, an acyltransferase of the DGAT1 acyl-CoA:diacylglycerol acyltransferase family in the oleaginous yeast *Yarrowia lipolytica*. New insights into the storage lipid metabolism of oleaginous yeasts. *Appl. Microbiol. Biotechnol.* 93, 1523–1537. doi:10.1007/s00253-011-3506-x

Berg, L., Strand, T. A., Valla, S., and Brautaset, T. (2013). Combinatorial mutagenesis and selection to understand and improve yeast promoters. *Biomed Res. Int.* 2013, 926985. doi:10.1155/2013/926985

Blazeck, J., Garg, R., Reed, B., and Alper, H. S. (2012). Controlling promoter strength and regulation in *Saccharomyces cerevisiae* using synthetic hybrid promoters. *Biotechnol. Bioeng.* 109, 2884–2895. doi:10.1002/bit.24552

Blazeck, J., Hill, A., Liu, L., Knight, R., Miller, J., Pan, A., et al. (2014). Harnessing *Yarrowia lipolytica* lipogenesis to create a platform for lipid and biofuel production. *Nat. Commun.* 5, 3131. doi:10.1038/ncomms4131

Bloom, J., Ehrenreich, I., Loo, W., Lite, T., and Kruglyak, L. (2013). Finding the sources of missing heritability in a yeast cross. *Nature* 494, 234–237. doi:10.1038/nature11867

Blount, B. A., Weenink, T., and Ellis, T. (2012a). Construction of synthetic regulatory networks in yeast. *FEBS Lett.* 586, 2112–2121. doi:10.1016/j.febslet.2012.01.053

Blount, B. A., Weenink, T., Vasylechko, S., and Ellis, T. (2012b). Rational diversification of a promoter providing fine-tuned expression and orthogonal regulation for synthetic biology. *PLoS ONE* 7:e33279. doi:10.1371/journal.pone.0033279

Branduardi, P., Dato, L., and Porro, D. (2014). Molecular tools and protocols for engineering the acid-tolerant yeast *Zygosaccharomyces bailii* as a potential cell factory. *Methods Mol. Biol.* 1152, 63–85. doi:10.1007/978-1-4939-0563-8_4

Bullard, J. H., Purdom, E., Hansen, K. D., and Dudoit, S. (2010). Evaluation of statistical methods for normalization and differential expression in mRNA-Seq experiments. *BMC Bioinformatics* 11:94. doi:10.1186/1471-2105-11-94

Caspeta, L., Chen, Y., Ghiaci, P., Feizi, A., Buskov, S., Hallström, B. M., et al. (2014). Altered sterol composition renders yeast thermotolerant. *Science* 346, 75L–78. doi:10.1126/science.1258137

Celinska, E., Bialas, W., Borkowska, M., and Grajek, W. (2015). Cloning, expression, and purification of insect (*Sitophilus oryzae*) alpha-amylase, able to digest granular starch, in *Yarrowia lipolytica* host. *Appl. Microbiol. Biotechnol.* 99, 2727–2739. doi:10.1007/s00253-014-6314-2

Chan, G. F., Gan, H. M., Ling, H. L., and Rashid, N. A. (2012). Genome sequence of *Pichia kudriavzevii* M12, a potential producer of bioethanol and phytase. *Eukaryotic Cell* 11, 1300–1301. doi:10.1128/EC.00229-12

Chang, J. J., Ho, F. J., Ho, C. Y., Wu, Y. C., Hou, Y. H., Huang, C. C., et al. (2013). Assembling a cellulase cocktail and a cellodextrin transporter into a yeast host for CBP ethanol production. *Biotechnol. Biofuels* 6, 19. doi:10.1186/1754-6834-6-19

Cho, J. Y., and Jeffries, T. W. (1999). Transcriptional control of ADH genes in the xylose-fermenting yeast *Pichia stipitis*. *Appl. Environ. Microbiol.* 65, 2363–2368.

Chubukov, V., Mukhopadhyay, A., Petzold, C. J., Keasling, J. D., and Martín, H. G. (2016). Synthetic and systems biology for microbial production of commodity chemicals. *NPJ Syst. Biol. Appl.* 2, 16009. doi:10.1038/npjsba.2016.9

Cong, L., Ran, F. A., Cox, D., Lin, S., Barretto, R., Habib, N., et al. (2013). Multiplex genome engineering using CRISPR/Cas systems. *Science* 339, 819–823. doi:10.1126/science.1231143

Dandi, N. D., Dandi, B. N., and Chaudhari, A. B. (2013). Bioprospecting of thermo- and osmo-tolerant fungi from mango pulp-peel compost for bioethanol production. *Antonie Van Leeuwenhoek* 103, 723–736. doi:10.1007/s10482-012-9854-4

Dominguez, A., Fermiñán, E., Sánchez, M., González, F. J., Pérez-Campo, F. M., García, S., et al. (1998). Non-conventional yeasts as hosts for heterologous protein production. *Int. Microbiol.* 1, 131–142.

Dueber, J. E., Mirsky, E. A., and Lim, W. A. (2007). Engineering synthetic signaling proteins with ultrasensitive input/output control. *Nat. Biotechnol.* 25, 660–662. doi:10.1038/nbt1308

Dujon, B., Sherman, D., Fischer, G., Durrens, P., Casaregola, S., Lafontaine, I., et al. (2004). Genome evolution in yeasts. *Nature* 430, 35–44. doi:10.1038/nature02579

Dulermo, T., and Nicaud, J. M. (2011). Involvement of the G3P shuttle and β-oxidation pathway in the control of TAG synthesis and lipid accumulation in *Yarrowia lipolytica*. *Metab. Eng.* 13, 482–491. doi:10.1016/j.ymben.2011.05.002

Dunlop, M. J. (2011). Engineering microbes for tolerance to nextgeneration biofuels. *Biotechnol. Biofuels* 4, 32. doi:10.1186/1754-6834-4-32

Elowitz, M. B., and Leibler, S. A. (2000). Synthetic oscillatory network of transcriptional regulators. *Nature* 403, 335–338. doi:10.1038/35002125

Faber, K. N., Haima, P., Harder, W., Veenhuis, M., and Geert, A. B. (1994). Highly-efficient electrotransformation of the yeast *Hansenula polymorpha*. *Curr. Genet.* 25, 305–310. doi:10.1007/BF00351482

Garcia Sanchez, R., Karhumaa, K., Fonseca, C., Sànchez Nogué, V., Almeida, J. R., Larsson, C. U., et al. (2010). Improved xylose and arabinose utilization by an industrial recombinant *Saccharomyces cerevisiae* strain using evolutionary engineering. *Biotechnol. Biofuels* 3, 13. doi:10.1186/1754-6834-3-13

Gardner, T. S., Cantor, C. R., and Collins, J. J. (2000). Construction of a genetic toggle switch in *Escherichia coli*. *Nature* 403, 339–342. doi:10.1038/35002131

Gellissen, G., and Hollenberg, C. P. (1997). Application of yeasts in gene expression studies: a comparison of *Saccharomyces cerevisiae*, *Hansenula polymorpha* and *Kluyveromyces lactis* – a review. *Gene* 190, 87–97. doi:10.1016/S0378-1119(97)00020-6

Gellissen, G., Kunze, G., Gaillardin, C., Cregg, J. M., Berardi, E., Veenhuis, M., et al. (2005). New yeast expression platforms based on methylotrophic *Hansenula polymorpha* and *Pichia pastoris* and on dimorphic *Arxula adeninivorans* and *Yarrowia lipolytica* a comparison. *FEMS Yeast Res.* 5, 1079–1096. doi:10.1016/j.femsyr.2005.06.004

Gilbert, L. A., Larson, M. H., Morsut, L., Liu, Z., Brar, G. A., Torres, S. E., et al. (2013). CRISPR-mediated modular RNA-guided regulation of transcription in eukaryotes. *Cell* 154, 442–451. doi:10.1016/j.cell.2013.06.044

Gonzalez-Siso, M. I., Tourino, A., Vizoso, A., Pereira-Rodriguez, A., Rodriguez-Belmonte, E., Becerra, M., et al. (2015). Improved bioethanol production in an engineered *Kluyveromyces lactis* strain shifted from respiratory to fermentative metabolism by deletion of NDI1. *Microb. Biotechnol.* 8, 319–330. doi:10.1111/1751-7915.12160

Hartner, F. S., Ruth, C., Langenegger, D., Johnson, S. N., Hyka, P., Lin-Cereghino, G. P., et al. (2008). Promoter library designed for fine-tuned gene expression in *Pichia pastoris*. *Nucleic Acids Res.* 36, e76. doi:10.1093/nar/gkn369

Horwitz, A. A., Walter, J. M., Schubert, M. G., Kung, S. H., Hawkins, K., Platt, D. M., et al. (2015). Efficient multiplexed integration of synergistic alleles and metabolic pathways in yeasts via CRISPR-Cas. *Cell Syst.* 1, 1–9. doi:10.1016/j.cels.2015.02.001

Hoshida, H., Murakami, N., Suzuki, A., Tamura, R., Asakawa, J., Abdel-Banat, B. M. A., et al. (2014). Non-homologous end joining-mediated functional marker selection for DNA cloning in the yeast *Kluyveromyces marxianus*. *Yeast* 31, 29–46. doi:10.1002/yea.2993

Jeffries, T. W., and Jin, Y. S. (2000). Ethanol and thermotolerance in the bioconversion of xylose by yeasts. *Adv. Appl. Microbiol.* 47, 221–268. doi:10.1016/S0065-2164(00)47006-1

Jeffries, T. W., and Van Vleet, J. R. (2009). *Pichia stipitis* genomics, transcriptomics, and gene clusters. *FEMS Yeast Res.* 9, 793–807. doi:10.1111/j.1567-1364.2009.00525.x

Juretzek, T., Le Dall, M., Mauersberger, S., Gaillardin, C., Barth, G., and Nicaud, J. (2001). Vectors for gene expression and amplification in the yeast *Yarrowia lipolytica*. *Yeast* 18, 97–113. doi:10.1002/1097-0061(20010130)18:2<97::AID-YEA652>3.0.CO;2-U

Katahira, S., Fujita, Y., Mizuike, A., Fukuda, H., and Kondo, A. (2004). Construction of a xylan-fermenting yeast strain through co display of xylanolytic enzymes on the surface of xylose-utilizing *Saccharomyces cerevisiae* cells. *Appl. Environ. Microbiol.* 70, 5407–5414. doi:10.1128/AEM.70.9.5407-5414.2004

Kiel, C., and Serrano, L. (2011). Challenges ahead in signal transduction: MAPK as an example. *Curr. Opin. Biotechnol* 23, 305–314. doi:10.1016/j.copbio.2011.10.004

Kooistra, R., Hooykaas, P. J., and Steensma, H. Y. (2004). Efficient gene targeting in *Kluyveromyces lactis*. *Yeast* 21, 781–792. doi:10.1002/yea.1131

Kumar, S., Dheeran, P., Singh, S. P., Mishra, I. M., and Adhikari, D. K. (2014). Bioprocessing of bagasse hydrolysate for ethanol and xylitol production using thermotolerant yeast. *Bioprocess Biosyst. Eng.* 38, 39–47. doi:10.1007/s00449-014-1241-2

Kurylenko, O. O., Ruchala, J., Hryniv, O. B., Abbas, C. A., Dmytruk, K. V., and Sibirny, A. A. (2014). Metabolic engineering and classical selection of the methylotrophic thermotolerant yeast *Hansenula polymorpha* for improvement of high temperature xylose alcoholic fermentation. *Microb. Cell Fact.* 13, 122. doi:10.1186/s12934-014-0122-3

Kwon, Y. J., Ma, A. Z., Li, Q., Wang, F., Zhuang, G. Q., and Liu, C. Z. (2011). Effect of lignocellulosic inhibitory compounds on growth and ethanol fermentation of newly-isolated thermotolerant *Issatchenkia orientalis*. *Bioresource Technol.* 102, 8099–8104. doi:10.1016/j.biortech.2011.06.035

Laplaza, J. M., Torres, B. R., Jin, Y. S., and Jeffries, T. W. (2006). Sh ble and Cre adapted for functional genomics and metabolic engineering of *Pichia stipitis*. *Enzyme Microb. Technol.* 38, 741–747. doi:10.1016/j.enzmictec.2005.07.024

Leandro, M. J., Sychrová, H., Prista, C., and Loureiro-Dias, M. C. (2011). The osmotolerant fructophilic yeast *Zygosaccharomyces rouxii* employs two plasmamembrane fructose uptake systems belonging to a new family of yeast sugar transporters. *Microbiology* 157, 601–608. doi:10.1099/mic.0.044446-0

Ledesma-Amaro, R., Dulermo, T., and Nicaud, J. M. (2015). Engineering *Yarrowia lipolytica* to produce biodiesel from raw starch. *Biotechnol. Biofuels* 8, 148. doi:10.1186/s13068-015-0335-7

Li, M. Z., and Elledge, S. J. (2007). Harnessing homologous recombination in vitro to generate recombinant DNA via SLIC. *Nat. Methods* 4, 251–256. doi:10.1038/nmeth1010

Madhavan, A., and Sukumaran, R. K. (2014). Promoter and signal sequence from filamentous fungus can drive recombinant protein production in the yeast *Kluyveromyces lactis*. *Bioresour. Technol.* 165, 302–308. doi:10.1016/j.biortech.2014.03.002

Madzak, C. (2015). *Yarrowia lipolytica*: recent achievements in heterologous protein expression and pathway engineering. *Appl. Microbiol. Biotechnol.* 99, 4559–4577. doi:10.1007/s00253-015-6624-z

Madzak, C., Otterbein, L., Chamkha, M., Moukha, S., Asther, M., Gaillardin, C., et al. (2005). Heterologous production of a laccase from the basidiomycete *Pycnoporus cinnabarinus* in the dimorphic yeast *Yarrowia lipolytica*. *FEMS Yeast Res.* 5, 635–646. doi:10.1534/genetics.114.169060

Medema, M. H., Raaphorst, R., Takano, E., and Breitling, R. (2012). Computational tools for the synthetic design of biochemical pathways. *Nat. Rev. Microbiol* 10, 191–202. doi:10.1038/nrmicro2717

Miklenic, M., Stafa, A., Bajic, A., Zunar, B., Lisnic, B., and Svetec, I.-K. (2013). Genetic transformation of the yeast *Dekkera/Brettanomyces bruxellensis* with non-homologous DNA. *J. Microbiol. Biotechnol.* 23, 674–680. doi:10.4014/jmb.1211.11047

Morin, N., Cescut, J., Beopoulos, A., Lelandais, G., Le Berre, V., Uribelarrea, J. L., et al. (2011). Transcriptomic analyses during the transition from biomass production to lipid accumulation in the oleaginous yeast *Yarrowia lipolytica*. *PLoS ONE* 6:e27966. doi:10.1371/journal.pone.0027966

Mukherjee, V., Steensels, J., Lievens, B., Van de Voorde, I., Verplaetse, A., Aerts, G., et al. (2014). Phenotypic evaluation of natural and industrial *Saccharomyces* yeasts for different traits desirable in industrial bioethanol production. *Appl Microbiol Biotechnol* 98, 9483–9498. doi:10.1007/s00253-014-6090-z

Näätsaari, L., Mistlberger, B., Ruth, C., Hajek, T., Hartner, F. S., and Glieder, A. (2012). Deletion of the *Pichia pastoris* KU70 homologue facilitates platform strain generation for gene expression and synthetic biology. *PLoS ONE* 7:e39720. doi:10.1371/journal.pone.0039720

Nielsen, J. (2015). Yeast cell factories on the horizon. *Science* 349, 1050–1051. doi:10.1371/journal.pone.0039720

Oberoi, H. S., Babbar, N., Sandhu, S. K., Dhaliwal, S. S., Kaur, U., Chadha, B. S., et al. (2012). Ethanol production from alkali-treated rice straw via simultaneous saccharification and fermentation using newly isolated thermotolerant *Pichia kudriavzevii* HOP-1. *J. Ind. Microbiol. Biotechnol.* 39, 557–566. doi:10.1007/s10295-011-1060-2

Pan, P., and Hua, Q. (2012). Reconstruction and in silico analysis of metabolic network for an oleaginous yeast, *Yarrowia lipolytica*. *PLoS One* 7:e51535. doi:10.1371/journal.pone.0051535

Papapetridis, I., Dijk, M., Dobbe, A., Metz, B., Pronk, J. T., and Maris, A. J. A. (2016). Improving ethanol yield in acetate-reducing *Saccharomyces cerevisiae* by cofactor engineering of 6-phosphogluconate dehydrogenase and deletion of ALD6. *Microb. Cell Fact.* 15, 67. doi:10.1186/s12934-016-0465-z

Park, C. S., Chang, C. C., Kim, J. Y., Ogrydziak, D. M., and Ryu, D. D. (1997). Expression, secretion, and processing of rice alpha-amylase in the yeast *Yarrowia lipolytica*. *J. Biol. Chem.* 272, 6876–6881. doi:10.1074/jbc.272.11.6876

Peralta-Yahya, P. P., Zhang, F., del Cardayre, S. B., and Keasling, J. D. (2012). Microbial engineering for the production of advanced biofuels. *Nature* 488, 320–328. doi:10.1038/nature11478

Petschacher, B., and Nidetzky, B. (2008). Altering the coenzyme preference of xylose reductase to favor utilization of NADH enhances ethanol yield from xylose in a metabolically engineered strain of *Saccharomyces cerevisiae*. *Microb. Cell Fact.* 7, 9. doi:10.1186/1475-2859-7-9

Piskur, J., Rozpedowska, E., Polakova, S., Merico, A., and Compagno, C. (2006). How did *Saccharomyces* evolve to become a good brewer? *Trends Genet.* 22, 183–186. doi:10.1016/j.tig.2006.02.002

Pomraning, K. R., Wei, S., Karagiosis, S. A., Kim, Y. M., Dohnalkova, A. C., Arey, B. W., et al. (2015). Comprehensive metabolomic, lipidomic and microscopic profiling of *Yarrowia lipolytica* during lipid accumulation identifies targets

for increased lipogenesis. *PLoS ONE* 10:e0123188. doi:10.1371/journal.pone.0123188

Qiao, K., Imam Abidi, S. H., Liu, H., Zhang, H., Chakraborty, S., Watson, N., et al. (2015). Engineering lipid overproduction in the oleaginous yeast *Yarrowia lipolytica*. *Metab. Eng.* 29, 56–65. doi:10.1016/j.ymben.2015.02.005

Qin, X., Qian, J., Yao, G., Zhuang, Y., Zhang, S., and Chu, J. (2011). GAP promoter library for fine-tuning of gene expression in *Pichia pastoris*. *Appl. Environ. Microbiol.* 77, 3600–3608. doi:10.1128/AEM.02843-10

Ramezani-Rad, M., Hollenberg, C. P., Lauber, J., Wedler, H., Griess, E., Wagner, C., et al. (2003). The *Hansenula polymorpha* (strain CBS4732) genome sequencing and analysis. *FEMS Yeast Res.* 4, 207–215. doi:10.1016/S1567-1356(03)00125-9

Redding-Johanson, A. M., Batth, T. S., Chan, R., Krupa, R., Szmidt, H. L., Adams, P. D., et al. (2011). Targeted proteomics for metabolic pathway optimization: application to terpene production. *Metab. Eng.* 13, 194–203. doi:10.1016/j.ymben.2010.12.005

Ruyters, S., Mukherjee, V., Verstrepen, K. J., Thevelein, J. M., Willems, K. A., and Lievens, B. (2015). Assessing the potential of wild yeasts for bioethanol production. *J Ind Microbiol Biotechnol* 42, 39–48. doi:10.1007/s10295-014-1544-y

Saraya, R., Krikken, A. M., Kiel, J. A., Baerends, R. J., Veenhuis, M., and van der Klei, I. J. (2012). Novel genetic tools for *Hansenula polymorpha*. *FEMS Yeast Res.* 12, 271–278. doi:10.1111/j.1567-1364.2011.00772.x

Schifferdecker, A. J., Siurkus, J., Anderson, M. R., Joerck-Ramberg, D., Ling, Z., Zhou, N., et al. (2016). Alcohol dehydrogenase gene ADH3 activates glucose alcoholic fermentation in genetically engineered *Dekkera bruxellensis* yeast. *Appl. Microbiol. Biotechnol.* 100, 3219–3231. doi:10.1007/s00253-015-7266-x

Sharma, N. K., Behera, S., Arora, R., and Kumar, S. (2016). Enhancement in xylose utilization using *Kluyveromyces marxianus* NIRE-K1 through evolutionary adaptation approach. *Bioprocess. Biosyst. Eng* 39, 835–843. doi:10.1007/s00449-016-1563-3

Sheng, J., and Feng, X. (2015). Metabolic engineering of yeast to produce fatty acid-derived biofuels: bottlenecks and solutions. *Front. Microbiol* 6:554. doi:10.3389/fmicb.2015.00554

Sherman, D. J., Martin, T., Nikolski, M., Cayla, C., Souciet, J. L., Durrens, P., et al. (2009). Génolevures: protein families and synteny among complete hemi ascomycetous yeast proteomes and genomes. *Nucleic Acids Res.* 37, D550–D554. doi:10.1093/nar/gkn859

Shetty, R. P., Endy, D., and Knight, T. F. (2008). Engineering BioBrick vectors from BioBrick parts. *J. Biol. Eng.* 2, 5. doi:10.1186/1754-1611-2-5

Souciet, J. L., Dujon, B., Gaillardin, C., Johnston, M., Baret, P. V., Cliften, P., et al. (2009). Comparative genomics of protoploid Saccharomycetaceae. *Genome Res.* 19, 1696–1709. doi:10.1101/gr.091546.109

Swinnen, S., Klein, M., Carrillo, M., Mcinnes, J., Nguyen, T., Nevoigt, E., et al. (2013). Re-evaluation of glycerol utilization in *Saccharomyces cerevisiae*: characterization of an isolate that grows on glycerol without supporting supplements. *Biotechnol. Biofuels* 6, 157. doi:10.1186/1754-6834-6-157

Tai, M., and Stephanopoulos, G. (2013). Engineering the push and pull of lipid biosynthesis in oleaginous yeast *Yarrowia lipolytica* for biofuel production. *Metab. Eng.* 15, 1–9. doi:10.1016/j.ymben.2012.08.007

Teo, W. S., and Chang, M. W. (2014). Development and characterization of AND-gate dynamic controllers with a modular synthetic GAL1 core promoter in *Saccharomyces cerevisiae*. *Biotechnol. Bioeng.* 111, 144–151. doi:10.1002/bit.25001

Tsai, C. S., Kwak, S., Turner, T. L., and Jin, Y. S. (2015). Yeast synthetic biology toolbox and applications for biofuel production. *FEMS Yeast Res.* 15, 1–15. doi:10.1111/1567-1364.12206

Unkles, S. E., Valiante, V., Mattern, D. J., and Brakhage, A. A. (2014). Synthetic biology tools for bioprospecting of natural products in eukaryotes. *Chem. Biol.* 21, 502–508. doi:10.1016/j.chembiol.2014.02.010

Van den Berg, J. A., Van der Laken, K. J., Van Ooyen, A. J., Renniers, T. C., Rietveld, K., Schaap, A., et al. (1990). *Kluyveromyces* a host for heterologous gene expression: expression and secretion of prochymosin. *Biotechnology* 8, 135–139. doi:10.1038/nbt0290-135

Van Ooyen, A. J., Dekker, P., Huang, M., Olsthoorn, M. M., Jacobs, D. I., Coluss, P. A., et al. (2006). Heterologous protein production in the yeast *Kluyveromyces lactis*. *FEMS Yeast Res.* 6, 381–392. doi:10.1111/j.1567-1364.2006.00049.x

Verbeke, J., Beopoulos, A., and Nicaud, J. M. (2013). Efficient homologous recombination with short length flanking fragments in Ku70 deficient *Yarrowia lipolytica* strains. *Biotechnol. Lett.* 35, 571–576. doi:10.1007/s10529-012-1107-0

Vogl, T., Hartner, F. S., and Glieder, A. (2013). New opportunities by synthetic biology for biopharmaceutical production in Pichia pastoris. *Curr. Opin. Biotechnol.* 24, 1094–1101. doi:10.1016/j.copbio.2013.02.024

Voronovsky, A. Y., Rohulya, O. V., Abbas, C. A., and Sibirny, A. A. (2009). Development of strains of the thermotolerant yeast *Hansenula polymorpha* capable of alcoholic fermentation of starch and xylan. *Metab. Eng.* 11, 234–242. doi:10.1016/j.ymben.2009.04.001

Vroom, J. A., and Wang, C. L. (2008). Modular construction of plasmids through ligation-free assembly of vector components with oligonucleotide linkers. *BioTechniques* 44, 924–926. doi:10.2144/000112808

Wagner, J. M., and Alper, H. S. (2016). Synthetic biology and molecular genetics in non-conventional yeasts: current tools and future advances. *Fungal Genet. Biol.* 89, 126–136. doi:10.1016/j.fgb.2015.12.001

Wei, H., Wang, W., Alahuhta, M., Vander Wall, T., Baker, J. O., Decker, S. R., et al. (2014). Engineering towards a complete heterologous cellulase secretome in Yarrowia lipolytica reveals its potential for consolidated bioprocessing. *Biotechnol. Biofuels* 7, 148. doi:10.1186/s13068-014-0148-0

Wei, N., Quarterman, J., Kim, S. R., Cate, J. H. D., and Jin, Y. (2013). Enhanced biofuel production through coupled acetic acid and xylose consumption by engineered yeast. *Nat. Commun* 4, 2580. doi:10.1038/ncomms3580

Wolf, K. (2012). *Nonconventional Yeasts in Biotechnology: A Handbook.* Seiten: Springer Science & Business Media.

Xuan, Y., Zhou, X., Zhang, W., Zhang, X., Song, Z., and Zhang, Y. (2009). An upstream activation sequence controls the expression of AOX1 gene in *Pichia pastoris. FEMS Yeast Res.* 9, 1271–1282. doi:10.1111/j.1567-1364.2009.00571.x

Yanase, S., Hasunuma, T., Yamada, R., Tanaka, T., Ogino, C., Fukuda, H., et al. (2010). Direct ethanol production from cellulosic materials at high temperature using the thermotolerant yeast *Kluyveromyces marxianus* displaying cellulolytic enzymes. *Appl. Microbiol. Biotechnol.* 88, 381–388. doi:10.1007/s00253-010-2784-z

Zalatan, J. G., Lee, M. E., Almeida, R., Gilbert, L. A., Whitehead, E. H., La Russa, M., et al. (2015). Engineering complex synthetic transcriptional programs with CRISPR RNA scaffolds. *Cell* 160, 339–350. doi:10.1016/j.cell.2014.11.052

Zhang, C., and Hua, H. (2015). Applications of Genome-Scale Metabolic Models in Biotechnology and Systems Medicine. *Front. Physiol.* 6, 413. doi:10.3389/fphys.2015.00413

Zhang, B., Li, L., Zhang, J., Gao, X., Wang, D., and Hong, J. (2013). Improving ethanol and xylitol fermentation at elevated temperature through substitution of xylose reductase in *Kluyveromyces marxianus. J. Ind. Microbiol. Biotechnol.* 40, 305–316. doi:10.1007/s10295-013-1230-5

Zhou, Y. J., Buijs, N. A., Zhu, Z., Qin, J., Siewers, V., and Nielsen, J. (2016). Production of fatty acid-derived oleochemicals and biofuels by synthetic yeast cell factories. *Nat. Commun* 7, 11709. doi:10.1038/ncomms11709

Zinjarde, S., Apte, M., Mohite, P., and Kumar, A. R. (2014). *Yarrowia lipolytica* and pollutants: interactions and applications. *Biotechnol. Adv.* 32, 920–933. doi:10.1016/j.biotechadv.2014.04.008

Conflict of Interest Statement: The authors declare that the research was conducted in the absence of any commercial or financial relationships that could be construed as a potential conflict of interest.

21

Saccharification of Agricultural Lignocellulose Feedstocks and Protein-Level Responses by a Termite Gut-Microbe Bioreactor

Swapna Priya Rajarapu[†] and Michael E. Scharf*

Department of Entomology, Purdue University, West Lafayette, IN, USA

Edited by:
Jalel Labidi,
University of the Basque
Country, Spain

Reviewed by:
Bing-Zhi Li,
Tianjin University, China
Shishir P. S. Chundawat,
Rutgers University, USA
Zheng-Jun Li,
Beijing University of Chemical
Technology, China

*Correspondence:
Michael E. Scharf
mscharf@purdue.edu

†Present address:
Swapna Priya Rajarapu,
Department of Entomology, Ohio
State University-OARDC, Wooster,
OH, USA

Specialty section:
This article was submitted to
Bioenergy and Biofuels,
a section of the journal
Frontiers in Energy Research

This study investigated saccharification and protein-level responses to the candidate biofuel feedstocks corn stover (CS) and soybean residue (SR) by the gut of a lower termite. The focus termite was *Reticulitermes flavipes*, which is a highly efficient digester of wood lignocellulose that houses a mixture of prokaryotic and eukaryotic microbes in its gut. Our specific objectives were to (i) measure saccharification potential of the CS and SR feedstocks by termite gut protein extracts, (ii) identify specific proteins in the termite gut responding to feeding on CS and SR diets, and (iii) evaluate gut lignocellulase and accessory enzyme activity responses to CS and SR feeding. Cellulose paper was the control diet. Although CS was saccharified at higher levels, termite gut protein extracts saccharified both CS and SR irrespective of feedstock loading. Consumption of the CS and SR feedstocks by termites resulted in surprisingly few differences in gut protein profiles, with the main exception being elevated myosin abundance with SR feeding. Activity of potential lignocellulases and accessory enzymes was generally similar between CS and SR fed guts as well; however, cellobiohydrolase/exoglucanase activity was higher with CS feeding and glutathione peroxidase activity with SR feeding. These findings have significance from two perspectives. First, SR feeding/digestion appears to cause physiological stress in the termite gut that likely would extend to other types of microbial environments including those within industrial bioreactors. Second, because termites can survive on exclusive CS and SR diets and their guts exhibit clear CS and SR saccharification activity, this validates the *R. flavipes* system as a potential source for CS and SR degrading enzymes; in particular, cellobiohydrolases/exoglucanases and glutathione peroxidases from this system may play roles in CS and SR breakdown.

Keywords: *Reticulitermes flavipes*, second-generation feedstock, corn stover, soybean residue, cellulase, ligninase

Citation:
Rajarapu SP and Scharf ME (2017)
Saccharification of Agricultural
Lignocellulose Feedstocks and
Protein-Level Responses by a Termite
Gut-Microbe Bioreactor.
Front. Energy Res. 5:5.

INTRODUCTION

Lignocellulosic biomass offers promise for biofuel production due to its abundance and availability. Lignocellulose is a complex structure with its three main components of lignin, cellulose, and hemicellulose. It is crucial to break the lignin and hemicellulose polymers encapsulating the cellulose polymer, which is a source for many renewable bioproducts. In nature, organisms with physiological abilities to degrade lignocellulose occur across the tree of life (Cragg et al., 2015). Physiological degradation of lignocellulose is conferred mainly by cellulases and hemicellulases belonging to the glycoside hydrolase (GH) superfamily, and other accessory enzymes (e.g., Horn et al., 2012; Levasseur et al., 2013). Breakdown and/or dissociation of lignin is important to increase access to cellulose and hemicellulose. Cellulose depolymerization occurs with the synergistic action of endoglucanases, cellobiohydrolases (or exoglucanases), and β-glucosidases. Hemicellulose is degraded not only by endo- and exo-xylanases from GH family 5 (Aspeborg et al., 2012) but also in the focus termite of the current study by GH9 endoglucanases that target internal glucose residues (Scharf et al., 2011b). In recent years, wood feeding insects have gained attention for biofuel production due to their inherent ability to enzymatically degrade recalcitrant lignocellulose. Specifically, two groups of wood-feeding insects, wood-boring beetles, and termites (both higher and lower) have been studied for the purpose of deciphering lignocellulose unlocking mechanisms (Geib et al., 2008; Sun and Scharf, 2010; Scharf, 2015a).

Termites along with their gut symbiota efficiently degrade lignocellulose. Higher termites including *Nasutitermes takasagoensis* and lower termites including *Coptotermes formosanus, C. gestroi,* and *Reticulitermes flavipes* have been extensively studied from genes to function to dissect their lignocellulose digestive physiology (Tokuda et al., 1997; Cairo et al., 2011; Zhang et al., 2012; Scharf, 2015a). There is ample evidence of termites having endogenous cellulases that aid in cellulose digestion independent of symbiont action (Watanabe et al., 1998; Zhou et al., 2007), but clearly a synergistic collaboration exists between the termite and its symbiota that achieves efficient lignocellulose digestion (Scharf et al., 2011a,b). Several GH families have been identified in termites with activity toward various cellulosic substrates (Tokuda et al., 1997; Zhou et al., 2007, 2010; Scharf et al., 2010). In contrast, little is known about lignin breakdown mechanisms of wood-feeding insects (Geib et al., 2008), which is a limiting step in biofuel production (Yang and Wyman, 2008).

Reticulitermes flavipes termites provide a model system for studying lignocellulose digestive mechanisms (Scharf, 2015b). In addition to secreting host-derived digestive enzymes, this lower termite harbors a diversity of eukaryotic (protist) and prokaryotic (bacteria) symbionts in its hindgut, which is analogous to a fermentation chamber (Watanabe and Tokuda, 2010; Brune, 2014). Comparison of metatranscriptome profiles for wood and paper (cellulose)-fed termites has shed much light on putative lignocellulose degrading enzymes and detoxification enzymes potentially involved in lignin and metabolite degradation (Raychoudhury et al., 2013). In addition, detoxification and antioxidant enzymes including phenoloxidases, laccases, esterase, cytochrome P450s, catalases, superoxide dismutases, and glutathione peroxidases (GPX) were previously found upregulated in termites fed on lignin phenolics (Tartar et al., 2009; Coy et al., 2010; Sethi et al., 2013). *In vitro* digestion of a complex lignocellulose substrate (pine sawdust) using recombinant enzymes corresponding to termite lignocellulases conclusively demonstrated their role in lignocellulose saccharification (Sethi et al., 2013). In an extension of the above research to investigate breakdown of second-generation agricultural feedstocks, the gut metatranscriptomes of termites fed on corn stover (CS) and soybean residue (SR) were characterized (Rajarapu et al., 2015). This prior study revealed decreases in microbial abundance with CS and SR feeding relative to cellulose paper, but few differences in functional enzymatic profiles.

The present study investigated protein-level impacts of CS and SR feeding within the *R. flavipes* gut. Our objectives were as follows: (i) to determine saccharification potential of CS and SR by termite gut homogenates *in vitro*, (ii) define impacts of CS and SR feeding on gut protein expression, and (iii) identify candidate CS and SR digestive mechanisms in termite guts using enzyme activity assays. Our hypotheses in relation to these objectives were that the termite gut can saccharify CS and SR *in vitro*, protein profiles of termite guts fed on CS and SR would be similar, and key enzyme activities would be optimized to CS and SR diets according to the "diet-adaptation hypothesis" (Karl and Scharf, 2015). Our findings demonstrate the ability of termite guts to saccharify CS and SR. Additionally, we identified apparently deleterious physiological effects of SR feeding/digestion on total gut protein profiles, along with candidate enzymes that potentially can synergize the saccharification of CS and SR feedstocks.

MATERIALS AND METHODS

Feedstocks and Chemicals

Corn stover was "specialty 3557 variety" and SR was "Williams 82 variety." Both CS and SR were grown without chemical insecticides and transgenic toxins. CS and SR were donated by Dr. Nathan Mosier of Purdue University and Dr. Karen Hudson of the USDA-ARS. CS contained 45% glucan, 29% xylan, and 7.1 g lignin/100 g cell wall. SR contained 44% glucan, 33% xylan, and 6.4 g lignin/100 g cell wall (Rajarapu et al., 2015). All substrates used for enzyme activity assays were from Sigma-Aldrich (USA) and Carbosynth Limited (UK). Rat monoclonal myosin antibody was purchased from Abcam and was a gift from Dr. Chris Mattison, USDA-ARS. Soybean Kuntiz trypsin inhibitor was purchased from Cayman Chemicals (USA).

Termites and Feeding Assays

Reticulitermes flavipes workers from lab colonies were used in all studies and were maintained at 22°C in darkness with 70% relative humidity. Feeding assays were performed with the CS and SR feedstocks as described previously (Rajarapu et al., 2015). Feedstocks were ground to fine powder in a coffee grinder (Cuisinart) to prepare "cookies." 40 termites per treatment were fed these diet cookies for 7 days at 28°C. Each treatment was replicated three times. Feeding bioassays with Kunitz protease inhibitor were performed similarly; fifteen termites were fed six different concentrations (10–0.001 mg) of protease inhibitor on filter paper for 7 days. Controls were fed filter paper without protease inhibitor. Filter paper was moistened every other day with 50 µl sodium phosphate buffer (0.1M, pH 7.0).

Dissections and Protein Extraction

After 7-day feeding assays, whole guts were collected along with salivary glands in 100mM sodium phosphate buffer (pH 7.0). Gut tissues were homogenized in ice-cold phosphate buffer and centrifuged at $10,000 \times g$ for 10 min at 4°C. Supernatant was used for enzyme activity assays and for separation by sodium dodecyl sulfate-polyacrylamide gel electrophoresis (SDS-PAGE). Protein content was estimated at 562 nm using the Bicinchoninic acid method with bovine serum albumin standard following the manufacturer's protocol (Thermo Scientific) using a BioTek PowerWave microplate reader (Winooski, VT, USA).

In Vitro Saccharification Assays

Approximately 50 worker termites were dissected per experimental replicate to collect guts. Whole guts from replicate termite lab colonies were isolated and homogenized in 1.5 ml phosphate buffer (0.1M, pH 7.0), centrifuged at $10,000 \times g$ for 10 min at 4°C, and the clarified supernatant was saved for use in *in vitro* assays (Scharf et al., 2011a). Feedstocks were milled to 0.5 mm particle size in a UDY cyclone sample mill for use in assays. Saccharification assays were done in 750 μl volumes containing 150 μl gut supernatant (i.e., five gut equivalents) and 600 μl HEPES buffer (0.1M, pH 7.0). The assays were contained within vented 1.5 ml Eppendorf tubes in a shaking incubator set to 37°C and 200 rpm. Tubes were vented by melting a pinhole in the assay tube lids. All assay conditions are based on extensive preliminary work showing circum-neutral gut pH in R. flavipes, digestive enzyme pH optima around 7, and enzyme stability at 37°C (Zhou et al., 2007, 2010; Tartar et al., 2009; Coy et al., 2010; Scharf et al., 2010, 2011a; Sethi et al., 2013). A range of feedstock amounts was compared (0.2, 0.4, 0.8, 1, and 2% w/v). Assays were incubated for 4 h and reactions stopped with a final concentration of 4mM ethylenediaminetetraacetic acid (EDTA). EDTA was added to inhibit metalloenzymes, like laccases (Coy et al., 2010), and it also stabilizes color formation for the glucose detection reagent, which includes hydrogen peroxide that could be acted upon by gut metalloenzymes (Scharf et al., 2011a). Liberated glucose was quantified using a colorimetric glucose detection kit that utilizes glucose mutarotase biochemistry (Wako Chemicals USA; Catalog No. 298-65701) against a glucose standard curve (5–0.3125mM). Two experimental replicates were performed for each feedstock using termites from the same colony. Blanks that included feedstock without gut homogenate for each w/v% were run in parallel and used to correct for non-enzymatic saccharification (negligible). Gut preparations were also tested for the presence glucose and it was negligible (Scharf et al., 2011a and results not shown).

SDS-PAGE and Protein Identification by LC-MS/MS

Gut homogenates from termites fed on paper, CS, and SR were combined 1:1 with sample loading buffer, loaded at 40 μg/lane and separated on 8% or 4–20% SDS-PAGE gels at a constant voltage of 120 V for 1.0 or 1.5 h. Gels were stained in 0.3% Coomassie blue for 1–2 h and destained in 40% methanol with 10% acetic acid for 1 h and 10% Methanol + 5% acetic acid overnight. Stained gels were photographed using ChemiDoc™ XRS + system (Bio-Rad, Hercules, SA). Coomassie stained bands were quantified using ImageJ (Rasband, 2014). Bands of interest were cut from gels and sequenced at the Indiana University–Purdue University

proteomics core, Indianapolis, IN, USA. Briefly, gel bands were destained, reduced with 10mM dithiothreitol in 10mM ammonium bicarbonate, and then alkylated with 55mM iodoacetamide (prepared in 10mM ammonium bicarbonate). Alkylated samples were digested with trypsin (Promega) overnight at 37°C. Digested peptides were extracted from gel slices first with 50% ACN (acetonitrile) + 49.9%water + 0.1% TFA (trifluoro acetic acid), and then 99.9% ACN + 0.1% TFA. Trypsin-cleaved peptides were injected onto a C18 column (Thermo-Fisher Scientific LTQ and Surveyor HPLC system). Peptides were eluted with a linear gradient of 5–35% ACN (in water with 0.1% FA) developed over 60 min at room temperature, at a flow rate of 50 μl min⁻¹, and effluent was electro-sprayed into the LTQ mass spectrometer. Blanks were run prior to samples to ensure there was no significant signal from solvents or the column. Band identification was done using the Sequest™ algorithm against the translated host and symbiont metatranscriptomes of R. flavipes (Genbank accession Nos. PRJNA275308, FL634956-FL640828, and FL641015-FL645753) (Tartar et al., 2009; Rajarapu et al., 2015) and the predicted proteome of Zootermopsis nevadensis (UP000027135).

Immunoblotting

For western blotting, 8% SDS-gels were run at the same conditions mentioned above and proteins transferred to nitrocellulose membrane at 100 V for 1 h. For dot blots, five serial dilutions of gut protein (20–0.032 μg) per sample were spotted directly onto nitrocellulose membranes and allowed to dry. Membranes were blocked with 3% non-fat dry milk overnight, incubated with primary rat myosin antibody in a ratio of 1:4,000 for 1 h, washed 3× for 10 min with Tris buffered saline (TBS), incubated with horse radish peroxidase conjugated anti-rat G goat antibody secondary antibody (Bio-Rad) in a ratio of 1:5,000 for 30 min, washed 3× for 10 min with TBS; and stained with SuperSignal™ West Pico chemiluminescent substrate (Life Technologies). Blots were visualized using a Chemi-Doc XRS system (Bio-Rad).

Enzyme Assays
Carbohydrate Degrading Enzymes
Cellulose and Hemicellulose Substrates

Assays to estimate the activity of potential cellulases and hemicellulases were performed using seven model substrates (**Table 1**). Endoglucanase and hemicellulase activities were assayed using carboxymethyl cellulose, beech xylan, and wheat arabinan using the dinitrosalicylate (DNSA) detection method as described (Zhou et al., 2007; Karl and Scharf, 2015). Reaction mixtures containing 10 μl gut supernatant and 90 μl of substrate solutions were incubated for 1 h at 30°C. Reactions were stopped by adding 100 μl DNSA solution and incubated at 100°C for 20 min. Absorbance was measured at 540 nm after cooling plates on ice for 15 min. Glucose released was determined against a glucose standard curve (5–0.625mM).

p-Nitrophenyl Substrate Assays

All p-nitrophenyl carbohydrate substrates were assayed as described earlier with slight modifications (Karl and Scharf, 2015). Briefly, stock solutions of p-nitrophenyl substrates were

TABLE 1 | Enzyme activities assayed and their respective substrates.

Enzyme	Substrate	Solubility	Final assay concentration[a]	Reference
Carbohydrate-active enzymes				
Cellulase	Carboxymethyl cellulose	Nanopure water	1%	Zhou et al. (2007)
Hemicellulase	Wheat arabinoxylan	Nanopure water	2%	Zhou et al. (2007)
	Beechwood xylan	Nanopure water	2%	Zhou et al. (2007)
β-Glucosidase	p-Nitrophenyl (pNP) glucopyranoside	Nanopure water	4mM	Karl and Scharf (2015)
β-Cellobiohydrolase/exoglucanase	pNP-cellobioside	Dimethyl sulfoxide	4mM	Karl and Scharf (2015)
β-Arabinosidase	pNP-arabinoside	Dimethyl sulfoxide	4mM	Karl and Scharf (2015)
β-Mannosidase	pNP-mannoside	Dimethyl formamide	4mM	Karl and Scharf (2015)
Candidate lignase and detoxification/antioxidant enzymes				
Phenoloxidase	Pyrogallol	Nanopure water	4mM	Karl and Scharf (2015)
Esterases	pNP-acetate	Acetonitrile	4mM	Karl and Scharf (2015)
	Naphthyl propionate	Acetonitrile	25mM	Wheeler et al. (2010)
Feruloyl esterase	pNP-trans ferulate	Dimethyl sulfoxide	1mM	Mastihuba et al. (2002)
Glutathione-S-transferase	1-Chloro-3,4-dinitrobenzene	Acetone	1mM	Tang et al. (1998)
Superoxide dismutase	Nitro blue tetrazolium chloride	Nanopure water	0.05mM	Ewing and Janero (1995)
Glutathione peroxidase	Cumene hydroperoxide, glutathione	NA	–	Cayman International

[a]*Final assay concentration in sodium acetate buffer (0.1M, pH 7.0).*

prepared in their respective solvents (**Table 1**), and 4mM working solutions were prepared fresh in sodium acetate buffer (0.1M, pH 7.0). Reaction mixtures had 10 µl of gut supernatant + 90 µl of substrate solution in a reaction volume that was made to 200 µl with sodium acetate buffer (0.1M, pH 7.0). Reactions were read kinetically for 20 min at 410 nm. Mean velocities were calculated using Gen5 data analysis software (BioTek) and were used for calculating specific activity as micromole p-nitrophenol released per minute per milligram of protein. The molar extinction coefficient of p-nitrophenol is 0.6605mM OD^{-1} (Zhou et al., 2007).

Accessory Enzymes

Esterases

Esterase activity was measured using two different substrates, naphthyl propionate and p-nitrophenyl acetate. Naphthyl propionate assays were performed as described earlier (Wheeler et al., 2010). Reaction mixtures consisted of 10 µl gut supernatant, 2 µl of 25mM naphthyl propionate in ACN and 188 µl sodium phosphate buffer (100mM, pH 7.5). Reactions were incubated for 10 min at room temperature and stopped with 50 µl of 0.3% Fast Blue BB dissolved in nanopure water containing 3.5% SDS. Reactions proceeded at 30°C for 15 min and were read as endpoints at 600 nm. Specific activity was measured as the amount of α-naphthol released against a naphthol standard curve (5–0.625mM). p-Nitrophenyl acetate assays were performed as detailed in the preceding section. Feruloyl esterase activity was measured using p-nitrophenyl trans-ferulate as described by Mastihuba et al. (2002) with slight modifications. The assay was adapted to a 96-well microplate wherein the reaction mixture consisted of 10 µl gut supernatant and 190 µl of 1mM substrate. Working solution of 1mM substrate was prepared in an emulsion of phosphate buffer with Tween-20 (3 µl ml^{-1}) and DMSO (12 µl ml^{-1}). The phosphate buffer emulsion was added to the appropriate amount of substrate stock and mixed immediately to prevent precipitation. The substrate solution was prepared fresh and used within an hour. Absorbance was read kinetically at

410 nm for every 10 s for 5 min. Substrate reactions and enzyme blanks were run in parallel.

Phenoloxidase, Superoxide Dismutase, Glutathione Peroxidase, and Glutathione-S-Transferase

Phenoloxidase activity was measured using the substrate pyrogallol as described previously (Coy et al., 2010). Superoxide dismutase was assayed as detailed previously (Ewing and Janero, 1995). Briefly, the reaction mixture consisted of 25 µl of gut homogenate, 50µM nitro-blue tetrazolium (NBT) chloride, and 98µM NADH in 50mM sodium phosphate buffer (pH 7.4) containing 0.1mM EDTA. Reactions were initiated by the addition of 25 µl of 33µM phenazine methylsulfate in 50mM phosphate buffer. Absorbance of reduced NBT was measured kinetically at 560 nm for 5 min, with the extinction coefficient for reduced NBT of 15×10^6 mM^{-1} cm^{-1}. Glutathione-S-transferase activity was measured using 1-chloro-2,4-dinitrobenzene (CDNB) as the substrate (Tang et al., 1998). Assay mixtures of 1mM CDNB, 5mM reduced glutathione in sodium phosphate buffer (100mM, pH 7.4) were prepared fresh. The reaction was started by adding 225 µl assay mixture to 10 µl gut supernatant. Absorbance of the resultant conjugate, dinitrophenyl glutathione, was read kinetically at 324 nm for 10 min. The dinitrophenol extinction coefficient used was 9.5 mM^{-1} cm^{-1}. Glutathione peroxidase activity was measured using a glutathione peroxidase assay kit (Cayman Chemical, Ann Arbor, MI, USA) following manufacturer protocols.

Statistics

Significant differences in glucose liberation from saccharification assays were determined by non-parametric Kruskal–Wallis tests. Statistical differences in protein band densities were determined by comparing means of three replicates by Kruskal–Wallis tests. Activities of different enzymes were tested for significant differences using one way ANOVA with Tukey mean separation tests to identify diet effects. Statistics were performed at α of 0.05 in JMP statistical software.

RESULTS

In Vivo Feeding and *In Vitro* Saccharification Assays

Termites consumed similar amounts of cellulose paper, CS, and SR diets but their body weights did not change significantly on any of the diets ($p > 0.05$; Figure S1 in Supplementary Material). Termite gut homogenates were able to saccharify CS and SR at significant detectable levels; however, CS was saccharified at 150–220× greater levels than SR across the loading rates tested (**Figures 1A,B**). Background glucose levels in termite gut homogenates were negligible and therefore did not interfere with glucose detection. Glucose release was not significantly affected by the % w/v of feedstock present in assays for both CS and SR ($p > 0.05$; Figure S2 in Supplementary Material). There was also an unexpected reduction in detectable glucose levels in incubations lasting longer than 4 h for both feedstocks (Figures S2 and S3A in Supplementary Material), and co-incubation of assays with the antimicrobial compound sodium azide did not significantly impact glucose release levels over time (Figure S3B in Supplementary Material).

Protein Profiles of Termite Guts Fed on CS and SR

sodium dodecyl sulfate-polyacrylamide gel electrophoresis (SDS-PAGE) protein profiles of termite guts fed on paper, CS, and SR were similar, except for 250, 15, and 10 kDa protein bands that were uniquely associated with SR feeding (**Figure 2A**) These PAGE results were verified by densitometry analyses of band intensities from independent biological replicate gels (**Figure 2B**), which showed only the 250 kDa band (myosin heavy chain) was significantly different across treatments. Tandem mass spectrometry was performed to identify protein bands of interest.

Proteins were identified by querying against termite transcript and protein databases previously generated from our lab and elsewhere. Identity was confirmed based on coverage scores of sequenced peptides relative to reference proteins, with higher coverage scores indicating higher confidence of identity (**Table 2**). The significant SR-associated proteins were identified as myosin heavy chain (250 kDa), Cu–Zn superoxide dismutase (15 kDa), and lipocalin/fatty acid binding protein (10 kDa). Bands near the predicted molecular weight of putative lignocellulose degrading enzymes that were relatively abundant after CS and SR feeding were also sequenced and corroborated with previous proteomic studies in *R. flavipes* (Sethi et al., 2013) (**Table 2**). However, densitometry analyses of SDS-PAGE separated proteins showed only the 250 kDa (myosin heavy chain) band was significantly different across treatments.

Myosin Immunoblotting

The identity of the 250 kDa SR-abundant protein identified above as myosin by tandem MS was further verified by immunoblotting (**Figure 3**). Western and dot blots with a myosin-specific antibody together showed more intense signals in guts from termites that had fed on SR. Myosin patterns in guts of termites fed with soybean Kunitz protease inhibitor were different than those of SR-fed guts (Figure S4 in Supplementary Material), indicating the observed myosin result is not linked to soybean-associated protease inhibition.

Impacts of CS and SR Feeding on Activity of Candidate Lignocellulase and Accessory Enzymes

In vitro gut lignocellulase activities were mostly similar across the CS, SR, and paper-feeding treatments (**Table 3**). However, among

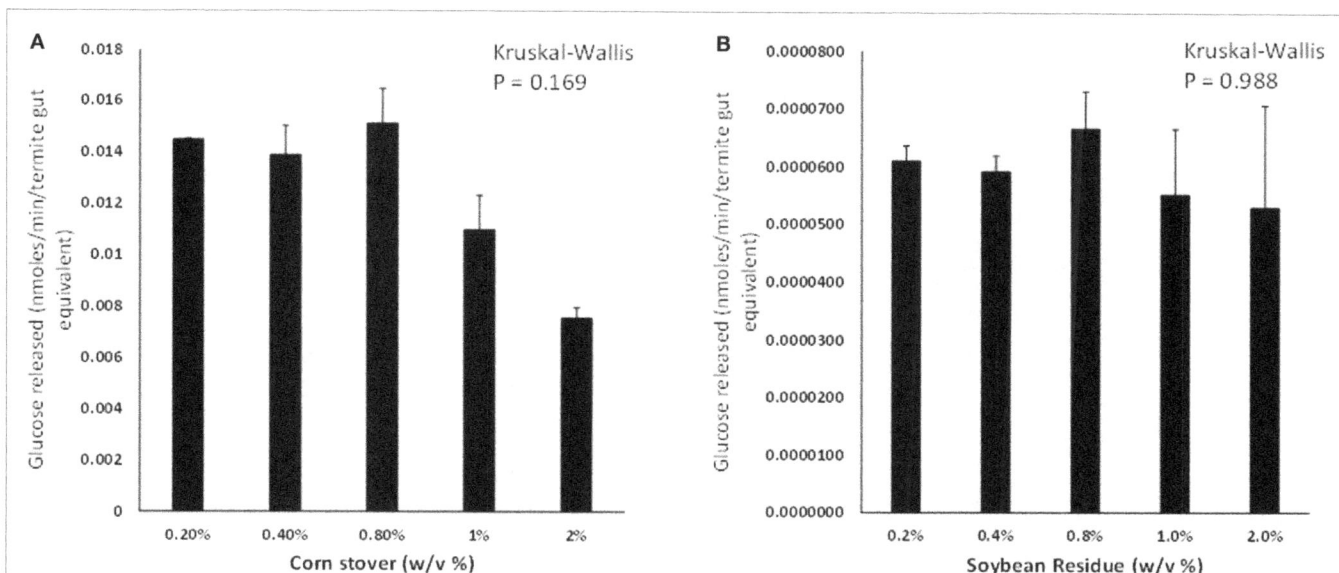

FIGURE 1 | Glucose liberation (nanomoles of glucose liberated per minute per termite gut equivalent) from (A) corn stover and (B) soybean residue by termite gut protein extracts *in vitro*. Data points represent the average ± SE of two experimental replicates using termite gut preparations from the same colony. Results of global Kruskal–Wallis analyses are shown indicating no effects of feedstock w/v% loading rates.

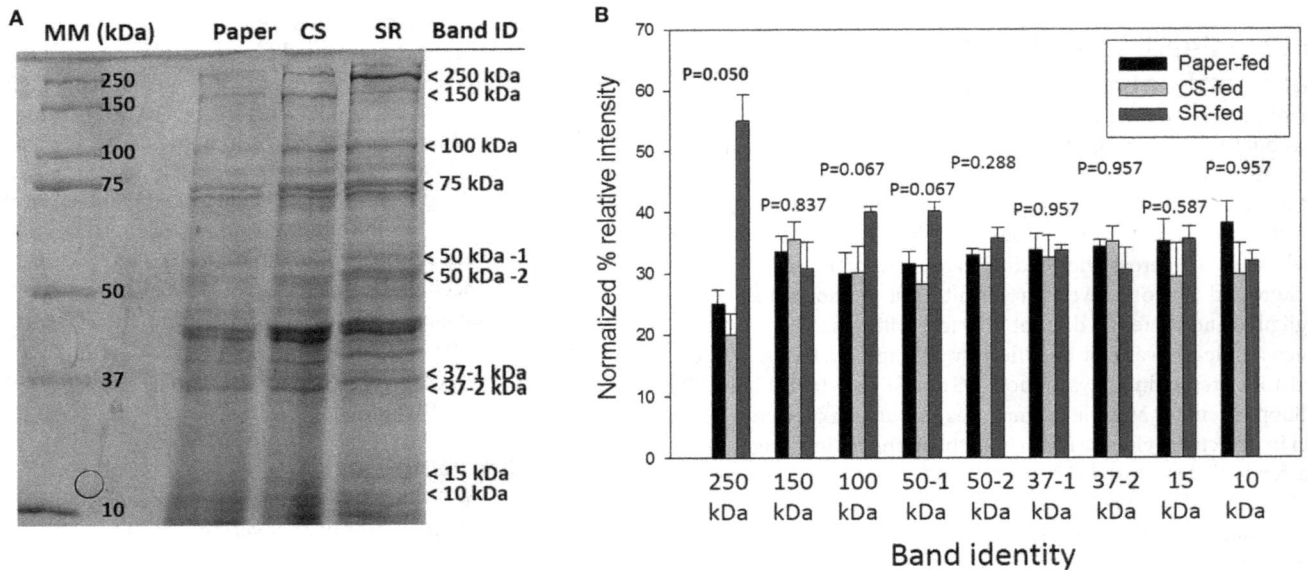

FIGURE 2 | (A) Sodium dodecyl sulfate-polyacrylamide gel electrophoresis (SDS-PAGE) of whole gut protein extracts from termites fed on paper, corn stover (CS) or soybean residue (SR). A representative 4–20% gradient gel stained with 0.1% Coomassie blue R-250 is shown. Arrows indicate bands sequenced for identification (**Table 1**). **(B)** Densitometry quantification represented as mean density of three biological replicates (±SEM) of SDS-PAGE separated proteins indicated by arrows in panel **(A)**.

TABLE 2 | Identification of sodium dodecyl sulfate-polyacrylamide gel electrophoresis protein bands by tandem MS analysis.

Molecular mass (kDa)	Paper	Corn stover (CS)	Soybean residue (SR)
250			Myosin (62.49)
75	Hexamerin II (179)	Hexamerin II (174.5)	Hexamerin II (184.4)
50 (1)	Endogenous cellulase (276.5) β-Actin (105)	α-Tubulin (125.9)	α-Tubulin (133.8)
50 (2)		Endogenous cellulase (416.9) β-Actin (273.8)	Endogenous cellulase (512.2) β-Actin (359.2)
37	Endogenous cellulase (215.1) β-Actin (144.6) Aldo-keto reductase (131.9) Glyceraldehyde-3-phosphate (111.3)	Endogenous cellulase (236.2) Aldo-keto reductase (164.6) β-Actin (117) Glyceraldehyde-3-phosphate (97.4)	Endogenous cellulase (131) β-Actin (95) Aldo-keto reductase (212.3) Glyceraldehyde-3-phosphate dehydrogenase (128.1)
15			Cu-Zn superoxide dismutase (87.10)
10			Lipocalin/fatty acid binding protein (86.17)

Respective peptide sequences for each band are provided in Table S1 in Supplementary Material. Numbers in parentheses are peptide identification numbers.

carbohydrate-active enzymes, cellobiohydrolase/exoglucanase (CBH) activity was higher in CS relative to SR and paper-fed guts ($p = 0.0006$, one way ANOVA). β-Mannosidase activity was lower in SR relative to paper-fed guts ($p = 0.047$, one way ANOVA), and there were no differences in β-xylosidase and β-arabinosidase activities across feeding treatments. Among the accessory enzymes studied, esterases had lower activity in SR relative to paper-fed guts ($p = 0.0003$, one way ANOVA); whereas, glutathione peroxidase had higher activity relative to both paper- and CS-fed guts ($p = 0.0125$, one way ANOVA). No differences were observed for other accessory enzyme activities investigated.

DISCUSSION

Saccharification is the major limiting step in biofuel production from lignocellulosic biomass. Existing pretreatment strategies to degrade the lignocellulose matrix include biological, physical, chemical, and physio-chemical methods that are expensive and in many cases produce hazardous wastes. Due to these disadvantages, there has been a need to develop efficient and environmentally favorable pretreatment methods (Yang and Wyman, 2008). Enzymatic strategies modeled after enzymatic mechanisms of organisms specializing on wood diets have utility for pretreatment of lignocellulosic biomasses (Ke et al., 2011). The termite studied here, *R. flavipes*, has been an excellent model system for identifying potential enzymes involved in the breakdown of both forest and agricultural feedstocks *via* the process of diet adaptation (Scharf, 2015a). Through preceding work, we studied the metatranscriptome profiles of *R. flavipes* guts fed on CS and SR relative to paper in an attempt to identify novel transcripts responding to the respective diets; however, while results showed shifts in gut microbe profiles, few differences in the functional profiles of termite guts fed on CS and SR relative to paper were identifiable (Rajarapu et al., 2015). In the current

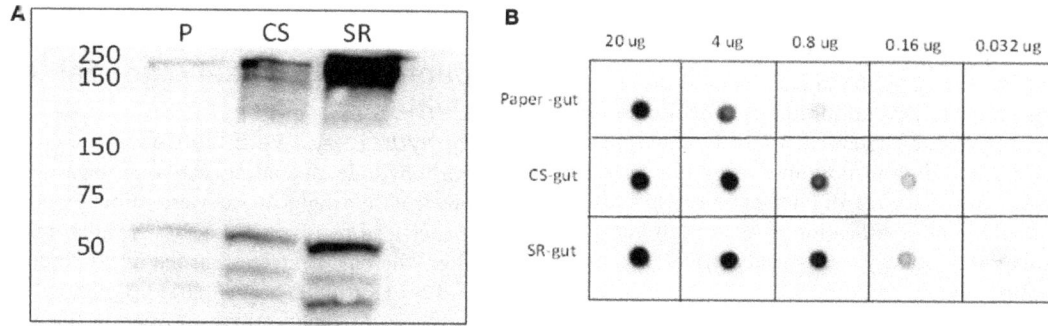

FIGURE 3 | (A) Western blots and **(B)** dot blots of whole gut protein extracts from termites fed on paper (P), corn stover (CS) and soybean residue (SR). Blots were incubated with primary rat monoclonal myosin antibody (1:4,000) and secondary goat anti-rat horse radish peroxidase-conjugated antibody (1:5,000). Hybridizations were visualized *via* chemiluminescence. Numbers across the top in panel **(B)** indicate protein loadings per dot.

TABLE 3 | Activities of carbohydrate-active and accessory enzymes (μmol/min/mg) in whole-gut homogenates prepared from termites fed for 7 days on paper, corn stover (CS), or soybean residue (SR).

Activity	Substrate	Paper-fed[1]	CS-fed[1]	SR-fed[1]
Carbohydrate-targeted activities				
Endoglucanase	Carboxymethylcellulose	2.4 ± 0.4[a]	2.34 ± 0.8[a]	2.10 ± 1.8[a]
Cellobiohydrolase/exoglucanase	pNP cellobioside	19.28 ± 1.5[a]	**29.64 ± 3.1[b]**	4.05 ± 2.0[a]
β-Glucosidase	pNP glucopyranoside	583.48 ± 11.7[a]	650.40 ± 38.5[a]	538.75 ± 46.5[a]
Xylanase	Beechwood xylan	3.56 ± 0.0[a]	3.44 ± 0.0[a]	3.41 ± 0.0[a]
Endoarabinanase	Wheat arabino xylan	3.66 ± 0.1[a]	3.55 ± 0.0[a]	3.44 ± 0.1[a]
β-Arabinosidase	pNP-arabinoside	2.34 ± 0.6[a]	4.34 ± 0.4[a]	2.82 ± 0.5[a]
β-Mannosidase	pNP-mannoside	14.53 ± 2.1[b]	12.30 ± 1.8[b]	**7.41 ± 0.4[a]**
Candidate detoxification and antioxidant "accessory" activities				
Phenoloxidase	Pyrogallol	17,399.1 ± 7,155.3[a]	15,771.0 ± 5,703.3[a]	10,521.52 ± 4,590.1[a]
Esterase	pNP-acetate	604.70 ± 75.9[a]	522.0 ± 52.1[a]	405.82 ± 76.7[a]
	Naphthyl propionate	4,160.0 ± 137.1[b]	4,425.0 ± 11.3[b]	**3,227.5 ± 92.4[a]**
Feruloyl esterase	pNP trans ferulate	2,130.21 ± 198.2[a]	1,774.0 ± 384.1[a]	1,571.72 ± 242.8[a]
Superoxide dismutase	Nitroblue tetrazolium chloride	164.28 ± 20.3[a]	133.02 ± 24.7[a]	109.3 ± 38.7[a]
Glutathione-S-transferase	1-Chloro-3,4-dinitrobenzene	113.98 ± 12.9[a]	150.75 ± 14.4[a]	162.6 ± 18.6[a]
Glutathione peroxidase	Cumene hydroperoxide	725.99 ± 118.0[a]	1,283.28 ± 450.1[a]	**2,733.83 ± 328.6[b]**

[1]Activities are expressed as mean μmol/min/mg ± SEM (n = 3).
[a,b]Values within rows with the same superscript letters are not significantly different (one way ANOVA, α = 0.05).
Bold text denotes values that are significant.

study, we evaluated protein-level saccharification and changes in response to feeding on the same CS and SR feedstocks. Similar to our prior metratranscriptome study, termites remained active and survived well on the diets for a 7-day period, displaying their flexibility to accept novel lignocellulosic diets. Saccharification assays showing gut-dependent release of glucose from the CS and SR feedstocks agree with the high degrees of survivability seen in bioassays. Overall gut protein profiles and activity of potential lignocellulose degrading enzymes were mostly similar across the CS, SR, and paper-feeding treatments, but some key differences as detailed below were still observed.

Termite Guts Saccharify Feedstocks *In Vitro*

Findings showed a clear ability of the *R. flavipes* gut to saccharify CS and SR *in vitro*. These findings are supported by *in vivo* bioassays in which termites had essentially 100% survival for 1 week when held with CS and SR as their only food sources [present study and Rajarapu et al. (2015)]. However, in *in vitro* studies, CS released considerably more glucose than SR, which correlates with the apparent proteolytic impacts of SR on protein composition (discussed in the next section), as well as disappearance of gut microbiota in response to SR feeding (Rajarapu et al., 2015). There also was an unexpected reduction in detectable glucose levels in incubations lasting longer than 4 h for both feedstocks, but co-incubation of assays with the antimicrobial compound sodium azide did not significantly impact glucose release levels over time. These findings point toward conversion of the released glucose and/or inhibition of the saccharification process with longer incubation times, rather than microbial consumption of released glucose in assay reactions. Released glucose can be further converted to byproducts not detectable by the colorimetric method we used to detect glucose (Popoff and Theander, 1972, 1976) or glucose might be conjugated with secondary plant chemicals present in the feedstocks. Production of inhibitory products during pretreatment is another major concern in biofuel production. Several inhibitory compounds have been identified in physical and chemical pretreatments (Jönsson et al., 2013). Additionally, lignin monomers have the potential to inhibit cellulases (Qin et al., 2016). Further research studying metabolite profiles of the saccharification assay mixture can shed more light on potential inhibitory compounds formed during of CS and SR processing.

Gut Protein Profiles

The rationale behind protein profiling studies was that differentially expressed proteins identified would potentially be host (termite) proteins responsive to dietary components, or symbiont proteins showing differential expression in correlation with fluctuating symbiont populations. Overall gut protein profiles of termites fed on CS and SR assessed by SDS-PAGE were very similar except for minor differences in 10 and 15 kDa bands (putatively identified as a lipocalin and a superoxide dismutase), and a 250-kDa band identified as myosin that was unique to SR-fed guts. The lack of changes in carbohydrate-active proteins here is in strong agreement with prior metatranscriptome sequencing performed after CS and SR feeding that also did not identify changes in carbohydrate-active enzyme encoding transcripts (Rajarapu et al., 2015). Myosin belongs to a large superfamily of proteins involved in muscle contraction and cytoskeletal structure (Weiss and Leinwand, 1996). While there is little information on the functional roles of myosin in insects, significant upregulation of myosin in the soldier castes of *R. flavipes* and *C. formosanus* termites was previously observed, suggesting a role for myosin in soldier musculature or cell differentiation (Tarver et al., 2012). Also, in *Drosophila melanogaster*, the *myosin heavy chain* protein is involved in cell shape and motility during development (Young et al., 1993). Given the structural role of myosin, it is possible that abundance of myosin in SR-fed termite guts might be due to breakdown of termite tissue or symbiota in the hindgut, as suggested by previous research (Rajarapu et al., 2015). Alternatively, myosin liberation might result from direct or indirect effects of soybean anti-herbivory defense mechanisms, including protease inhibitors. We tested this hypothesis by feeding commercially available Kunitz trypsinase inhibitor to termites in a 7-day bioassay. Total protein profiles of termite guts fed on protease inhibitor were different than those of guts fed a SR diet (Figure S4 in Supplementary Material) suggesting that complex proteolysis-related mechanisms, or their inhibition (Chu et al., 2015), might be involved in increasing the myosin detectability in SR fed termite guts. Nonetheless, this investigation of protein profiles demonstrates an effect of SR digestion on termite gut tissue and/or microbial physiology, suggesting possible implications for SR processing by microbes in bioreactor environments. This "degradation" hypothesis may explain the discrepancy between saccharification potential of CS and SR by termite gut protein preparations, but not in most model substrate activities as detailed in the following section. Another possible explanation is

the relative recalcitrance of SR and CS feedstocks relative to each other and the model substrates tested.

Impacts on Potential Lignocellulase Activities

Carbohydrate-Active Enzymes

All carbohydrate degrading activities investigated except cellobiohydrolase/exoglucanase were similar across the CS, SR, and paper-feeding treatments. These results are similar to earlier findings wherein *R. flavipes* was fed five different diets including CS and SR (Karl and Scharf, 2015). Cellulose is synergistically degraded in *R. flavipes* by endogenous (host) endoglucanase and β-glucosidase activities, with additive effects by protist (symbiont) CBH/exoglucanase activity (Scharf et al., 2011a; Sethi et al., 2013). Cellulose exists in crystalline and amorphous forms due to differences in hydrogen bonds within the polymer. Endoglucanases hydrolyze internal glycosidic bonds of the cellulose polymer whereas CBHs (also known as exoglucanases) act on amorphous cellulose. Nevertheless, CBH from yeast, fungi, and marine invertebrates can also hydrolyze crystalline cellulose (Teeri, 1997). β-glucosidase acts on the hydrolyzates of CBH, releasing monomeric glucose units. Cellulose substrates from different sources vary in amorphous content and crystallinity, as well as size and shape of crystallites (Montanari et al., 2005). The higher CBH activity measured in CS-fed guts might be due to qualitative differences in cellulose composition between CS and SR. Additionally, CBH (GHF7) enzymes in *R. flavipes* are primarily produced by gut symbionts (Tartar et al., 2009), and higher activity of CBH in CS-fed guts therefore suggests significant participation by symbionts in CS saccharification. In contrast, endoglucanase and β-glucosidase, which are primarily host-derived (Tartar et al., 2009; Scharf et al., 2010; Zhou et al., 2010), had similar activity across guts from termites fed on different diets. An alternative explanation for the observed differences in CBH activity between feedstocks is that, as noted above, protein degradation/inhibition was more substantial with SR feeding and/or SR impacts on symbiont mortality were more significant.

Access to cellulose in lignocellulose is increased with degradation of hemicellulose surrounding the cellulose polymer. Hemicellulose is a heterogeneous polysaccharide frequently containing mannose, xylose, glucose, galactose, and arabinose. Interestingly, gut hemicellulase activity was similar in termites fed on CS and SR, except for β-mannosidase that was reduced with SR feeding. In lower termites like *R. flavipes*, β-mannosidases are likely contributed by hindgut protists (Brune, 2014). A decrease in β-mannosidase activity could be thus due to a decrease in protist numbers in termite guts fed on SR (Rajarapu et al., 2015). It was hypothesized that xylanase, endoarabinase, and β-arabinosidase activities would be higher in CS-fed termite guts, as hemicellulose in the Poaceae is generally composed of a xylanopyranosyl backbone with arabinofuranosyl side branches. However, there were no significant differences observed in these activities across feeding treatments.

Detoxification and Antioxidant (Accessory) Enzymes

Unlike cellulose and hemicellulose, lignin is a polymer of phenyl propane units linked by C–C or C–O bonds between the three monomers coumaryl, coniferyl, and sinapyl alcohol (Chen, 2014). There is significant independent evidence suggesting that termites can degrade lignin polymers, but the mechanisms of degradation remain elusive (Cookson, 1987, 1988; Geib et al., 2008). From previous studies, enzyme-coding transcripts responding to lignin diets were identified, which included esterase, feruloyl esterase, laccase/phenoloxidase, glutathione-S-transferase, superoxide dismutase, and glutathione peroxidase (Sethi et al., 2013), which were activities tested in this study. Among these activities, glutathione peroxidase activity was higher in SR-fed relative to paper- and CS-fed guts; whereas, naphthyl propionate-specific esterases had lowest activity in SR-fed guts. Lignin is esterified onto the sugar side chains of hemicellulose within the lignocellulose structure. Esterases can thus play a role in breaking the bonds between the lignin polymer and hemicellulose. However, the CS and SR tested in this study have similar lignin quantities around 6–7 mg/100 mg cell wall (Rajarapu et al., 2015), and thus, the difference in esterase activity measured between CS and SR feeding treatments might be due to qualitative differences in lignin composition between CS and SR, or possibly due to aromatics distinct to soybeans (e.g., isoflavones, anthocyanins, phenolics). Alternatively, esterases in *R. flavipes* are produced by both host and symbiota (Karl and Scharf, 2015). Consequently, the decrease in esterase activity with SR feeding might be due to a loss of gut symbiota. Along with esterases, peroxidases might be involved in lignin degradation either directly by radicalizing bonds in the lignin polymer or indirectly by quenching the potential oxidative stress generated by lignin disassociation/depolymerization (Sethi et al., 2013). Lastly, GPXs are cytosolic enzymes involved in detoxification of hydrogen peroxide with assistance from reduced glutathione. GPX could therefore neutralize oxidative stress created during saccharification. Alternatively, it is possible that GPX could potentially saccharify SR *in vitro*, but this is not consistent with the known mechanism of action for GPX enzymes.

CONCLUSION

This study assessed the simultaneous enzymatic pretreatment and saccharification of agricultural feedstocks *in vivo* and *in vitro* by the termite gut. Our overall objective was to identify termite gut proteins that might specifically degrade lignocellulose in CS and SR to fermentable glucose. Our findings demonstrate the ability of the lower termite gut to saccharify CS and SR, verifying the utility of this system as a source of enzymes for enzymatic pretreatment and/or saccharification of agricultural feedstocks. Additionally, we identified lignocellulase activities that might be responsible for degrading both CS and SR feedstocks, also supporting the diet-adaptation hypothesis as proposed earlier (Karl and Scharf, 2015). Among the potential lignocellulase and accessory activities assayed, CBH and glutathione peroxidase activity were higher, respectively, in termite guts fed on CS and SR. The cytoskeletal/structural protein myosin was also more abundant in termite guts fed on SR. Myosin-dependent degradation of SR does not seem likely, and its increased presence is more likely indicative of a physiological stress response to deleterious soybean secondary chemicals known to interfere with a number of physiological processes (Weiss and Leinwand, 1996). This outcome highlights the potential effects of SR digestion on biological systems that likely would extend to fermentation chamber microenvironments.

AUTHOR CONTRIBUTIONS

SR performed the research, analyzed the data, and wrote the manuscript. MS designed the research, interpreted the data, and edited the manuscript.

ACKNOWLEDGMENTS

The authors thank Indiana University Purdue University proteomics core for sequencing and identification of protein bands and Jesse Hoteling for maintaining termite colonies.

FUNDING

This work was supported by National Science Foundation grant no. CBET 1233484 awarded to MS, and the O.W. Rollins/Orkin Endowment in the Department of Entomology at Purdue University.

REFERENCES

Aspeborg, H., Coutinho, P. M., Wang, Y., Brumer, H., and Henrissat, B. (2012). Evolution, substrate specificity and subfamily classification of glycoside hydrolase family 5 (GH5). *BMC Evol. Biol.* 12:186. doi:10.1186/1471-2148-12-186

Brune, A. (2014). Symbiotic digestion of lignocellulose in termite guts. *Nat. Rev. Microbiol.* 12, 168–180. doi:10.1038/nrmicro3182

Cairo, J., Leonardo, F. C., Alvarez, T. M., Ribeiro, D. A., Büchli, F., Costa-Leonardo, A. M., et al. (2011). Functional characterization and target discovery of glycoside hydrolases from the digestome of the lower termite *Coptotermes gestroi*. *Biotechnol. Biofuels* 4, 50. doi:10.1186/1754-6834-4-50

Chen, H. (ed.) (2014). "Chemical composition and structure of natural lignocellulose," in *Biotechnology of Lignocellulose* (Beijing: Springer), 25–71.

Chu, C. C., Zavala, J. A., Spencer, J. L., Curzi, M. J., Fields, C. J., Drnevich, J., et al. (2015). Patterns of differential gene expression in adult rotation-resistant and wild-type western corn rootworm digestive tracts. *Evol. Appl.* 8, 692–704. doi:10.1111/eva.12278

Cookson, L. (1987). 14C-lignin degradation by three Australian termite species. *Wood Sci. Technol.* 21, 11–25.

Cookson, L. (1988). The site and mechanism of 14C-lignin degradation by *Nasutitermes exitiosus*. *J. Insect Physiol.* 34, 409–414. doi:10.1016/0022-1910(88)90111-4

Coy, M., Salem, T., Denton, J., Kovaleva, E., Liu, Z., Barber, D., et al. (2010). Phenol-oxidizing laccases from the termite gut. *Insect Biochem. Mol. Biol.* 40, 723–732. doi:10.1016/j.ibmb.2010.07.004

Cragg, S. M., Beckham, G. T., Bruce, N. C., Bugg, T. D., Distel, D. L., Dupree, P., et al. (2015). Lignocellulose degradation mechanisms across the Tree of Life. *Curr. Opin. Chem. Biol.* 29, 108–119. doi:10.1016/j.cbpa.2015.10.018

Ewing, J. F., and Janero, D. R. (1995). Microplate superoxide dismutase assay employing a nonenzymatic superoxide generator. *Anal. Biochem.* 232, 243–248. doi:10.1006/abio.1995.0014

Geib, S. M., Filley, T. R., Hatcher, P. G., Hoover, K., Carlson, J. E., del Mar Jimenez-Gasco, M., et al. (2008). Lignin degradation in wood-feeding insects. *Proc. Natl. Acad. Sci. U.S.A.* 105, 12932–12937. doi:10.1073/pnas.0805257105

Horn, S. J., Vaaje-Kolstad, G., Westereng, B., and Eijsink, V. G. (2012). Novel enzymes for the degradation of cellulose. *Biotechnol. Biofuels* 5, 45. doi:10.1186/1754-6834-5-45

Jönsson, L. J., Alriksson, B., and Nilvebrant, N.-O. (2013). Bioconversion of lignocellulose: inhibitors and detoxification. *Biotechnol. Biofuels* 6, 16. doi:10.1186/1754-6834-6-16

Karl, Z. J., and Scharf, M. E. (2015). Effects of five diverse lignocellulosic diets on digestive enzyme biochemistry in the termite *Reticulitermes flavipes*. *Arch. Insect Biochem. Physiol.* 90, 89–103. doi:10.1002/arch.21246

Ke, J., Laskar, D. D., Singh, D., and Chen, S. (2011). *In situ* lignocellulosic unlocking mechanism for carbohydrate hydrolysis in termites: crucial lignin modification. *Biotechnol. Biofuels* 4, 17. doi:10.1186/1754-6834-4-17

Levasseur, A., Drula, E., Lombard, V., Coutinho, P. M., and Henrissat, B. (2013). Expansion of the enzymatic repertoire of the CAZy database to integrate auxiliary redox enzymes. *Biotechnol. Biofuels* 6, 41. doi:10.1186/1754-6834-6-41

Mastihuba, V. R., Kremnický, L. R., Mastihubová, M., Willett, J., and Côté, G. L. (2002). A spectrophotometric assay for feruloyl esterases. *Anal. Biochem.* 309, 96–101. doi:10.1016/S0003-2697(02)00241-5

Montanari, S., Roumani, M., Heux, L., and Vignon, M. R. (2005). Topochemistry of carboxylated cellulose nanocrystals resulting from TEMPO-mediated oxidation. *Macromolecules* 38, 1665–1671. doi:10.1021/ma048396c

Popoff, T., and Theander, O. (1972). Formation of aromatic compounds from carbohydrates: part 1. Reaction of D-glucuronic Acid, D-glacturonic Acid, D-xylose, and L-arabinose in slightly acidic, aqueous solution. *Carbohyd. Res.* 22, 135–149. doi:10.1016/S0008-6215(00)85733-X

Popoff, T., and Theander, O. (1976). Formation of aromatic compounds from carbohydrates. *Acta Chem. Scand* 30, 397–402. doi:10.3891/acta.chem.scand.30b-0397

Qin, L., Li, W.-C., Liu, L., Zhu, J.-Q., Li, X., Li, B.-Z., et al. (2016). Inhibition of lignin-derived phenolic compounds to cellulase. *Biotechnol. Biofuels* 9, 1–10. doi:10.1186/s13068-016-0485-2

Rajarapu, S. P., Shreve, J. T., Bhide, K. P., Thimmapuram, J., and Scharf, M. E. (2015). Metatranscriptomic profiles of Eastern subterranean termites, *Reticulitermes flavipes* (Kollar) fed on second generation feedstocks. *BMC Genomics* 16:332. doi:10.1186/s12864-015-1502-8

Rasband, W. S. (2014). *ImageJ*. Bethesda, MD: U. S. National Institutes of Health. Available at: http://imagej.nih.gov/ij/

Raychoudhury, R., Sen, R., Cai, Y., Sun, Y., Lietze, V. U., Boucias, D., et al. (2013). Comparative metatranscriptomic signatures of wood and paper feeding in the gut of the termite *Reticulitermes flavipes* (Isoptera: Rhinotermitidae). *Insect Mol. Biol.* 22, 155–171. doi:10.1111/imb.12011

Scharf, M. E. (2015a). Omic research in termites: an overview and a roadmap. *Front. Genet.* 6:76. doi:10.3389/fgene.2015.00076

Scharf, M. E. (2015b). Termites as targets and models for biotechnology. *Annu. Rev. Entomol.* 60, 77–102. doi:10.1146/annurev-ento-010814-020902

Scharf, M. E., Karl, Z. J., Sethi, A., and Boucias, D. G. (2011a). Multiple levels of synergistic collaboration in termite lignocellulose digestion. *PLoS ONE* 6:e21709. doi:10.1371/journal.pone.0021709

Scharf, M. E., Karl, Z. J., Sethi, A., Sen, R., Raychoudhury, R., and Boucias, D. G. (2011b). Defining host-symbiont collaboration in termite lignocellulose digestion: "The view from the tip of the iceberg". *Commun. Integr. Biol.* 4, 761–763. doi:10.4161/cib.17750

Scharf, M. E., Kovaleva, E. S., Jadhao, S., Campbell, J. H., Buchman, G. W., and Boucias, D. G. (2010). Functional and translational analyses of a beta-glucosidase gene (glycosyl hydrolase family 1) isolated from the gut of the lower termite *Reticulitermes flavipes*. *Insect Biochem. Mol. Biol.* 40, 611–620. doi:10.1016/j.ibmb.2010.06.002

Sethi, A., Slack, J. M., Kovaleva, E. S., Buchman, G. W., and Scharf, M. E. (2013). Lignin-associated metagene expression in a lignocellulose-digesting termite. *Insect Biochem. Mol. Biol.* 43, 91–101. doi:10.1016/j.ibmb.2012.10.001

Sun, J.-Z., and Scharf, M. (2010). Exploring and integrating cellulolytic systems of insects to advance biofuel technology. *Insect Sci.* 17, 163. doi:10.1111/j.1744-7917.2010.01348.x

Tang, J., Siegfried, B. D., and Hoagland, K. D. (1998). Glutathione-S-transferase and in vitro metabolism of atrazine in freshwater algae. *Pestic. Biochem. Physiol.* 59, 155–161. doi:10.1006/pest.1998.2319

Tartar, A., Wheeler, M. M., Zhou, X., Coy, M. R., Boucias, D. G., and Scharf, M. E. (2009). Parallel metatranscriptome analyses of host and symbiont gene expression in the gut of the termite *Reticulitermes flavipes*. *Biotechnol. Biofuels* 2, 25. doi:10.1186/1754-6834-2-25

Tarver, M. R., Florane, C. B., Mattison, C. P., Holloway, B. A., and Lax, A. (2012). Myosin gene expression and protein abundance in different castes of the Formosan subterranean termite (*Coptotermes formosanus*). *Insects* 3, 1190–1199. doi:10.3390/insects3041190

Teeri, T. T. (1997). Crystalline cellulose degradation: new insight into the function of cellobiohydrolases. *Trends Biotechnol.* 15, 160–167. doi:10.1016/S0167-7799(97)01032-9

Tokuda, G., Watanabe, H., Matsumoto, T., and Noda, H. (1997). Cellulose digestion in the wood-eating higher termite, *Nasutitermes takasagoensis* (Shiraki): distribution of cellulases and properties of endo-β-1, 4-glucanase. *Zoolog. Sci.* 14, 83–93. doi:10.2108/zsj.14.83

Watanabe, H., Noda, H., Tokuda, G., and Lo, N. (1998). A cellulase gene of termite origin. *Nature* 394, 330–331. doi:10.1038/28527

Watanabe, H., and Tokuda, G. (2010). Cellulolytic systems in insects. *Annu. Rev. Entomol.* 55, 609–632. doi:10.1146/annurev-ento-112408-085319

Weiss, A., and Leinwand, L. A. (1996). The mammalian myosin heavy chain gene family. *Annu. Rev. Cell Dev. Biol.* 12, 417–439. doi:10.1146/annurev.cellbio.12.1.417

Wheeler, M. M., Tarver, M. R., Coy, M. R., and Scharf, M. E. (2010). Characterization of four esterase genes and esterase activity from the gut of the termite *Reticulitermes flavipes*. *Arch. Insect Biochem. Physiol.* 73, 30–48. doi:10.1002/arch.20333

Yang, B., and Wyman, C. E. (2008). Pretreatment: the key to unlocking low-cost cellulosic ethanol. *Biofuels Bioprod. Biorefin.* 2, 26–40. doi:10.1002/bbb.49

Young, P., Richman, A., Ketchum, A., and Kiehart, D. (1993). Morphogenesis in *Drosophila* requires nonmuscle myosin heavy chain function. *Genes Dev.* 7, 29–41. doi:10.1101/gad.7.1.29

Zhang, D., Lax, A. R., Henrissat, B., Coutinho, P., Katiya, N., Nierman, W. C., et al. (2012). Carbohydrate-active enzymes revealed in *Coptotermes formosanus* (Isoptera: Rhinotermitidae) transcriptome. *Insect Mol. Biol.* 21, 235–245. doi:10.1111/j.1365-2583.2011.01130.x

Zhou, X., Kovaleva, E. S., Wu-Scharf, D., Campbell, J. H., Buchman, G. W., Boucias, D. G., et al. (2010). Production and characterization of a recombinant beta-1,4-endoglucanase (glycohydrolase family 9) from the termite *Reticulitermes flavipes*. *Arch. Insect Biochem. Physiol.* 74, 147–162. doi:10.1002/arch.20368

Zhou, X., Smith, J. A., Oi, F. M., Koehler, P. G., Bennett, G. W., and Scharf, M. E. (2007). Correlation of cellulase gene expression and cellulolytic activity throughout the gut of the termite *Reticulitermes flavipes*. *Gene* 395, 29–39. doi:10.1016/j.gene.2007.01.004

Conflict of Interest Statement: The authors declare that the research was conducted in the absence of any commercial or financial relationships that could be construed as a potential conflict of interest.

Permissions

The contributors of this book come from diverse backgrounds, making this book a truly international effort. This book will bring forth new frontiers with its revolutionizing research information and detailed analysis of the nascent developments around the world.

We would like to thank all the contributing authors for lending their expertise to make the book truly unique. They have played a crucial role in the development of this book. Without their invaluable contributions this book wouldn't have been possible. They have made vital efforts to compile up to date information on the varied aspects of this subject to make this book a valuable addition to the collection of many professionals and students.

This book was conceptualized with the vision of imparting up-to-date information and advanced data in this field. To ensure the same, a matchless editorial board was set up. Every individual on the board went through rigorous rounds of assessment to prove their worth. After which they invested a large part of their time researching and compiling the most relevant data for our readers.

The editorial board has been involved in producing this book since its inception. They have spent rigorous hours researching and exploring the diverse topics which have resulted in the successful publishing of this book. They have passed on their knowledge of decades through this book. To expedite this challenging task, the publisher supported the team at every step. A small team of assistant editors was also appointed to further simplify the editing procedure and attain best results for the readers.

Apart from the editorial board, the designing team has also invested a significant amount of their time in understanding the subject and creating the most relevant covers. They scrutinized every image to scout for the most suitable representation of the subject and create an appropriate cover for the book.

The publishing team has been an ardent support to the editorial, designing and production team. Their endless efforts to recruit the best for this project, has resulted in the accomplishment of this book. They are a veteran in the field of academics and their pool of knowledge is as vast as their experience in printing. Their expertise and guidance has proved useful at every step. Their uncompromising quality standards have made this book an exceptional effort. Their encouragement from time to time has been an inspiration for everyone.

The publisher and the editorial board hope that this book will prove to be a valuable piece of knowledge for researchers, students, practitioners and scholars across the globe.

List of Contributors

Fernanda Vargas e Silva and Luiz Olinto Monteggia
Institute of Hydraulic Research, Federal University of Rio Grande do Sul, Porto Alegre, Brazil

P. Chiranjeevi and S. Venkata Mohan
Bioengineering and Environmental Sciences (BEES), CSIR-Indian Institute of Chemical Technology (CSIR-IICT), Hyderabad, India
Academy of Scientific and Innovative Research (AcSIR), India

Mads Ville Markussen, Piotr Oleskowicz-Popiel, Jens Ejbye Schmidt and Hanne Østergård
Department of Chemical and Biochemical Engineering, Technical University of Denmark, Kgs Lyngby, Denmark

Siri Pugesgaard
Department of Agroecology, Aarhus University, Tjele, Denmark

Anne E. Harman-Ware, Robert Sykes and Mark Davis
National Bioenergy Center, National Renewable Energy Laboratory, Golden, CO, USA

Gary F. Peter
School of Forest Resources and Conservation, University of Florida, Gainesville, FL, USA

Sebastian Teir
VTT Technical Research Centre of Finland Ltd., Espoo, Finland

Toni Auvinen
Outotec Dewatering Technology Center, Lappeenranta, Finland

Arshe Said
Department of Energy Technology, School of Engineering, Aalto University, Espoo, Finland

Tuukka Kotiranta and Heljä Peltola
Outotec Research Center, Pori, Finland

Yaqin Qiao
Key Laboratory of Algal Biology, Institute of Hydrobiology, Chinese Academy of Sciences, Wuhan, China
University of Chinese Academy of Sciences, Beijing, China

Junfeng Rong
SINOPEC Research Institute of Petroleum Processing, Beijing, China

Hui Chen, Chenliu He and Qiang Wang
Key Laboratory of Algal Biology, Institute of Hydrobiology, Chinese Academy of Sciences, Wuhan, China

Qusai Al Abdallah and Jarrod R. Fortwendel
Department of Clinical Pharmacy, University of Tennessee Health Science Center, Memphis, TN, USA

B. Tracy Nixon
Department of Biochemistry and Molecular Biology, The Pennsylvania State University, University Park, PA, USA

Güray Güven
Giner, Inc., Newton, MA, USA

Samet Şahin and Eileen H. Yu
Chemical Engineering and Advanced Materials, Merz Court, Newcastle University, Newcastle upon Tyne, UK

Arcan Güven
Pennsylvania State University College of Medicine, Hershey, PA, USA

Anand Javee and Steffi James Pallissery
Biotechnology Division, National Institute for Interdisciplinary Science and Technology (NIIST), Council of Scientific and Industrial Research (CSIR), Trivandrum, India

Sujitha Balakrishnan Sulochana and Muthu Arumugam
Biotechnology Division, National Institute for Interdisciplinary Science and Technology (NIIST), Council of Scientific and Industrial Research (CSIR), Trivandrum, India
Academy of Scientific and Innovative Research (AcSIR), New Delhi, India

Vincent Anayochukwu Ani
Department of Electronic Engineering, University of Nigeria, Nsukka, Nigeria

Bruno Panella, Luigi Consalvo De Giorgi, Mario De Salve, Cristina Bertani and Mario Malandrone
Department of Energy, Politecnico di Torino, Torino, Italy

Simone P. Souza and Luiz A. Horta Nogueira
Interdisciplinary Center for Energy Planning, University of Campinas (UNICAMP), Campinas, SP, Brazil

Helen K. Watson
School of Agricultural, Earth and Environmental Sciences, University of KwaZulu-Natal, Durban, KZN, South Africa

Lee Rybeck Lynd
Dartmouth College, Thayer School of Engineering, Dartmouth, NH, USA

Mosad Elmissiry
New Partnership for Africa's Development (NEPAD), Johannesburg, GT, South Africa

Luís A. B. Cortez
Faculty of Agricultural Engineering, University of Campinas (UNICAMP), Campinas, SP, Brazil

Renato Baciocchi, Giulia Costa and Daniela Zingaretti
Department of Civil Engineering and Computer Science Engineering, University of Rome "Tor Vergata", Rome, Italy

Alessandra Polettini, Raffaella Pomi and Alessio Stramazzo
Department of Civil and Environmental Engineering, University of Rome "La Sapienza", Rome, Italy

Jaya Shankar Tumuluru
Idaho National Laboratory, Idaho Falls, ID, USA

Anya Skatova
Horizon Digital Economy Research, University of Nottingham, Nottingham, UK
Warwick Business School, University of Warwick, Coventry, UK

Benjamin Bedwell
Horizon Digital Economy Research, University of Nottingham, Nottingham, UK

Benjamin Kuper-Smith
Horizon Digital Economy Research, University of Nottingham, Nottingham, UK
Institute of Neurology, University College London, London, UK

Christos N. Markides
Clean Energy Processes (CEP) Laboratory, Department of Chemical Engineering, Imperial College London, London, UK

Julia Hansson
Climate and Sustainable Cities, IVL Swedish Environmental Research Institute, Stockholm, Sweden
Division of Physical Resource Theory, Department of Energy and Environment, Chalmers University of Technology, Göteborg, Sweden

Roman Hackl
Climate and Sustainable Cities, IVL Swedish Environmental Research Institute, Stockholm, Sweden

Maria Taljegard
Division of Energy Technology, Department of Energy and Environment, Chalmers University of Technology, Göteborg, Sweden

Selma Brynolf and Maria Grahn
Division of Physical Resource Theory, Department of Energy and Environment, Chalmers University of Technology, Göteborg, Sweden

Cristiano E. Rodrigues Reis and Bo Hu
Department of Bioproducts and Biosystems Engineering, University of Minnesota, Saint Paul, MN, USA

Didi Adisaputro
Department of Chemical and Biological Engineering, University of Sheffield, Sheffield, UK
Department of Energy Security, Indonesian Defence University, Bogor, Indonesia

Bastian Saputra
Department of Chemical and Biological Engineering, University of Sheffield, Sheffield, UK

Aravind Madhavan
Biotechnology Division, National Institute for Interdisciplinary Science and Technology, Council of Scientific and Industrial Research, Trivandrum, India
Rajiv Gandhi Centre for Biotechnology, Trivandrum, India

Anju Alphonsa Jose, Parameswaran Binod, Raveendran Sindhu and Rajeev K. Sukumaran
Biotechnology Division, National Institute for Interdisciplinary Science and Technology, Council of Scientific and Industrial Research, Trivandrum, India

Ashok Pandey
Biotechnology Division, National Institute for Interdisciplinary Science and Technology, Council of Scientific and Industrial Research, Trivandrum, India
Center for Innovative and Applied Bioprocessing, Mohali, Punjab, India

Galliano Eulogio Castro
Dpt. Ingeniería Química, Ambiental y de los Materiales Edificio, Universidad de Jaén, Jaén, Spain

Swapna Priya Rajarapu and Michael E. Scharf
Department of Entomology, Purdue University, West Lafayette, IN, USA

Index